31st UIT (Italian Union of Thermo-Fluid-Dynamics) Heat Transfer Conference 2013

Journal of Physics: Conference Series Volume 501

Como, Italy
25 – 27 June 2013

ISBN: 978-1-63439-281-5
ISSN: 1742-6588

Printed from e-media with permission by:

Curran Associates, Inc.
57 Morehouse Lane
Red Hook, NY 12571

Some format issues inherent in the e-media version may also appear in this print version.

Printed by Curran Associates, Inc. (2014)

For permission requests, please contact the Institute of Physics
at the address below.

Institute of Physics
Dirac House, Temple Back
Bristol BS1 6BE UK

Phone: 44 1 17 929 7481
Fax: 44 1 17 920 0979

techtracking@iop.org

Additional copies of this publication are available from:

Curran Associates, Inc.
57 Morehouse Lane
Red Hook, NY 12571 USA
Phone: 845-758-0400
Fax: 845-758-2634
Email: curran@proceedings.com
Web: www.proceedings.com

TABLE OF CONTENTS

NATURAL, FORCED AND MIXED CONVECTION

ENTRANCE AND TEMPERATURE DEPENDENT PROPERTY EFFECTS IN THE LAMINAR FORCED CONVECTION IN STRAIGHT DUCTS WITH UNIFORM WALL TEMPERATURE 1
S Del Giudice, S Savino, C Nonino

INVERSE ESTIMATION OF THE LOCAL HEAT TRANSFER COEFFICIENT IN CURVED TUBES: A NUMERICAL VALIDATION 10
F Bozzoli, L Cattani, C Corradi, M Mordacci, S Rainieri

ON THE THERMAL INSTABILITY IN A HORIZONTAL RECTANGULAR POROUS CHANNEL HEATED FROM BELOW BY A CONSTANT FLUX 20
A Barletta, E Rossi Di Schio, L Storesletten

THERMOCONVECTIVE INSTABILITY AND LOCAL THERMAL NON-EQUILIBRIUM IN A POROUS LAYER WITH ISOFLUX-ISOTHERMAL BOUNDARY CONDITIONS 30
M Celli, A Barletta, L Storesletten

MIXED CONVECTION IN HORIZONTAL CHANNELS PARTIALLY FILLED WITH ALUMINIUM FOAM HEATED FROM BELOW AND WITH EXTERNAL HEAT LOSSES ON UPPER PLATE 39
B Buonomo, G Ferraro, O Manca, L Marinelli, S Nardini

CONTROL OF NATURAL CIRCULATION LOOPS BY ELECTROHYDRODYNAMIC PUMPING 49
W Grassi, D Testi, D Della Vista

HEAT TRANSFER CHARACTERISTICS IN FORCED CONVECTION THROUGH A RECTANGULAR CHANNEL WITH 60° TILTED STAGGERED RIBS 61
M Cucchi, D Fustinoni, P Gramazio, L P M Colombo, A Niro

CONDUCTION

ESTIMATION OF TRIAL PARAMETERS FOR PULSE PHASE THERMOGRAPHY WITH LOW POWER HEAT SOURCES 69
L Vitali, D Fustinoni, P Gramazio, A Niro

RADIATION

PERFORMANCES OF INFRARED EMITTERS APPLIED TO THE POROUS THIN MATERIALS DRYING 77
P Di Marco, S Filippeschi, L Pietrasanta

MULTI-PHASE FLUID DYNAMICS, HEAT TRANSFER AND INTERFACE PHENOMENA

PORE-SCALE SIMULATION OF LAMINAR FLOW THROUGH POROUS MEDIA 87
M Piller, D Casagrande, G Schena, M Santini

A NEW SPOOL PIECE FOR HORIZONTAL TWO-PHASE FLOW MEASUREMENT 100
G Monni, M De Salve, B Panella

EXPERIMENTAL AND NUMERICAL INVESTIGATION OF A PHASE CHANGE ENERGY STORAGE SYSTEM 111
G Casano, S Piva

EXPERIMENTAL STUDY ON THE FLOW PATTERNS AND THE TWO-PHASE PRESSURE DROPS IN A HORIZONTAL IMPACTING T-JUNCTION 121
C Bertani, M Malandrone, B Panella

SINGLE AND TWO-PHASE FLOW FLUID DYNAMICS IN PARALLEL HELICAL COILS 131
M De Salve, M Orio, B Panella

MEASUREMENT OF THE OIL HOLDUP FOR A TWO-PHASE OIL-WATER FLOW THROUGH A SUDDEN CONTRACTION IN A HORIZONTAL PIPE 141
L P M Colombo, M Guilizzoni, G M Sotgia, S Bortolotti, L Pavan

PREDICTION OF TWO-PHASE CHOKED-FLOW THROUGH SAFETY VALVES 148
G Arnulfo, C Bertani, M De Salve
EXPERIMENTAL ANALYSIS OF R134A FLOW BOILING INSIDE A 5 PPI COPPER FOAM 158
A Diani, S Mancin, L Rossetto
EXPERIMENTAL SIZING AND ASSESSMENT OF TWO-PHASE PRESSURE DROP CORRELATIONS FOR A CAPILLARY TUBE WITH TRANSCRITICAL AND SUBCRITICAL CARBON DIOXIDE FLOW 168
R Trinchieri, G Boccardi, N Calabrese, G P Celata, G Zummo
EXPERIMENTAL INVESTIGATION OF FREE SURFACE VORTICES AND DEFINITION OF GAS ENTRAINMENT OCCURRENCE MAPS 181
G Caruso, L Cristofano, M Nobili, D Vitale Di Maio

COMPUTATIONAL FLUID DYNAMICS AND HEAT TRANSFER

A DYNAMIC MIXED SUBGRID-SCALE MODEL FOR LARGE EDDY SIMULATION ON UNSTRUCTURED GRIDS: APPLICATION TO TURBULENT PIPE FLOWS 191
P Lampitella, E Colombo, F Inzoli
MICROTOMOGRAPHY-BASED CFD ANALYSIS OF TRANSPORT IN OPEN-CELL ALUMINUM METAL FOAMS 204
P Ranut, E Nobile, L Mancini
APPEARANCE OF QUASIPERIODICITY WITHIN A PERIOD DOUBLING ROUTE TO CHAOS OF A SWAYING THERMAL PLUME 215
D Angeli, M A Corticelli, A Fichera, A Pagano
CHARACTERIZATION OF A MINI-CHANNEL HEAT EXCHANGER FOR A HEAT PUMP SYSTEM 227
A Arteconi, G Giuliani, M Tartuferi, F Polonara
STUDY OF LAMINAR SEPARATION BUBBLE ON LOW REYNOLDS NUMBER OPERATING AIRFOILS: RANS MODELLING BY MEANS OF AN HIGH-ACCURACY SOLVER AND EXPERIMENTAL VERIFICATION 237
A Crivellini, V D'Alessandro, D Di Benedetto, S Montelpare, R Ricci
ON THE INFLUENCE OF CURVATURE AND TORSION ON TURBULENCE IN HELICALLY COILED PIPES 249
M Ciofalo, M Di Liberto, G Marotta

MICRO AND NANO SCALES

ELECTRO-OSMOTIC FLOWS INSIDE TRIANGULAR MICROCHANNELS 261
P Vocale, M Geri, G L Morini, M Spiga
USE OF THE μPIV TECHNIQUE FOR AN INDIRECT DETERMINATION OF THE MICROCHANNEL CROSS-SECTION PASSAGE GEOMETRY 270
G Puccetti, B Pulvirenti, G L Morini
DROPWISE CONDENSATION ON SUPERHYDROPHOBIC NANOSTRUCTURED SURFACES: LITERATURE REVIEW AND EXPERIMENTAL ANALYSIS 280
A Bisetto, D Torresin, M K Tiwari, D Del Col, D Poulikakos
DYNAMIC COUPLING OF KINETIC AND CONTINUUM SOLVERS FOR MICRO GAS FLOW COMPUTATIONS 290
O Rovenskaya, G Croce

HEAT TRANSFER IN NUCLEAR PLANTS

THERMAL-HYDRAULICS OF INTERNALLY HEATED MOLTEN SALTS AND APPLICATION TO THE MOLTEN SALT FAST REACTOR 300
C Fiorina, A Cammi, L Luzzi, K Mikityuk, H Ninokata, M E Ricotti
PRELIMINARY EXPERIMENTAL EVALUATION OF THERMAL CONDUCTIVITY OF CERAMIC PEBBLE BEDS 310
D Aquaro, R Lo Frano

HEAT TRANSFER AND EFFICIENCY IN ENERGY SYSTEMS, ENVIRONMENTAL TECHNOLOGIES AND BUILDINGS

SENSITIVITY ANALYSIS ON THE EFFECT OF KEY PARAMETERS ON THE PERFORMANCE OF PARABOLIC TROUGH SOLAR COLLECTORS 320
L S W Muhlen, B Najafi, F Rinaldi, R Marchesi

BOREHOLE HEAT EXCHANGERS: HEAT TRANSFER SIMULATION IN THE PRESENCE OF A GROUNDWATER FLOW 333
A Angelotti, L Alberti, I La Licata, M Antelmi

ENERGY AND ECONOMICAL COMPARISON OF POSSIBLE CULTURES FOR A TOTAL-INTEGRATED ON-FIELD BIODIESEL PRODUCTION 342
G Allesina, S Pedrazzi, S Tebianian, A Muscio, P Tartarini

QUALITATIVE THERMAL CHARACTERIZATION AND COOLING OF LITHIUM BATTERIES FOR ELECTRIC VEHICLES 354
A Mariani, F D'Annibale, G Boccardi, G P Celata, C Menale, R Bubbico, F Vellucci

CO AND NO EMISSIONS FROM PELLET STOVES: AN EXPERIMENTAL STUDY 367
D Petrocelli, A M Lezzi

Author Index

Preface

Organized by

Sponsored by

Journal of Physics: Conference Series **501** (2014) 011001 doi:10.1088/1742-6596/501/1/011001

31st UIT Heat Transfer Conference

Como, Italy, 25–27 June, 2013

The annual Conference of the "Unione Italiana di Termofluidodinamica" (UIT) aims at promoting cooperation in the field of heat transfer and thermal sciences, by bringing together scientists and engineers working in related areas.

The 31st UIT Conference was held in Moltrasio (Como), Italy, 25–27 June, 2013 at the Grand Hotel Imperiale.

The response has been enthusiastic, with more than 70 quality contributions from 224 authors on heat transfer related topics: natural, forced and mixed convection, conduction, radiation, multi-phase fluid dynamics and interface phenomena, computational fluid dynamics, micro- and nano-scales, efficiency in energy systems, environmental technologies and buildings.

To encourage the debate, the Conference Program has scheduled ample poster sessions and invited lectures from the best experts in the field along with a few of the most talented researchers. Keynote Lectures were given by Professor Roberto Mauri (University of Pisa), Professor Lounés Tadrist (Polytech Marseille) and Professor Maurizio Quadrio (Politecnico di Milano).

This special volume collects a selection of the scientific contributions discussed during this conference; these works give a good overview of the state-of-the art Italian research in the field of Heat Transfer related topics.

I would like to thank sincerely the authors for presenting their works at the conference and in this special issue. I would also like to extend my thanks to the Scientific Committee and the authors for their accurate review process of each paper for this special issue. Special thanks go to the organizing committee and to our sponsors.

As a professor of Politecnico di Milano, let me say I am very proud to have been the chair of this conference in the 150th anniversary of my university.

Professor Alfonso Niro

Journal of Physics: Conference Series **501** (2014) 011001 doi:10.1088/1742-6596/501/1/011001

About UIT (Unione Italiana Termofluidodinamica)

The Italian Union of Thermal-Fluid Dynamics (UIT) was founded in Bologna on December 19, 1984, following a proposal by Professor Enrico Lorenzini.

Mission

The Italian Union of Thermal-Fluid Dynamics (UIT) has the fundamental aim to contribute to the progress of the Italian research and application of thermal-fluid dynamics, by promoting and enhancing the scientific, professional and social values among its members, as well as the formation of new generations of young scientists towards the research work in the field of thermal sciences, heat transfer and their applications.

In order to achieve these objectives UIT undertakes:

- the organization of a heat transfer conference, on yearly basis
- the organization of international Conferences, Seminars, Courses, etc.
- the organization of a summer school, on yearly basis, for PhD students and young researchers
- the institution of a prize, on yearly basis, for the best Italian Masters thesis dealing with thermal-fluid dynamics
- the institution of a prize, on a two-year basis, for the best Italian PhD thesis dealing with thermal-fluid dynamics.

Governance

The governing bodies of the Italian Union of Thermal-Fluid Dynamics (UIT) are:

- the assembly (Assemblea)
- the steering committee (Comitato Direttivo)
- the president (Presidente)
- the auditors (Revisori dei conti)
- the arbiters (Probiviri)
- the secretary (Segretario).

The assembly of the UIT is made up by all ordinary members (natural persons, institutions and associations), honorary members, distinguished members and meets at least once a year. The assembly establishes the activity of the association, votes its budget and balance, appoints the governing bodies. At present more than 200 members from universities, national laboratories and industries take part in the UIT. The scientific management of the UIT, subject to the limitations of the constitution (Statuto), is handled by a steering committee chosen from its membership. It is made up of seven officers (president, vice-president, secretary, and four members). The steering committee is renewed every three years. President, vice-President and secretary cannot be appointed for the same chair more than twice on end.

Journal of Physics: Conference Series **501** (2014) 011001 doi:10.1088/1742-6596/501/1/011001

UIT 2013

Steering committee
Vincenzo Naso, President
Giuseppe Vella, Vice-President
Sara Rainieri, Secretary
Gian Piero Celata
Davide Del Col
Paolo Di Marco
Alfonso Niro

Auditors
Mario De Salve
Marco Spiga
Luca Tagliafico

Past-Presidents
Enrico Lorenzini
(First President and Founder)
Sergio Faggiani †
Claudio Pisoni †
Gian Piero Celata

Past-Secretaries
Sandro Salvigni
Piero Vigni †
Gian Piero Celata
Paolo Di Marco

Distinguished members
Gaetano Alfano †
Vittorio Betta
Maurizio Cumo
Giovanni Del Tin
Giovanni De Comelli †
Sergio Faggiani †
Enrico Lorenzini
Elio Oliveri
Claudio Pisoni †
Sandro Salvigni
Giorgio Sotgia

Honorary members
Arthur E Bergles
Michael W Collins
Marino Di Marzo
Frank P Incropera
Antonio C M Sousa

Journal of Physics: Conference Series **501** (2014) 011001 doi:10.1088/1742-6596/501/1/011001

Scientific committee of the 31st UIT heat transfer conference

Alfonso Niro, *Chair*	Politecnico di Milano
Giovanni Sebastiano Barozzi	Università degli Studi di Modena e Reggio Emilia
Gian Paolo Beretta	Università degli Studi di Brescia
Gian Piero Celata	ENEA
Giovanni Cesini	Università Politecnica delle Marche
Michele Ciofalo	Università degli Studi di Palermo
Elvio Cossali	Università degli Studi di Bergamo
Paolo Di Marco	Università degli Studi di Pisa
Marino Di Marzo	University of Maryland, USA
Walter Grassi	Università degli Studi di Pisa
Fabio Inzoli	Politecnico di Milano
Enrico Lorenzini	Università degli Studi di Bologna
Oronzio Manca	Seconda Università degli Studi di Napoli
Gian Luca Morini	Università degli Studi di Bologna
Adriano Muzzio	Politecnico di Milano
Vincenzo Naso	Università degli Studi di Napoli
Hisashi Ninokata	Professor Emeritus of the Tokyo Institute of Technology, Japan Visiting Professor at the Politecnico di Milano, Italy
Enrico Nobile	Università degli Studi di Trieste
Bruno Panella	Politecnico di Torino
Stefano Piva	Università degli Studi di Ferrara
Giorgio Sotgia	Politecnico di Milano
Antonio C M Sousa	University of Brunswick, Canada
John Richard Thome	EPFL, Lausanne, Switzerland
Akio Omyiama	Kobe University, Japan
Giuseppe Vella	Università degli Studi di Palermo
Evelyn Wang	MIT, Cambridge Mass, USA
Iztok Zun	Slovenian Academy of Engineering, Slovenia

Keynote lecturers of 31st UIT heat transfer conference

Roberto Mauri	Universià degli Studi di Pisa
Lounés Tadrist	Polytech Marseille
Maurizio Quadrio	Politecnico di Milano

Journal of Physics: Conference Series **501** (2014) 011001 doi:10.1088/1742-6596/501/1/011001

Organizing committee of 31st UIT heat transfer conference

Alfonso Niro	Chair
Sergio Arosio	Co-chair
Guido Parissenti	
Luigi Vitali	
Damiano Fustinoni	
Pasqualino Gramazio	
Susanna Ciminà	
Luigi Colombo	
Manfredo Guilizzoni	
Emanuela Colombo	

UIT conferences

1	1983	Trieste	16	1998	Siena
2	1984	Bologna	17	1999	Ferrara
3	1985	Palermo	18	2000	Cernobbio (CO)
4	1986	Genova	19	2001	Modena
5	1987	Torino	20	2002	Maratea (PZ)
6	1988	Bari	21	2003	Udine
7	1989	Firenze	22	2004	Genova
8	1990	Ancona	23	2005	Parma
9	1991	Pisa	24	2006	Napoli
10	1992	Genova	25	2007	Trieste
11	1993	Milano	26	2008	Palermo
12	1994	L'Aquila	27	2009	Reggio Emilia
13	1995	Bologna	28	2010	Brescia
14	1996	Roma	29	2011	Torino
15	1997	Torino	30	2012	Bologna

31 2013 MOLTRASIO (CO)

Peer review statement

All papers published in this volume of *Journal of Physics: Conference Series* have been peer reviewed through processes administered by the proceedings Editors. Reviews were conducted by expert referees to the professional and scientific standards expected of a proceedings journal published by IOP Publishing.

Journal of Physics: Conference Series **501** (2014) 012001 doi:10.1088/1742-6596/501/1/012001

Entrance and temperature dependent property effects in the laminar forced convection in straight ducts with uniform wall temperature

Stefano Del Giudice, Stefano Savino and Carlo Nonino

Dipartimento di Ingegneria Elettrica, Gestionale e Meccanica, Università degli Studi di Udine, Via delle Scienze 206, 33100 Udine, Italy

E-mail: stefano.delgiudice@uniud.it, carlo.nonino@uniud.it, stefano.savino@uniud.it

Abstract. The paper reports the results of a parametric investigation on the effects of temperature dependent viscosity and thermal conductivity on forced convection in simultaneously developing laminar flows of liquids in straight ducts of constant cross-sections. Uniform temperature boundary conditions are specified at the duct walls. Viscosity is assumed to vary with temperature according to an exponential relation, while a linear dependence of thermal conductivity on temperature is assumed. The other fluid properties are held constant. Two different cross-sectional geometries, namely circular and flat ducts, are considered. A finite element procedure is employed for the solution of the parabolized momentum and energy equations. Computed axial distributions of the local Nusselt number are presented for different values of the entrance Prandtl number and of the viscosity and thermal conductivity Pearson numbers. Moreover, a superposition method is proved to be applicable in order to obtain an approximate value of the local Nusselt number by separately considering the effects of temperature dependent viscosity and those of temperature dependent thermal conductivity.

1. Introduction

Temperature dependence of fluid properties can play an important role in the simultaneous development of the velocity and temperature fields in duct flows [1, 2]. If the fluid is a liquid, the relative variations of viscosity with temperature are the most relevant, while those of thermal conductivity are, in general, much smaller, and those of density and specific heat capacity are almost negligible [3, 4, 5]. As far as velocity distribution and pressure drop are concerned, the main effects of temperature dependent fluid properties can be retained even if only viscosity is allowed to vary with temperature, while the other properties are assumed constant [1, 2]. Instead, heat transfer characteristics, namely the Nusselt number, are also influenced by thermal conductivity [6, 7, 8]. However, while viscosity exhibits a similar behaviour for all liquids, i.e., it can be assumed to decrease with increasing temperature according to a nearly exponential relation, thermal conductivity can both increase or decrease with increasing temperature, depending on the nature of the fluid considered. For this reason, since parametric analyses valid for all liquids are rather difficult, most of the studies reported in the literature are based on the assumption of temperature dependences of viscosity and thermal conductivity given by specific relations of empirical nature, leading to results which cannot be considered general and applicable to all liquids [2, 5, 9, 10, 11].

The authors have already carried out systematic studies on entrance and temperature dependent viscosity effects in duct flows with different thermal boundary conditions applied at the solid walls [12, 13], but they did not consider the effects of temperature dependent thermal conductivity. More recently, they proposed a simplified approach to the analysis of the effects of temperature dependent viscosity and thermal conductivity on forced convection in simultaneously developing laminar flows of liquids in straight ducts with uniform wall heat flux boundary conditions [14]. A superposition method, where the effects of temperature dependent viscosity and those of temperature dependent thermal conductivity were considered separately, was demonstrated to be applicable in order to obtain accurate values of characteristic heat transfer parameters. A similar approach was also proposed by the authors in the context of forced convection in laminar microchannel flows of liquids with temperature dependent properties and non-negligible viscous dissipation [15].

In this paper, the effects of temperature dependent viscosity and thermal conductivity on forced convection in simultaneously developing laminar flows of liquids in straight ducts with uniform wall temperature boundary conditions are analysed. Also in this case a superposition method, where the effects of temperature dependent viscosity and those of temperature dependent thermal conductivity are considered separately, is demonstrated to be applicable to predict the value of the Nusselt number.

A finite element procedure is employed for the step-by-step solution of the parabolized momentum and energy equations in a domain corresponding to the cross-section of the duct [16]. In most situations of practical interest, because of the high value of the ratio between the total length and the hydraulic diameter, such an approach is very advantageous with respect to that based on the steady-state solution of the elliptic form of the governing equations in a three-dimensional domain corresponding to the whole duct.

2. Statement of the problem

The laminar forced convection in the entrance region of straight ducts of constant cross-sections with uniform wall temperature t_w is studied parametrically. The hypotheses made here are that viscous dissipation effects are negligible and that liquid heating begins at the duct inlet, where the velocity boundary layer also starts developing. Therefore, at the entrance of the duct, uniform values t_e and $u_e = \overline{u}$ of the temperature t and of the axial velocity component u are specified, being $\overline{u}$ the average axial velocity.

Since the fluids considered here are liquids, the dynamic viscosity μ is assumed to decrease with increasing temperature, and μ_e is its values at t_e. Exponential type relations are usually employed to represent the temperature dependence of viscosity. The one adopted in this paper is the widely used exponential formula [2, 10, 11, 17, 18, 19]

$$\mu = \mu_e \exp[-\,\beta\,(t - t_e)] \tag{1}$$

where $\beta = -(\mathrm{d}\mu/\mathrm{d}t)/\mu = \text{const}$ is positive. The thermal conductivity k is also assumed to vary with temperature and k_e is its value at t_e. However, while viscosity exhibits a similar behaviour for all liquids, thermal conductivity can both increase or decrease with increasing temperature, depending on the fluid considered. Therefore, the following linear relation [4, 7, 8, 9] is assumed in this paper

$$k = k_e\,[1 + \alpha\,(t - t_e)] \tag{2}$$

where $\alpha = (\mathrm{d}k/\mathrm{d}t)/k_e = \text{const}$ can be both positive or negative. By means of simple manipulations, equation (1) can be cast in the following dimensionless form

$$\frac{\mu}{\mu_e} = \exp(-\,\mathrm{Pn}_\mu\,T) = \left(\frac{\mu_w}{\mu_e}\right)^T \tag{3}$$

Journal of Physics: Conference Series **501** (2014) 012001 doi:10.1088/1742-6596/501/1/012001

where $T = (t - t_e)/(t_w - t_e)$ is the dimensionless temperature, μ_w is the value of the dynamic viscosity at t_w, while $\text{Pn}_\mu = \beta\,(t_w - t_e) = -\ln(\mu_w/\mu_e)$ is the viscosity Pearson number, which represents the ratio of the characteristic process temperature difference $(t_w - t_e)$ to the characteristic temperature difference $1/\beta$ that can produce appreciable viscosity variations. Similarly, equation (2) can be cast in the following dimensionless form

$$\frac{k}{k_e} = 1 + \text{Pn}_k\, T = 1 + \left(\frac{k_w}{k_e} - 1\right) T \tag{4}$$

where k_w is the value of the thermal conductivity at t_w and $\text{Pn}_k = \alpha\,(t_w - t_e) = k_w/k_e - 1$ is the thermal conductivity Pearson number, which represents the ratio of the characteristic process temperature difference $(t_w - t_e)$ to the characteristic temperature difference $1/\alpha$ that can produce appreciable thermal conductivity variations.

It is worth noting that the local Reynolds number $\text{Re} = \rho\,\overline{u}\,D_h/\mu$, the local Prandtl number $\text{Pr} = \mu\,c/k$ and the local Péclet number $\text{Pe} = \text{Re}\,\text{Pr} = \rho\,c\,\overline{u}\,D_h/k$ all vary with temperature because of the variations of μ, of the ratio μ/k and of k, respectively. In these expressions D_h is the hydraulic diameter of the duct, while ρ and c are the density and the specific heat of the liquid. Therefore, since ρ and c are assumed to be constant, for a given duct flow we have $\mu/\mu_e = \text{Re}_e/\text{Re}$ and $k/k_e = \text{Pe}_e/\text{Pe}$, where Re_e and Pe_e are the Reynolds and Péclet numbers evaluated at the entrance temperature t_e. Moreover, since the viscosity of liquids decreases with increasing temperature $(\beta > 0)$, we have $\text{Pn}_\mu > 0$ and $\text{Re}_e/\text{Re} < 1$ in the case of fluid heating $(t_w > t_e)$ and $\text{Pn}_\mu < 0$ and $\text{Re}_e/\text{Re} > 1$ in the case of fluid cooling $(t_w < t_e)$, while $\text{Pn}_\mu = 0$ and $\text{Re}_e/\text{Re} = 1$ refer to constant viscosity fluids $(\beta = 0)$. Instead, since the thermal conductivity can either increase $(\alpha > 0)$ or decrease $(\alpha < 0)$ with increasing temperature, we have $\text{Pn}_k > 0$ and $\text{Pe}_e/\text{Pe} > 1$ with $t_w > t_e$ and $\text{Pn}_k < 0$ and $\text{Pe}_e/\text{Pe} < 1$ with $t_w < t_e$ in the first case, and $\text{Pn}_k < 0$ and $\text{Pe}_e/\text{Pe} < 1$ with $t_w > t_e$ and $\text{Pn}_k > 0$ and $\text{Pe}_e/\text{Pe} > 1$ with $t_w < t_e$ in the second case, while $\text{Pn}_k = 0$ and $\text{Pe}_e/\text{Pe} = 1$ refer to constant thermal conductivity fluids $(\alpha = 0)$.

Numerical results concerning axial distributions of the local Nusselt number

$$\text{Nu} = \frac{h\,D_h}{k_e} \tag{5}$$

are presented in the following. In this expression h is the local convective heat transfer coefficient defined as

$$h = \frac{q''_w}{t_w - t_b} \tag{6}$$

where q''_w is the wall heat flux (with $q''_w > 0$ in fluid heating and and $q''_w < 0$ in fluid cooling) and t_b is the bulk temperature. It is worth noting that in the above definition of the Nusselt number we use the value of the thermal conductivity at the duct entrance, with the purpose of showing how the heat transfer coefficient is actually affected by the temperature dependence of thermophysical properties.

3. Mathematical model

If the effects of axial diffusion can be neglected and there is no recirculation in the longitudinal direction, steady-state flow and heat transfer in straight ducts of constant cross-sections are governed by the continuity and the parabolized Navier-Stokes and energy equations [20, 21]. Since the inverse of the Reynolds number is representative of the relative importance of diffusive and advective components of the axial momentum flow rate, while the inverse of the Péclet number is representative of the relative importance of conductive and advective components of the axial heat flow rate, the parabolic approximation of the Navier-Stokes and energy equations can be considered adequate, except for the immediate neighbourhood of the inlet, for values of the Reynolds and Péclet numbers larger than 50 [1, 22]. According to the assumption of parabolic flows, all the derivatives in the axial direction are neglected in the diffusive terms of momentum and energy equations [21]. Moreover, in the axial momentum equation, reference can be made to the gradient of the cross-section averaged pressure. In fact, except for very close to the entrance, the hydrodynamically developing flow problem is similar to a boundary layer problem and, therefore, the pressure in the axial momentum equation can be considered a function only of the axial coordinate x [1]. With reference to incompressible fluids with temperature dependent viscosity and thermal conductivity, in the hypotheses of negligible body forces and viscous dissipation, the above mentioned equations can be written in the following form, valid for axisymmetric geometries

$$\frac{\partial u}{\partial x} + \frac{1}{r}\frac{\partial}{\partial r}(r\,v) = 0 \tag{7}$$

$$\rho\,u\,\frac{\partial u}{\partial x} = \frac{1}{r}\frac{\partial}{\partial r}\left(\mu\,r\,\frac{\partial u}{\partial r}\right) - \rho\,v\,\frac{\partial u}{\partial r} - \frac{d\overline{p}}{dx} \tag{8}$$

$$\rho\,u\,\frac{\partial v}{\partial x} = \frac{2}{r}\left[\frac{\partial}{\partial r}\left(\mu\,r\,\frac{\partial v}{\partial r}\right) - \mu\,\frac{v}{r}\right] - \rho\,v\,\frac{\partial v}{\partial r} - \frac{\partial p}{\partial r} \tag{9}$$

$$\rho\,c\,u\,\frac{\partial t}{\partial x} = \frac{1}{r}\frac{\partial}{\partial r}\left(k\,r\,\frac{\partial t}{\partial r}\right) - \rho\,c\,v\,\frac{\partial t}{\partial r} \tag{10}$$

In the above equations, r is the radial coordinate, v represents the radial components of velocity, p is the deviation from the hydrostatic pressure and $\overline{p}$ is its average value over the cross-section.

At the entrance of the duct, the inlet conditions $u = u_e = \overline{u}$, $v = 0$ and $t = t_e$ are imposed. The solution domain can be bounded by rigid walls or by symmetry axes. On rigid boundaries, the no-slip conditions, that is, $u = v = 0$, are imposed together with the temperature (Dirichlet) boundary condition $t = t_w$. Instead, symmetry conditions are $\partial u/\partial r = 0$, $v = 0$ and $\partial t/\partial r = 0$ at the symmetry axis.

A finite element procedure for the analysis of the forced convection of fluids with temperature dependent properties in the entrance region of straight ducts and microchannels [12, 13, 16, 23, 24, 25] is used to solve the model equations. The adopted procedure is based on a segregated approach which implies the sequential solution of the momentum and energy equations on a one-dimensional domain corresponding to the cross-section of the duct. A marching method is then used to move forward in the axial direction. The pressure-velocity coupling is dealt with using an improved projection algorithm already employed by one of the authors (C.N.) for the solution of the Navier-Stokes equations in their elliptic form [26]. Since the description reported in Reference [16] can be easily adapted to the case of a fluid with temperature dependent properties considered here, only the details concerning the estimation of the gradient of the cross-section averaged pressure $d\overline{p}/dx$, which is necessary to solve the momentum equation in the axial direction, are reported in this paper. With reference to the flow in straight ducts, integration of the axial momentum equation over the cross-section A gives [1, 16]

$$-\frac{d\overline{p}}{dx} = \frac{d}{dx}\left(\frac{1}{A}\int_A \rho\,u^2 dA\right) - \frac{1}{A}\int_C \mu\,\frac{\partial u}{\partial n}\,dC \tag{11}$$

where C is the perimeter of the cross-section and n is the outward normal.

The adopted procedure has already been validated, with reference to both constant and temperature dependent property fluids, by comparing heat transfer and pressure drop results with existing literature data for simultaneously developing laminar flows in straight ducts and microchannels [12, 16, 23, 24].

4. Numerical results and approximations

In this study, two cross-sectional geometries, namely circular ducts with outer radius r_0 and flat ducts with plate spacing a, are considered, representing the limiting situations corresponding to the minimum and maximum values of the ratio between the perimeter and the area of the cross-section. Both these cross-sectional geometries are axisymmetric, since the latter can be considered the limiting case of a concentric annular duct with finite spacing $a = r_o - r_i$ and the ratio r_o/r_i between the outer and the inner radiuses tending to one, which implies $r_i, r_o \to \infty$. However, in practice, reference is made to equations (7) to (10) for the solution of circular duct problems, while to solve parallel plate problems it is convenient to refer to the equations valid for one-dimensional plane problems, i.e., to the limiting forms of equations (7) to (10) corresponding to $r \to \infty$. The computational domains have been defined taking into account existing symmetries. Therefore, the circular and flat cross-sections correspond to one-dimensional domains of lengths equal to the external radius r_o and to the plate spacing a, respectively.

The one-dimensional domains are discretized by means of three-node Lagrangian parabolic elements whose sizes gradually increase with increasing distance from the walls. A total of 40 elements and 81 nodal points are used in the discretization of the domain corresponding to the circular cross-section, and a total of 80 elements and 161 nodal points in that of the domain corresponding to the cross-section of the parallel plate channel. The adopted meshes are finer near the walls to allow an accurate representation of the steep velocity and temperature gradients taking place in that region as the flow develops. Several preliminary test were carried out to verify that these discretizations are fine enough to give mesh-independent results. In all the computations, the axial step is gradually increased from the starting value $\Delta x = 0.0001\, D_h$ to the maximum value $\Delta x = 0.05\, D_h$. As the initial value of the axial step is very small, the strong variations of the axial pressure gradient arising near the entrance of the duct can be adequately captured.

The same value $\mathrm{Re}_e = 450$ of the entrance Reynolds number is assumed in all the computations. Instead, the values $\mathrm{Pr}_e = 5$ and 100 of the Prandtl number at t_e are selected to take into account the behaviours of different liquids. The corresponding values of the Péclet number are $\mathrm{Pe}_e = 2250$ and 45000. The values of the dynamic viscosity Pearson number $\mathrm{Pn}_\mu = 0, \pm 0.75$ and ± 1.5 are considered to account for reasonable viscosity temperature dependences. Thus, for the assumed Re_e, the minimum and maximum values $\mathrm{Re} = \mathrm{Re}_w$ of the local Reynolds number are 100 (fluid cooling) and 2017 (fluid heating), respectively. For each value of Pn_μ four nonzero values of Pn_k are selected (two positive and two negative), giving the corresponding values of the ratio $\mathrm{Pn}_\mu/\mathrm{Pn}_k = \beta/\alpha = \pm 10$ and ± 20. Therefore, on the whole, the six values $\mathrm{Pn}_k = \pm 0.0375, \pm 0.075$ and ± 0.15 are considered, besides the value $\mathrm{Pn}_k = 0$. Therefore, the local Péclet number $\mathrm{Pe} = \mathrm{Pe}_w$ varies between 1957 ($\mathrm{Pr}_e = 5$ and $\mathrm{Pn}_k = 0.15$) and 52941 ($\mathrm{Pr}_e = 100$ and $\mathrm{Pn}_k = -0.15$). Finally, for the considered values of Pr_e, Pn_μ and Pn_k, the minimum and maximum values $\mathrm{Pr} = \mathrm{Pr}_w$ of the local Prandtl number are 0.97 and 527, respectively.

It must be pointed out that, even if the numerical results reported in the following were obtained for $\mathrm{Re}_e = 450$, they are much more general than they appear, since they are valid in the laminar regime for any value of the entrance Reynolds number larger than 50. In fact, for given values of the entrance Prandtl number Pr_e and of the Pearson numbers Pn_μ and Pn_k, the axial distributions of Nu are independent of the entrance Reynolds number Re_e provided that the appropriate dimensionless axial coordinate $X^* = x/(D_h\, \mathrm{Pe}_e)$ is employed. The validity of the above statement, which is well established for constant property flows ($\mathrm{Pn}_\mu = \mathrm{Pn}_k = 0$) with negligible axial diffusion of momentum and negligible axial heat conduction in the absence of recirculation in the longitudinal direction [1, pp. 44 and 58], was verified by means of sample numerical tests under the variable viscosity and thermal conductivity assumptions in the ranges

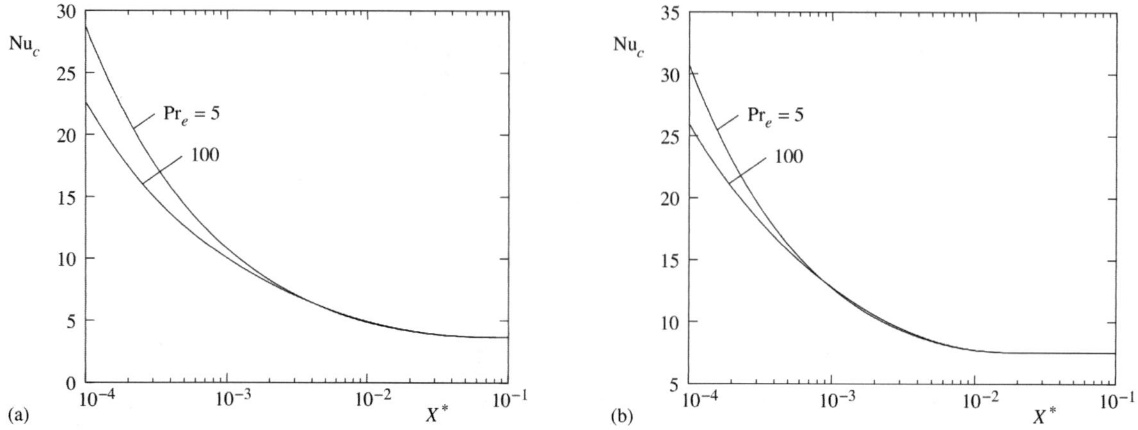

Figure 1. Axial distributions of the Nusselt number Nu_c for constant property flows with $\mathrm{Pr}_e = 5$ and $\mathrm{Pr}_e = 100$ in the entrance region of (a) circular and (b) flat ducts.

$50 \leq \mathrm{Re}_e \leq 1500$, $5 \leq \mathrm{Pr}_e \leq 100$, $-1.5 \leq \mathrm{Pn}_\mu \leq 1.5$ and $-0.15 \leq \mathrm{Pn}_k \leq 0.15$.

To illustrate the effects on heat transfer of temperature dependent properties (viscosity and thermal conductivity), the local Nusselt number $\mathrm{Nu}_{\mu k}$, obtained for given nonzero values of Pn_μ and Pn_k, is compared with the corresponding local Nusselt number Nu_c computed for simultaneously developing constant property flows ($\mathrm{Pn}_\mu = \mathrm{Pn}_k = 0$). Axial distributions of Nu_c for both cross-sectional geometries considered here are presented in figure 1, while axial distributions of the ratio $\mathrm{Nu}_{\mu k}/\mathrm{Nu}_c$ with different values of Pn_μ and Pn_k are presented in the following.

As can be seen in figure 1, for a given cross-sectional geometry Pr_e significantly affects Nu_c for low values of X^* while, as expected, the same asymptotic value of Nu_c is reached for fully developed conditions. The computed asymptotic values 3.65680 and 7.54075 for circular and flat ducts, respectively, match very well the corresponding literature values 3.65679 and 7.54070 [1].

Axial distributions of the ratio $\mathrm{Nu}_{\mu k}/\mathrm{Nu}_c$ for circular and flat ducts with different values of Pr_e, Pn_μ and Pn_k are presented in figures 2 and 3 (solid lines). As expected, the effects of Pr_e are significant only for low values of X^*, resulting in larger differences from 1 of the value of $\mathrm{Nu}_{\mu k}/\mathrm{Nu}_c$ for larger values of Pr_e. Near the entrance the effects of temperature dependent viscosity prevail over those due to the temperature dependence of thermal conductivity. In fact, the values of $\mathrm{Nu}_{\mu k}/\mathrm{Nu}_c$ are always higher than 1 for $\mathrm{Pn}_\mu > 0$ (fluid heating) and lower than 1 for $\mathrm{Pn}_\mu < 0$ (fluid cooling), no matter whether Pn_k is positive or negative. Instead, when fully developed conditions are approached, the effects of temperature dependence of viscosity vanish since t_b tends to t_w. Thus, the asymptotic value of $\mathrm{Nu}_{\mu k}/\mathrm{Nu}_c$ is higher than 1 for $\mathrm{Pn}_k > 0$ and lower than 1 for $\mathrm{Pn}_k < 0$, being equal to $k_w/k_e = 1 + \mathrm{Pn}_k$. As can be seen, the effects of temperature dependent thermal conductivity are comparable with those of temperature dependent viscosity ($\mathrm{Pn}_k = 0$). Thus, if thermal conductivity significantly varies with temperature, heat transfer results obtained under the assumptions of temperature dependent viscosity ($\mathrm{Pn}_\mu \neq 0$) and constant thermal conductivity ($\mathrm{Pn}_k = 0$) are inadequate to predict actual heat transfer rates. In other words, the Nusselt number Nu_μ, computed under such assumptions, can be considerably different from $\mathrm{Nu}_{\mu k}$.

However, it will be shown that the values of Nu_μ, together with those of the Nusselt number Nu_k, computed under the assumptions of temperature dependent thermal conductivity ($\mathrm{Pn}_k \neq 0$) and constant viscosity ($\mathrm{Pn}_\mu = 0$), can be used to predict the corresponding values

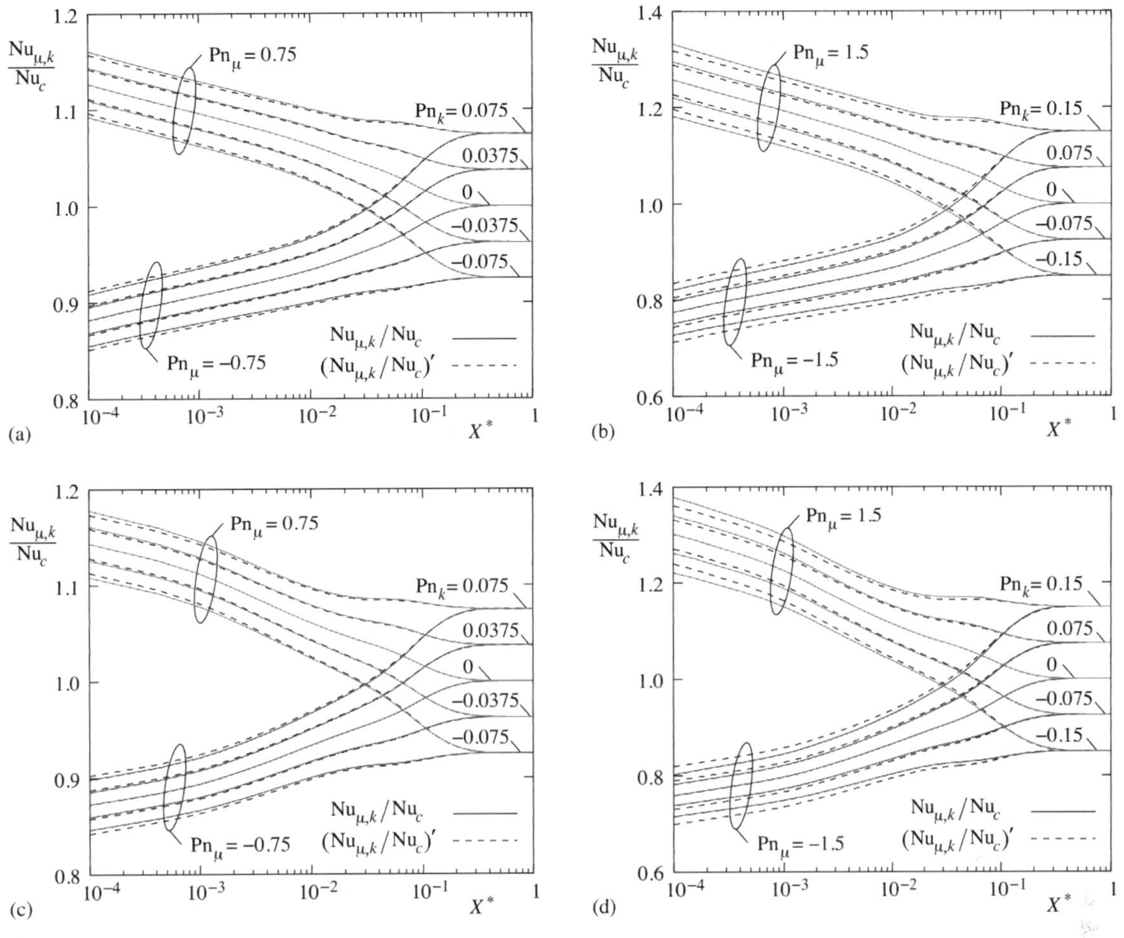

Figure 2. Axial distributions of the ratios $\mathrm{Nu}_{\mu k}/\mathrm{Nu}_c$ and $(\mathrm{Nu}_{\mu k}/\mathrm{Nu}_c)'$ for simultaneously developing laminar flows in circular ducts with different values of Pn_k and: $\mathrm{Pr}_e = 5$ and (a) $|\mathrm{Pn}_\mu| = 0.75$, (b) $|\mathrm{Pn}_\mu| = 1.5$; $\mathrm{Pr}_e = 100$ and (c) $|\mathrm{Pn}_\mu| = 0.75$, (d) $|\mathrm{Pn}_\mu| = 1.5$.

of $\mathrm{Nu}_{\mu k}$ with an accuracy which can be considered satisfactory in most situations. In fact, if the effects of temperature dependence of viscosity and thermal conductivity are considered separately to compute Nu_μ and Nu_k, respectively, a superposition method is applicable in order to obtain approximate values of the Nusselt number $\mathrm{Nu}_{\mu k}$ according to the relation

$$\frac{\mathrm{Nu}_{\mu k} - \mathrm{Nu}_c}{\mathrm{Nu}_c} \cong \frac{\mathrm{Nu}_\mu - \mathrm{Nu}_c}{\mathrm{Nu}_c} + \frac{\mathrm{Nu}_k - \mathrm{Nu}_c}{\mathrm{Nu}_c} \tag{12}$$

The applicability of such a superposition method can be justified by considering that temperature dependent viscosity effects are mostly due to the temperature variations over the duct cross-section, while those of temperature dependent thermal conductivity mainly come from the temperature variations between the inlet section and the considered axial location. With reference to the ratios $\mathrm{Nu}_{\mu k}/\mathrm{Nu}_c$, $\mathrm{Nu}_\mu/\mathrm{Nu}_c$ and $\mathrm{Nu}_k/\mathrm{Nu}_c$, equation (12) can be recast in the form

$$\frac{\mathrm{Nu}_{\mu k}}{\mathrm{Nu}_c} \cong \left(\frac{\mathrm{Nu}_{\mu k}}{\mathrm{Nu}_c}\right)' = \frac{\mathrm{Nu}_\mu}{\mathrm{Nu}_c} + \frac{\mathrm{Nu}_k}{\mathrm{Nu}_c} - 1 \tag{13}$$

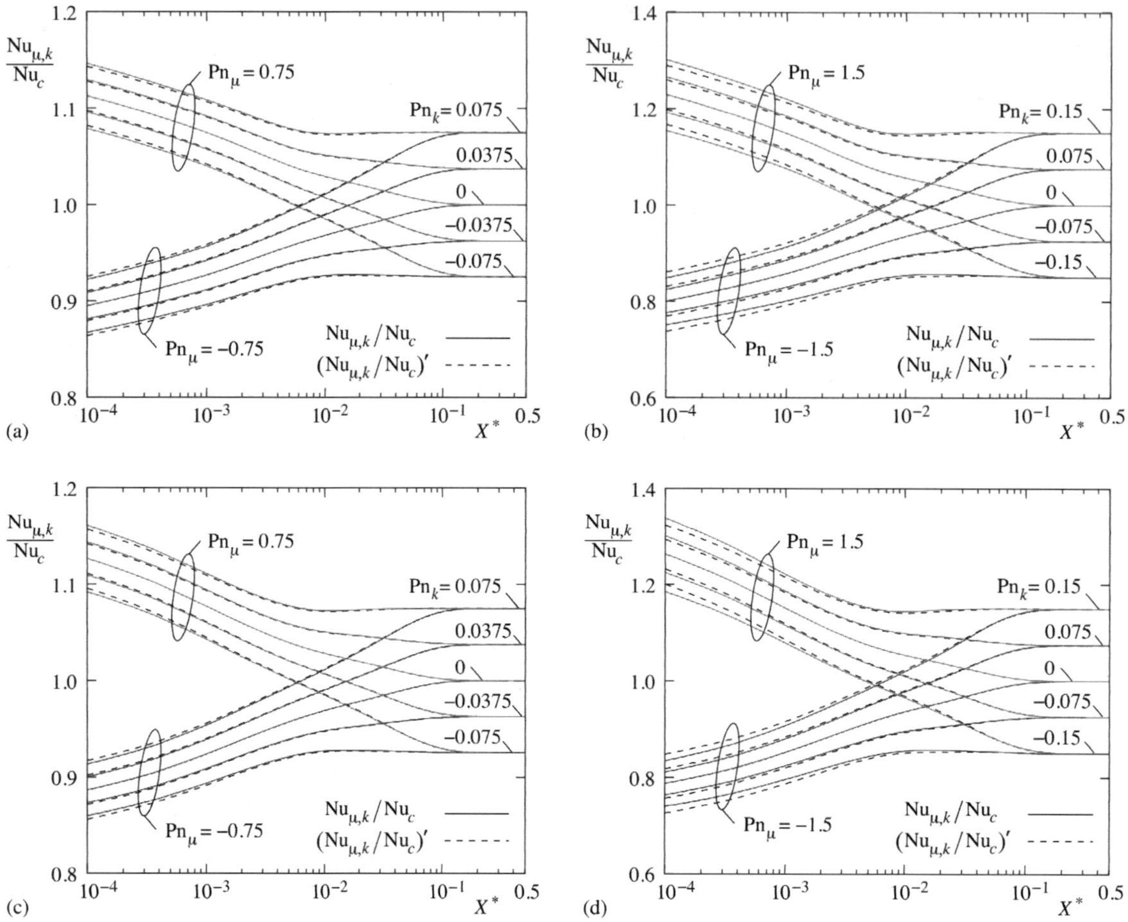

Figure 3. Axial distributions of the ratios $\mathrm{Nu}_{\mu k}/\mathrm{Nu}_c$ and $(\mathrm{Nu}_{\mu k}/\mathrm{Nu}_c)'$ for simultaneously developing laminar flows in flat ducts with different values of Pn_k and: $\mathrm{Pr}_e = 5$ and (a) $|\mathrm{Pn}_\mu| = 0.75$, (b) $|\mathrm{Pn}_\mu| = 1.5$; $\mathrm{Pr}_e = 100$ and (c) $|\mathrm{Pn}_\mu| = 0.75$, (d) $|\mathrm{Pn}_\mu| = 1.5$.

where $(\mathrm{Nu}_{\mu k}/\mathrm{Nu}_c)'$ is the approximate value of $\mathrm{Nu}_{\mu k}/\mathrm{Nu}_c$ given by the superposition method. To prove the validity of equation (13), axial distributions of the ratio $(\mathrm{Nu}_{\mu k}/\mathrm{Nu}_c)'$ for circular and flat ducts for different values of Pn_μ and Pn_k, are compared in figures 2 and 3 with the corresponding distributions of $\mathrm{Nu}_{\mu k}/\mathrm{Nu}_c$. As can be seen, the dashed curves, representing axial distributions of the ratio $(\mathrm{Nu}_{\mu k}/\mathrm{Nu}_c)'$, are very close, even for the highest/lowest values of Pn_μ and Pn_k considered here, to the solid ones representing the computed axial distributions of $\mathrm{Nu}_{\mu k}/\mathrm{Nu}_c$. Therefore, the validity of the approximation given by equation (13) can be considered confirmed.

5. Conclusions

The effects of temperature dependent viscosity and thermal conductivity on forced convection in simultaneously developing laminar flows of liquids in straight ducts of constant cross-sections have been investigated parametrically. Uniform temperature boundary conditions have been specified at the duct walls. Viscosity has been assumed to vary with temperature according to an exponential relation, while a linear dependence of thermal conductivity on temperature has been considered. The other fluid properties have been held constant. Two different cross-sectional geometries have been considered, namely circular and flat ducts. An efficient finite element procedure has been employed for the solution of the parabolized momentum and energy equations.

In this parametric study, computed axial distributions of the local Nusselt number have been presented for different values of the entrance Prandtl number and of the viscosity and thermal conductivity Pearson numbers. Moreover, a superposition method has been proved to be applicable in order to obtain an approximate value of the local Nusselt number by separately considering the effects of temperature dependent viscosity and those of temperature dependent thermal conductivity.

Acknowledgments
This work was funded by MIUR (PRIN/COFIN 2009 project).

References

[1] Shah R K and London A L 1978 (New York: Academic Press)
[2] Kakaç S 1987 *Handbook of Single-Phase Convective Heat Transfer* ed S Kakaç, R K Shah and W Aung (New York: Wiley) chapter 18
[3] Herwig H 1985 *Int. J. Heat Mass Transfer* **28** 423
[4] Herwig H, Voigt M and Bauhaus F J 1989 *Int. J. Heat Mass Transfer* **32** 1907
[5] Herwig H and Mahulikar S P 2006 *Int. J. Thermal Sci.* **45** 977
[6] Mahulikar S P and Hervig H 2006 *Phys. Fluids* **18** 073601
[7] Hooman K, Hooman F and Famouri M 2009 *Int. Commun. Heat Mass Transfer* **36** 192
[8] Hooman K and Ejlali A 2010 *Int. Commun. Heat Mass Transfer* **37** 34
[9] Nóbrega J M, Pinho F T, Oliveira P J and Carneiro O S 2004 *Int. J. Heat Mass Transfer* **47** 1141
[10] Shannon R L and Depew C A 1969 *J. Heat Transfer* **91** 251
[11] Joshi S D and Bergles A E 1980 *AIChE Symp. Ser.* **76** 270
[12] Nonino C, Del Giudice S and Savino S 2006 *Int. J. Heat Mass Transfer* **49** 4469
[13] Del Giudice S, Savino S and Nonino C 2011 *J. Heat Transfer, Trans. ASME* **133** 101702 p 1-11.
[14] Del Giudice S, Savino S and Nonino C 2010 *Proc. 28th UIT Heat Transfer Congress (Brescia, Italy)* (UIT) p 231
[15] Del Giudice S, Savino S and Nonino C 2012 *Proc. 10th ICNMM2012 (Puerto Rico, USA)* (ASME)
[16] Nonino C, Del Giudice S and Comini G 1988 *Numer. Heat Transfer* **13** 451
[17] Lin C R and Chen C K 1994 *J. Phys. D: Appl. Phys.* **27** 29
[18] Costa A and Macedonio G 2002 *Geophysics Research Letters* **29** Art. 1402
[19] Costa A and Macedonio G 2003 *Nonlinear Processes in Geophysics* **10** 545
[20] Patankar S V and Spalding D B 1972 *Int. J. Heat Mass Transfer* **15** 1787
[21] Hirsh C 1988 vol 1 (New York: Wiley) p 70
[22] Javeri V 1977 *Wärme- und Stoffübertragung* **10** 137
[23] Del Giudice S, Nonino C and Savino S 2007 *Int. J. Heat and Fluid Flow* **28** 15
[24] Nonino C, Del Giudice S and Savino S 2007 *J. Heat Transfer, Trans. ASME* **129** 1187
[25] Nonino C, Del Giudice S and Savino S 2010 *Heat Transfer Eng.*, **31** 682
[26] Nonino C 2003 *Numer. Heat Transfer* B **44** 61

Journal of Physics: Conference Series **501** (2014) 012002 doi:10.1088/1742-6596/501/1/012002

Inverse estimation of the local heat transfer coefficient in curved tubes: a numerical validation

F Bozzoli[1], L Cattani[1] , C Corradi[2], M Mordacci[2] and S Rainieri[1,2]

[1] Department of Industrial Engineering, University of Parma, Parco Area delle Scienze 181/A I-43124 Parma, Italy
[2] SITEIA.PARMA Interdepartmental Centre, University of Parma, Parco Area delle Scienze 181/A, I-43124 Parma, Italy

E-mail address of the corresponding Author (Sara Rainieri): sara.rainieri@unipr.it

Abstract. Wall curvature represents one of the most used passive techniques to enhance convective heat transfer. The effectiveness of wall curvature is due to the fact that it gives origin to the centrifugal force: this phenomenon induces local maxima in the velocity distribution that locally increase the temperature gradients at the wall by then maximizing the heat transfer. This fact brings to a significant variation of the wall temperature and of the wall heat flux along the circumferential coordinate. The convective heat transfer coefficient is consequently not uniformly distributed along the tube's perimeter and is characterized by higher values at the extrados wall surface in comparison to the ones at the intrados wall surface. Therefore, for predicting the overall performance of heat transfer apparatuses that involve the use of curved tubes, it becomes important to know the local distribution of the convective heat transfer coefficient not only along the axis of the heat transfer section, but also on the internal tube's surface along the cross section circumference. The present paper is intended to the assessment of a procedure developed to evaluate the local convective heat transfer coefficient, along the circumferential coordinate, at the internal wall of a coiled pipe.

1. Introduction

Among the most used passive techniques to enhance convective heat transfer, wall curvature is found. The effectiveness of wall curvature is due to the fact that it gives origin to the centrifugal force: this phenomenon induces local maxima in the velocity distribution that locally increase the temperature gradients at the wall by then maximizing the heat transfer [1-3]. The asymmetrical distribution of the velocity field over the tube's cross-section leads to a significant variation of the wall temperature and of the wall heat flux along the circumferential coordinate: the convective heat transfer coefficient is then not uniformly distributed along the tube's perimeter but it presents higher values at the extrados wall surface in comparison to the ones registered at the intrados wall surface.

This irregular distribution could be critical in some industrial application, such as in the ones that involve a thermal process. For instance, in food pasteurization, the irregular temperature field induced by the wall curvature could reduce the bacteria heat-killing or could locally overheat the product. Therefore, in order to predict the overall performance of heat transfer apparatuses that involve the use of curved tubes it is necessary to know the local distribution of the convective heat transfer coefficient not only along the axis of the heat transfer section, but also on the internal tube's surface along the cross section circumference.

Although many Authors have investigated the forced convective heat transfer in coiled tubes, most of them have presented the results only in terms of the Nusselt number averaged along the wall circumference: only few Authors have studied the phenomenon locally. Jayakumar et al.[4] numerically analyzed the turbulent heat transfer in helically coiled tubes and presented the local Nusselt number at various cross sections along the curvilinear coordinate. The results showed that, on any cross section, the highest Nusselt number is on the outer side of the coil, while the lowest one is expected on the inner sides. Moreover, the Authors proposed a correlation for predicting the local Nusselt number as a function of average Nusselt number and angular location for both constant temperature and constant heat flux boundary conditions. Bai et al. [5] studied experimentally the turbulent heat transfer in helically coiled tubes using deionized water as working fluid. As expected they found out that the local heat transfer coefficient was not evenly distributed along the periphery on the cross section and that in particular, at the outside surface of the coil it was up-to four times than that at the inside surface. One of the most promising way to estimate the local convective heat transfer coefficient on the interior wall surface of a solid domain is found in the solution of the inverse heat conduction problem (IHCP) in the wall, starting from the temperature distribution acquired on the external wall surface [6]. However IHCPs generally present some complications due to the fact that they are ill-posed: this entails that they show a great sensitiveness to small variations in the input data. In order to find a solution to this problem many methods based on a numerical approach have been improved: among these techniques the conjugate gradient iterative method based on the adjoint problem [7], the Laplace transform method [6], the direct sensitivity coefficient method [8], the space marching method [9], the sequential function specification method [10], the maximum entropy method [11], the Tikhonov regularization method [12] and the filtering technique [13-14] are found.

Tikhonov method is perhaps the most common used regularization technique: it is based on the minimization of an object function in order to reformulate the IHCP as a well-posed problem. The object function is expressed by the sum of the squared difference between measured and estimated temperature discrete data and of a regularization parameter times a term that expresses the smoothness of the unknown quantity. This regularization scheme, originally suggested by Tikhonov and Arsenin [12], enables to overcome the problem's instability in case of particularly critical signal to noise ratio and it proved to be very successful, although the selection of the regularization parameter requires some care.

2. Problem's definition

A typical condition in which curved tubes are tested is the one in which the convective heat transfer within the fluid that flows inside the tube occurs under the uniform heat flux boundary conditions, such as the one considered by Rainieri et al. [1,15] in their experimental investigations where a heat flux was dissipated by Joule effect directly within the tube wall.

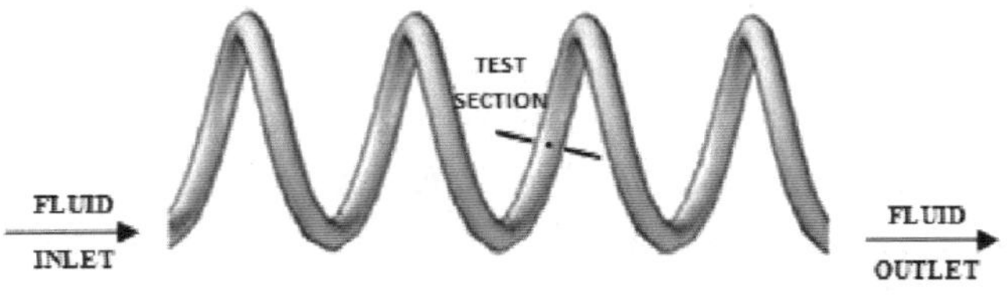

Figure 1: Coiled tube.

In order to evaluate the local actual value of the convective heat transfer coefficient at the fluid internal wall interface on a given cross section (as highlighted in figure 1) the following procedure could be followed: the temperature distribution is acquired on the external wall surface on a given test section and then the IHCP in the wall domain is solved by considering the convective heat transfer coefficient distribution on the internal wall surface to be unknown.

The temperature on the external side could be acquired by using thermocouples, located along the circumference of the text section or more suitably by adopting an infrared thermographic system.

In order to test the above described parameter estimation procedure a simplified 2-D numerical model of the test section (sketched in figure 2) was formulated by assuming that along the axis of the tube the temperature gradient is almost negligible.

In the 2-D solid domain the steady state energy balance equation is expressed in the form:

$$\frac{k}{r}\frac{\partial}{\partial r}\left(r\frac{\partial T}{\partial r}\right)+\frac{k}{r^2}\frac{\partial}{\partial \phi}\left(\frac{\partial T}{\partial \phi}\right)+q_g=0 \tag{1}$$

where q_g is the heat generated per unit volume and k is the tube wall thermal conductivity.

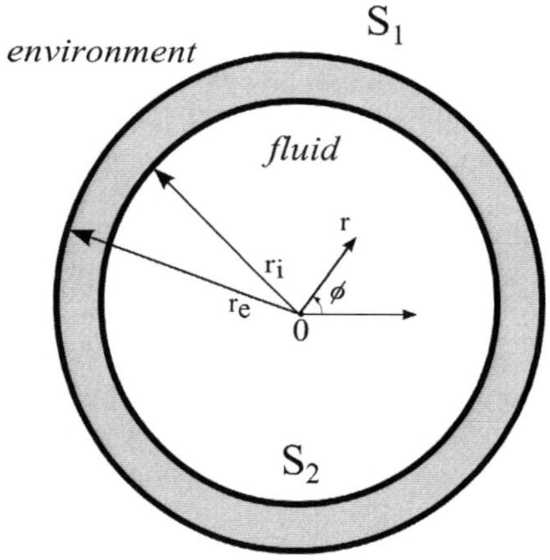

Figure 2. Geometrical domain with coordinate system.

The following two boundary conditions completed the energy balance equation:

$$k\frac{\partial T}{\partial r}=\alpha_e\cdot\left(T_e-T(\phi)\right) \tag{2}$$

that is applied on surface S_1 and where α_e is the overall heat transfer coefficient between the tube wall and the surrounding environment with the temperature T_e ;

$$-k\frac{\partial T}{\partial r}=h_{int}\cdot\left(T_b-T(\phi)\right) \tag{3}$$

that is applied on surface S_2 and where T_b is the bulk fluid temperature and h_{int} is the local convective heat transfer coefficient at the fluid- internal wall interface.

From the knowledge of the total wall heat flux and by measuring the temperature of the fluid at the inlet section, it is possible to calculate the bulk fluid temperature on the tube's cross-section by imposing that the energy balance is satisfied.

3. Parameter estimation procedure

The parameter estimation procedure is embedded in the inverse solution of the problem expressed by equations (1-3). The temperature distribution on the external surface of the section, $T(\phi, r = r_e, h_{int})$, can be easily computed numerically, by imposing a trial convective heat transfer coefficient distribution on the internal wall side h_{int}.

In the inverse formulation of problem, this computed temperature distribution is forced to match the experimental temperature distribution $\Theta(\phi)$, by tuning the convective heat transfer coefficient distribution on the internal wall side, $h_{int}(\phi)$ .The matching of the two temperature distributions (the computed and the experimentally acquired) could be easily performed under a least square approach. To overcome the ill-posed nature of the problem and in particular its ill-conditioned character the Tikhonov regularization method is adopted [12]. This approach, successfully applied in literature [16] allows to reformulate the problem as a well-posed problem by minimizing the following objective function:

$$J(h_{int}(\phi)) = \|T(\phi, r = r_e, h_{int}) - \Theta(\phi)\|_2^2 + \lambda^2 \|D^n\, h_{int}\|_2^2 \quad (4)$$

where λ is the regularization parameter, D is the derivative operator and $T(\phi, r = r_e, h_{int})$ is the distribution of external surface temperature derived from direct numerical solution of the problem obtained by imposing a convective heat transfer coefficient distribution on the internal wall side h_{int}. Often D is the zero, first or second derivative operator: in this work the second derivative formulation was chosen in order to preserve local variation of heat transfer coefficient distribution. Under this assumption the object function becomes:

$$J(h_{int}(\phi)) = \|T(\phi, r = r_e, h_{int}) - \Theta(\phi)\|_2^2 + \lambda^2 \left\| \frac{d^2 h_{int}(\phi)}{d\phi^2} \right\|_2^2 \quad (5)$$

The above function represents a trade-off between two optimization processes: agreement between the data and solution and smoothness or stability of the solution. Thus, an appropriate choice of λ should give an optimal balance. The importance of the choice of the regularization parameter was analyzed by Bazàn and Borges [17]. Choosing an high regularization parameter means to impose too much regularization to the solution prejudicing the fitting of the data and obtaining a great residual; the absence of regularization or a too small regularization parameter will bring to a good fitting but also to a solution affected by the data errors. This idea has led to the development of the L-curve, which was first proposed by Hansen and O'Leary [18]. This method defines the ideal regularization parameter by locating the 'corner' on a plot of the function of the norm of the second derivative of computed convective heat transfer coefficient, $\|d^2 h_{int}(\phi)/d\phi^2\|_2$, versus the norm of the difference between experimental and computed temperature values, $\|T(\phi, r = r_e, h_{int}) - \Theta(\phi)\|_2$.

In the present analysis in order to limit the number of degrees of freedom in the inverse problem approach, the convective heat transfer distribution was simplified by considering a continuous piecewise linear function composed of six sections as follows:

$$h_{int}(\phi) = \begin{cases} b_0 + \phi \cdot \left(\frac{b_1 - b_0}{\pi/3}\right), & 0 \le \phi < \frac{\pi}{3} \\ b_1 + \left(\phi - \frac{\pi}{3}\right) \cdot \left(\frac{b_2 - b_1}{\pi/3}\right), & \frac{\pi}{3} \le \phi < \frac{2\pi}{3} \\ b_2 + \left(\phi - \frac{2\pi}{3}\right) \cdot \left(\frac{b_3 - b_2}{\frac{\pi}{3}}\right), & \frac{2\pi}{3} \le \phi \le \pi \\ b_3 + (\phi - \pi) \cdot \left(\frac{b_4 - b_3}{\frac{\pi}{3}}\right), & \pi \le \phi \le \frac{4\pi}{3} \\ b_4 + \left(\phi - \frac{4\pi}{3}\right) \cdot \left(\frac{b_5 - b_4}{\frac{\pi}{3}}\right), & \frac{4\pi}{3} \le \phi \le \frac{5\pi}{3} \\ b_5 + \left(\phi - \frac{5\pi}{3}\right) \cdot \left(\frac{b_0 - b_5}{\frac{\pi}{3}}\right), & \frac{5\pi}{3} \le \phi \le 2\pi \end{cases} \quad (6)$$

where b_0 b_1 b_2 b_3 b_4 b_5 values become the six unknowns of the inverse problem.

4. Results and discussion

The objective of this paper was to validate the above described procedure of estimation of the local convective heat transfer coefficient. The physical parameters used in this work correspond to the ones of a stainless steel type AISI 304. For what concern the geometrical parameters an internal radius r_i of 7 mm and an external radius r_e of 8 mm were considered. The validation of the procedure was performed by adopting synthetic data. By imposing a known distribution of h_{int} and by solving the governing equations (1-3), a synthetic temperature distribution on the external wall surface was obtained. For the local convective heat transfer coefficient, two different distributions ${h_{int}}^{(A)}$ and ${h_{int}}^{(B)}$ according to the data of Jayakumar et al. [4] and shown in figures 3-4, were considered.

The two distributions were derived by numerical simulations performed under the turbulent flow regime and they have been chosen in this work because they represent two extreme cases: the former distribution, that corresponds to a representative axial location in the thermal entry region, is almost flat and it is characterized by a maximum to minimum value ratio of about 1.3, the latter, that corresponds to the thermally fully developed region, shows a significant variation along the curvilinear coordinate and it is characterized by a maximum to minimum value ratio of about 5. Then the synthetic temperature distribution on the external wall surface, deliberately spoiled by random noise, was used as the input data of the inverse problem. In particular, a white noise characterized by a standard deviation of 0.1 was considered. By solving the inverse problem, i.e. by forcing the external surface temperature $T(\phi, r = r_e, h_{int})$ to match the synthetic temperature distribution $\Theta(\phi)$, it's possible to restore the convective heat transfer coefficient distribution on the internal wall side $h_{int}(\phi)$, that is the only term unknown.

The $h_{int}(\phi)$ distribution found was then compared to the distribution used to obtain the synthetic temperature $\Theta(\phi)$ in order to evaluate the effectiveness of the parameter estimation procedure.

Given the symmetry of the two distributions of the convective heat transfer coefficient, the number of the unknown variables reduces to four, being $b_2 = b_4$ and $b_1 = b_5$. This enabled to further reduce the computational cost of the minimization algorithm.

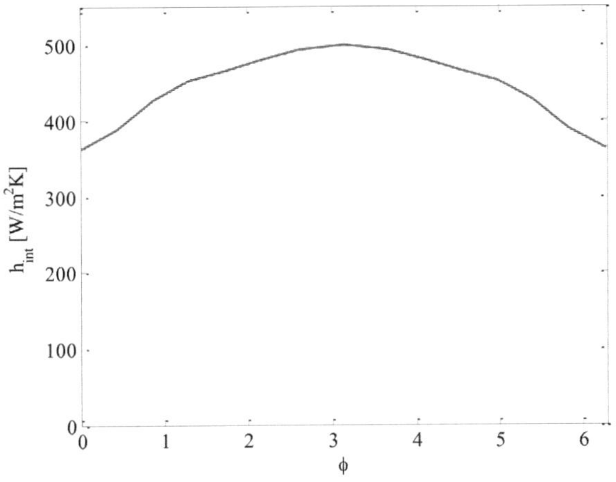

Figure 3. First distribution of the local convective heat transfer coefficient $h_{int}{}^{(A)}$ (Jayakumar et al. [4]).

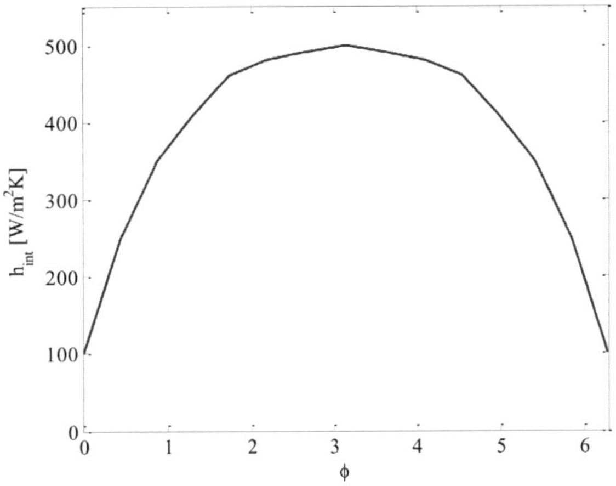

Figure 4. Second distribution of the local convective heat transfer coefficient $h_{int}{}^{(B)}$ (Jayakumar et al. [4]).

The direct problem was solved by finite element method implemented in Comsol Multiphysics® environment with a mesh of about 2400 triangular elements while the minimization of Tikhonov target function was run within the Matlab Optimization Toolbox® by using as stopping criterion a relative tolerance on object function lower than 1e-004. The algorithm adopted in the minimization process is the Nelder-Mead algorithm [19] that represents one of the best known solution for multidimensional unconstrained optimization.

In figure 5 the L-curve for the distribution $h_{int}{}^{(A)}$ is shown. Following the above described L-curve method, the optimal value $\lambda = 0.0003$, corresponding to the 'corner' of the curve, was identified.

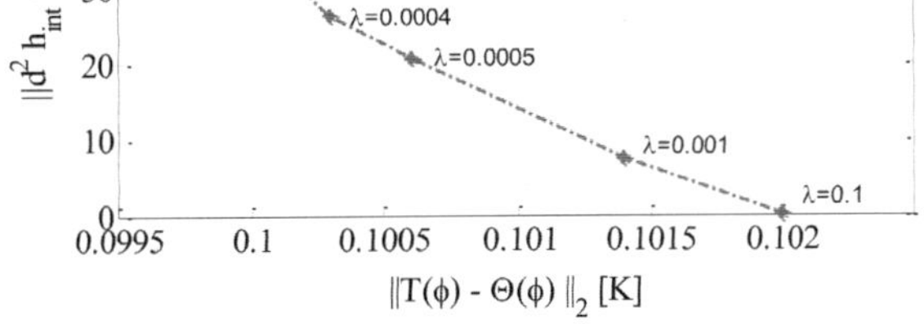

Figure 5. L-curve for distribution $h_{int}{}^{(A)}$.

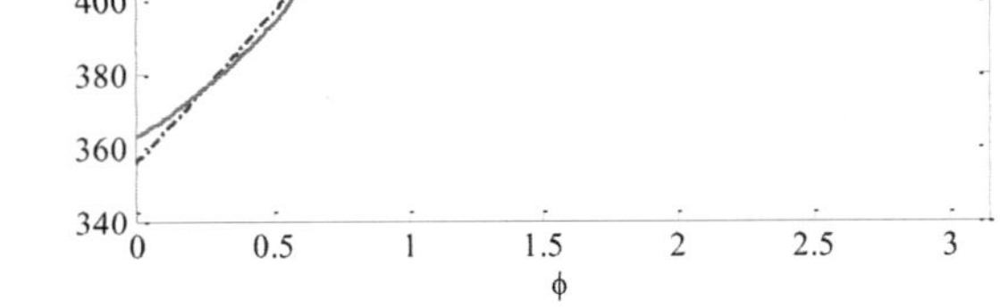

Figure 6. Restored and exact local heat transfer coefficient distribution ($h_{int}{}^{(A)}$, $\lambda = 0.0003$).

In figure 6 the heat transfer coefficient distribution, restored by adopting the regularization parameter $\lambda = 0.0003$, is reported and compared to the exact value.

The restored values matched with a good approximation the original local heat transfer distribution h_{int} (A), by demonstrating the robustness of the procedure presented in this paper.

In figure 7 the restored temperature distribution resulting from the minimization procedure, implemented by considering λ = 0.0003, is shown. The data confirm that, by adopting the optimal regularization parameter, the Tikhonov scheme enables to filter out the noise from the raw input signal by restoring the smoothest approximate solution compatible with the experimental data within a given noise level [13,14].

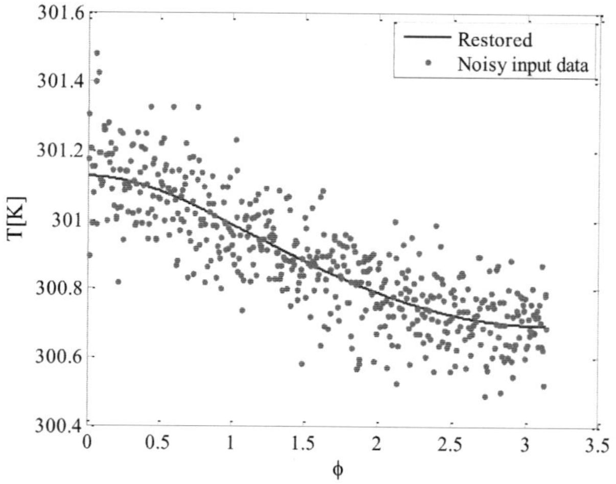

Figure 7. Restored temperature distribution (h_{int} (A) , λ = 0.0003).

It is well known that the smoothness is controlled by the choice of the regularization parameter: if the regularization parameter is not the correct one the problem's ill-conditioned character could lead to an incorrect solution. For instance by considering the values λ = 0 and λ = 0.1, corresponding to the two extreme values of the L-curve, the local heat transfer coefficient obtained from minimization of the function given by equation (4) doesn't match the exact distribution as it is shown in figures 8 and 9.

In both cases (λ = 0 and λ = 0.1) the parameter estimation procedure fails by proving that the selection of the regularization parameter is a critical task within Tikhonov regularization: if it is too small the filtering of the raw signal is too weak, while if it is too large the regularized solution is over-smoothed.

The same procedure was applied to the second local heat transfer coefficient distribution ${h_{int}}^{(B)}$: the L-curve is reported for this case in figure 10 and the optimal value of λ chosen, corresponding to the 'corner' of the curve resulted λ = 0.0002.

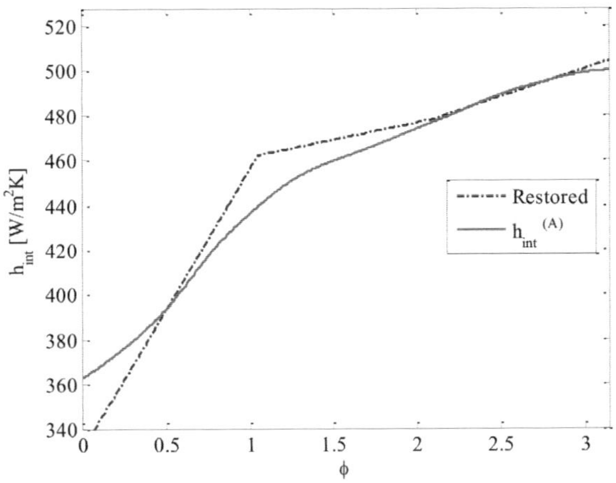

Figure 8. Restored and exact local heat transfer coefficient ($h_{int}{}^{(A)}$, $\lambda = 0$)

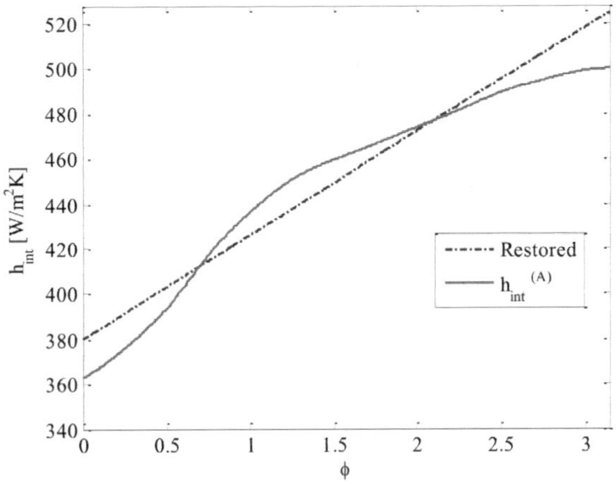

Figure 9. Restored and exact local heat transfer coefficient ($h_{int}{}^{(A)}$, $\lambda = 0.1$).

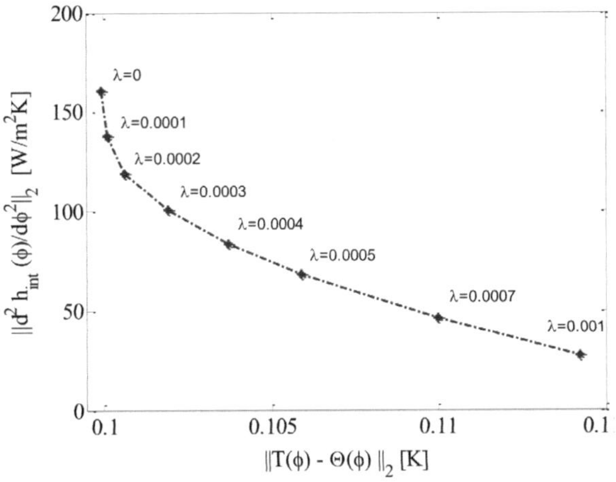

Figure 10. L-curve for distribution $h_{int}{}^{(B)}$

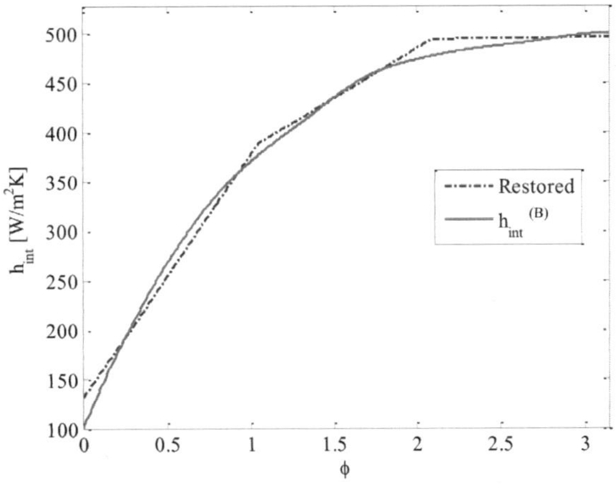

Figure 11. Restored and exact local heat transfer coefficient ($h_{int}{}^{(B)}$, $\lambda = 0.0002$).

The restored values of local heat transfer coefficient are reported in figure 11. Also in this case there is a good matching between the restored and the exact distributions when the regularization parameter is identified by means of the L-curve method. In order to highlight the effect of the choice of the regularization parameter a residual analysis could be performed by plotting the estimation error, defined as follows:

$$E = \frac{\left\| \left(h_{int}(\phi)\right)_{restored} - \left(h_{int}(\phi)\right)_{exact} \right\|_2^2}{\left\| \left(h_{int}(\phi)\right)_{exact} \right\|_2^2} \tag{7}$$

versus the regularization parameter λ.

The estimation error is reported in figure 12 against the regularization parameter λ for the representative case corresponding to the distribution $h_{int}{}^{(B)}$: as expected, the data show that the minimum error corresponds to the optimal λ value, that for this case was identified equal to 0.0002 by the L-curve method. Moreover, these data confirm the necessity of the regularization scheme for enhancing the parameter estimation accuracy.

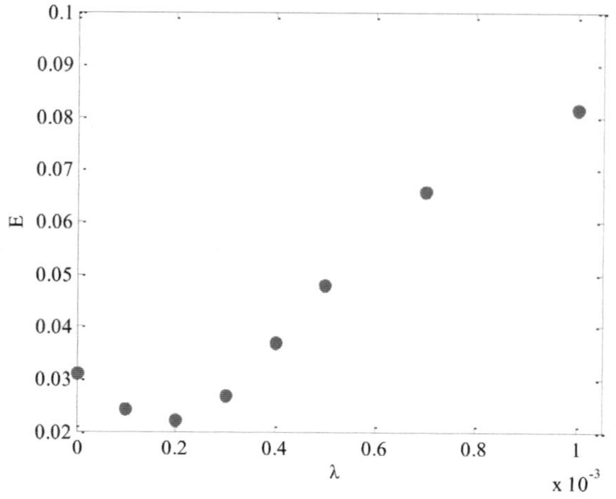

Figure 12. Estimation error for the distribution $h_{int}^{(B)}$.

5. Conclusions

In the present paper a procedure to estimate the local convective heat transfer coefficient along the circumferential coordinate at the internal wall of a coiled pipe and based on the solution of the Inverse Heat Conduction Problem (IHCP) was presented and validated. This problem could be particularly interesting in applications that require the knowledge of the local heat transfer distribution at the fluid-wall interface, such as in food industries.

The validation of the procedure was here performed throughout its application to numerical data.

A simplified 2-D numerical model of the test section was formulated by assuming that along the axis of the tube the temperature gradient is almost negligible. A synthetic temperature distribution on the external wall surface was then obtained by imposing a given distribution of the local convective heat transfer coefficient on the internal side of the wall. Regarding the distribution of the local convective heat transfer coefficient, the results of Jayakumar et al. [4] were considered. The synthetic temperature distribution on the external wall surface, deliberately spoiled by random noise, was then used as the input data of the inverse problem. The IHCP solution was derived by the Tikhonov regularization technique coupled to the L-curve method.

The results showed that the restored values of the convective heat transfer coefficient match with a good approximation with the exact distribution, by demonstrating the robustness of the here addressed parameter estimation procedure. The results enabled also to highlight the criticality associated to the choice of the regularization parameter successfully identified by the L-curve method. A necessary further step of the research is the application of the estimation procedure to real experimental data.

Nomenclature

Symbol	Quantity	SI Unit
E	Estimation error, equation (7)	-
h	Convective heat transfer coefficient	W/m^2K
k	Thermal conductivity	W/mK
q_g	Internal heat generation per unit volume	W/m^3
r	Radial coordinate	m
T	Temperature	K

α	Overall heat transfer coefficient	W/m^2K
ϕ	Angular coordinate	rad
λ	Regularization parameter	-

Subscripts

b	bulk
e	external
i,int	internal

References

[1] Rainieri S, Bozzoli F, Cattani L and Pagliarini G 2013 Compound convective heat transfer enhancement in helically coiled wall corrugated tubes *Int J Heat Mass Tran* **59** 353
[2] Rainieri S, Bozzoli F and Pagliarini G 2012 Experimental investigation on the convective heat transfer in straight and coiled corrugated tubes for highly viscous fluids: Preliminary results *Int J Heat Mass Tran* **55** 498
[3] Rainieri S, Bozzoli F, Schiavi L and Pagliarini G 2011 Numerical analysis of convective heat transfer enhancement in swirl tubes *Int J Numer Method H* **21**(5) 559
[4] Jayakumar JS, Mahajani SM,. Mandal JC, Iyer KN and Vijayan PK 2010 CFD analysis of single-phase flows inside helically coiled tubes *Comput Chem Eng* **34** 430
[5] Bai B, Guo L, Feng Z and Chen X 1999 Turbulent Heat Transfer in a Horizontal Helically Coiled Tube *Heat Tran Asian Res* **5** 28
[6] Beck JV, Blackwell B and St.Clair Jr. C R 1985 *Inverse Heat Conduction – Ill-posed problems* (New York: John Wiley & Sons, Inc.)
[7] Alifanov O M 1994 Inverse Heat Transfer Problem (Berlin:Springer)
[8] Tseng A A, Chen T C and Zhao F Z 1995 Direct sensitivity coefficient method for solving two-dimensional inverse heat conduction problems by finite-element scheme 1995 *Numer Heat Tr* **27** 291
[9] D'Souza N 1965 Numerical soultion of one-dimensional inverse transient heat conduction by finite difference method *ASME Paper* No. 75-WA/HT-81
[10] Chantasiriwan S 1999 Comparison of three sequential function specification algorithms for the inverse heat conduction problem *Int Commun Heat Mass* **26** 115
[11] Sun Kyoung Kim and Woo Il Lee 2002 Solution of inverse heat conduction problems using maximum entropy method *Int J Heat Mass Tran* **45** 381
[12] Tikhonov A N and Arsenin V Y 1977 *Solution of Ill-posed Problems* (Washington, DC: Winston & sons)
[13] Rainieri S, Bozzoli F and Pagliarini G 2008 Characterization of an uncooled infrared thermographic system suitable for the solution of the 2-D inverse heat conduction problem *Exp Therm Fluid Sci* **32**(8) 1492-1498
[14] Bozzoli F, Pagliarini G and Rainieri S 2013 Experimental validation of the filtering technique approach applied to the restoration of the heat source field *Exp Therm Fluid Sci* **44** 858
[15] Rainieri S, Bozzoli F, Cattani L and Pagliarini G 2012 Experimental investigation on the convective heat transfer enhancement for highly viscous fluids in helical coiled corrugated tubes *J Phys Conf Ser* **395**
[16] Orlande HRB, Olivier F, Maillet D and Cotta RM 2011 *Thermal measurements and inverse techniques* (New York : Taylor and Francis Group)
[17] Bazán FSV and Borges LS 2010 GKB-FP: an algorithm for large-scale discrete ill-posed problems *Bit Numer Math* **50** 481
[18] Hansen P and O'Leary D 1993 The use of the L-curve in the regularization of discrete ill-posed problems *SIAM J Sci Comput* **14** 1487
[19] Nelder J A and Mead R 1965 A simplex method for function minimization *Comp J* **7** 308

Journal of Physics: Conference Series **501** (2014) 012003 doi:10.1088/1742-6596/501/1/012003

On the thermal instability in a horizontal rectangular porous channel heated from below by a constant flux

Antonio Barletta[1], Eugenia Rossi di Schio[1] and Leiv Storesletten[2]
[1]Department of Industrial Engineering, Alma Mater Studiorum Università di Bologna, Viale Risorgimento 2, 40136 Bologna, Italy
[2]Department of Mathematics, University of Agder, Postboks 422, 4604 Kristiansand, Norway
E-mail: antonio.barletta@unibo.it

Abstract. The onset of thermoconvective instability in a rectangular horizontal channel filled with a fluid-saturated porous medium is studied. The channel is heated from below with a constant flux. The top wall is maintained at a uniform constant temperature, while the lateral boundaries are permeable and perfectly conducting. The stability of the basic motionless state is analysed with respect to small-amplitude disturbances. The eigenvalue problem for the neutral stability condition is solved analytically for the normal modes. A closed-form expression is obtained for the implicit neutral stability relation between the Darcy-Rayleigh number and the longitudinal wave number. The critical condition, *viz.* the absolute minimum of the Darcy-Rayleigh number for the instability, is determined for different aspect ratios of the rectangular cross-section. The preferred modes under critical conditions are detected. It is found that the selected patterns of instability at the critical Rayleigh number may be two-dimensional, for slender or square cross-sections of the channel. On the other hand, instability is three dimensional when the critical width-to-height ratio, 1.350517, is exceeded.

1. Introduction

In the last decades, many authors investigated the onset of convection in a fluid saturated porous medium heated from below, starting from the pioneering papers by Horton and Rogers [1], and by Lapwood [2]. Nowadays, this topic is termed, in a wide sense, either Horton-Rogers-Lapwood problem, or Darcy-Bénard problem. Surveys on the main results obtained in this field were written by Rees [3], Tyvand [4], Nield and Bejan [5], Straughan [6], Barletta [7]. The Darcy-Bénard problem was originally formulated for an infinitely wide horizontal saturated porous layer in a motionless state, bounded by two horizontal, isothermal and impermeable walls maintained at different temperatures. In later investigations, various types of thermal boundary conditions, involving all possible combinations of isothermal and isoflux conditions, were analysed. Furthermore, the condition of a finite lateral width of the channel was analysed as well. The case of a horizontal porous channel with rectangular cross-section was studied by Sutton [8], Beck [9], and further explored by Nilsen and Storesletten [10].

Recently, Barletta and Storesletten [11] studied the onset of convection in a horizontal rectangular channel, bounded above and below with impermeable isothermal walls at different temperatures, but with partially conducting sidewalls. These authors carried out a three-dimensional linear stability analysis both for a channel with a longitudinal infinite length and for a finite longitudinal length. The linear disturbance equations were solved by Galerkin's method of

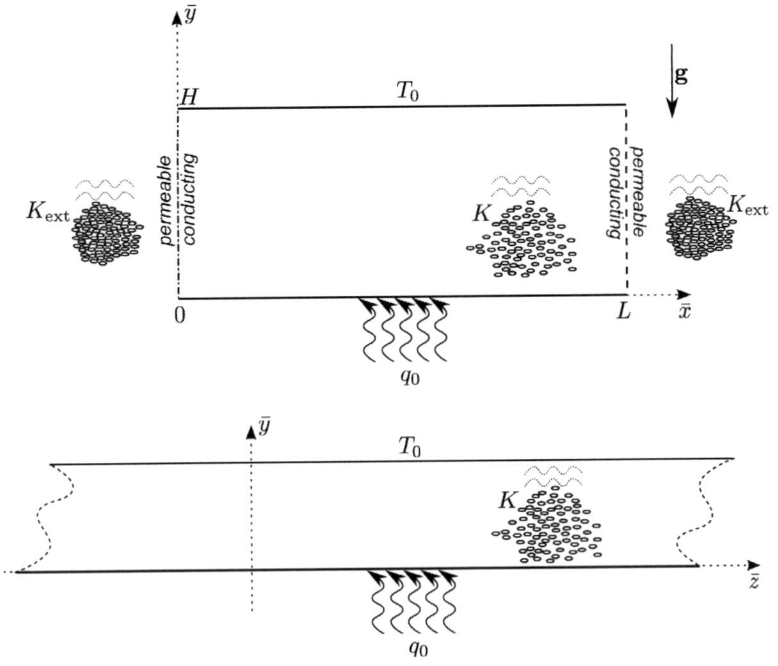

Figure 1. Sketch of the rectangular porous channel.

weighted residuals, as well as by a sixth-order Runge-Kutta method combined with the shooting method. The thermal behaviour of the vertical sidewalls was modelled through a Biot number associated with the convection to an external, thermally stratified, fluid environment.

The aim of the present paper is to extend the analysis presented in Ref. [11]. We will consider a horizontal rectangular porous channel, having infinite length in the longitudinal direction. The lower horizontal wall is impermeable and subject to an upward uniform heat flux, while the upper horizontal wall is impermeable and isothermal. The permeable side boundaries are assumed to be in perfect thermal and mechanical contact with an external porous reservoir, saturated by the same fluid and thermally stratified in the vertical direction, so that Dirichlet temperature and pressure conditions can be prescribed. The local momentum balance is modelled by Darcy's law, while the local energy balance is formulated by assuming local thermal equilibrium between the solid and the fluid phase. With the fluid that saturates the porous channel initially at rest, a three-dimensional linear stability analysis will be carried out. The governing equations for the linear disturbances of the basic state will be solved analytically. The neutral stability curves, as well as the critical values of the Darcy-Rayleigh number, will be determined for assigned values of the transverse aspect ratio.

2. Mathematical model

Let us devise a horizontal channel filled with a fluid saturated porous medium. In the following, the dimensional fields, coordinates, time and nabla operator are denoted by an overline. The channel is rectangular with height H and width L, and we choose a Cartesian coordinate system with the $\bar{y}-$ axis in the vertical direction and the $\bar{x}-$axis in the horizontal direction perpendicular to the channel $\bar{z}-$axis. The horizontal channel boundaries are at $\bar{y} = 0$ and $\bar{y} = H$, and the vertical boundaries at $\bar{x} = 0$ and $\bar{x} = L$ (see figure 1).

The components of the seepage velocity $\bar{\mathbf{u}}$, relative to the Cartesian set of coordinates $(\bar{x}, \bar{y}, \bar{z})$, are denoted as $(\bar{u}, \bar{v}, \bar{w})$, respectively. Let us assume that the porous medium is homogeneous and isotropic, that the effect of viscous dissipation is negligible, and that the solid and fluid phases are in local thermal equilibrium. We assume the validity of Darcy's law and of the Oberbeck-Boussinesq approximation. Then, the local balance equations of mass, momentum and energy are written as

$$\begin{aligned}
&\bar{\nabla} \cdot \bar{\mathbf{u}} = 0, \\
&\frac{\mu}{K}\,\bar{\mathbf{u}} = -\bar{\nabla}\bar{p} + \varrho\, g\, \beta\left(\bar{T} - T_0\right)\mathbf{e}_y, \\
&\sigma\,\frac{\partial \bar{T}}{\partial \bar{t}} + \bar{\mathbf{u}} \cdot \bar{\nabla}\bar{T} = \alpha\,\bar{\nabla}^2\bar{T},
\end{aligned} \tag{1}$$

where $\mathbf{g} = -g\,\mathbf{e}_y$ is the gravitational acceleration with $\mathbf{e}_y$ the unit vector along the vertical $\bar{y}-$axis, μ is the dynamic viscosity, K is the permeability, $\bar{p}$ is the dynamic pressure, *viz.* the local difference between the pressure and the hydrostatic pressure. Moreover, ρ is the reference density, β is the thermal expansion coefficient, $\bar{t}$ is time, $\bar{T}$ is temperature and T_0 is the reference temperature, α is the average thermal diffusivity, σ is the ratio between the average volumetric heat capacity of the fluid saturated porous medium and the volumetric heat capacity of the fluid.

The channel horizontal walls are impermeable, with the lower wall ($\bar{y} = 0$) subject to a positive uniform heat flux $q_0 = k\,\Delta T/H$, with k denoting the average thermal conductivity of the saturated porous medium. On the other hand, the upper wall ($\bar{y} = H$) is isothermal with temperature T_0. The lateral boundaries ($\bar{x} = 0, L$) are assumed to be perfectly permeable and in perfect thermal contact with an external porous reservoir saturated by the same fluid. The external reservoir is in a steady state thermally stratified in the vertical direction. The temperature field and the dynamic pressure field in the external reservoir are given by

$$\bar{T}_{\text{ext}}(\bar{y}) = T_0 + \Delta T\left(1 - \frac{\bar{y}}{H}\right), \tag{2a}$$

$$\bar{p}_{\text{ext}}(\bar{y}) = p_0 + \rho g \beta \Delta T\,\bar{y}\left(1 - \frac{\bar{y}}{2H}\right), \tag{2b}$$

where p_0 is a reference pressure. The permeability K_{ext} of the porous reservoir external to the rectangular channel is assumed to be much smaller than K, so that the steady state in the reservoir remains stable in spite of the possible onset of convection instability inside the porous channel.

On account of equations (2), the boundary conditions can be expressed as follows:

$$\begin{aligned}
&\bar{v} = 0, \quad -k\,\frac{\partial \bar{T}}{\partial \bar{y}} = q_0, \qquad \text{on} \quad \bar{y} = 0, \quad 0 < \bar{x} < L, \\
&\bar{v} = 0, \quad \bar{T} = T_0, \qquad \text{on} \quad \bar{y} = H, \quad 0 < \bar{x} < L, \\
&\bar{p} = \bar{p}_{\text{ext}}(\bar{y}), \quad \bar{T} = \bar{T}_{\text{ext}}(\bar{y}), \qquad \text{on} \quad \bar{x} = 0, L, \quad 0 < \bar{y} < H.
\end{aligned} \tag{3}$$

The dimensionless quantities are defined such that

$$\begin{aligned}
&(\bar{x}, \bar{y}, \bar{z}) = (x,\, y,\, z)\,H, \quad \bar{\mathbf{u}} = (\bar{u}, \bar{v}, \bar{w}) = (u,\, v,\, w)\,\frac{\alpha}{H} = \mathbf{u}\,\frac{\alpha}{H}, \\
&\bar{\nabla} = \frac{1}{H}\nabla, \quad \bar{T} = T_0 + \frac{q_0\,H}{k}\,T, \quad \bar{p} = p_0 + \frac{\mu\,\alpha}{K}\,p, \\
&\bar{t} = \frac{\sigma\,H^2}{\alpha}\,t, \quad s = \frac{L}{H}.
\end{aligned} \tag{4}$$

Journal of Physics: Conference Series **501** (2014) 012003 doi:10.1088/1742-6596/501/1/012003

The governing equations (1) can thus be rewritten in a dimensionless form as

$$\nabla \cdot \mathbf{u} = 0, \tag{5a}$$

$$\mathbf{u} = -\nabla p + R\, T\, \mathbf{e}_y, \tag{5b}$$

$$\frac{\partial T}{\partial t} + \mathbf{u} \cdot \nabla T = \nabla^2 T, \tag{5c}$$

where R is the Darcy-Rayleigh number, given by

$$R = \frac{\varrho g \beta q_0 K H^2}{k \mu \alpha}. \tag{6}$$

The boundary conditions (2) and (3) can be rewritten as

$$\begin{aligned} & v = 0, \quad \frac{\partial T}{\partial y} = -1, \quad \text{on} \quad y = 0, \quad 0 < x < s, \\ & v = 0, \quad T = 0, \quad \text{on} \quad y = 1, \quad 0 < x < s, \\ & p = R\, y \left(1 - \frac{y}{2}\right), \quad T = 1 - y, \quad \text{on} \quad x = 0, s, \quad 0 < y < 1. \end{aligned} \tag{7}$$

3. Basic solution and linear disturbances

A stationary solution of equations (5) and (7) is given by

$$u_b = v_b = w_b = 0, \quad T_b = 1 - y, \quad p_b = R\, y \left(1 - \frac{y}{2}\right). \tag{8}$$

where b stands for *basic solution.* We can adopt a pressure-temperature formulation of equations (5) by substituting equation (5b) into (5a) and (5c), thus yielding

$$\begin{aligned} & \nabla^2 p = R \frac{\partial T}{\partial y}, \\ & \nabla^2 T = \frac{\partial T}{\partial t} - \nabla p \cdot \nabla T + R\, T \frac{\partial T}{\partial y}. \end{aligned} \tag{9}$$

The boundary conditions now become

$$\begin{aligned} & \frac{\partial p}{\partial y} = R\, T, \quad \frac{\partial T}{\partial y} = -1, \quad \text{on} \quad y = 0, \quad 0 < x < s, \\ & \frac{\partial p}{\partial y} = 0, \quad T = 0, \quad \text{on} \quad y = 1, \quad 0 < x < s, \\ & p = R\, y \left(1 - \frac{y}{2}\right), \quad T = 1 - y, \quad \text{on} \quad x = 0, s, \quad 0 < y < 1. \end{aligned} \tag{10}$$

We perturb the basic solution (8) with small amplitude disturbances, defined by

$$p = p_b + \varepsilon\, P, \qquad T = T_b + \varepsilon\, \theta. \tag{11}$$

Following the usual linear stability analysis, we substitute equation (11) into (9) and we neglect the terms of order ε^2. Thus, we obtain

$$\begin{aligned} & \nabla^2 P = R \frac{\partial \theta}{\partial y}, \\ & \nabla^2 \theta = \frac{\partial \theta}{\partial t} - R\, \theta + \frac{\partial P}{\partial y}, \end{aligned} \tag{12}$$

to be subjected to the boundary conditions

$$\begin{aligned}
&\frac{\partial P}{\partial y} = R\,\theta, \quad \frac{\partial \theta}{\partial y} = 0, \qquad \text{on} \quad y = 0, \quad 0 < x < s, \\
&\frac{\partial P}{\partial y} = 0, \quad \theta = 0, \qquad \text{on} \quad y = 1, \quad 0 < x < s, \\
&P = 0, \quad \theta = 0, \qquad \text{on} \quad x = 0, s, \quad 0 < y < 1.
\end{aligned} \tag{13}$$

4. Normal modes

Due to the principle of exchange of stabilities, the study of the neutral stability condition is relative to the time-independent solutions of equations (12). Following the usual normal mode decomposition, we may write

$$\begin{aligned}
P(x, y, z) &= f_n(y) \sin\!\left(\frac{n\pi x}{s}\right) \cos(az), \\
\theta(x, y, z) &= h_n(y) \sin\!\left(\frac{n\pi x}{s}\right) \cos(az),
\end{aligned} \tag{14}$$

where $n = 1, 2, \ldots$, and a is the dimensionless wave number in the $z-$direction. We substitute equation (14) into (12) so that we obtain

$$f_n'' - \left(a^2 + \frac{n^2\pi^2}{s^2}\right) f_n - R\,h_n' = 0, \tag{15a}$$

$$h_n'' - \left(a^2 + \frac{n^2\pi^2}{s^2} - R\right) h_n - f_n' = 0, \tag{15b}$$

with the boundary conditions

$$\begin{aligned}
&f_n' = R\,h_n, \quad h_n' = 0, \qquad \text{on} \quad y = 0, \\
&f_n' = 0, \quad h_n = 0, \qquad \text{on} \quad y = 1.
\end{aligned} \tag{16}$$

In equations (15) and (16), primes denote derivatives with respect to y. Let us define

$$\varphi_n(y) = -f_n'(y) + R\,h_n(y), \tag{17a}$$

$$b_n^2 = a^2 + \frac{n^2\pi^2}{s^2}, \tag{17b}$$

then (15) and (16) can be reformulated as

$$\varphi_n'' - b_n^2 \varphi_n + R\,b_n^2 h_n = 0, \tag{18a}$$

$$h_n'' - b_n^2 h_n + \varphi_n = 0, \tag{18b}$$

with the boundary conditions

$$\begin{aligned}
&\varphi_n = 0, \quad h_n' = 0, \qquad \text{on} \quad y = 0, \\
&\varphi_n = 0, \quad h_n = 0, \qquad \text{on} \quad y = 1.
\end{aligned} \tag{19}$$

Journal of Physics: Conference Series **501** (2014) 012003 doi:10.1088/1742-6596/501/1/012003

5. Analytical solution

The solution of equations (18) and (19) can be determined by employing the usual method of the characteristic equation. Since the boundary value problem is homogeneous, the solution is to be determined up to an arbitrary scale factor. It is not restrictive to set the latter so that

$$h_n(0) = 1. \tag{20}$$

Then, we may write

$$\begin{aligned} \varphi_n(y) &= b_n\sqrt{R}\left[A_n \sin(\lambda_n y) + \frac{1}{2}\cos(\lambda_n y) + \frac{A_n\lambda_n}{\eta_n}\sinh(\eta_n y) - \frac{1}{2}\cosh(\eta_n y)\right], \\ h_n(y) &= A_n \sin(\lambda_n y) + \frac{1}{2}\cos(\lambda_n y) - \frac{A_n\lambda_n}{\eta_n}\sinh(\eta_n y) + \frac{1}{2}\cosh(\eta_n y), \end{aligned} \tag{21}$$

where

$$\begin{aligned} \lambda_n &= \sqrt{b_n\left(\sqrt{R} - b_n\right)}, \\ \eta_n &= \sqrt{b_n\left(\sqrt{R} + b_n\right)}, \\ A_n &= \frac{\eta_n\left(\cosh\eta_n - \cos\lambda_n\right)}{2\left(\eta_n \sin\lambda_n + \lambda_n \sinh\eta_n\right)}. \end{aligned} \tag{22}$$

Equations (21) and (22) ensure that (18) and (20) are satisfied, together with the boundary conditions, given by Eq. (19): $\varphi_n(0) = \varphi_n(1) = 0$ and $h'_n(0) = 0$. On the other hand, the last boundary condition, $h_n(1) = 0$, leads to the dispersion relation

$$\frac{\lambda_n \sinh\eta_n \cos\lambda_n + \eta_n \cosh\eta_n \sin\lambda_n}{\eta_n \sin\lambda_n + \lambda_n \sinh\eta_n} = 0. \tag{23}$$

Equations (17b), (22) and (23) allow one to plot the neutral stability curves in the parametric plane (a, R), for any given aspect ratio s.

6. Discussion of the results

The dispersion relation (23) can be displayed in the parametric plane (b_n, R). Obviously, if we do not specify the modal number n and the transverse aspect ratio s, the parameter b_n is nothing but a redefined wave number that can vary continuously over the set of positive real numbers. In this case, for the sake of simplicity, we denote b_n with b and plot equation (23) in the plane (b, R). This plot is shown in figure 2. There are two disconnected curves, each one displaying a minimum. The lower curve has a minimum for

$$b = 2.326215, \qquad R = 27.09763, \tag{24}$$

while the upper curve has a minimum for

$$b = 5.468656, \qquad R = 132.3633. \tag{25}$$

We note that the lower curve is that describing the transition to instability. When R is below this curve, for a given b, we have in fact linear stability. Then, the pair (b, R) given by equation (24) is the critical pair for the onset of instability. We mention that the same critical value of R was found by Ribando and Torrance [12], and more recently by Wang [13] on studying

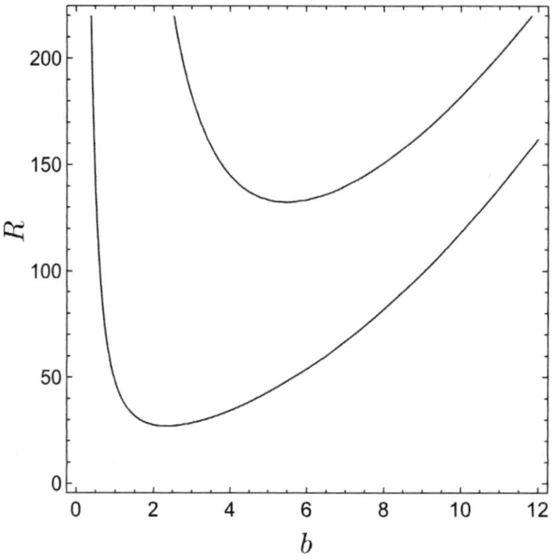

Figure 2. Neutral stability curves in the plane (b, R).

the stability of a saturated rectangular porous box with lateral impermeable and adiabatic boundaries, upper horizontal wall maintained at constant temperature, and lower horizontal wall heated by a constant flux.

Due to the lateral confinement, with a given aspect ratio s, the longitudinal wave number a is related to b_n through equation (17b), so that we can plot the neutral stability curves with respect to a for prescribed modal numbers n. These plots are displayed in figures 3-8, relative to the aspect ratios $s = 1/2,\ 3/4,\ 1,\ 3/2,\ 2,\ 5/2$. The aspect ratio $s = 1$ (figure 5) is for a square channel, while $s < 1$ (figures 3 and 4) is for a slender rectangular channel, and $s > 1$ (figures 6, 7 and 8) is for a shallow rectangular channel.

Figures 3-5 show that the most unstable modes are those with $n = 1$ and that the onset of convection takes place for vanishingly small wave numbers, meaning $z-$independent or two-dimensional modes. In particular, for $s = 1/2,\ 3/4,\ 1$ the critical wave number is $a_c = 0$ and the critical values of R are $R_c = 57.33,\ 35.83,\ 29.21$, respectively. This behaviour changes slightly in figure 6 where, even if the least stable modes are still for $n = 1$, we have a nonvanishing critical wave number $a_c = 1.012$ and a critical Darcy-Rayleigh number $R_c = 27.10$, *viz.* just the same value of R reported in (24). We mention that the value $a_c = 1.012$ arises just by taking the value of b from equation (25) and employing (17) with $n = 1$ and $s = 3/2$. The transition from a condition with $a_c = 0$ to one with $a_c > 0$ arises with a value of s intermediate between $s = 1$ (figure 5) and $s = 3/2$ (figure 6). A precise evaluation of this transition aspect ratio yields

$$s = 1.350517. \tag{26}$$

Below this aspect ratio, the onset of convection is two-dimensional ($z-$independent), while for higher aspect ratios the selected modes at neutral stability are three-dimensional. Relative to a shallow channel, figures 7 and 8 show that the modes with $n = 1$ are not the most unstable over the whole range of a, but only for sufficiently large wave numbers: $a > 0.6290$ and $a > 1.401$ for $s = 2$ and $s = 5/2$, respectively. For smaller wave numbers, the modes with $n = 2$ prevail. Anyway, the absolute minimum of R for neutral stability takes place for $n = 1$ in both cases, and this is for a nonvanishing critical wave number. Another interesting feature displayed by figures 5-8 is that, on increasing the aspect ratio s, the neutral stability curves with $n = 1, 2, 3$ become

Journal of Physics: Conference Series **501** (2014) 012003 doi:10.1088/1742-6596/501/1/012003

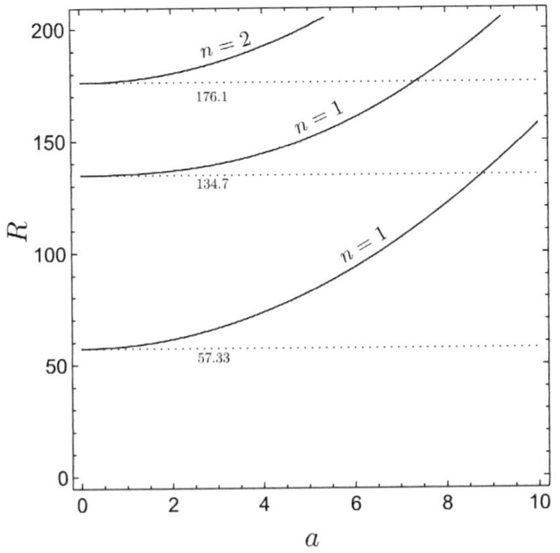

Figure 3. Neutral stability curves in the plane (a, R) for $s = 1/2$. The dotted lines denote the minimum of each branch.

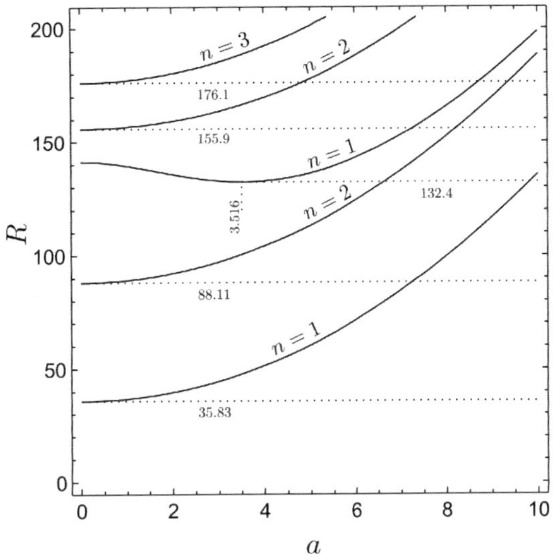

Figure 4. Neutral stability curves in the plane (a, R) for $s = 3/4$. The dotted lines denote the minimum of each branch.

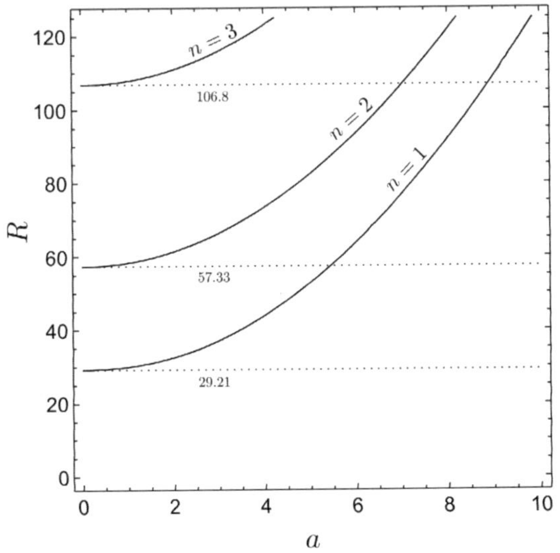

Figure 5. Neutral stability curves in the plane (a, R) for $s = 1$. The dotted lines denote the minimum of each branch.

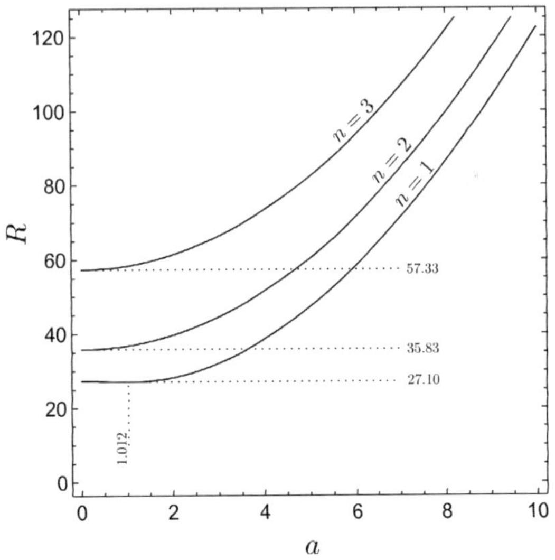

Figure 6. Neutral stability curves in the plane (a, R) for $s = 3/2$. The dotted lines denote the minimum of each branch.

closer and closer, meaning an easier transition between different modes at slightly supercritical conditions.

7. Conclusions

The stability versus small-amplitude disturbances of the steady motionless state in a horizontal rectangular channel containing a fluid-saturated porous material has been investigated. A special case is described where the heating from below is modelled by a uniform flux and the cooling

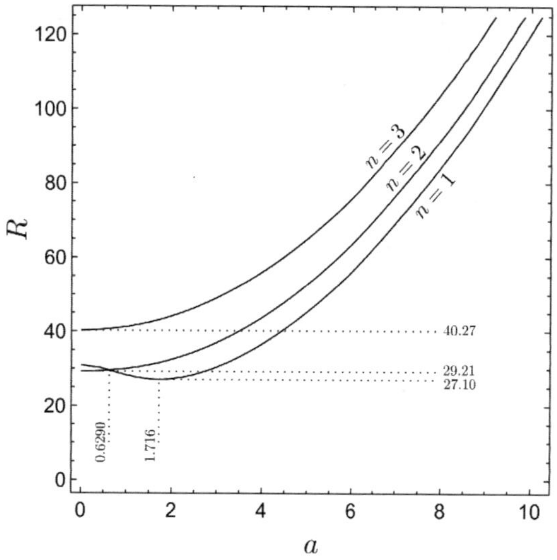

Figure 7. Neutral stability curves in the plane (a, R) for $s = 2$. The dotted lines denote the minimum of each branch and the intersection between different branches.

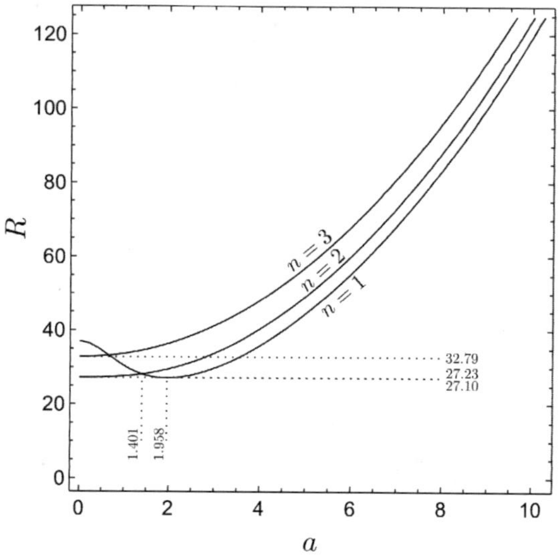

Figure 8. Neutral stability curves in the plane (a, R) for $s = 5/2$. The dotted lines denote the minimum of each branch and the intersection between different branches.

from above is modelled by a uniform temperature. The lateral confinement is by thermal and mechanical contact with a saturated porous reservoir in a steady state with vertical thermal stratification.

Darcy's law and the Oberbeck-Boussinesq approximation are adopted for modelling the buoyant flow. The dimensionless parameters governing the onset of the instability are the Darcy-Rayleigh number, R, and the width-to-height aspect ratio, s. An analytical solution of the disturbance equations has been found for the normal modes. These modes are labelled by the natural number, n, that yields the number of nodes in the sinusoidal temperature distribution along the transverse horizontal direction, namely along the $x-$axis.

A study of the neutral stability curves has been carried out for different aspect ratios ranging from $s = 1/2$ to $s = 5/2$. The main feature emerged from this study is that the modes with $n = 1$ are preferred at onset of convection. However, the critical condition, defined as the absolute minimum of R leading to instability, may be displayed either with two-dimensional instability patterns or three-dimensional patterns. The former are rolls invariant along the longitudinal $z-$direction. The three-dimensional patterns are preferred at the onset of instability when the aspect ratio is sufficiently large, *viz.* exceeding the threshold value $s = 1.350517$. Under these conditions, the absolute minimum of R leading to instability is the critical value $R = 27.09763$.

Opportunities for future developments of this study are foreseen. The effect of a finite longitudinal length along the $z-$axis is yet to be analysed. With this respect, specific assumptions about the planes, $z = constant$, bounding longitudinally the channel are to be made. Another possible future development is relative to the role played by the type of lateral confinement. We have shown that permeable and perfectly conducting side-boundaries lead to an analytical solution of the linear stability problem. The closely related case of impermeable and perfectly conducting sidewalls does not allow an analytical solution, and linear stability can be discussed only by means of a numerical solution of the governing equations for the disturbances. This challenging problem is a natural extension of the present analysis.

Acknowledgments
This work was financially supported by Italian government, MIUR grant PRIN-2009KSSKL3.

References
[1] Horton C W and Rogers F T 1945 *J Appl Phys* **16** 367–370
[2] Lapwood E R 1948 *Math Proc Cambridge* **44** 508–521
[3] Rees D A S 2000 *Handbook of Porous Media* ed Vafai K and Hadim H A (New York: CRC Press) chap 12, pp 521–558
[4] Tyvand P A 2002 *Transport Phenomena in Porous Media II* ed Ingham D B and Pop I (New York: Pergamon) chap 4, pp 82–112
[5] Nield D A and Bejan A 2006 *Convection in Porous Media* 3rd ed (New York: Springer)
[6] Straughan B 2008 *Stability and Wave Motion in Porous Media* (New York: Springer)
[7] Barletta A 2011 *Heat Transfer in Multi-Phase Materials* ed Öchsner A and Murch G E (New York: Springer) pp 381–414
[8] Sutton F M 1970 *Phys Fluids* **13** 1931–1934
[9] Beck J L 1972 *Phys Fluids* **15** 1377–1383
[10] Nilsen T and Storesletten L 1990 *J Heat trans-T ASME* **112** 396–401
[11] Barletta A and Storesletten L 2012 *Int J Therm Sci* **55** 1–15
[12] Ribando R J and Torrance K E 1976 *J Heat trans-T ASME* **98** 42–48
[13] Wang C Y 1999 *Phys Fluids* **11** 1673–1675

Thermoconvective instability and local thermal non-equilibrium in a porous layer with isoflux-isothermal boundary conditions

Michele Celli[1], Antonio Barletta[1] and Leiv Storesletten[2]

[1]Department of Industrial Engineering, Alma Mater Studiorum Universitá di Bologna, Viale Risorgimento 2, 40136 Bologna, Italy

[2]Department of Mathematics, University of Agder, Postboks 422, 4604 Kristiansand, Norway

E-mail: michele.celli3@unibo.it

Abstract. The effects of lack of local thermal equilibrium between the solid phase and the fluid phase are taken into account for the convective stability analysis of a horizontal porous layer. The layer is bounded by a pair of plane parallel walls which are impermeable and such that the lower wall is subject to a uniform flux heating, while the upper wall is isothermal. The local thermal non-equilibrium is modelled through a two-temperature formulation of the energy exchange between the phases, resulting in a pair of local energy balance equations: one for each phase. Small-amplitude disturbances of the basic rest state are envisaged to test the stability. Then, the standard normal mode procedure is adopted to detect the onset conditions of convective rolls. Beyond the Darcy-Rayleigh number, playing the role of order parameter for the transition to instability, the relevant dimensionless parameters are the inter-phase heat transfer parameter and the thermal conductivity ratio. The disturbance governing equations, formulated as an eigenvalue problem, are solved numerically by a shooting method. Results are reported for the neutral stability curves and for the critical values for the onset of instability.

1. Introduction

The onset of thermoconvective instability in a horizontal fluid-saturated porous layer heated from below has been extensively studied in the last decades. The attention deserved by this subject is due to the many applications of this interdisciplinary research field, ranging from geophysical research to biophysical applications as well as to petroleum and heat transfer engineering. A detailed discussion of the literature on this subject can be found in the books by Nield and Bejan [1], and by Straughan [2], as well as in the reviews by Tyvand [3], Rees [4], and Barletta [5]. Cornerstone studies of the onset of thermoconvective instability in porous layers are reported by Horton and Rogers [6] and Lapwood [7]. They presented the first linear stability analysis of what is now well-known as either the Horton-Rogers-Lapwood problem, or the Darcy-Bènard problem. The Darcy-Bènard problem is closely linked to the classical Rayleigh-Bènard problem for clear fluids. It consists of a basic motionless state with a uniform temperature drop across a porous layer with a heated isothermal lower wall and a cooled isothermal upper wall.

A variant of the Darcy-Bènard problem will be investigated in this paper. Darcy's law for momentum transfer in the porous medium is assumed to be valid, and a two-temperature model, a temperature for the fluid phase and a temperature for the solid phase, is employed. Our

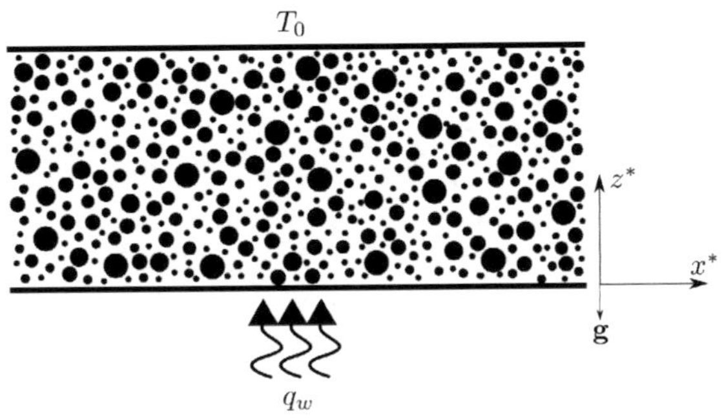

Figure 1. Sketch of the porous layer.

aim is to analyse the effect of the local thermal non-equilibrium model on the onset of the thermoconvective instability. An inter-phase heat transfer coefficient is defined and the local energy balance equations, one for the fluid phase and one for the solid phase, contain heat exchange terms proportional to the local temperature difference between the two phases [1, 8, 9].

We aim to extend two previous analyses [10, 11] of the Darcy-Bènard problem with local thermal non-equilibrium. Banu and Rees [10] considered a porous layer bounded by impermeable isothermal walls, while Barletta and Rees [11] studied the case where both the boundary walls are impermeable and isoflux.

In our paper we will assume, in analogy with [10, 11], that the boundary walls are impermeable. On the other hand, unlike in the studies reported in [10, 11], we will analyse asymmetric temperature conditions, where the upper wall is kept at constant temperature, while a uniform heat flux is prescribed on the lower wall. The uniform heat flux condition will be formulated according to the Amiri-Vafai-Kuzay Model A [11, 12, 13]. The latter model is valid when the boundary wall has a finite thickness and a high thermal conductivity. A linear stability analysis of the motionless basic state, where the fluid is thermally stratified in the vertical direction, is performed. The normal mode method is employed to investigate the onset conditions of the thermoconvective instability. A one-dimensional eigenvalue problem is thus obtained. The analysis is focused on determining the neutral stability curves and the critical values of the Darcy-Rayleigh number. Results are obtained numerically and are compared with those valid for the limiting case of local thermal equilibrium between the phases.

2. Mathematical model

We consider a plane porous layer bounded by impermeable boundaries at $z^* = 0$ and $z^* = L$. A uniform upward heat flux, q_w, is prescribed at $z^* = 0$ and a uniform temperature, T_0, is prescribed at $z^* = L$, see Figure 1. The porous medium is homogeneous and isotropic. Darcy's law and the Oberbeck-Boussinesq approximation can be employed. The viscous dissipation effect is negligible and no internal heat source exists. The local thermal non-equilibrium between the solid and fluid phases is modeled through an inter-phase heat transfer coefficient h.

2.1. Governing equations

The local thermal non-equilibrium model prescribes an energy balance equation for each phase: one for the solid phase, and the other for the fluid phase. The set of dimensional governing equations is given by a local mass balance equation, a local momentum balance equation and two local energy balance equations, namely

$$\nabla^* \cdot \mathbf{v}^* = 0, \tag{1a}$$

$$\frac{\mu}{K}\,\nabla^* \times \mathbf{v}^* = \rho_f g\,\beta\,\nabla^* \times \left[(T_f^* - T_0)\,\mathbf{e}_z\right], \tag{1b}$$

$$(1-\varphi)\,\frac{\partial T_s^*}{\partial t^*} = (1-\varphi)\alpha_s \nabla^{*2} T_s^* + \frac{h}{(\rho c)_s}\left(T_f^* - T_s^*\right), \tag{1c}$$

$$\varphi\,\frac{\partial T_f^*}{\partial t^*} + \mathbf{v}^* \cdot \nabla^*\, T_f^* = \varphi \alpha_f \nabla^{*2} T_f^* - \frac{h}{(\rho c)_f}\left(T_f^* - T_s^*\right), \tag{1d}$$

where the subscripts f, s denote the fluid and solid phase, respectively, while the asterisks denote dimensional variables and operators. We denoted as $\mathbf{v}^*$ the velocity field with Cartesian components (u^*, v^*, w^*), while $T_{f,s}^*$ are the fluid and solid phase temperatures. Moreover, $\alpha_{f,s}$ are the thermal diffusivities of the two phases, *viz.* $\alpha_{f,s} = k_{f,s}/(\rho c)_{f,s}$, t^* is time, μ is the dynamic viscosity, K is permeability, ρ is density, g is the modulus of the gravitational acceleration $\mathbf{g}$, β is the thermal expansion coefficient, T_0 is the reference temperature, $\mathbf{e}_z$ is the unit vector for the direction z, φ is porosity, and c is the heat capacity per unit mass. The set of boundary conditions can be expressed as [11, 12, 13]

$$\begin{aligned} z^* = 0: \quad & -\varphi k_f\,\frac{\partial T_f^*}{\partial z^*} - (1-\varphi)k_s\,\frac{\partial T_s^*}{\partial z^*} = q_w, \quad T_f^* = T_s^*, \quad w^* = 0, \\ z^* = L: \quad & T_f^* = T_s^* = T_0, \quad w^* = 0. \end{aligned} \tag{2}$$

A dimensionless formulation of equations (1) and (2) is now introduced by the rescaling

$$\mathbf{x}^* = \mathbf{x}\,L, \quad t^* = t\,\frac{L^2}{\alpha_f}, \quad \mathbf{v}^* = \mathbf{v}\,\frac{\alpha_f}{L} = (u, v, w)\,\frac{\alpha_f}{L}, \quad T_{s,f}^* = T_0 + T_{s,f}\,\Delta T. \tag{3}$$

Here, $k_m = (1-\varphi)\,k_s + \varphi\,k_f$ is the average thermal conductivity, and $\Delta T = q_w L/k_m$ is the reference temperature difference. The dimensionless set of governing equations can now be written as

$$\nabla \cdot \mathbf{v} = 0, \tag{4a}$$

$$\nabla \times \mathbf{v} = \varphi\,\frac{1+\gamma}{\gamma}\,R\,\nabla \times (T_f\,\mathbf{e}_z), \tag{4b}$$

$$\lambda\,\frac{\partial T_s}{\partial t} = \nabla^2 T_s + H\gamma\,(T_f - T_s), \tag{4c}$$

$$\frac{\partial T_f}{\partial t} + \frac{1}{\varphi}\mathbf{v} \cdot \nabla\, T_f = \nabla^2 T_f - H\,(T_f - T_s), \tag{4d}$$

where the Darcy-Rayleigh number R and the other dimensionless parameters employed in equations (4) are defined as

$$\lambda = \frac{\alpha_f}{\alpha_s}, \quad H = \frac{hL^2}{\varphi k_f}, \quad \gamma = \frac{\varphi k_f}{(1-\varphi)k_s}, \quad R = \frac{g\,\beta\,\Delta T K L}{\alpha_m \nu}, \tag{5}$$

with $\alpha_m = k_m/(\rho c)_f$. On the other hand, the dimensionless boundary conditions are given by

$$\begin{aligned} z = 0: \quad & -\gamma\,\frac{\partial T_f}{\partial z} - \frac{\partial T_s}{\partial z} = 1 + \gamma, \quad T_f = T_s, \quad w = 0, \\ z = 1: \quad & T_f = T_s = 0, \quad w = 0. \end{aligned} \tag{6}$$

3. The basic state and its stability

A stationary basic state with vanishing velocity field is allowed as a solution of equations (4) and (6),

$$\mathbf{v}_B = 0, \quad T_{s,B} = T_{f,B} = 1 - z, \tag{7}$$

where B stands for basic state. One may note, on account of equation (7), that the temperatures of the fluid phase and of the solid phase are exactly the same and thus, for the basic state, local thermal equilibrium between the two phases occurs.

We now introduce small amplitude disturbances of the basic state defined by (7), such that

$$\mathbf{v} = \mathbf{v}_B + \epsilon\,\tilde{\mathbf{V}}, \quad T_s = T_{s,B} + \epsilon\,\tilde{\Phi}, \quad T_f = T_{f,B} + \epsilon\,\tilde{\Theta}. \tag{8}$$

Here, ϵ is a perturbation parameter, small enough as to neglect the nonlinear terms $O(\epsilon^2)$, while the perturbation velocity $\tilde{\mathbf{V}}$ has Cartesian components $(\tilde{U}, \tilde{V}, \tilde{W})$, $\tilde{\Phi}$ is the perturbation temperature of the solid phase and $\tilde{\Theta}$ is the perturbation temperature of the fluid phase. A linear system of equations governing the disturbances is obtained by substituting equations (7) and (8) into (4), namely

$$\boldsymbol{\nabla} \cdot \tilde{\mathbf{V}} = 0, \tag{9a}$$

$$\boldsymbol{\nabla} \times \tilde{\mathbf{V}} = \varphi\,\frac{1+\gamma}{\gamma}\,R\,\boldsymbol{\nabla} \times \left(\tilde{\Theta}\,\mathbf{e}_z\right), \tag{9b}$$

$$\lambda\,\frac{\partial \tilde{\Phi}}{\partial t} = \nabla^2 \tilde{\Phi} + H\gamma\left(\tilde{\Theta} - \tilde{\Phi}\right), \tag{9c}$$

$$\frac{\partial \tilde{\Theta}}{\partial t} - \frac{1}{\varphi}\,\tilde{W} = \nabla^2 \tilde{\Theta} - H\left(\tilde{\Theta} - \tilde{\Phi}\right). \tag{9d}$$

Since the basic state, equation (7), is invariant by rotations around the z-axis, a two-dimensional formulation of the problem, based on the coordinates (x, z), is allowed without any loss of generality. Thus, a streamfunction can be defined

$$\tilde{U} = \varphi\,\frac{\partial \tilde{\Psi}}{\partial z}, \quad \tilde{W} = -\,\varphi\,\frac{\partial \tilde{\Psi}}{\partial x}. \tag{10}$$

The governing equations (9), on account of (10), are rewritten as

$$\nabla^2 \tilde{\Psi} + \frac{1+\gamma}{\gamma}\,R\,\frac{\partial \tilde{\Theta}}{\partial x} = 0, \tag{11a}$$

$$\lambda\,\frac{\partial \tilde{\Phi}}{\partial t} = \nabla^2 \tilde{\Phi} + H\gamma\left(\tilde{\Theta} - \tilde{\Phi}\right), \tag{11b}$$

$$\frac{\partial \tilde{\Theta}}{\partial t} + \frac{\partial \tilde{\Psi}}{\partial x} = \nabla^2 \tilde{\Theta} - H\left(\tilde{\Theta} - \tilde{\Phi}\right). \tag{11c}$$

From equations (6)-(8), the boundary conditions associated with (11) are

$$z = 0: \quad \gamma\,\frac{\partial \tilde{\Theta}}{\partial z} + \frac{\partial \tilde{\Phi}}{\partial z} = 0, \quad \tilde{\Phi} = \tilde{\Theta}, \quad \tilde{\Psi} = 0,$$
$$z = 1: \quad \tilde{\Phi} = \tilde{\Theta} = 0, \quad \tilde{\Psi} = 0. \tag{12}$$

The normal mode method is here employed. The disturbances are expressed in terms of plane waves, namely

$$\left\{ \begin{array}{c} \tilde{\Psi}(x,z,t) \\ \tilde{\Phi}(x,z,t) \\ \tilde{\Theta}(x,z,t) \end{array} \right\} = \left\{ \begin{array}{c} i\,\Psi(z) \\ \Phi(z) \\ \Theta(z) \end{array} \right\} e^{i(ax - \omega t)}. \tag{13}$$

Journal of Physics: Conference Series **501** (2014) 012004 doi:10.1088/1742-6596/501/1/012004

Functions $\Psi(z)$, $\Phi(z)$ and $\Theta(z)$ express the amplitude of the normal modes, a is the real-valued wave number, while ω is a complex parameter such that its real part represents the angular frequency of the wave. A positive imaginary part of ω means a perturbation exponentially growing in time and, hence, an unstable behaviour. A negative imaginary part of ω yields an exponential damping of the perturbation, corresponding to linear stability. Hereafter, we will be interested in the stability/instability threshold, *viz.* in the neutral stability. At neutral stability, the parameter ω is real-valued. Moreover, the principle of exchange of stabilities implies that the neutrally stable modes are non-travelling, *i.e.* that ω is zero. Therefore, on account of (13), equations (11) and (12) yield a system of real ordinary differential equations, namely

$$\Psi'' - a^2\Psi + a\,\frac{1+\gamma}{\gamma}\,R\Theta = 0, \tag{14a}$$

$$\Phi'' - a^2\Phi + H\gamma\,(\Theta - \Phi) = 0, \tag{14b}$$

$$\Theta'' - a^2\Theta - H\,(\Theta - \Phi) + a\Psi = 0, \tag{14c}$$

with the boundary conditions

$$\gamma\,\Theta'(0) + \Phi'(0) = 0, \qquad \Phi(0) = \Theta(0), \qquad \Psi(0) = 0,$$
$$\Phi(1) = \Theta(1) = 0, \qquad \Psi(1) = 0. \tag{15}$$

We now define

$$\hat{\Psi} = \gamma\Psi, \qquad \Lambda = \gamma\Theta + \Phi, \tag{16}$$

so that the governing equations (14) and the boundary conditions, (15), can be rewritten as

$$\hat{\Psi}'' - a^2\hat{\Psi} + a\,\frac{1+\gamma}{\gamma}\,R\,(\Lambda - \Phi) = 0, \tag{17a}$$

$$\Phi'' - a^2\Phi + H\,[\Lambda - (1+\gamma)\,\Phi] = 0, \tag{17b}$$

$$\Lambda'' - a^2\Lambda + a\hat{\Psi} = 0, \tag{17c}$$

$$\Lambda'(0) = 0, \qquad \Lambda(0) = (1+\gamma)\,\Phi(0), \qquad \hat{\Psi}(0) = 0,$$
$$\Lambda(1) = \Phi(1) = 0, \qquad \hat{\Psi}(1) = 0. \tag{17d}$$

Equations (17) yield an eigenvalue problem, where R is the eigenvalue to be determined for assigned values of (a, H, γ). This problem can be solved numerically according to the procedure described in the next Section 3.1.

3.1. Numerical solution

Equations (17) can be solved numerically by means of a Runge-Kutta solver combined with a shooting method. More precisely, equations (17) are integrated numerically with the initial conditions

$$\hat{\Psi}(0) = 0, \quad \hat{\Psi}'(0) = \eta, \quad \Lambda(0) = 1, \quad \Lambda'(0) = 0, \quad \Phi(0) = \frac{1}{1+\gamma}, \quad \Phi'(0) = \xi. \tag{18}$$

The additional condition $\Lambda(0) = 1$ serves to break the overall scale invariance of the solution. On the other hand, the unknown parameters η and ξ, are determined, together with the eigenvalue R, by means of a shooting method set up to satisfy the three target conditions at $z = 1$, namely

$$\hat{\Psi}(1) = 0, \quad \Lambda(1) = \Phi(1) = 0. \tag{19}$$

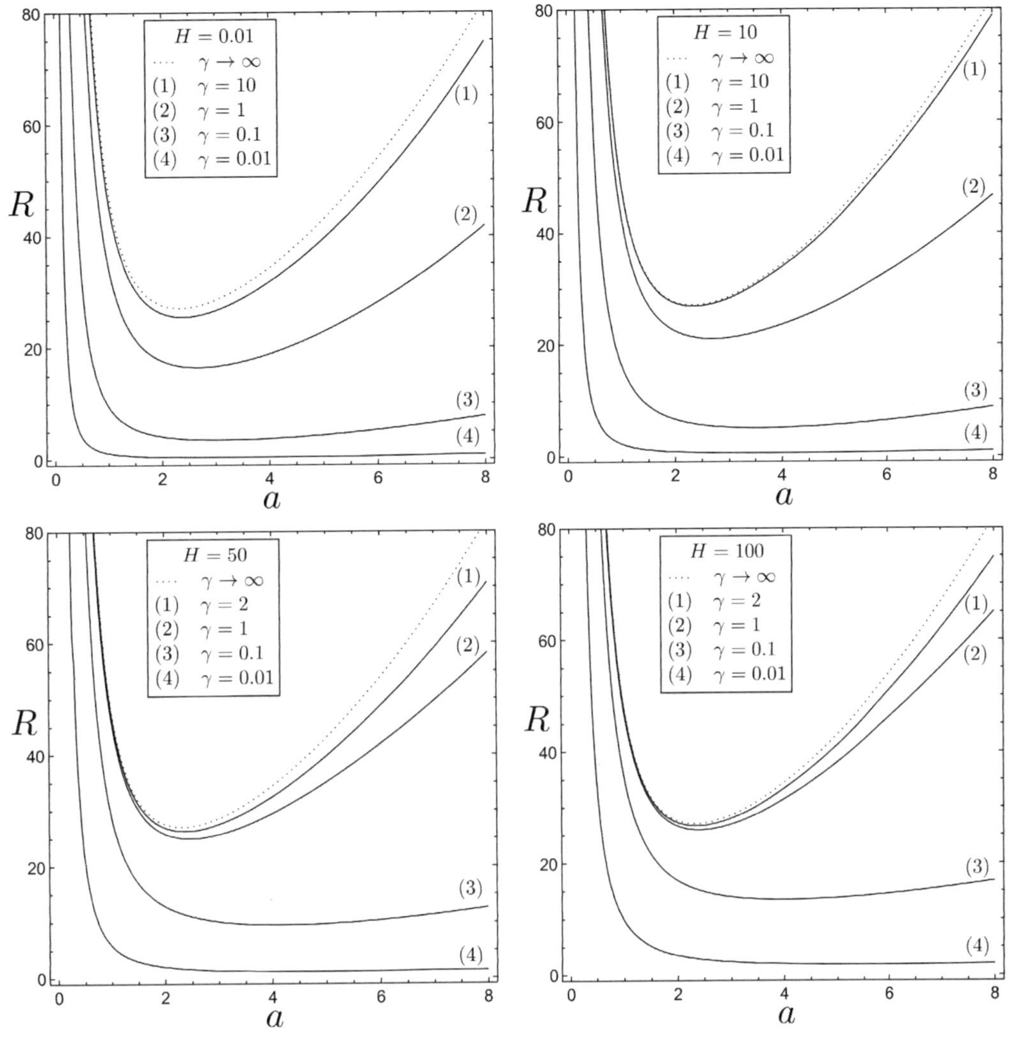

Figure 2. Neutral stability curves.

This numerical technique is accomplished inside the *Mathematica 9* (© Wolfram Research, Inc.) environment [14]. The Runge-Kutta solver is implemented by the built-in function `NDSolve` and the shooting method by the built-in function `FindRoot`. The `FindRoot` function allows one to solve numerically the constraints equation (19). The overall numerical procedure requires the assignment of the input data (a, H, γ). Once the input data are fixed, a neutral stability curve $R(a)$ for every pair (H, γ) is obtained. Instability occurs when $R > R(a)$, while linear stability is for $R < R(a)$. The absolute minimum of $R(a)$ gives the critical values (a_{cr}, R_{cr}) for the onset of the convective instability.

4. Discussion of the results

We solved the eigenvalue problem defined by equations (17) for a wide range of values of the pair (H, γ). Figure 2 contains different frames, each frame refers to a different value of H. Starting from the upper left corner, the values of H are: 0.01, 10, 50 and 100. Inside each frame, each curve is relative to a different value of γ. Figure 2 shows that the onset condition of the instability is very sensitive to the quantity γ. As γ decreases, the system becomes more and more unstable. Very small values of γ, on account of (5), imply a solid phase much more conductive than the fluid phase. This is in fact a condition very close to the behaviour of a gas saturated metallic foam.

As shown by Figure 2, the effect of a decreasing H is also destabilising. We mention that larger and larger values of the product γH lead to the asymptotic condition of local thermal equilibrium. Local thermal equilibrium can be attained by taking the limit $H \to \infty$ with a finite nonvanishing value of γ. From equation (17b), this limit leads to $\Lambda = (1+\gamma)\,\Phi$ and, as a consequence of equations (17), to $\Phi = \Theta$. Thus, in this limit, we can rewrite (17) as

$$\hat{\Psi}'' - a^2\hat{\Psi} + aR\Lambda = 0, \tag{20a}$$

$$\Lambda'' - a^2\Lambda + a\hat{\Psi} = 0, \tag{20b}$$

$$\Lambda'(0) = 0, \qquad \hat{\Psi}(0) = 0,$$
$$\Lambda(1) = 0, \qquad \hat{\Psi}(1) = 0. \tag{20c}$$

On the other hand, on account of equation (14b), by taking the limit $\gamma \to \infty$ with a finite nonvanishing value of H, one is lead to $\Phi = \Theta$. Thus, the governing equations (14) reduce to

$$\Psi'' - a^2\Psi + aR\Theta = 0, \tag{21a}$$

$$\Theta'' - a^2\Theta + a\Psi = 0, \tag{21b}$$

$$\Theta'(0) = 0, \qquad \Psi(0) = 0,$$
$$\Theta(1) = 0, \qquad \Psi(1) = 0. \tag{21c}$$

The eigenvalue problem given by equations (20) is completely equivalent to that given by (21). In other words, the local thermal equilibrium condition is attained either by taking the limit $H \to \infty$ with a finite γ, or by taking the limit $\gamma \to \infty$ with a finite H. We mention that the solution of the eigenvalue problem equations (20) or (21) was found by Ribando and Torrance [15], and analysed in detail by Wang [16]. These authors considered a horizontal porous layer with an isoflux lower boundary and an isothermal upper boundary, assuming that both these boundaries are impermeable and that local thermal equilibrium between the solid and the fluid phase holds. Data on the neutral stability condition and on the critical values of R and a for the onset of convection with local thermal equilibrium have been reported by [1, 15, 16]. In particular, we were able to recover the critical values for local thermal equilibrium,

$$R_{cr} = 27.097628, \qquad a_{cr} = 2.3262146, \tag{22}$$

by solving numerically equations (20) or (21), with a suitably simplified version of the procedure described in Section 3.1.

The trend of R_{cr} versus γ and that of a_{cr} versus γ, with different finite values of H, are shown in Figures 3 and 4, respectively. Hence, these figures display the effects of a departure from the local thermal equilibrium asymptotic condition ($\gamma H \gg 1$). Interestingly enough, Figure 3 suggests that R_{cr} tends to zero when γ tends to zero, for every finite value of H. We already noted above that this regime may be relevant for the study of heat transfer in gas-saturated metallic foams. In fact, a vanishingly small R_{cr} means that whatever small value of the heat flux q_w prescribed on the lower wall leads to a destabilisation of the basic motionless state through the onset of convective rolls. Another evident feature of Figures 3 and 4 is that the critical values of R and a approach monotonically the local thermal equilibrium asymptotes (dashed lines) when $\gamma \to \infty$.

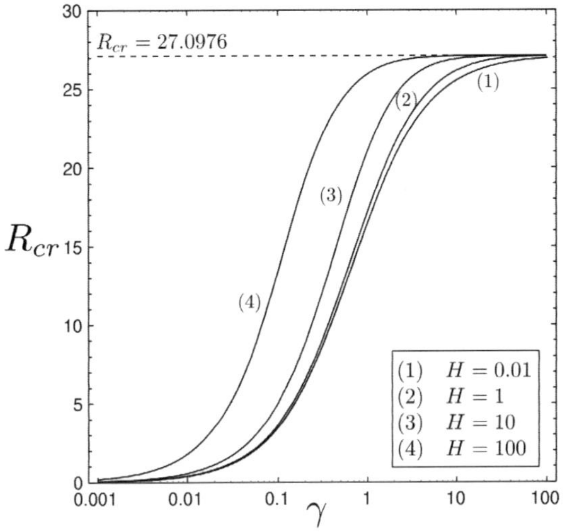

Figure 3. Plots of R_{cr} versus γ.

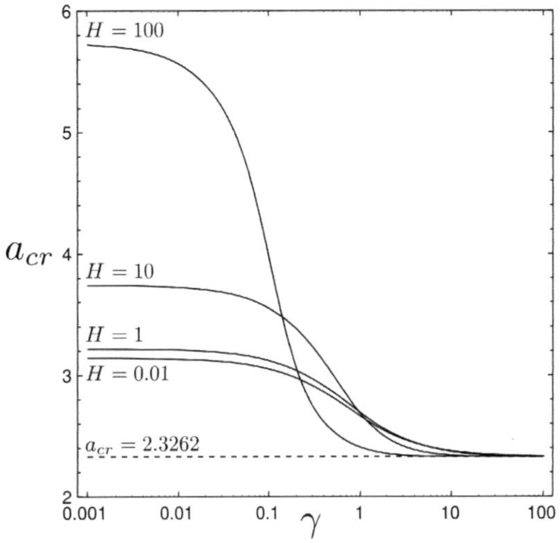

Figure 4. Plots of a_{cr} versus γ.

5. Conclusions

The onset of convection in a horizontal porous layer saturated by a fluid and heated from below by a uniform heat flux has been studied. The assumption of local thermal equilibrium has been relaxed. Different local values of the fluid temperature and of the solid temperature are assumed, with an inter-phase energy exchange effect. The local thermal non-equilibrium thus yields a mathematical model with two energy balance equations. These equations, together with the local mass balance equation and Darcy's law with buoyancy force, have been employed to test the stability of the motionless and local thermal equilibrium basic state of the layer. The disturbance equations, in a streamfunction/temperature formulation, have been linearised and solved for the normal modes. We thus obtained an ordinary differential eigenvalue problem that allowed us to study the neutral stability condition, as well as to determine the critical values of the Darcy-Rayleigh number, R, and of the wave number, a, for the onset of convective rolls.

We pointed out that two dimensionless parameters are crucial in the transition to instability: the thermal conductivity ratio, γ, and the inter-phase heat transfer parameter, H. When the product γH becomes very large, the condition of local thermal equilibrium is attained. On the other hand, small values of γ and H mean significant departures from local thermal equilibrium. In particular, lower and lower values of both γ and H result into a destabilisation of the basic state. A significant effect has been revealed when $\gamma \ll 1$, a condition approached when the solid phase is much more conductive than the fluid phase. In fact, when $\gamma \ll 1$, the critical value of R tends to zero. This feature means that, in this parametric regime, the basic motionless state is always unstable to small-amplitude disturbances. This result is of special interest for all the applications involving a highly-conductive solid, such as the gas-saturated metallic foams. We mention that recently attention in the heat transfer community has been deserved by the properties of the metallic foams with respect to the design of heat exchangers [17].

Acknowledgments

This work was financially supported by Italian government, MIUR grant PRIN-2009KSSKL3.

References

[1] Nield D A and Bejan A 2013 *Convect. Porous Media* 4th ed (New York: Springer-Verlag)

[2] Straughan B 2008 *Stability and Wave Motion in Porous Media* Appl. Math. Sci. Ser. (New York: Springer-Verlag)
[3] Tyvand P A 2002 Onset of Rayleigh-Bénard convection in porous bodies *Transp. Phenom. Porous Media II* ed Ingham D B and Pop I (Oxford: Pergamon) pp 82–112
[4] Rees D A S 2000 Stability of Darcy-Bénard convection *Handb. Porous Media* ed Vafai K (Redding, CT: Begell House) pp 521–558
[5] Barletta A 2011 Thermal instabilities in a fluid saturated porous medium *Heat Transf. Multi-Phase Mater.* ed Öchsner A and Murch G E (New York: Springer-Verlag) pp 381–414
[6] Horton C W and Rogers F T 1945 *J. Appl. Phys.* **16** 367–370
[7] Lapwood E R 1948 *Proc. Camb. Philos. Soc.* **44** 508–521
[8] Kuznetsov A V 1998 Thermal nonequilibrium forced convection in porous media *Transp. Phenom. Porous Media* ed Ingham D B and Pop I (Oxford: Pergamon) pp 103–129
[9] Rees D A S and Pop I 2005 Local thermal non-equilibrium in porous medium convection *Transp. Phenom. Porous Media III* ed Ingham D B and Pop I (Oxford: Pergamon) pp 147–173
[10] Banu N and Rees D A S 2002 *Int. J. Heat Mass Tran.* **45** 2221–2228
[11] Barletta A and Rees D A S 2012 *Int. J. Heat Mass Tran.* **55** 384–394
[12] Amiri A, Vafai K and Kuzay T M 1995 *Numer. Heat Transfer A* **27** 651–664
[13] Yang K and Vafai K 2010 *Int. J. Heat Mass Tran.* **54** 4316–4325
[14] Wolfram S 2003 *The Mathematica Book* 5th ed (Champaign, IL: Wolfram Media)
[15] Ribando R J and Torrance K E 1976 *ASME J. Heat Trans.* **98** 42–48
[16] Wang C Y 1999 *Phys. Fluids* **11** 1673–1675
[17] Mahjoob S and Vafai K 2008 *Int. J. Heat Mass Tran.* **51** 3701–3711

Journal of Physics: Conference Series **501** (2014) 012005 doi:10.1088/1742-6596/501/1/012005

Mixed convection in horizontal channels partially filled with aluminium foam heated from below and with external heat losses on upper plate

Bernardo Buonomo, Giuseppe Ferraro, Oronzio Manca, Lorenzo Marinelli and Sergio Nardini
Dipartimento di Ingegneria Industriale e dell'Informazione Seconda Università degli studi di Napoli via Roma 29, 81031 Aversa (CE), Italy

E-mail: bernardo.buonomo@unina2.it; giufero@gmail.com; oronzio.manca@unina2.it; lorenzod83@hotmail.it; sergio.nardini@unina2.it

Abstract. In this work mixed convection in a horizontal channel partially filled with a porous medium and the lower wall heated at uniform heat flux is studied both experimentally and numerically. The experimental test section is made of a horizontal wall and a parallel wall with heat losses toward the external ambient. A simplified two dimensional numerical model is considered to evaluate the dynamic and thermal fields inside the heated channel. The investigation allows to evaluate the effect of the aluminium foam on the mixed convection in the heated channel by wall temperature measurements and flow visualization. Results are given for heated channel without and with foam in terms of wall temperature profiles for different Reynolds number value, from 100 to 300, wall heat flux and for aluminium foam with 10 and 20 pore per inch. Experimental and numerical results are given in terms of wall temperature profiles and both allow to show that the foam determines lower wall temperature profiles along the fluid flow than the ones in the clean channel cases. Moreover, the foams with different pores per inch present different local thermal behaviours. The simplified numerical model allows to estimate the possible heat losses toward the external ambient.

1. Introduction

In recent years mixed convection in horizontal channels has been studied diffusively for its presence in many applications in different fields. Solar components, cooling of electronic equipments and chemical vapor deposition are just some of the fields concerned, as indicated in [1-5]. Low velocity forced flow in horizontal parallel plates channels, heated from below, can be affected by secondary flow which are induced by buoyancy and a three dimensional mixed convection flows is determined. Moreover, the local heat transfer increases due to the transition from a two-dimensional laminar flow to a three-dimensional one [1-3,6,7]. The onset point of the secondary flows is important because it delineates the region after which the two-dimensional laminar flow becomes three-dimensional and a transition motion from laminar to turbulent flow is observed. The unstable and stable vortex flow associated to the secondary flow is advantageous for some application such as cooling of electronic equipments [4]. On the contrary in other technological applications such as in the chemical vapor deposition processes used to grow thin crystal films on silicon substrates, the presence of the vortex flow determines a non-uniform deposition on the substrates [5]. The secondary flow and the enhancement of the heat transfer, in the so-called Poiseuille-Rayleigh-Benard flow, depends strongly on the Rayleigh number, the Reynolds number, the Prandtl number of the channel flow, the boundary conditions, and the geometry of the channel, as well explained in [1]. The relevant literature on mixed convection in horizontal channels heated from below has been reviewed in [1-3,6-11].

Porous media with high thermal conductivity and porosity, such as metal foams, have emerged as an effective method of heat transfer enhancement due to their large surface area to volume ratio and to intense mixing of the flow. In the last two decades metal foams have been applied in heat transfer and their use in thermal engineering is growing quite quickly. Several investigations have been accomplished on this topic, as reviewed in [12]. The possibility to employ a metal foam to improve heat transfer in a horizontal channel has been investigated in several studies as reported in [12,13].

Investigations on mixed convection in horizontal channels filled or partially filled with a metal foam heated from below seems very few as indicated in [14]. Several investigations have been focused on instability in a fluid saturated porous medium in a heated channel, as recently reported in [15-17]. Moreover, mixed convection in horizontal channels partially filled with porous blocks intermittently inserted in transverse to the channel axis on the channel bottom wall have been extensively investigated, as reported for example in [18-20]. Some experimental investigations on mixed convection in vertical channel partially filled with metal foams have been carried out in [21,22]. Two porous plates are set along the channel walls and fill partially the channel. More recently, an experimental investigation on mixed convection in horizontal channel partially filled with an aluminum foam has been accomplished in [23].

It seems that mixed convection with porous medium inserted in horizontal channels heated from below has not been adequately investigated experimentally or numerically, though it has a lot of potential engineering applications like electronic equipment cooling, solar collectors, thermal energy storage, chemical vapour deposition (CVD) and thermal hydraulics of nuclear reactors. In this work mixed convection in a horizontal channel partially filled with a porous medium and the lower wall heated at uniform heat flux is experimentally and numerically studied. Results in terms of wall temperature profiles as a function of the pores per inch (PPI), Reynolds number values and wall heat flux are presented. The experimental results are compared with the ones carried out with a simplified two-dimensional numerical model. The comparison is accomplished in terms of wall temperature profiles.

2. Experimental apparatus

A schematic layout of the experimental set-up is shown in figure 1. The experimental apparatus consists of a convergent channel at the entrance of the main horizontal channel with the lower wall partially heated and a parallel unheated glass plate above, a diffuser, a flow control valve and an AC fan. The horizontal channel is 1.40 m long, 0.475 m width and the height can range from 10 mm to 100 mm. It is made up by a central part, long 0.400 m which represents the test section, with the lower wall uniformly heated, a parallel unheated glass plate above and two vertical unheated glass walls, as reported in figure 2. The upstream and downstream extension parts of the main channel have lower and vertical walls made of wood, figure 1. The blower attached to the channel through the diffuser provided a variable mass flow rate. The apparatus was located within a room, sealed to eliminate extraneous air currents.

The heated wall consists of two 400x530 mm^2 sandwiched phenolic fiberboard plates and is 4.8 mm thick. The upper glass plate is 500 x 600 mm^2 and 6 mm thick. The side walls of the heated part of the main channel are made of glass, which are machined to an accuracy of ±0.3 mm, in order to take pictures of the flow motion. The plate spacing is measured to an accuracy of ±0.25 mm by a dial-gauge equipped caliper. In some runs, the heated part of the channel is partially filled by a porous medium layer on the lower wall. The porous medium is an aluminum foam layer and it has a thickness equal to 20 mm whereas the length and the width were the same of the channel. The investigated aluminum foams can have 10 or 20 Pores Per Inches (PPI) with porosity equal to 0.97 and 0.95, respectively. The lower wall was made of two plates.

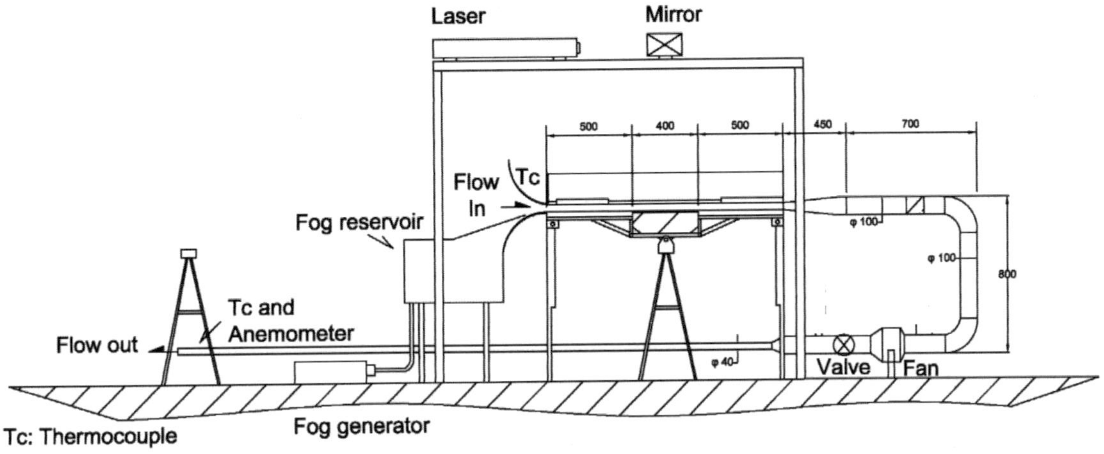

Figure 1. Layout of the experimental apparatus, dimensions in mm.

The heated plate facing the channel has the surface adjacent to the internal air coated with a 35 μm thick nickel plated copper layer. The low emissivity of nickel (0.05) minimized radiation effects on heat transfer. The rear plate is 2.6 mm thick. Its back surface is coated with a 17.5 μm thick copper layer, which is the heater. It is an electrical resistance obtained cutting the copper layer in a serpentine shape. Its runs are 19.6 mm wide with a gap of nearly 0.5 mm between each one, giving the heater a total length of 9.0 m. Its expected electrical resistance is 0.50 Ω. The lower wall is heated by passing a direct electrical current through the heater. In order to reduce conductive heat losses, a 150 mm Polystyrene block is affixed to the rear face of each principal wall. The narrow gaps between the runs, together with the relatively high thickness (4.8 mm) of the resulting low-conductive fiberglass are suitable to maintain a nearly uniform heat flux at the plate surface. Direct electrical current through the heaters is accomplished by using Agilent Tar*E3632A*dc power supply.

The electrical power supplied is evaluated by measuring the voltage drop across the plate and the current passing through it. An HP-3465A digital multi-meter measures the voltage drop, while the current is calculated by the measured voltage drop across a reference resistance. To avoid electrical contact resistances, thick copper bars soldered both to the electric supply wire and to the ends of heater are bolted together. The dissipated heat flux is evaluated to an accuracy of ±2%.

Wall temperatures are measured by 0.50 mm OD ungrounded iron-constantan thermocouples embedded in each fiberboard plate and in contact with the outer layer. They are located at twelve longitudinal stations at three different z values. Fifteen thermocouples are affixed to the rear surface of the plates and embedded in the Polystyrene to enable the evaluation of conductive heat losses. The

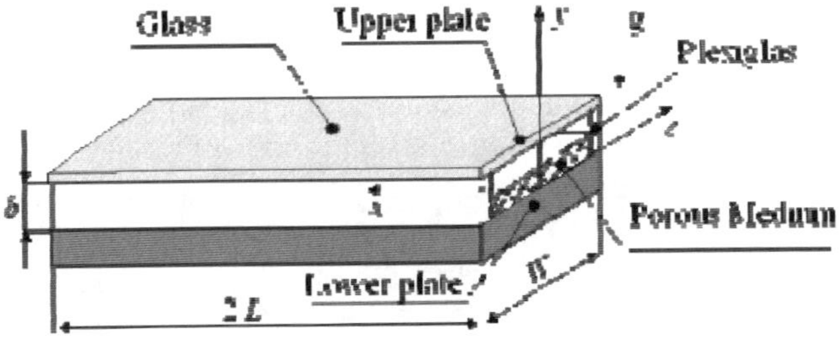

Figure 2. View of the test section.

ambient air temperature is measured by a shielded thermocouple placed near the leading edge of the channel. A Isotech TRU Model 938 ice point is used as a reference for thermocouples junctions. Their voltages are recorded by an Agilent 34980A multifunction measurement unit and a personal computer is used for the data collection and reduction. Calibration of the temperature measuring system shows an estimated precision of the thermocouple-readout system of ±0.1 °C.

Mass flow rate is evaluated by measuring the velocity with a hot wire anemometer Dantec Mini CTA 54T30 with a 55P11 probe. The sensor is located at 2500 mm from the inlet section of a circular duct, with a diameter of 30 mm, in order to have a fully developed laminar flow, figure 1. Furthermore, the hot wire probe is placed at about 6.0 m away from the outlet rectangular section of the test section. The hot wire probe is calibrated at 15°C, 20°C and 25°C in the velocity range from 0.010 m/s to 1.10 m/s. The maximum uncertainty in this range is about ±4%, the uncertainty on the measurement of the duct, diameter is ±1% and the uncertainty of the location of the sensor is ±2%.

It was observed that the steady-state condition was reached after 4–6 h.

3. Data reduction

The governing parameters in the problem are the buoyancy parameter Grashof, *Gr*, Rayleigh, *Ra*, and Richardson number, *Ri*, the Reynolds number, *Re*, and the Prandtl number, *Pr*. The Grashof and the Reynolds numbers are defined as:

$$Gr = \frac{g\beta \bar{q}_c b^4}{\nu^2 k}\ ; \quad Ra = GrPr\ ; \quad Re = \frac{u_i b}{\nu}\ ; \quad Ri = \frac{Gr}{Re^2}\ ; \quad Pr = \frac{\nu}{a} \tag{1}$$

where q_c is the average convective heat flux:

$$\bar{q}_c = \frac{1}{L}\int_0^L q_c(x)dx\ . \tag{2}$$

thermo-physical properties were evaluated at the reference temperature $T_r = (\bar{T}_w + T_0)/2$ with:

$$\bar{T}_w = \frac{1}{L}\int_0^L T_w(x)dx \tag{3}$$

where $\bar{T}_w$ is the average wall temperature along the heated lower plate.

Local convective heat flux, $q_c(x)$, was not uniform because of radiation and conduction heat losses. Experimental data were reduced by first introducing, in the equations presented above, the local convective heat flux.

$$q_c(x) = q_\Omega(x) - q_k(x) - q_r(x) \tag{4}$$

where $q_\Omega(x)$ is the local heat flux due to Ohmic dissipation, assumed uniform along x, $q_k(x)$ is the local conduction heat losses from the plate and $q_r(x)$ is the local radiative heat flux from the plate. For each run, the terms $q_k(x)$ were calculated by means of a numerical procedure, a three-dimensional distribution of the temperature being assumed in the Polystyrene.

Therefore, q_k on the wall was a function of both x and z coordinates, and its values were averaged along z. The predicted temperatures for some configurations of the system were previously compared with those measured by thermocouples embedded in the Polystyrene insulation and the relationship was very good, the maximum deviation being 3%.

A two-dimensional radiative cavity was made of the two plates, considered as diffuse-grey surfaces and two black edge sections at room temperature. In all cases, the radiative heat losses were not greater than 2% of the Ohmic dissipated power. The $q_r(x)$ terms were calculated for each temperature distribution of the wall, ambient temperature and plate spacing, by dividing each plate into sixteen strips along its length. Each strip was assumed at the span wise average temperature.

Journal of Physics: Conference Series **501** (2014) 012005 doi:10.1088/1742-6596/501/1/012005

The uncertainty in the calculated quantities was determined according to the standard single sample analysis recommended by Moffat [24]. The uncertainty of the Rayleigh and Reynolds numbers were 6% and 7%, respectively.

4. Model description and numerical procedure

A preliminary simplified numerical simulation is accomplished in a two-dimensional, steady state and incompressible flow. The domain is made of a principal channel, partially filled with an aluminum foam plate, and two channels with adiabatic walls, one upstream the principal channel and the other downstream, figure 3. In this problem the geometrical parameters are the distance between the horizontal walls, the thickness of the aluminum foam plate and the heated plate length. The principal channel is made up of a uniformly heated horizontal wall at uniform heat flux and a parallel glass plate located above which allows heat transfer with the external ambient. An upper external reservoir over the glass plate, four times the channel height, simulates the natural convection toward the external ambient. Viscous dissipation, heat generation and pressure work are all assumed to have negligible effect on the velocity and temperature fields, therefore they are neglected. All the thermophysical properties of the fluid and the solid matrix of the porous medium are assumed constant except for the variation in density of the air with temperature (Boussinesq approximation) giving rise to the buoyancy forces. The thermo-physical properties of the fluid and the solid matrix of the porous medium are evaluated at the ambient temperature, T_0, which is equal to 300 K in all cases. In the porous medium region, the generalized flow model, known as the Brinkman-Forchheimer-extended Darcy model, is used in the governing equations. It is assumed that fluid and solid phase of porous medium are in local thermal equilibrium. Governing equations for the considered problem are not given in the present paper but the reader can refer to [25, 26].

Thermo-physical properties of the upper glass plate are: ρ=2800 kg m^{-3}, k=0.81 W m^{-1} K^{-1}, c_p=800 J kg^{-1} K^{-1}. The lower heated plate has ρ= 1206 kg m^{-3}, k=2.0 W m^{-1}K^{-1}, c_p= 361 J kg^{-1}K^{-1}. The geometrical parameter in the simulations are the same of the experimental section along its longitudinal section. For the aluminum foam it is assumed a permeability equal to 1.17x10^{-7} m^2 and 1.12x10^{-7} m^2 for 10 and 20 PPI, respectively.

The commercial CFD code Ansys-Fluent [26] is employed to solve the governing equations. The SIMPLE scheme is chosen to couple pressure and velocity. The porous medium model is active in the porous region and the LTE is employed. The convergence criteria of 10^{-6} for the residual of continuity equation and velocity components and 10^{-8} for the residuals of energy are assumed.

A grid dependence test is accomplished to realize the most convenient grid size by monitoring local wall temperature for wall heat flux equal to 100 W/m^2 and a Reynolds number equal to 200. Three

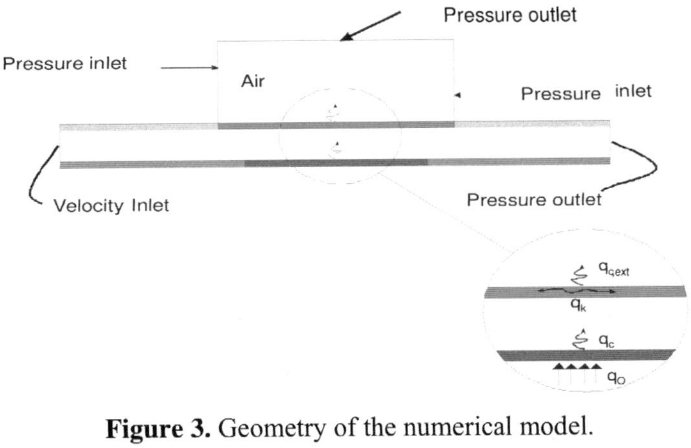

Figure 3. Geometry of the numerical model.

different uniform grids are tested with 112x1120, 160x1600, 226x2260 nodes in the channel. The 160x1600 node grid is the best compromise between accuracy and computational time. This mesh size was employed for all considered cases.

It should be underlined that the two-dimensional geometrical model has been employed just to have some indication on the main flow and the effect of the aluminum foam. It is not able to detect the transversal secondary motions that are present in the channel [23].

5. Results and discussion

The experiments were carried out for $100 \leq Re \leq 300$ and $1.0\text{x}10^5 \leq Ra \leq 1.8\text{x}10^6$ with a corresponding Richardson number values from 18 to 220 and for two wall heat flux equal to 100 and 200 Wm^{-2}. The longitudinal aspect ratio was 10 both in the clean cases, without porous medium, and in the cases with the foam. The transversal aspect ratio was 12.45 in all cases.

Thermal behaviours in all cases are evaluated by wall temperature profiles. In figure 4, experimental wall temperature profiles along the axis for all considered cases, with and without foam (clean case) and two different wall heat flux values, 100 W m^{-2} and 200 W m^{-2} and Re = 300, show that temperature increases along the flow up to x about 300 mm, where a maximum temperature value is attained, and after it decreases. The comparison between the channel without foam or with foam allows to observe that the presence of aluminum foam determines a wall temperatures reduction. For the cases with the aluminum foam with 10 PPI lower temperature values than the ones for 20 PPI are detected in the inlet region whereas the opposite is observed in the outlet zone also in accordance with the results given in [23]. In fact, in the inlet zone the fluid penetrates in 10 PPI foams better than in the 20 PPI and the surface heat transfer on the heated wall is favorite for 10 PPI. The inversion of the values along the axial direction is due to a better contact between the 20 PPI foams layer and the heated wall than the one between the foam with 10 PPI and the bottom heated wall. Consequently, a better heat transfer is obtained in the case of the 20 PPI foam which presents lower temperature values than the ones for 10 PPI foams for x>100. The effect is more evident for the highest considered wall heat flux, as can be noted in figure 4b.

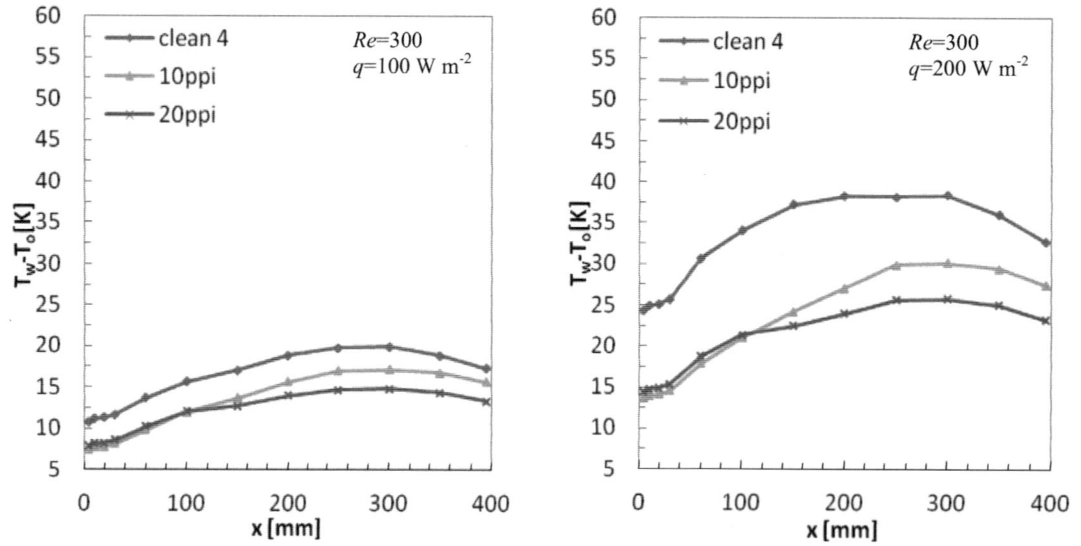

(a) **(b)**

Figure 4. Wall temperature profiles along the heated plate for channel with (10 and 20 PPI) and without (clean 4) aluminum foam plate, for Re=300 and: a) q=100 W m^{-2} and b) q=200 W m^{-2}.

In figures 5-7 the comparisons between experimental and numerical results are reported for Reynolds number values equal to 100, 200 and 300 and two wall heat flux values, 100 W m^{-2} and q=200 W m^{-2}.

For the all *Re* values considered, it is noted that the numerical temperature values are higher than the experimental ones but the trends are very similar on a quality level. In fact for the two different

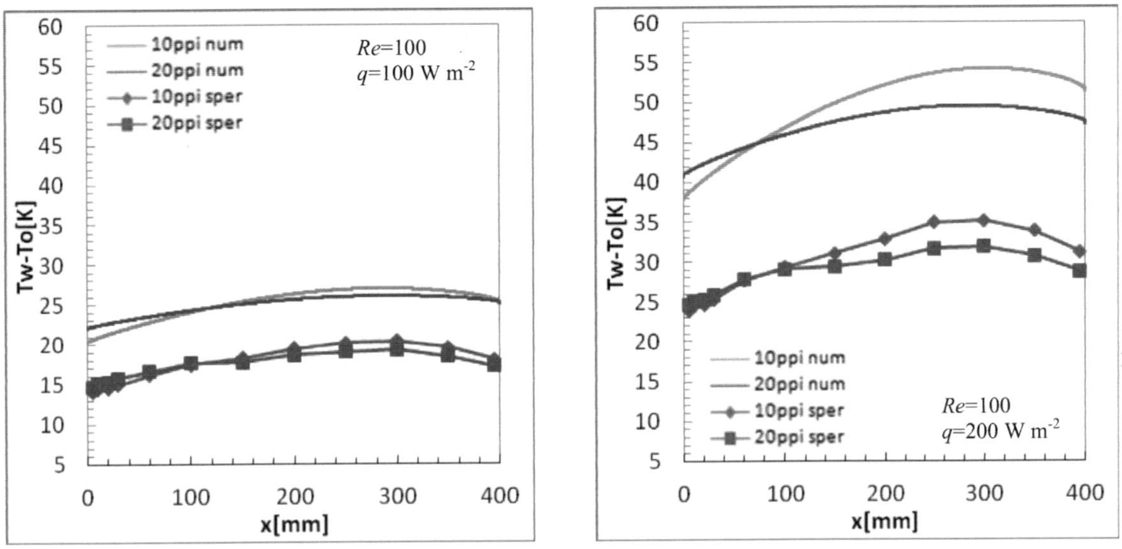

(a) **(b)**

Figure 5. Comparison between experimental and numerical wall temperature profiles for *Re*=100 and: a) q=100 W m^{-2} and b) q=200 W m^{-2}.

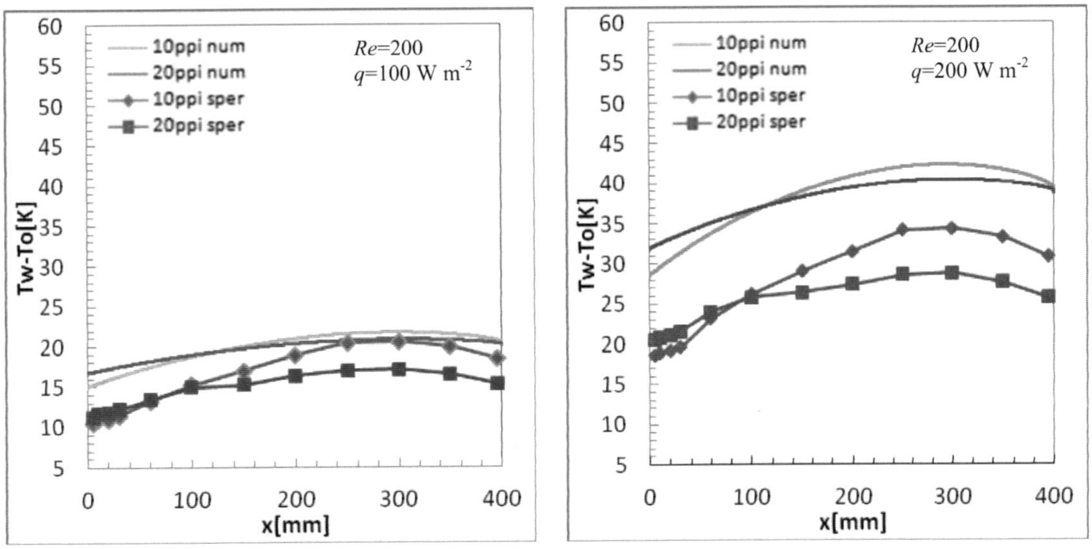

(a) **(b)**

Figure 6. Comparison between experimental and numerical wall temperature profiles for *Re*=200 and: a) q=100 W m^{-2} and b) q=200 W m^{-2}.

PPI values, in all cases the profiles present an intersection between the curves and the curve slopes are very similar both in the experimental and in numerical results. In figure 5, for *Re*=100 the differences between the experimental and numerical profiles are the greatest. These differences decrease increasing the *Re* value, as can be noted in figures 6 and 7 for *Re*=200 and 300, respectively. Moreover, for higher wall heat flux cases the differences between the wall temperature profiles with different PPI value increase. These two last observations suggest that the secondary motions become weaker increasing the Reynolds numbers as well as decreasing the wall heat flux. A two dimensional numerical simulation can give a good estimation of the main fluid flow inside the heated channel.

In figure 8 air temperature and stream function fields are presented for the cases clean and with aluminum foam, 10 PPI and 20 PPI, for *Re*=100 and 300 with q=100 W m^{-2}. For *Re*=100 the buoyancy pushes the fluid flow toward the upper wall in the outflow zone of the heated channel. This determines a heat loss toward the external ambient. For *Re*=300 the heat losses seem significantly decreased because of the higher velocity of the forced flow which reduces the buoyancy effect. The flow does not move adjacent the upper plate in the outlet zone. For the heated channel partially filled with aluminum foam and *Re*=100, the heated fluid reaches closer the upper plate before the middle zone of the heated channel and the heat losses are higher than in the clean case. Moreover, in the case with 10 PPI the external plume is stronger. For *Re*=300, the forced flow with higher velocity sweeps in advance the heated fluid and it does not allow to the hottest fluid to reach completely the upper plate. The heat losses toward the external ambient are significantly reduced.

6. Conclusions

Experimental and numerical results, given in terms of wall temperature profiles, showed that the foam determined lower temperature values along the wall than the ones in clean channel cases in accordance also with [23]. Different behaviours of local wall temperatures was detected for the foam with different PPI, for lower PPI at the inlet zone of the heated plate lower temperature were attained whereas the opposite was noted in the other part of the heated channel. Similar trends on a quality level were observed between numerical and experimental wall temperature profiles. A two dimensional numerical model can give a good evaluation of main flow inside a heated channel from below in mixed convection.

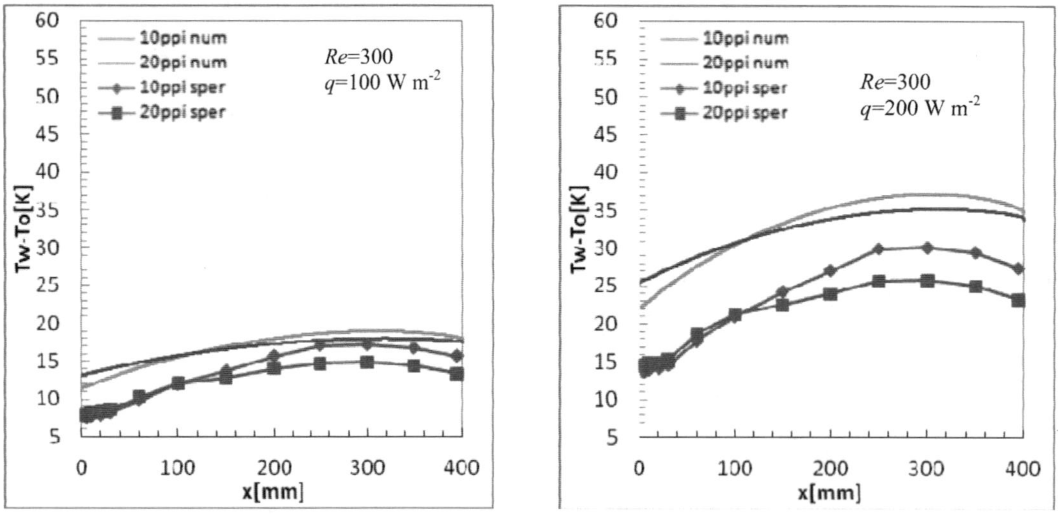

(a) (b)

Figure 7. Comparison between experimental and numerical wall temperature profiles for *Re*=300 and: a) q=100 W m^{-2} and b) q=200 W m^{-2}.

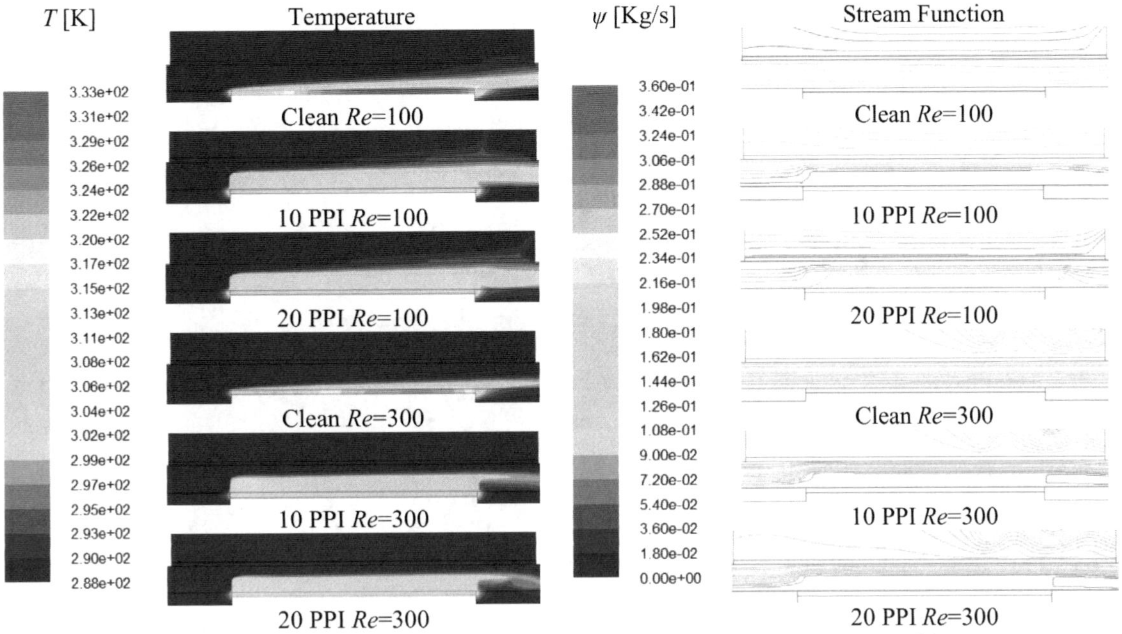

(a) **(b)**

Figure 8. Numerical results, fields for q= 100 W m^{-2} and Re=100 and 300: a) air temperature fields and b) stream function fields

Nomenclature

b	channel spacing, m
c_p	specific heat
g	acceleration due to the gravity, ms^{-2}
Gr	Grashof number, Eq.(1)
h	heat transfer coefficient, W m^{-2} K^{-1}
k	thermal conductivity, Wm^{-1}K^{-1}
L	length of the plate, m
Pr	Prandtl number, Eq. (1)
q	heat flux, Wm^{-2}
Ra	Rayleigh number, = Gr Pr, Eq. (1)
Re	Reynolds number, Eq. (1)
Ri	Richardson number, Eq. (1)
T	temperature, K
u_i	average velocity at inlet section of the channel, m s^{-1}
x	horizontal coordinate distance, m
y	vertical coordinate distance, m
z	coordinate along the width of the plates, m
W	width of the plate, m
Greek symbols	
β	volumetric coefficient of expansion, K^{-1}
ν	kinematic viscosity, m^{2}s^{-1}
Subscript	
c	convective
k	conductive
i	inflow

0 ambient air
r radiative
w wall
Ω Ohmic dissipation

Acknowledgments
This work was financially supported by Italian Government, MIUR grant PRIN-2009KSSKL3.

References
[1] Nicolas X 2002 *Int. J. Thermal Sciences* **41** 961
[2] Lin T F 2003 *Int. J. Heat Fluid Flow* **24** 299
[3] Rahli O, Bennacer R, Bouhadef K and Ameziani D E 2011 *Num. Heat Transfer Pt. A Appl.* **59** 349
[4] Incropera F P 1988 *J. Heat Transfer* **110** 1097
[5] Hitchman M L and Jensen K 1993 *Chemical Vapor Deposition Principles and Applications* (San Diego: Academic Press)
[6] Stamatopoulo Ch, Mathioulakis D and Katsamouris A 2011 *Exp. Thermal Fluid Science* **35** 375
[7] Bennacer R, Rahli O and Hammed H B 2012 *Thermal Sciences* **4** 549
[8] Koffi F, Abid C, Medale M and Papini F 2011 *Heat Mass Transfer* **47** 1251
[9] Seychelles F, Mergui S and Nicolas X 2012 *J. Phys.: Conf. Series* **395** 012101
[10] Nicolas X, Zoueidi N and Xin S 2012 *Phys. Fluids* **24** 084101
[11] Medale M nad Haddad A 2012 *J. Phys.: Conf. Series* **395** 012095
[12] Zhao C Y 2012 *Int. J. Heat Mass Transfer* **55** 3618
[13] Mancin S, Zilio C, Diani A and Rossetto L 2012 *Exp. Thermal Fluid Science* **36** 224
[14] Kurtbas I and Celik N 2009 *Int. J. Heat Mass Transfer* **52** 1313
[15] Chung T J, Choi C K, Yoon D Y and Kim M C 2010 *Int. J. Heat Mass Transfer* **53** 5139
[16] Rahli O, Bennacer R, Bouhadef K, Ameziani D E and Ghorbel E 2010 *Def. Diff. Forum* **297-301** 1010
[17] Barletta A 2012 *Transp. Porous Med.* **92** 419
[18] Jaballah S, Bennacer R, Sammouda H and Belghith A 2008 *J. Porous Media* **11** 247
[19] Guerroudj N and Kahalerras H 2010 *Energy Conv. Manag.* **51** 505
[20] Jaballah S, Sammouda H and Bennacer R 2012 *J. Porous Media* **15** 51
[21] Kamath P M, Balaji C and Venkateshan S P 2011 *Int. J. Heat Mass Transfer* **54** 5231
[22] Kamath P M, Balaji C and Venkateshan S P 2013 *Int. J. Thermal Sciences* **64** 1
[23] Buonomo B, Ferraro G, Manca O, Marinelli L and Nardini S 2013 ASME 2013 Summer Heat Transfer Conference (Minneapolis Minnesota, US 14 - 19 July 2013) HT2013-17306
[24] Moffat R J 1988 *Exp. Thermal Fluid Science* **1** 3
[25] Buonomo B, Manca O, Mesolella P and Nardini S 2012 Atti 67° Congresso Annuale ATI (Trieste, IT 11-14 Settembre 2012) 3-121
[26] Buonomo B., D'Ambrosio S., Manca O. Nardini S 2010 Atti 65° Congresso Nazionale ATI (Chia Laguna Resort Domus de Maria Cagliari, IT 13 - 17 Settembre 2010) 11-191
[27] Ansys-Fluent Incorporated, Fluent 6.3.26, User Manual, 2011

Journal of Physics: Conference Series **501** (2014) 012006 doi:10.1088/1742-6596/501/1/012006

Control of natural circulation loops by electrohydrodynamic pumping

W Grassi, D Testi and D Della Vista

University of Pisa, DESTEC (Department of Energy, Systems, Territory and Construction Engineering), Largo Lucio Lazzarino, 56122, Pisa, Italy

E-mail: daniele.testi@ing.unipi.it

Abstract. The paper analyses the effect of electrohydrodynamic (EHD) pumping on the control of natural circulation loops (NCLs). The two major objectives of the investigation are: finding the optimal configuration of an EHD pump and demonstrating that the NCL flow direction can be inverted by exploiting the EHD phenomena. In the initial experimental set-up, we measured the static pressure rise given by an EHD pump made of three consecutive modules of point-ring electrodes for different dielectric fluids and electrode materials. When reversing the polarity of the applied DC voltage, we observed opposite pumping directions, suggesting the presence of two distinct EHD phenomena, inducing motion on opposite directions: ion-drag pumping and conduction pumping. The former was identified as a more efficient process compared to the latter. Based on these preliminary experiments, we built a NCL, operating with the fluid HFE-7100. Two oppositely mounted optimised pumping sections could be alternately activated, to promote clockwise or anticlockwise motion. In the first series of tests, alternately, the pumping sections were triggered prior to the heat input. In any case, the circulation followed the EHD pumping direction. In other tests, the electric field was applied when natural circulation was already present and the flow was reversed by means of opposite EHD pumping, at both polarities. Simply inverting the polarity of the applied voltage, we could alternate ion-drag and conduction pumping; in this way, we easily controlled the direction of motion by means of a single EHD pumping device.

1. Introduction on natural circulation loops

The main objectives of the present experimental investigation are: finding the optimal configuration for an electrohydrodynamic (EHD) pump and demonstrating that the flow direction in natural circulation loops (NCLs) can be inverted and managed by exploiting the EHD phenomena.

The problem of heat transfer and stability prediction and control in thermo-gravitational systems and particularly in NCLs has been extensively investigated in theoretical/numerical and experimental works. Research has been stimulated by important applications of the phenomenon in oceanography, geophysics and engineering systems such as nuclear reactor emergency cooling, solar thermosyphon collectors and other industrial equipment. The pioneering works of Keller [1] and Welander [2] proved that even a simplified one-dimensional model can predict that a steady-state condition can be broken by small perturbations, inducing oscillatory non-periodic motion of the fluid in the loop. Welander proposed a physical explanation of such instability, based on the creation of thermal anomalies (hot or cold "pockets") in correspondence of the heat source or sink that materially accelerate and decelerate around the loop. It was also pointed out that the flow is supposed to be stable when buoyancy and frictional forces are small or large enough. Modelling a rectangular loop, Chen [3] showed that its aspect ratio (height-to-width) is an important stability parameter and that the flow is least stable in a square loop. Vijayan et al. [4] predicted that the flow is more stable at larger length-to-diameter ratios of the loop.

The first experimental confirmation of the Welander instability in ordinary single-phase fluids was obtained by Creveling et al. [5] in a toroidal loop. Many subsequent experimental works were reviewed by Vijayan et al. [6]. Stability maps based on the one-dimensional theory were found to be highly conservative, predicting much larger unstable zones with respect to the experimental results. This discrepancy was attributed to the multidimensional nature of the phenomenon, with additional damping effects especially occurring during the low-velocity flow phase of the oscillation cycle. More recently, Marcelli et al. [7] performed a three-dimensional simulation reproducing experimental data, providing further insight on the complex instability mechanisms.

Fichera and Pagano [8] tackled the problem of stabilising the dynamics of NCLs by means of a feedback control strategy based on the physical model. An alternative approach that we propose as a mean for managing the thermosyphon flow is the superposition of a body force of electrical nature. As mentioned, we experimentally analysed the effect of optimised EHD pumps on the flow in a NCL. Two different mechanisms were employed: ion-drag pumping [9,10] and conduction pumping [11,12], both based on the Coulomb force. Ion-drag pumping is associated with the ion injection phenomenon, occurring at the metal-liquid interface, while conduction pumping is based on the process of dissociation of neutral species in the fluid bulk, enhanced by the electric field. Both mechanisms are illustrated in detail in the next section.

A previous application of EHD pumping on a NCL is reported by Shelestynsky et al. [13], though this case deals with a two-phase loop, in which vapour bubbles are pumped by polarisation forces (dielectrophoresis and electrostriction) acting at the liquid-vapour interface.

2. EHD pumping mechanisms

EHD pumping has shown extensive potential for low velocity and micro-pumping applications, such as cooling of microprocessors [14], microhydraulic actuators [15] and devices for thermal control in microgravity and space [16]. In spite of their low pumping efficiency and the necessity to operate only with dielectric fluids (as most refrigerants, solvents, esters and oils are), EHD pumps have a simple, lightweight, low-cost, easy-to-fabricate and reliable design, with no moving parts. Other advantages are: minimal maintenance, no vibrations, no acoustical noise and ease with which pumping can be controlled by modulation of the applied voltage. All these characteristics make EHD pumping suitable for the scope of flow management and control of NCLs. We now review the two main methods of pressure head generation by electric body force.

2.1. Conduction pumping

In weak electrolytes, neutral species partially dissociate in a reversible process by which positive and negative ions are formed and a thermodynamic dissociation/recombination equilibrium is reached. Liquid dielectrics have a very low concentration of ionic components and the recombination process generally involves only ion pairs. At the application of the electric field, the dissociation rate proportionally increases, while the recombination rate is independent of it [17]. An upper bound for the recombination rate constant was found by Langevin [18]. At electric fields above a threshold value (indicatively, around 10^5 V/m, depending on the liquid characteristics), the dissociation rate is so enhanced that ions are no longer in equilibrium with their parent electrolytes and their creation cannot keep pace with their removal by the field. The ions that are moving in any volume element of the liquid were created at some distance by irreversible dissociation. As a consequence, the flux of anions (or cations) cannot be uniform, increasing towards the anode (or cathode) [19]. The formation of heterocharge layers (i.e., layers constituted by charge of the opposite polarity from the one of the adjacent electrode) is an obvious consequence. The thickness of these layers is not only controlled by the kinetics of ionic dissociation, but is also proportional to the relaxation time of the working fluid and the strength of the local electric field [20]. While the charges are redistributed by the electric field, they induce a fluid motion from the liquid to the electrode side attracting them, which can be useful for pumping purposes. For instance, considering an electrode configuration as shown in figure 1, a net axial flow is induced, thanks to the contribution of the high-voltage (HV) electrode, where the electric field is particularly strong and the corresponding heterocharge layer is thick as well.

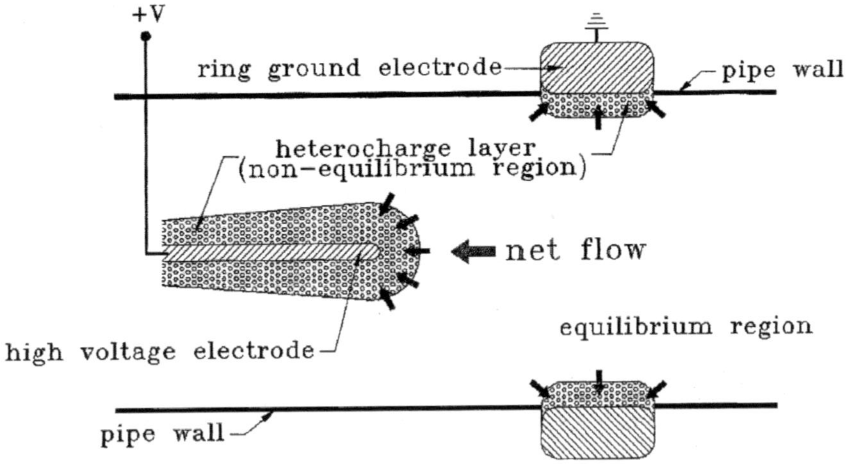

Figure 1. Axial motion generated by field-enhanced dissociation in a configuration with point and ring electrodes [20].

We refer to this pumping mechanism as conduction pumping, because the electrical current and the induced flow are due to charge carriers produced by dissociation of molecules within the fluid, as opposed to injection from electrodes. This latter, competing phenomenon of charge creation – which is described in the next subsection – generally develops at higher electric fields (in the order of 10^6-10^7 V/m, depending on the fluid properties and on the material of the sharper electrode) and rapidly becomes dominant.

2.2. Ion-drag pumping

Work has to be done in order to extract an electron from the surface of a metal. The potential energy barrier is called work function and depends on the material. The application of an electric field lowers the barrier of a significant quantity (s_1), as illustrated in figure 2 for the case of vacuum.

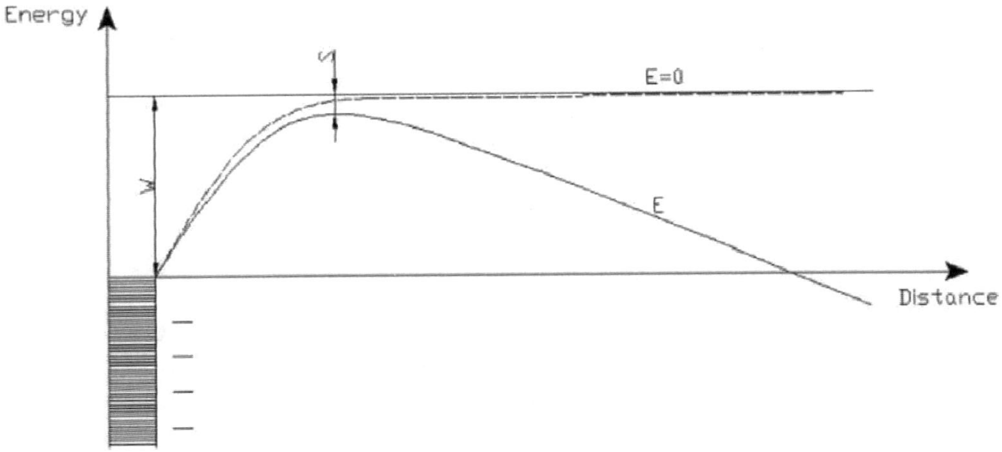

Figure 2. Potential barrier at the application of an electric field in vacuum [21].

When the electrode is wetted by an electrolytic solution, the barrier through which electrons must tunnel is thinner and lower than in vacuum [22] for two main reasons:

- condensed matter usually presents a free-electron band or at least numerous acceptor levels, whose energy is well below the one of an electron in vacuum;
- under an electric field, with the metal being the cathode, positive ions dissociated in the liquid congregate on the interface and form the so-called electric (or charged) double layer (see figure 3).

Within the double layer, the electric field is intensified by a factor η and a further lowering (s_2) of the potential energy barrier is obtained. At this point, the barrier is around 3 eV, but electrons do not have to cross through the whole of it, since they can reach acceptor levels which are 1 to 2 eV below, with the length of the tunnel accordingly shortened.

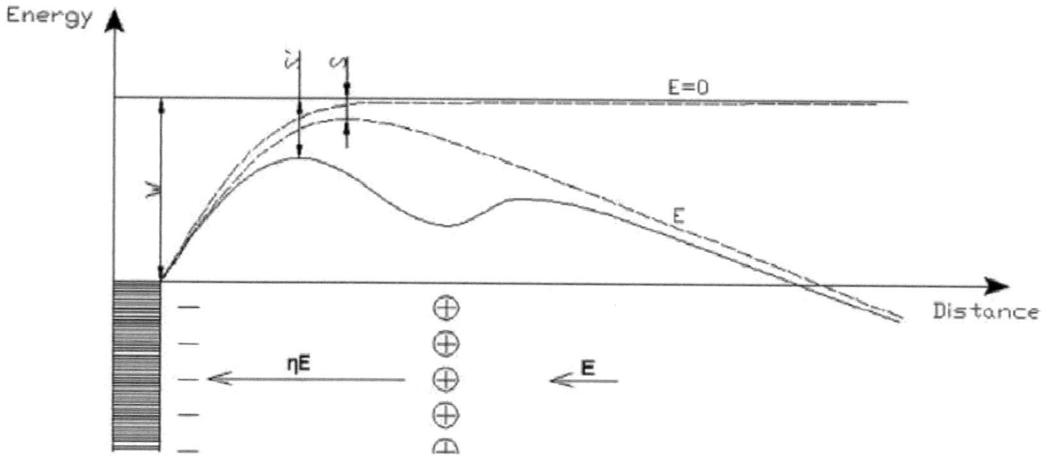

Figure 3. Potential barrier at the formation of the electric double layer [21].

In dielectric liquids, the concentration and composition of ionic impurities in the bulk determine the net charge in the double layer [23]. Moreover, neutral admixture molecules with high electro-acceptor qualities have a considerable probability of taking part in electrochemical reactions at the interface, by virtue of the lowering of the potential barrier promoted by the electric double layer. In this way, negative ions are created in the proximity of the cathode, as shown in figure 4. The same scheme can be applied to a positive electrode, with electrons following the opposite path (i.e., departing from molecules of high electro-donor qualities and reaching the metal), resulting in the creation of positive ions near the anode.

We are now dealing with homocharges (i.e., charges of the same polarity of the electrode, which we generally refer to as the emitter). For a sufficiently high electric field, the ions escape from the charged double layer and are pushed away from the emitter by electrical repulsion. The injected space charge gives rise to a plume-like motion towards the facing electrode, called collector [24]. This flow is stronger and in opposite direction with respect to the previously-described motion based on field-enhanced dissociation, so, when active, it prevails.

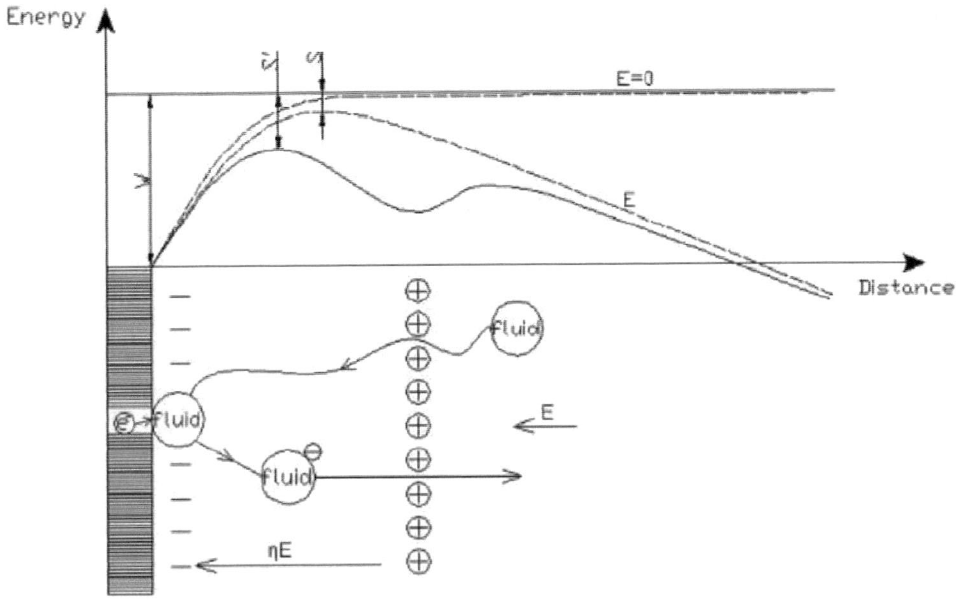

Figure 4. Electrochemical ionisation of neutral molecules at the metal/liquid interface [21].

The ion injection phenomenon occurs at the sharper electrode and, as already mentioned, is mainly controlled by the electrochemistry of the interface, which critically depends on the shape (i.e., radius of curvature) and composition of the emitter, as well as on the chemical properties of the dielectric fluid; in particular, the concentration and nature of the adsorbed species play a fundamental role for two different reasons:

- the ionic species created by the dissociation of electrolytes are responsible for the formation of the electric double layer (see again figure 3);
- the extrinsic neutral molecules of high electro-acceptor (or electro-donor) qualities may directly take part in the reduction (or oxidation) reactions of ionisation.

Obviously, the polarity set on the emitter plays a major role, since the rates of reduction reactions in the solution are totally unrelated to the rates of oxidation reactions that could occur at the application of the opposite polarity.

Granted that ionic layers are necessary for electron exchange to take place at liquid/metal contacts, the very nature of the injected carriers remains unknown. Identifying them is a very difficult task, since the amount of matter involved is extremely small by chemical standards. In a non-polar or weakly polar insulating liquid, these impurities have a strong influence on the rate of ion formation. Also, the rate easily changes over a wide range, by changing the concentration of an impurity [25]. Therefore, the creation of charge carriers at interfaces does not depend on the electric field alone, but more on the delicate balance between the competing effects of ion concentration and velocity, on one hand, and of kinetics of ion neutralisation, on the other. This explains why ion injection has always been found an unpredictable process [19] and related research heavily relies on experimentation.

As for the electrode, a participation at the ionisation reactions can be excluded for inert metals or for cathodic reductions, since metals are very poor electro-acceptors. In any case, the electrode has a role in the erratic behaviour of ion injection. In fact, the real structure of its surface is microscopically very rough. Even highly polished surfaces have a high density of micro-asperities; their characteristic dimensions are less than 2 μm, giving local electric-field amplification factors of about 10^2-10^3 [26].

From the fluid-dynamical viewpoint, as previously mentioned, the injected space charge is driven by the Coulomb force towards the opposite electrode. Along the path, the ions drag the fluid neutral molecules and set them into motion. This mechanism can be conveniently used for creating a submerged jet impinging on a surface and intensifying heat removal rates [27-32] or for pumping a fluid in a chosen direction; in this latter case, the method is known as ion-drag pumping [15,33-35]. As in conduction pumping, the pressure rise is easily controllable by adjusting the applied voltage.

The more evident difference between the two pumping mechanisms is the direction of flow for a given pair of electrodes. In a typical module of an ion-drag pump, the fluid moves from the sharper electrode, e.g. a point, to the collector, e.g. a ring. The opposite applies to conduction pumping, where the flow direction, with this particular design, would be from the ring to the point. Both methods were applied in the experimental campaign reported herein, in which EHD pumping was employed to control the flow of a NCL.

3. Experimental activity

3.1. EHD pump testing apparatus

Prior to application to a NCL, we intended to optimise the configuration of the EHD pump. For this sake, we built the pressure head generation testing apparatus illustrated in figure 5.

This basic testing equipment is made of a pumping section on the lower side and a static pressure head measuring section on the upper side, connected by flexible pipework. The EHD pump is a series of three consecutive electrical modules, each module containing a pair of electrodes in point-ring configuration (see figure 6). A microammeter measures the total current transiting through the modules, electrically connected in parallel. The measuring section is made of a digital camera and two vertical glass tubes with a graduated ruler positioned behind them. Pressure head is simply measured by recording the difference in height that is established between the free surfaces of the two columns of the tested fluid (whose density is known), when voltage is applied to the electrodes.

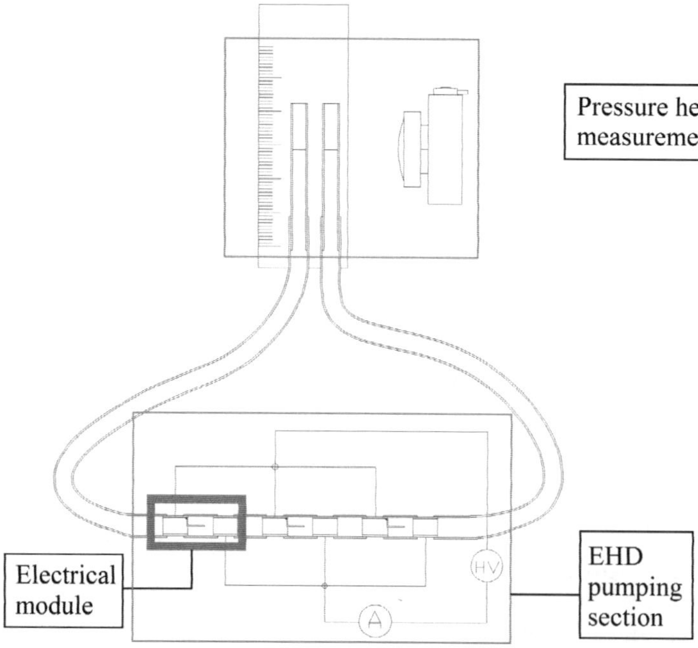

Figure 5. EHD pump testing apparatus.

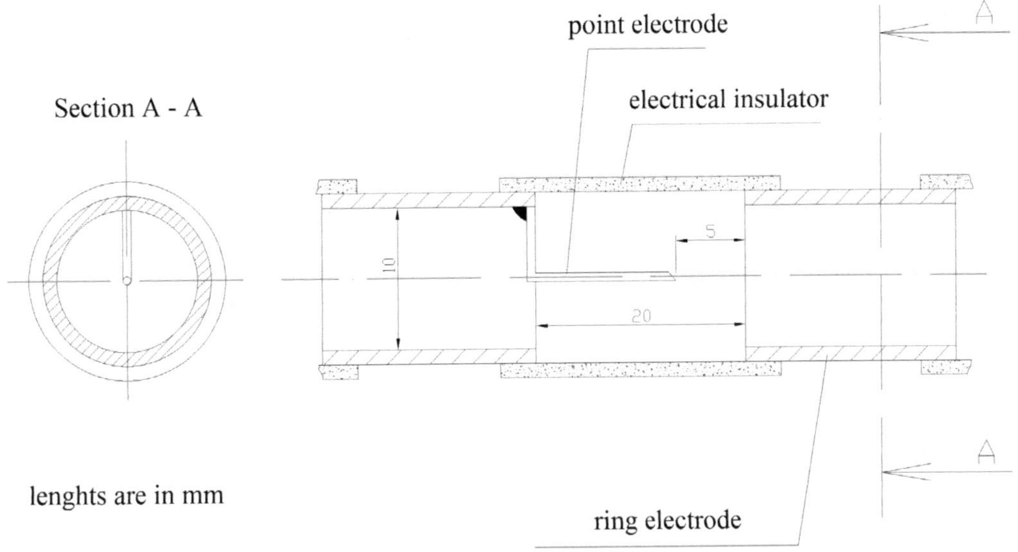

Figure 6. Detail of the EHD pumping module.

Tests were conducted at a DC voltage of 10 kV. The static pressure head generated by the EHD pump was measured in various configurations. Typically, a steady-state condition was reached after a transient phase about 15 minutes long, in which both electrical current and pressure head decreased. We changed the following factors:

- operating dielectric liquid: FC-72, Vertrel XF, HFE-7100 and H-Galden ZT85;
- electrode material: steel and copper/tin alloy;
- polarity of the applied voltage.

In no test FC-72 and Vertrel XF showed significant pumping effects, even though for different reasons. Electrical currents in FC-72 are very low (in the order of a microampere), suggesting that the electric field in the tested configuration is not intense enough to activate ion injection or field-enhanced dissociation in this non-polar liquid. On the other hand, in the moderately polar Vertrel XF, the transiting current is as high as 1 A even at an applied voltage of a few hundreds of volts. In this case, the current limitation of our HV power supply did not allow us to reach the 10 kV target. Nonetheless, the high electrical power consumption associated with this fluid makes it disadvantageous and impractical for pumping applications.

As for HFE-7100 and H-Galden ZT85, the test results are reported in figure 7. A negative pressure head indicates that the fluid received a thrust directed from the ring to the point electrodes. Then, we can observe that opposite pumping directions were always obtained when switching the polarity of the applied voltage. In particular, when the points were set to negative HV and the ring electrodes were earthed, the direction of the EHD thrust was from the points to the rings; conversely, when positive HV was applied to the points, the pumping force was weaker and its direction was reversed. This behaviour can be explained by the alternative presence of the two extensively-described EHD pumping mechanisms, which induce motion on opposite directions. Hence, we conclude that in our experiments ion-drag pumping prevailed when negative HV was applied to the points; on the other hand, the same mechanism was inactive at the positive polarity and conduction pumping was the dominant process.

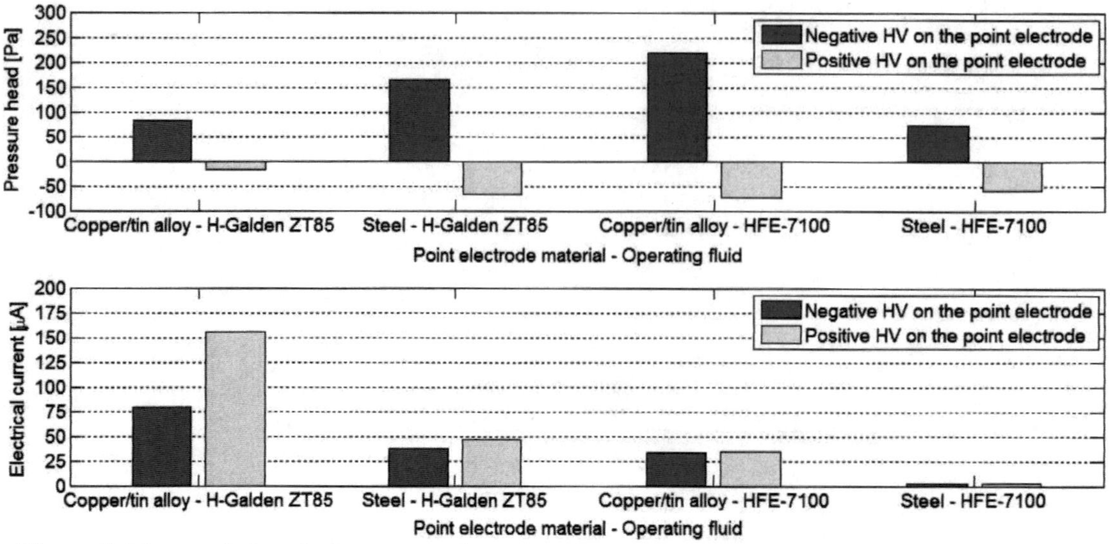

Figure 7. Measured electrical current and generated pressure head (steady-state values) for various configurations operated with H-Galden ZT85 and HFE-7100 (applied voltage: 10 kV).

Among the tested configurations, the one giving the highest static pressure head was: negative polarity on the points, HFE-7100 as working fluid and copper-tin alloy as electrode material. According to the above model, we assume that ion-drag pumping was occurring. Homocharges in the proximity of a HV electrode, typical of ion injection, reduce the intensity of the electric field on that region and, consequently, are expected to lower the transiting electrical current. When positive polarity was imposed on the points, not only the generated pressure head was lower, but we also measured higher electrical currents, coherently with the presence of heterocharges and conduction pumping. Thus, ion-drag pumping is a more efficient process than conduction pumping, generating a higher pressure head with a lower energy input; the main drawback is its less predictable behaviour and the possible deterioration of the working fluid (owing to the irreversible electrochemical reactions that may occur at the metal/liquid interface) that can only be ruled out by means of long-term tests.

The same order of magnitude of generated pressure head was obtained by Jeong and Seyed-Yagoobi [36] on R-123 with a similar experimental setup and an EHD pump design optimised for conduction pumping.

3.2. Experiments on an EHD-controlled NCL

Based on the results of the above-described tests, we built a natural circulation loop operating with HFE-7100. The physical properties of the liquid are reported in table 1.

The loop is a square of side 30 cm; the inner diameter of the pipe is 14 mm. A thermal power of 60 W was applied to the lower horizontal side and removed from the upper one; the vertical legs were adiabatic. We selected the size of the loop and the applied heat input according to the instrumentation already available in our laboratory and to the problems of heat removal associated with the use of dielectric fluids such as HFE-7100 (low thermal conductivity, low specific heat, low boiling point).

Temperature was measured at each of the four vertexes by Negative Temperature Coefficient (NTC) sensors (±0.2 K accuracy), previously calibrated in a thermostatic bath by a reference resistance thermometer. We also measured the volumetric flow rate of the circulating liquid by means of a transit time ultrasonic flow meter (±0.5% accuracy). A schematic of the experimental NCL is shown in figure 8.

Two pumping sections are present, each made by three consecutive electrical modules, as illustrated in figure 9. Every module is formed by a pair of electrodes obtained by a metallic grid; the high voltage electrode is properly cut and bent, in order to create four points, directed perpendicularly

Journal of Physics: Conference Series **501** (2014) 012006 doi:10.1088/1742-6596/501/1/012006

towards the opposite grid. The two pumping sections are oppositely mounted and can be alternately activated, to promote clockwise or anticlockwise motion. When the points are set at a negative HV and the grid is earthed, the expected pumping direction goes from the points to the grid (due to ion-drag pumping). The flow direction is determined by measuring the temperature difference between NTCs 1 and 2. The adiabaticity of the vertical legs of the loop is verified by sensors 3 and 4.

Table 1. Physical properties of HFE-7100 at 25°C and 1 bar (courtesy of 3M).

Chemical name	methoxy-nonafluorobutane
Chemical formula	$C_4F_9OCH_3$
Dielectric strength	11 MV/m
Electrical conductivity	$3.04 \cdot 10^{-8}$ S/m
Relative dielectric permittivity	7.39
Density	1.48 kg/dm^3
Specific heat	1.18 kJ/(kg·K)
Thermal conductivity	0.069 W/(m·K)
Thermal expansion coefficient	$1.53 \cdot 10^{-3}$ 1/K
Kinetic viscosity	0.369 cSt
Sonic velocity	580 m/s
Boiling point	61°C
Freezing point	-135°C

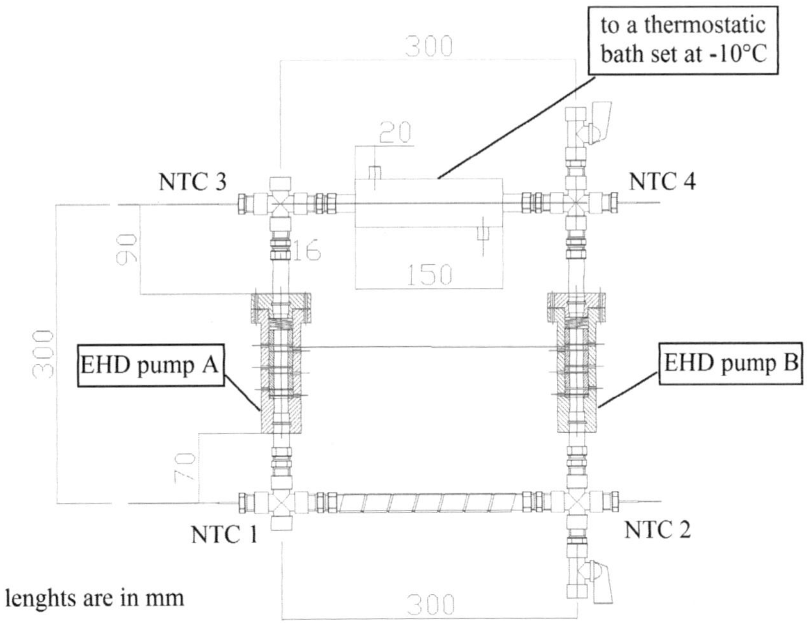

Figure 8. Schematic of the experimental EHD-controlled NCL.

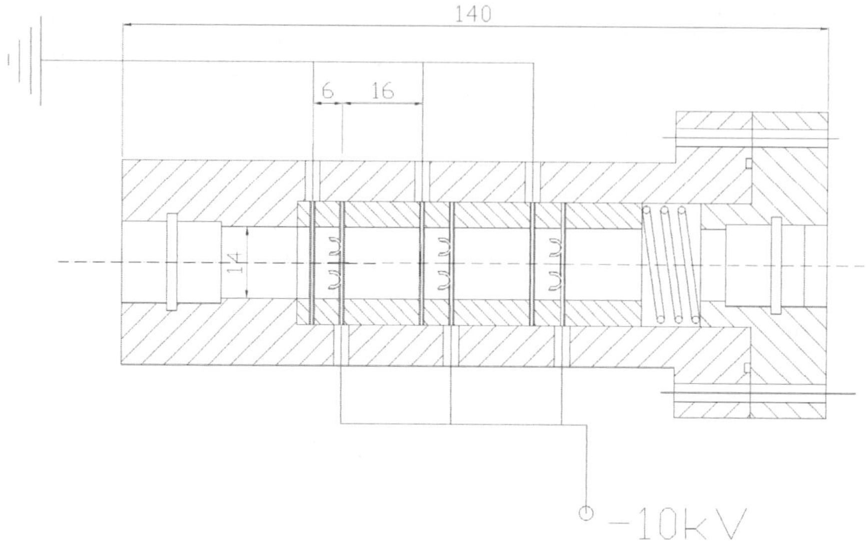

Figure 9. Detail of the EHD pump mounted on the NCL.

In the first series of tests, alternately, one of the pumping sections was activated prior to the heat input, applying negative polarity to the points. The circulation repeatedly followed the ion-drag pumping direction: 5 clockwise (activation of EHD pump B) and 5 anticlockwise tests (activation of EHD pump A).

In subsequent tests, the electric field was applied when natural circulation was already ongoing. In any case, the direction of motion was reversed by means of the ion-drag pumping mechanism. A typical ΔT-versus-time graph is showed in figure 10, ΔT being the temperature difference either between NTCs 1 and 4 for anticlockwise motion (negative values), or between NTCs 3 and 2 for clockwise flow. At the activation of EHD pump B (minute 6), the flow is reversed from anticlockwise to clockwise motion. Next (minute 10), the flow is forced to return to anticlockwise motion, switching off the electric field and, at the same time, giving a proper anticlockwise inclination (about 15 degrees) to the whole loop. At minute 16, pump B is reactivated, producing again the inversion of the flow. The measured flow rate goes from 15 cm/s in pure natural circulation to about 7 cm/s when pump B is active, owing to additional EHD-induced pressure drop.

In a final series of tests, we analysed the effect of switching the polarity of the 10 kV applied voltage on a single pumping section. According to the tests of pressure head generation, it means that we alternately activated the two EHD pumping mechanisms. In any case, the flow in the loop was reversed, following the local upward or downward EHD thrust. Therefore, it was experimentally verified that the direction of motion can be conveniently controlled by means of a single pumping section, simply inverting the polarity of the applied electric tension.

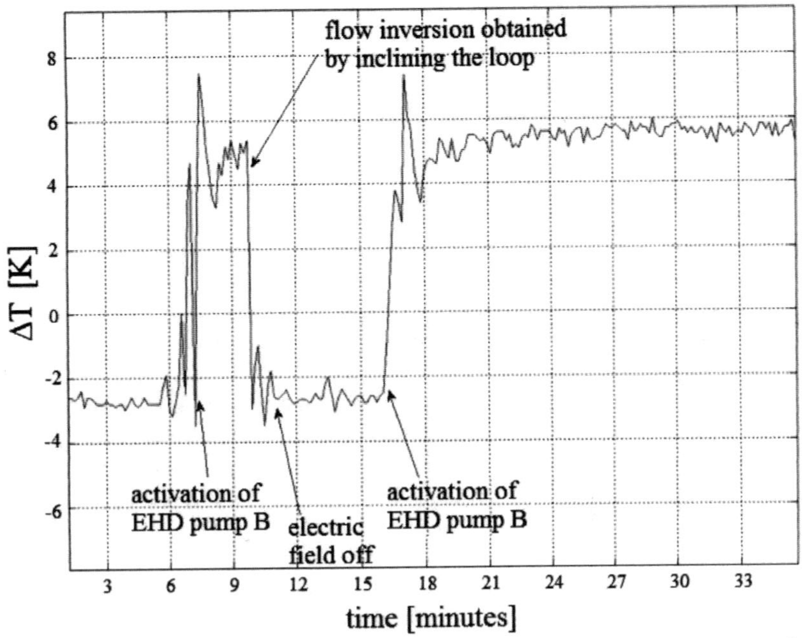

Figure 10. Typical ΔT-versus-time graph with and without ion-drag pumping.

4. Concluding remarks

The manuscript reports on experiments carried out on a natural circulation loop, controlled by electrohydrodynamic pumping devices. Optimization of EHD pumps and control of NCLs is a multi-disciplinary physical problem involving electrochemistry, fluid dynamics and heat transfer. The following major goals are achieved: first, the optimal configuration of an EHD pump is found by changing the working fluid, the high voltage electrode material and by reversing the applied DC voltage, allowing to identify ion-drag pumping as a more efficient process compared to conductive pumping; then, experiments are conducted on an EHD-controlled NCL, demonstrating that the NCL flow direction can be inverted and managed by exploiting the physical phenomena of ion-drag and conduction pumping.

Future progress of the research work includes a focus on stabilization of NCLs. We will attempt to develop a stability model of the NCL with and without the action of the electric field. The stability issues inherent in the considered thermal and geometrical configuration will be identified and characterized as, for instance, in Vijayan [37] or Mousavian et al. [38]. The stability analysis with EHD is expected to show suppression of instabilities typical of NCLs, as hinted by the present experimental investigation, in which instability phenomena seemed to be ruled out by the superimposed EHD-driven forced convection, establishing in any case the direction of flow in the loop.

Acknowledgements

The experimental work outlined in the present paper was carried out in the frame of the National Research Project PRIN 2008 entitled “Complex flow structures in thermogravitational systems, their prediction, promotion and heat transfer optimisation”, research package “Application of electrohydrodynamic techniques to heat transfer enhancement and optimization in single-phase thermogravitational systems”, co-funded by the University of Pisa and the Italian Ministry of Education, Universities and Research, which are gratefully acknowledged.

References

[1] Keller J B 1966 *J. Fluid Mech.* **26** 599
[2] Welander P 1967 *J. Fluid Mech.* **29** 17
[3] Chen K 1985 *ASME J. Heat Transfer* **107** 826
[4] Vijayan P K, Nayak A. K., Pilkhwal D S, Saha D and Venkat Raj V 1992 *Proc. 5th Int. Topical Meeting on Reactor Thermalhydraulics (Salt Lake City, UT)* vol 1 p 261
[5] Creveling H F, De Paz J F, Baladi J Y and Schoenhals R J 1975 *J. Fluid Mech.* **67** 65
[6] Vijayan P K, Sharma M and Saha D 2007 Exp. Thermal Fluid Science **31** 925
[7] Marcelli A, Galassi M C, Misale M, Devia F, Garibaldi P and D'Auria F 2009 *Proc. 64th ATI National Conf. (L'Aquila, Italy)* paper 08.06
[8] Fichera A and Pagano A 2003 *Int. J. Heat Mass Transfer* **46** 2425
[9] Stuetzer O M 1960 *J. Appl. Phys.* **31** 136
[10] Bryan J E and Seyed-Yagoobi J 1991 *IEEE Trans. Electr. Insul.* **26** 647
[11] Atten P and Seyed-Yagoobi J 2003 *IEEE Trans. Dielectr. Electr. Insul.* **10** 27
[12] Pearson M R and Seyed-Yagoobi J 2009 *IEEE Trans. Dielectr. Electr. Insul.* **16** 424
[13] Shelestynsky S J, Ching C Y and Chang J-S 2003 *Annual Report Conf. on Electrical Insulation and Dielectric Phenomena (Albuquerque, NM)* paper 7C-9
[14] Lin C-W and Jang J-Y 2005 *Sens. Actuators A* **122** 167
[15] Richter A, Plettner A, Hoffmann K A and Sandmaier H 1991 *Sens. Actuators A* **29** 159
[16] Babin B R, Peterson G P and Seyed-Yagoobi J 1993 *AIAA J. Thermophys. Heat Transfer* **7** 340
[17] Zhakin A I 1998 Classic Theories of Ion Recombination and Dissociation in Liquids *Electrohydrodynamics* ed A Castellanos (New York: Springer-Verlag Wien) pp 84-102
[18] Langevin P 1903 *Ann. Chim. Phys.* **28** 433 (in French)
[19] Felici N J 1971 *Direct Current* **2** 90
[20] Jeong S I, Seyed-Yagoobi J and Atten P 2003 *IEEE Trans. Ind. Appl.* **39** 355
[21] Grassi W, Testi D and Della Vista D 2007 *J. Enhanced Heat Transfer* **14** 161
[22] Felici N J 1985 *IEEE Trans. Electr. Insul.* **20** 233
[23] Jayaram S 1993 *Proc. IEEE Conf. on Electrical Insulation and Dielectric Phenomena* (*Pocono Manor, PA*) p 86
[24] Atten P, Malraison B and Zahn M 1997 *IEEE Trans. Dielectr. Electr. Insul.* **4** 710
[25] Stishkov Yu K and Buyanov A V 2002 *Proc. 14th Int. Conf. on Dielectric Liquids (Graz, Austria)* p 63
[26] Little R P and Whitney W T 1963 *J. Appl. Phys.* **34** 2430
[27] Grassi W, Testi D and Saputelli M 2005 *Int. J. Thermal Sciences* **44** 1072
[28] Grassi W, Testi D and Della Vista D 2006 *J. Electrostatics* **64** 574
[29] Grassi W and Testi D 2006 *Annals New York Academy Sciences* **1077** 527
[30] Testi D 2007 *J. Thermophys. Heat Transfer* **21** 431
[31] Grassi W and Testi D 2009 *Annals New York Academy Sciences* **1161** 452
[32] Testi D and Grassi W 2013 *J. Heat Transfer* **135** 082202-1
[33] Pickard W F 1963 *J. Appl. Phys.* **34** 246
[34] Pickard W F 1963 *J. Appl. Phys.* **34**, 251
[35] Babin B R, Peterson G P and Seyed-Yagoobi J 1993 *J. Thermophys. Heat Transfer* **7** 340
[36] Jeong S I and Seyed-Yagoobi J 2002 *J. Electrostatics* **56** 123
[37] Vijayan P K 2002 *Nuclear Eng. Design* **215** 139
[38] Mousavian S K, Misale M, D'Auria F and Salehi M A 2004 *Annals Nuclear Energy* **31** 1177

Journal of Physics: Conference Series **501** (2014) 012007 doi:10.1088/1742-6596/501/1/012007

Heat transfer characteristics in forced convection through a rectangular channel with 60° tilted staggered ribs

M Cucchi, D Fustinoni, P Gramazio, L P M Colombo and A Niro
Dipartimento di Energia, Politecnico di Milano
Campus Bovisa, via Lambruschini, 4 – I-20156 Milano, Italy

E-Mail: alfonso.niro@polimi.it

Abstract. In this paper we present new experimental results of investigation on average heat transfer characteristics of a forced air-flow through a rectangular channel with the lower and upper surfaces roughened by ribs; data for a rectangular channel with flat surfaces are presented for comparison as well. The channel cross-section is 120 mm wide and 12 mm high; the channel is operated with the lower and upper walls kept at fixed temperature whereas the sides are adiabatic. The ribs have a square cross section and are mounted 60° parallel-tilted (the angle is respect to main stream) in astaggered arrangement. The tested configurations differ each other for the rib side dimension, namely, 2 or 4 mm, and for their pitch-to-side ratio equal to 10, 20 and 40. Upstream the test channel, there is an entry-section consisting of a 800 mm long, rectangular duct with the same transverse dimensions as the test section but with flat and adiabatic walls. Air flow rates have been varied in order to have Reynolds numbers, based on the duct hydraulic diameter, ranging between 700 and 7500. The average Nusselt numbers are evaluated on the basis of the air-flow bulk-temperature at entrance and exit from the heated zone, as well as of the surface temperature measured by eight T-type thermocouples plugged into the heated walls. The test section is also equipped with static pressure taps placed at the heated zone ends. Results show an increase of the average Nusselt number, calculated as the ratio Nu/Nu_0, for the all tested ribbed channels ranging between 1.0 and 5.0.

Introduction

Heat transfer in forced convection inside rectangular channels is a very interesting matter for industry as it is encountered in critical heat transfer applications like gas-turbine blade cooling [1], and in devices largely used such as plate-fin compact heat exchangers. In designing these devices, high values of heat transfer area per unit volume are searched for; however, if this parameter is increased over a given value, thermal performances start worsening. In fact, the higher the surface-to-volume ratio the narrower the passages, so air velocity has to be lowered to maintain acceptable the pressure drops; however, narrow passages and low air velocities lead to laminar flows laminarization that is of course characterized by a rather poor convective coefficient which eventually defeats completely the area increase benefits. To overcome this limit, heat transfer is enhanced by configuring surfaces with a large variety of fins and ribs, which are an efficient and cost-effective solution. The literature review shows that, starting from the '70s, a continuously growing number of studies have been dealing with heat transfer over rib-roughened surfaces, as this is a very attractive solution for both gas-turbine blade cooling and compact heat exchangers.

A first crucial feature for this kind of roughness is the rib arrangement with respect to the main stream. The number of possible configurations, indeed, is very large but four arrangements have been particularly considered so far in literature, namely, transverse-, parallel-tilted-, cross-inclined- and V-shaped-ribs; in addition, ribs may be assembled on the opposite walls either staggered or in-line. In this paper, we focus particularly on the parallel-tilted and cross-tilted arrangements. A brief discussion of relevant works in literature is presented below making use of the following dimensionless geometric parameters: the channel aspect ratio *AR*, i.e., the base to height ratio (making reference to the channel cross section), the dimensionless pitch p/e , i.e., the pitch to rib height ratio, and the blockage factor e/D_h defined as the rib height to the channel hydraulic diameter ratio.

The first major studies were carried out starting from the first half of the '80s by Han [2-7], Park [8], and Kukreja [9]. It was observed that tilting the ribs with respect to the main flow direction results in higher heat transfer performance due to the presence of secondary flows. Actually, as a consequence of inclination, two counter-rotating vortices develop near the rib and slide along it. They interact with the main flow carrying part of the colder core towards the wall. Furthermore, the interaction between main and secondary flows affects the reattachment point, and hence the recirculation-zone between two successive ribs. Specifically, a great deal of work was devoted to the case with a tilt angle $\alpha = 45°$; it was found that the optimum dimensionless pitch p/e strongly depends on both e/D_h and *AR*. The most studied configuration presents AR=1 with all the sides heated.

Among the recent studies, there are the papers of Liu et al. [10] [11], Choi et al. [12], and Smulsky et al. [13]. Liu et al. [10] [11] carried out experiments on heat transfer characteristics in steam-cooled rectangular channels with parallel-tilted ribs on the opposite walls, for two tilt angles, i.e., 45° and 60°, several aspect ratios within 0.5 and 2.0, and Reynolds number ranging from 10000 to 140000. They found that thermal performances for α=45° are 15-25% larger than for α=60°. They also showed that heat transfer coefficients for steam are 12-25% higher than for air.

Choi et al. [12] studied the local heat transfer characteristics on two kinds of surfaces, namely, smooth or dimpled, with parallel-tilted ribs ($\alpha = 60°$). They found that ribs are more effective than dimples in enhancing heat transfer. Finally, Smulsky et al. [13] investigated the effect of different tilt angles of a single rib on a flat surface. The local heat transfer maximum for 50° is approximately 40% higher than that for 90°, while its position depends on rib side dimension.

Finally, in the last decade, rectangular channels with only one heated, ribbed surface have aroused great interest for applications in renewable energy technologies as solar air heaters. Reviews on this subject are reported in Bhushan and Singh [15], and Hans et al. [16] [17].

A closer look reveals that most of the work done up to now focused on fully turbulent flow regimes. Moreover, two aspects are very little investigated, namely, the impact of high values of the blockage factor, and the laminar-to-turbulent flow regime transition. Both these aspects are very relevant in designing compact heat exchangers, as well shown by the specific literature [18]. For this reason we started an experimental campaign aimed at investigating heat transfer coefficients and pressure drops in rectangular ducts with variously arranged square ribs, in a range of Reynolds numbers between 600 and 8000. In these conditions, it is possible to explore both the turbulence onset and the transition region. In this paper we present a selection of the available results while experimental runs are under completion.

Experimental setup

As schematically shown in figure 1(a), the experimental setup consists of two independent circuits, namely, the air-circuit containing the test section, and the heating-water circuit used to control the temperature of the channel walls.

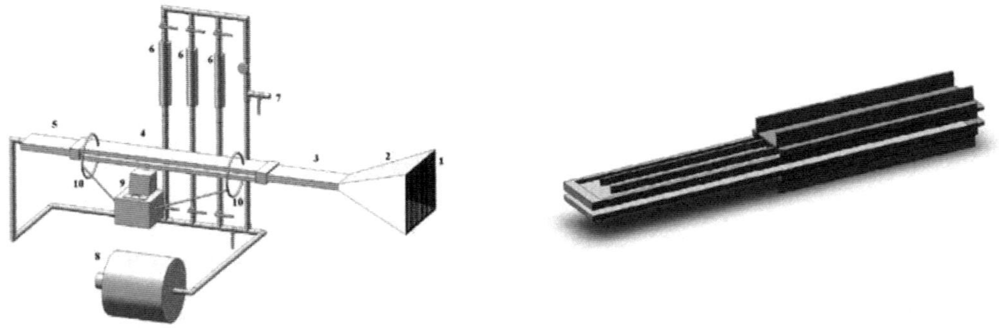

(a) (b)

Figure 1. Schematics of the experimental setup: 1. fine mesh screen cover; 2. convergent air inlet; 3. entry-section; 4. test-section; 5. exit-section; 6. rotameters; 7. by-pass; 8. blower; 9. heat bath; 10. water circuit piping (a), test section with the duct on the wall backside (b)

Through a convergent, room air flows into the circuit; a fine mesh screen covers the convergent inlet whereas at its outlet there is a flow straightener consisting in a matrix of staggered 4-mm-diameter, 40-mm long thin polystyrene tubes. Inside the convergent a thermo-resistance, i.e., a resistance thermometer, is mounted to measure the air temperature. From the straightener air flows into an entry-section that is a rectangular duct with the same cross dimensions as the tested channel; entry-section is 0.8-m long, i.e., near 40 times the channel hydraulic diameter, and its walls are made with 10-mm thick plexiglas plates and are not heated. At the end of this section, air enters the test-section that is a 120-mm wide, 12-mm height, 880-mm long duct; its lower and upper walls are two aluminium plates of 10-mm minimum thickness with a rib configured face.

As shown in figure 1(b), the backside of each plate is covered by a cap strongly tightened to the plate, so that they form a jacket where the heating water flows. Inside and outside the water jacket there are ribs which prevent the plate buckling. To check if temperature is uniform over the heated walls, four thermocouples are embedded in the lower wall and three in the upper one; each thermocouple is cemented into a 1.8-mm-wide, 0.5-mm-depth groove cut in the flat face; thermocouple locations are displayed in figure 2 whereas table 1 lists the exact position of their junctions. Eventually, the test-section sides are closed by 4-mm-thick glass plates to allow optical access inside the channel.

Table 1. Probe positions on ribbed channel wall.

Upper wall		Lower wall	
x (mm)	y (mm)	x (mm)	y (mm)
120	60	70	60
440	60	120	60
760	60	440	60
		760	60

(a)

(b)

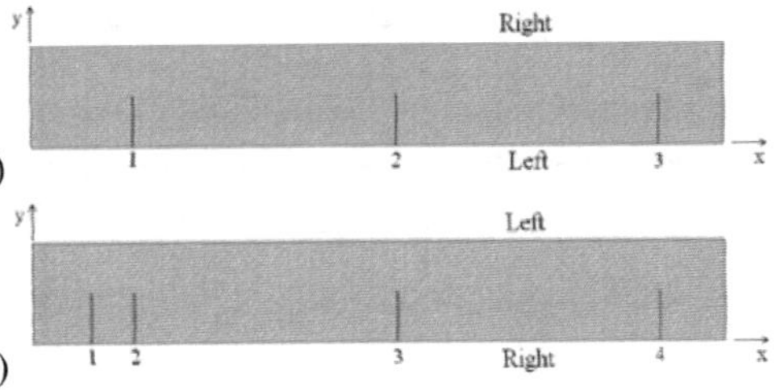

Figure 2. Thermocouples locations on upper (a) and lower (b) test channel walls.

At the test-section outlet, there is a short exit-section equipped with two turbolizer rows followed by a convergent that conveys air through a 20-mm-wide, 6-mm-high channel partially filled by a fine-meshed plastic net; after the screen, a PTR and a thermocouple are positioned on the centreline of this channel to measure the air bulk-temperature. Finally, inside the couplings between the entry-section and the test-section, and between this one and the exit-section there are two pressure taps, each consisting in a 1.5-mm-diameter hole (great care was devoted in eliminating any burr). Downstream the exit-section there are three rotameters, i.e., float-type flow-meters, connected in parallel with full

Journal of Physics: Conference Series **501** (2014) 012007 doi:10.1088/1742-6596/501/1/012007

scale of 6, 23.5 and 40 m³/h respectively, a metering valve, and a 7-stage, 30-kPa-head, 5.5-kW-power blower operating in suction mode. The exhausted air is discharged outside the laboratory. Air temperature is also measured upstream the flow-meters by means of a thermocouple plugged in the pipe.
The heating circuit is mainly composed by a heat bath which provides high mass flow-rate of water at constant temperature, and by the channels built into the upper and lower test-section walls (water and air stream in counterflow); two thermocouples are placed inside each water channel, near the inlet and outlet ports, respectively. The heat bath is the ThermoHaake B12 with a tank of 12-dm³, a 3-kW heater and a high precision controller; water temperature inside the tank is kept constant within 0.01 K.

Measurements, data processing and error analysis

The thermocouples used are T-type with 0.5-mm-diameter wires, whereas the PTR are 4-wires, 100-ohm, Platinum type, i.e., PT100, with dimensions of 2 mm x 4 mm. Both thermocouples and thermo-resistances were all preliminary calibrated over five points within the temperature range from 22 to 60 °C, by means of the ThermoHaake heat bath. The probes were immersed all together into the bath while devoting great care in their positioning; for each calibration point, 160 readings per probe were collected; the resulting standard deviation is of 0.02 K for the thermocouples, and of 0.01 K for the thermo-resistances. All temperature measurements are performed by means of an Agilent 34970A data logger equipped with a relay multiplexer and a 6½ digit multimeter. The channels are sequentially read, by waiting a 0.5-s settling time after each channel-locking, with an integration time of 400 ms which ensures a standard deviation of 0.01 K that is less than or equal to the probe uncertainties; consequently, reading cycles are performed every 20 s. Pressure drops are measured by connecting the two pressure taps, placed at the inlet and outlet of the test channel, to a differential micromanometer with a full scale of 250 Pa and a 0.125-Pa sensitivity. Air volume-flow-rate is measured by means of the aforementioned rotameters which have a 2% nominal accuracy; however, by means of a calibration performed by measuring pressure drops of laminar air-flows through a smooth circular tube, we found that their accuracy is better of 1%. Finally, in order to guarantee repeatability to measurements, we adopted a precise test procedures described in the following. First, we power the bath heater and water starts to circulate through the entire circuit included the ducts on the test-section wall backside. After 30 m, that is the time to stabilize the channel wall temperature to a prefixed value within a 0.01-K band, we power the blower and air starts to flow through the test section. Then we need to wait for other 30 m in order to attain regime conditions with all temperature time-fluctuations within 0.02-K band; in this conditions, channel wall-temperature is uniform within 0.1 K over the entire heated length. Eventually, we start to collect 20 reading-cycles which take 400 s; during this time, measurements of the air-volume-flow-rate and of reading-cycles was chosen first by collecting data for N=200 for some of the most representative regime conditions, and then by calculating their average standard deviation as a function of N; as a result, we found that at N=20 the average standard deviation becomes less than 0.01 K, and thus we assumed that 20 is the minimum reading-cycle number to be collected.
The average Nusselt over the test section and the apparent Darcy friction factor are calculated as follows

$$Nu = \frac{D_h}{k}\frac{\rho \dot{V} c_P}{A_S}\ln\frac{\theta_i}{\theta_o} \tag{1}$$

$$f = \frac{2\Delta p A^2 D_h}{\rho \dot{V}^2 \ell_{taps}} \tag{2}$$

Where $\dot{V}$ the air volume-flow-rate, ρ the density calculated at the temperature where $\dot{V}$ is measured, c_P the specific heat at constant pressure, k the thermal conductivity, A_S the total heated area, θ_i and θ_o the wall-to-air-bulk-temperature difference at the test section inlet and outlet respectively, D_h the hydraulic diameter, Δp the pressure drop, and ℓ_{taps} the distance between the pressure taps.

Journal of Physics: Conference Series **501** (2014) 012007 doi:10.1088/1742-6596/501/1/012007

The error analysis performed according to Moffat [19] gives an uncertainty less than 3% on the convective coefficient, and less than 7% on the apparent Darcy friction factor

Results

Experimental results here presented refer to a rectangular channel with parallel-tilted ribs on the lower and upper surfaces in a staggered arrangement, as schematically shown in figure 3. Ribs have a square-cross section and are made of linden wood, i.e., a material with a very low thermal conductivity, to minimize their fin effect on heat transfer. Tests have been carried out on many configurations differing for the rib side dimension and the pitch-to-side ratio as listed in table 2, whereas channel main dimensions are listed in table 3.

All data are for fixed and uniform wall-channel-temperature at the nominal value of 40°C, and for air-flows entering into the test-section at room temperature of near 22°C (elaborations, indeed, account for their actual values).

Figure 4(a) displays the average Nusselt number plotted versus the Reynolds number for all the tested configurations . Values obtained for the rectangular channel with flat surfaces, operated at the same conditions, are also plotted for comparison.

For the smooth rectangular-channel, the average Nusselt number remains nearly a constant for Reynolds numbers up to 2100, and then it starts to increase exhibiting a power-law dependence on *Re* with an exponent of about 0.8. The figures also reports as solid lines the values calculated for a flat rectangular-channel by the Shah and London correlation with Wibulsas correction for *Re*<2300, and by the Gnielinski correlation for *Re*≥2300 [20]; all predictions take into account thermal effects of the entry-region. As it can be seen, present data agree very well with predictions.

Table 2. Configurations parameters.

e [mm]	p [mm]	e/D_h	p/e	e/h
2	20	0.0917	10	0.1667
2	40	0.0917	20	0.1667
2	80	0.0917	40	0.1667
4	40	0.1833	10	0.3333
4	80	0.1833	20	0.3333
4	160	0.1833	40	0.3333

Table 3. Main channel parameters.

Channel height, *h*	12.00 mm
Channel lenght, *L*	880.00 mm
Channel width, *w*	120.00 mm
Hydraulic diameter, D_h	21.82 mm
Height to width ratio	0.10

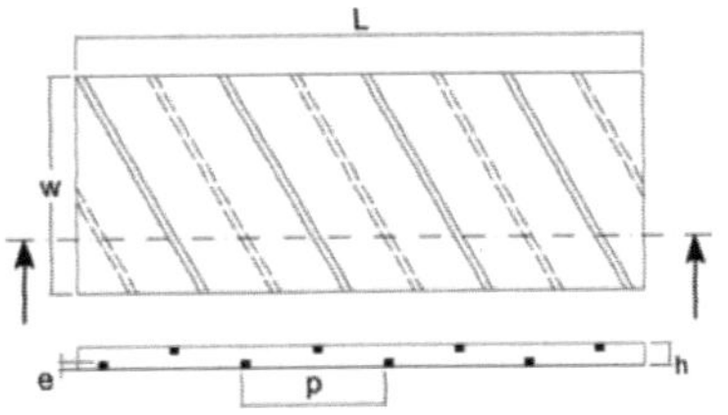

Figure 3. Tilted parallel staggered configuration

For all the tested ribbed configurations, however, the average Nusselt number displays an increasing trend even from the minimum value of the investigated Reynolds numbers, revealing a power-law dependence on *Re* yet with an exponent ranging between 0.80 and 0.95. The lack of a constant *Nu* region induces to conjecture that flow becomes turbulent at *Re*<600 in these channels; however, no comparison with other results in the open literature is possible because seemingly there are not data for this kind of geometries at *Re*<2000.

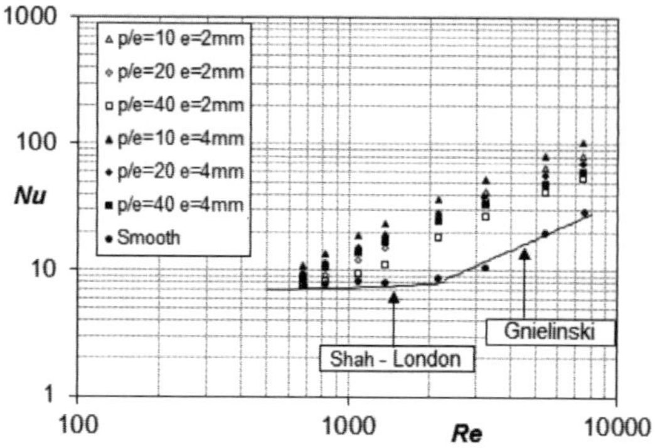

(a)

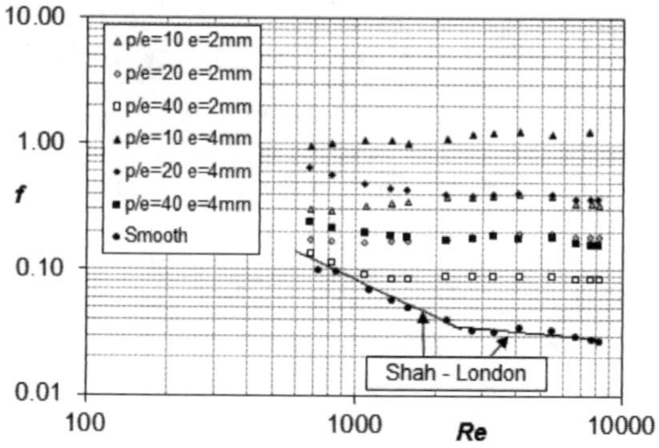

(b)

Figure 4. Average Nusselt number (a) and Darcy friction factor (b) versus Reynolds number

Among the tested configurations here reported, the highest heat transfer increase is for the *p/e*=10 and *e*=4mm, since it displays a mean value of the ratio Nu/Nu_0 equal to 3.2 (Nu_0 refers to the flat channel at the same operating conditions) with a maximum of 5.0 at *Re*=3250; whereas the configuration with *p/e*=40 and *e*=2mm is the less performing one being characterized by a mean Nu/Nu_0 ratio equal to 1.6. The Nusselt number ratio and the friction factor ratio decrease by increasing the pitch-to-side ratio, whereas they increase by increasing the rib side (or blockage values).
Figure 4bshows the experimental values of the apparent Darcy friction factor *f* plotted versus Reynolds number for all the tested configurations. For comparison, the figure also reports as solid lines the values f_0 calculated for a laminar and turbulent flow through a flat rectangular channel by means of Shah and London correlation [20]. From this figure, it is evident that *f* is substantially higher than f_0 for allconfigirations; more precisely, the worst configuration is for *p/e*=10 and *e*=4mm, as it displays a mean value of *f* 52 times larger than f_0, but f/f_0 is reduced to 2.2 for *p/e*=40 and *e*=2mm. These data display the typical trend of flows through a duct with walls characterized by an equivalent sand roughness, namely, the friction factor becomes independent of *Re* for sufficiently high values. It is noteworthy that at the highest of the two blockage values, corresponding to a rib height of 4 mm, the flow appears to be turbulent at any *Re*; at the lowest blockage, this behavior occurs for *p/e*=10 and 20, while the turbulence onset shifting to higher *Re* for *p/e*=40.

As a concluding remark, from an engineering point of view, it is interesting to compare performance of different ribbed channels, and indeed there are currently several criteria to carry out this comparison. The Criterion we used compares performances by taking into account at the same time both the heat transfer enhancement (Nu/Nu_0) and the pressure-drop penalization (f/f_0) at constant pumping power

$$\frac{Nu/Nu_0}{(f/f_0)^{1/3}} \tag{3}$$

Figure 11 shows a graph where the values calculated by this Criterion is plotted versus the Reynolds number for all the tested configurations. According to this Criterion, the highest value is 1.9 obtained at Re=3250 for the configuration with *p/e*=20 and *e*=2mmGenerally speaking, the most efficient configuration is for *p/e*=40 and *e*=4mm at low Reynolds number, and for *p/e*=20 and *e*=2mm at high one.

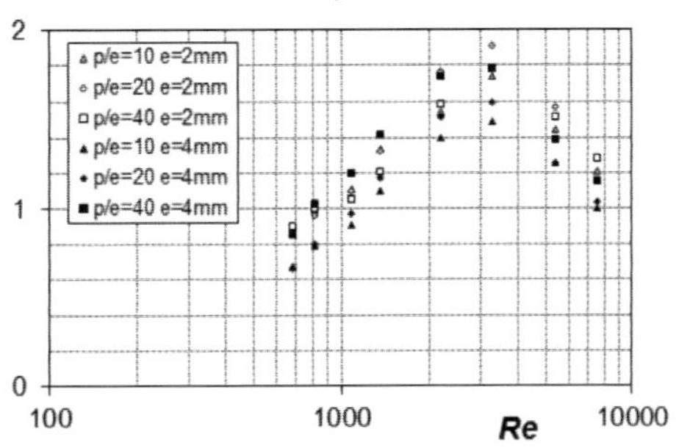

Figure 11. Tilted staggered ribs comparison at constant pumping power.

Conclusions

This paper reports about an experimental investigation on heat transfer characteristics of a forced air-flow through a rectangular channel with the lower and upper surfaces roughened by ribs with different configurations, at Reynolds numbers ranging between 700 and 7500. The enhancement effect is essentially due to the periodic streamline deflection induced by the ribs. Configurations differ each other for rib side, namely, 2 and 4 mm, and for the pitch-to-rib-side ratio equal to 10, 20 and 40. The Nusselt number averaged over the entire channel, and the Darcy friction factor are evaluated for each configuration and they have been plotted as a function of the Reynolds number; experimental results for a plate channel are also reported as reference values.

Heat transfer measurements show for all configurations a power-law dependence on Reynolds number with an exponent ranging between 0.80 and 0.95. As there is not a region at constant Nusselt number, we conjecture that flow is turbulent even from the lowest investigated values of the Reynolds number. Among the tested configurations, the one with p/e=10 and e=4mm shows the maximum thermal improvement with respect to the flat channel, and in fact it is characterized by a mean enhancement factor of 3.2, with a maximum of 5.0 at Re=3250, but in spite to this increment the apparent Darcy friction factor worsens of nearly 27.5 times. In order to compare performances by taking into account at the same time both heat transfer enhancement and pressure-drop penalization, we used a comparison criterion based on a fixed pumping power. The configuration withp/e=20 and e=2mm displays the better performances with a value of 1.9 at Re=3250.

Nomenclature

Symbol	Quantity	Unit
A_S	Total heated area	m^2
AR	Channel aspect ratio	-
D_h	Hydraulic diameter	m
e	Rib side dimension	mm
$\dot{V}$	Volume flow rate	$m^3\ s^{-1}$
c_P	Specific heat	$J\ kg^{-1}\ K^{-1}$
f	Darcy-Weisbach friction factor	-
h	Channel height	mm
k	Thermal conductivity	$W\ K^{-1}\ m^{-1}$
ℓ_{taps}	Distance between pressure taps	m
p	Ribs distance	mm
Δp	Pressure drop	Pa
Nu	Nusselt number	-
w	Channel width	mm

Journal of Physics: Conference Series **501** (2014) 012007 doi:10.1088/1742-6596/501/1/012007

Greek symbols

α	Rib angle respect to main stream	deg
θ	Wall-to-air-bulk-temperature difference	°C
ρ	Density	kg m^{-3}

Subscripts

i	Inlet
o	Outlet

References

[1] Gupta S, Chaube A and Verma P 2012 *J.Eng. Sci. Tech. Rev.* **1** 57-62
[2] Han J C and Park J S 1985 *J. Eng. Gas Turb. Power* **107** 628-635
[3] Han J C 1988 Heat *ASME J. Heat Trans.* **110** 321-328
[4] Han J C and Park J S 1988 *Int. J. Heat Mass Tran.* **31** 183-195
[5] Han J C, Ou S, Park J S and Lei C K 1989 *Int. J. Heat Mass Tran.* **32** 1619-30
[6] Han J C and Zhang Y M 1991 *J. Turbomach.* **113** 123-130
[7] Han J C, Zhang Y M and Lee C P 1991 *J. Heat Trans.* **113** 590-596
[8] Park J S, Han J C, Huang Y, Ou S and Boyle R J 1992 *Int. J. Heat Mass Tran.* **35** 2891-903
[9] Kukreja R T, Lau S C and McMillin R D 1993 *Int. J. Heat Mass Tran.* **36** 2013-20
[10] Liu J, Gao J and Gao T 2012 *J. Mec. Sci. Tech.* **4** 1291-1298
[11] Liu J, Gao J, Gao T and Shi X 2013 *App. Therm. Eng.* **50** 104-111
[12] Choi E Y, Choi Y D, Lee W S, Chung J T and Kwak J S 2013 *App. Therm. Eng.* **51** 435-441
[13] Smulsky Y I, Terekhov V I and Yarygina N I 2012 *Int. J. Heat Mass Tran.* **55** 726-733
[14] Won S Y and Ligrani P M 2004 *Int. J. Heat Mass Tran.* **47** 1573-1586
[15] Bhushan B and Singh R 2010 *Energy* **35** 202-212
[16] Hans V S, Saini R P and Saini J S 2010 *Sol. Energy* **84** 898-91
[17] Patil A K, Saini J S and Kumar K 2012 *Int. J. Ren. Energy Res.* **2** 1-15
[18] Webb R L 1994 *Principles of enhanced heat transfer* (New York: Winley-Interscience)
[19] Moffat R J 1988 Describing the uncertainties in experimental results *Exp. Therm. Fluid Sci.***1** 3-17
[20] Kakaç S, Shah R K and Aung W 1987 *Handbook of single-phase convective heat transfer*, (New York: Wiley-Interscience)

Journal of Physics: Conference Series **501** (2014) 012008 doi:10.1088/1742-6596/501/1/012008

Estimation of trial parameters for Pulse Phase Thermography with low power heat sources

L Vitali, D Fustinoni, P Gramazio and A Niro

Energy Department, Politecnico di Milano,
Campus Bovisa, Via Lambruschini 4, 20156 Milano – Italy

E-mail: luigi.vitali@mail.polimi.it

Abstract. Non-destructive Testing by Infrared Thermography (IR-NDT) is a widely adopted technique to reveal the presence of defects, i.e. discontinuity zones of thermal proprieties, inside materials. Pulsed Phase Thermography (PPT) is one of the most interesting techniques among IR-NDT: the specimen is heated by a thermal pulse and the sequence of thermograms of the surface cooling is transformed with the Discrete Fourier Transform (DFT). The resulting phase images (phasegrams) show little sensitivity to irregular heating and surface proprieties, and allow better defect identification by increasing the contrast. It is also possible to estimate the depth of the defect by correlating a characteristic frequency to the thermal diffusion length of the defect. The outcome of this analysis depends on the fine tuning of the technique and the appropriate choice of the parameters of the thermal pulse, namely length and power, as well as of the acquisition: frequency and observation time. While there are in literature a few guidelines for the choice of these parameters, a good knowledge of the technique and a certain degree of guessing is still required, especially when low heating power, longer pulses and small and deep defects are involved. This paper reports a method to estimate these parameters, partly based on theoretic considerations and partly on numerical simulations performed by means of a FEM commercial code on a 2D axial-symmetric model. Experimental results are also here presented, focusing on the difference between a thick plate and a thin one.

1. Introduction

Pulsed Phase Thermography (PPT) is a widely studied and used technique for the non destructive assessment of discontinuities within a huge range of materials and thicknesses, from concrete walls to metal or composite plates used in various engineering applications. Its main advantages are a low sensitivity to uneven heating and specimen emissivity, and a better identification of deeper defects compared to the time based Pulse Thermography techniques.

A lot of studies focusing on specific applications are available in the open literature, but only a few are about the correct choice of the trial parameters ([1], [2]). Indeed, these parameters are crucial for a successful test execution, especially when the PPT is carried out with halogen lamps whose pulse can not be shorter than a few seconds in order to adequately heat the specimen. The present work reports about a way to readily estimate these parameters, namely pulse length and power, observation time and frequency, via a combination of theoretical and numerical approaches, experimentally validated on polymethylmethacrylate specimens.

2. Theory

The 1D model of thermal pseudo-waves propagation (solution of a diffusion equation) is still useful as a basis to understand the behavior of the phenomenon under study [3], combined with knowledge about how thermal pseudo-waves reflect and refract on obstacles according to the Snell law [4].

Let us consider a very thick slab, and let us impose an harmonic heat flux boundary condition on one of its surfaces:

$$q(0,t) = q_0 \cos(2\pi f t - \varphi) \tag{1}$$

If we further assume that inside the slab there is a defect like a delamination at a depth L, the resulting surface temperature estimated for a 1D model is

$$\tilde{T}_0(t) = T_i + \theta_0 \, cos(2\pi f t - \varphi - \pi/4) + \hat{r}\theta_0 e^{-2L/\mu} cos\left(2\pi f t - \frac{2L}{\mu} - \varphi\right) \tag{2}$$

The factor μ is the *thermal diffusion length*, defined as

$$\mu = \sqrt{\frac{\alpha}{\pi f}} \tag{3}$$

It expresses how deep a thermal wave can travel inside a medium (for $z = \mu$ the amplitude is reduced by a factor *1/e*). $\hat{r}$ is the reflection coefficient

$$\hat{r} = \frac{e_1 - e_2}{e_1 + e_2} \tag{4}$$

It depends on the effusivity of both the specimen and the defect.

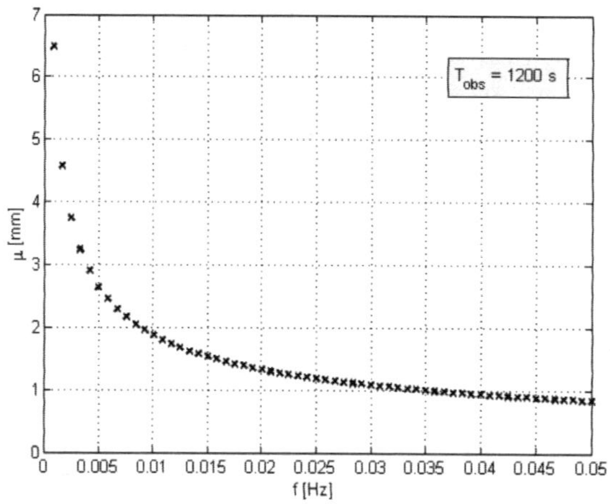

Figure 1. Thermal diffusion length, Plexiglas, fixed observation time

Although the 1D model is useful to understand the basic behavior of the phenomenon, it is quite inaccurate for prediction purposes, because the 2D and 3D effects, depending on the defect shape, cannot be ignored. In considering this, several analytical models have been proposed, especially in order to take into account the lateral size of the defect [5], and the effect of surface convection [3].

Journal of Physics: Conference Series **501** (2014) 012008 doi:10.1088/1742-6596/501/1/012008

3. Experiments

3.1. Experimental setup

The experimental setup is made by two reflectors, each one equipped with a 500W halogen lamp and a guillotine shutter in order to ensure a square thermal pulse to a good extent. The IR camera is the Raytheon Radiance HS (InSb, Focal Plane Array, detectable wavelength region 3-5 μm). Experiments are carried out using two 200 x 200 mm PMMA flat plates as specimens, 4 and 20 mm thick, respectively. On one side, flat-bottomed holes are drilled to simulate the defects, while the other is black painted to enhance its emissivity.

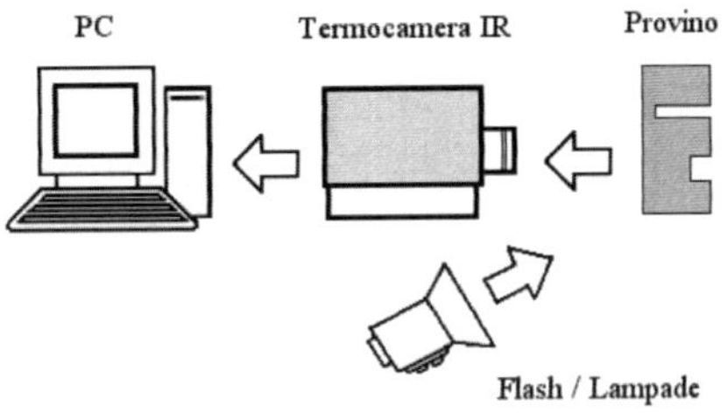

Figure 2. Experimental scheme

Table 1. Thermal proprieties

PMMA		
Density	ρ [kg/m^3]	1190
Specific heat capacity	c [J/(kg K)]	1470
Thermal conductivity	k [W/(m K)]	0.19
Thermal diffusivity	α [mm^2/s]	0.109

3.2. Inversion method

The inversion method used in this work was proposed by Gutierrez-Fajardo in [6]. It is observed that, for each defect depth, there is an optimal visibility frequency (f_{opt}) at which the defect appears the least distorted or affected by noise. In order to find this frequency, a correlation function R (7) between consecutive phasegrams $\Phi_k(f_k)$ (5) is calculated, according to the following equations:

$$\phi_k(f_k) = \tilde{\phi}_{ij}(f_k) = arg\{FFT[\tilde{T}_{ij}(t_n)]\} \tag{5}$$

$$r(A,B) = \frac{\sum_{i=1}^{Row}\sum_{j=1}^{Col}(A_{ij}-\bar{A})(B_{ij}-\bar{B})}{\sqrt{\left(\sum_{i=1}^{Row}\sum_{j=1}^{Col}(A_{ij}-\bar{A})^2\right)\left(\sum_{i=1}^{Row}\sum_{j=1}^{Col}(B_{ij}-\bar{B})^2\right)}} \tag{6}$$

$$R = \left[r(\phi_1,\phi_2),\ r(\phi_2,\phi_3), \dots\ r(\phi_i,\phi_{i+1}), \dots r\left(\phi_{\frac{N}{2}-1},\phi_{\frac{N}{2}}\right)\right] \tag{7}$$

Where $\tilde{T}_{ij}(t_n)$ is the thermographic image sequence.

The frequency at which R(f) is maximum (the lowest of the two for first five pairs of phasegrams, the average for the rest) goes into equation (8) to obtain the defect depth.

$$z_{exp} = C1\sqrt{\frac{\alpha}{\pi f_{Rmax}}} \tag{8}$$

C1 is a correlation constant obtained by a set of trials on known depth defects, and it could depend on the specimen material [2].
This criterion showed good results specially for the deepest defects that are associated to the lowest frequencies, because of a lower phase noise; and further, it has the advantage that the user has not to point at a sound zone.

3.3. Experimental procedure

3.3.1. Pulse. The power spectrum of an ideal δ(t) pulse is a constant: every frequency is excited with the same intensity. However, actual thermal pulses have a finite duration; while a 5 ms flash can be considered quite an ideal pulse for our purposes, pulse time τ is in the order of seconds when lamps are used, and it must be carefully selected in order to properly stimulate the highest frequencies associated with the most superficial flaws. Indeed, the longer the pulse times, the larger the power on low frequencies, and, as a consequence, long pulses are well suited for deeper investigations.

$$q(t) = \begin{cases} 0 & t \leq 0 \\ q_0 & 0 < t \leq \tau \\ 0 & t > \tau \end{cases} \tag{9}$$

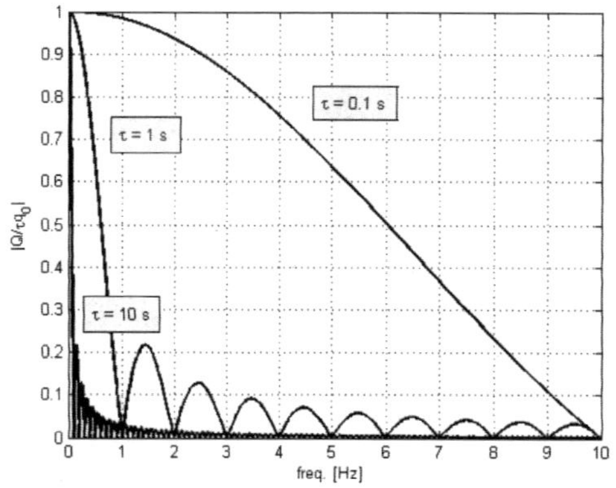

Figure 3. Energy distribution for various pulses, normalized

3.3.2. Postprocessing. Data acquired from the IR camera is automatically processed with a dedicated software developed by ThermALab following this procedure:

- Calibration from grayscale to temperature values; the calibration is carried out with a blackbody and it is normally non-linear.
- FFT for each pixel, obtaining N/2 phasegrams, where N is the number of analyzed thermograms.
- Median and Gaussian filters on each phasegram, in order to reduce crazy pixels and high frequency noise respectively.
- Morphological reconstruction [6]. This technique is based on the *image segmentation*, i.e., the division of an image into different regions (minima and maxima of phase) in order to use geodesic dilatation and erosion algorithms to extract both shape and extension of the defect from the phasegram image

- Correlation between phasegrams, according to equation (6).

4. Numerical simulations

4.1. Numerical model

A fast running numerical FEM model has been developed to quickly evaluate the trial parameters, as well as the constant *C1* in equation (3), while taking into account different materials or defect types, i.e., different reflection coefficients.

The model is characterized by a 2D axisymmetric geometry, and it is equivalent to a round plate with the defect located at its center. The points whose temperatures are extracted are the one on the center of the defect and that located 4.5 mm from the symmetry axis, respectively.

4.2. Validation

The model should reproduce the behavior of the sound zone. The analysis on the 20 mm thick specimen shows a good agreement between experimental and predicted data, even for the simplest possible model, whose boundary condition are adiabatic on every sides (figure 2a).

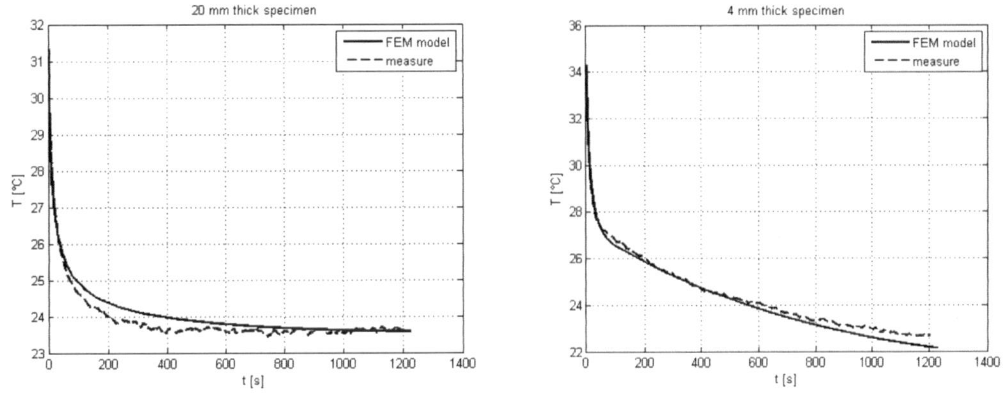

(a) (b)

Figure 4. numerical vs. experimental data on sound specimens, 20 mm thick (a), 4 mm thick (b)

The 4 mm thick specimen requires the heat losses by convection to be taken into account. To this end, a convective heat transfer boundary condition with $h=5$ W/(m^{2}K) was introduced. By means of the model, the FFT of the temperature is calculated, and the defect depth is evaluated by the frequency at which the phase difference shows a maximum. The agreement between the model results and the experimental ones is evaluated with the agreement on C1.

A validation of this model was carried out by comparisons with results from literature [2] of the behavior of the phase difference $\Delta\varphi$ between the defect centre and the sound zone, varying the depth and the size of the defect. This model clearly shows the expected results, namely the defect size only affects the magnitude of $\Delta\varphi$, whereas the correspondent frequency doesn't change. On the other end, an increase of depth causes a decrease of the frequency at which $\Delta\varphi$ peaks (figure 5).

4.3. Automatic regression

The simple FEM model here developed shows good results and has a fast execution time. So, a routine has been written to evaluate, given a set of defects, namely diameters and depths of flat-bottom holes, an estimate of the main parameters of interest. In particular, the correlation coefficient *C1*, whose evaluation requires normally an experimental campaign on a set of known flaws, the influence of the trial parameters on depth estimation errors and, given the IR camera sensitivity, an indication about which defects are detectable, by showing the times in which a temperature difference between the flow and the sound zone is measurable.

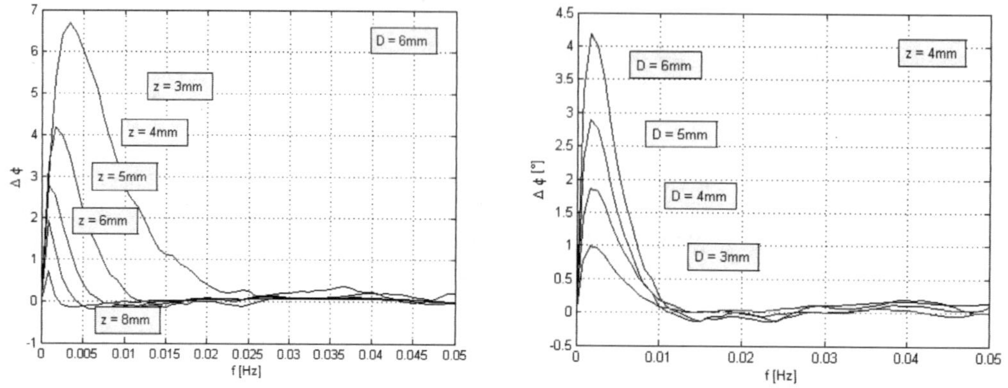

(a) (b)

Figure 5. Phase differences varying defect dimensions, 20 mm thick plate, fixed diameter (a), fixed depth (b)

Table 2. Automatic regression output

E	Tau	Mat	thickness	T_oss [s]	C1	σ	CONV	IR Cam sens.	
1350	20	plex/aria	4mm	1200	1,077	0,32715	SI	0,15	
Z	**D**	φ_{num}	z_{num}	$\Delta\varphi_{max}$	**n_freq**	**f_Δf max**	t_{start}	t_{end}	**Err [%]**
2	4	1,53	1,65	4,88	19	0,015	1	174	17,73
2	8	2,16	2,33	13,42	10	0,0075	1	361	-16,35
2	12	2,29	2,47	18,94	9	0,006667	1	504	-23,41
5	4	4,58	4,94	0,99	3	0,001667	0	0	1,27
5	8	4,58	4,94	3,96	3	0,001667	75	315	1,27
5	12	4,58	4,94	7,54	3	0,001667	55	577	1,27
7	4	6,48	6,98	0,46	2	0,000833	0	0	0,27
7	8	6,48	6,98	1,99	2	0,000833	0	0	0,27
7	12	6,48	6,98	4,30	2	0,000833	235	389	0,27

5. Estimation of parameters

5.1. Pulse

5.1.1. Pulse. The frequency of the first zero of the square pulse Fourier transform is $f_0 = 1/\tau_I$. Our criterion is that the desired frequency range should fall below the 50% of the main lobe in order to receive enough energy. Therefore, the automatic regression is run to find the maximum estimated frequencies associated with low-depth defects.

$$\tau_I = \frac{1}{2\max(f_{\Delta\varphi max})} \qquad (10)$$

5.1.2. Power. In [5] it is shown that the power of the thermal stimulus is proportional to the temperature contrast between the sound and the defect zones. As already observed, the program shows

Journal of Physics: Conference Series **501** (2014) 012008 doi:10.1088/1742-6596/501/1/012008

which defects are detectable for a certain power, and it can be used to evaluate whether a more powerful apparatus is required.

5.2. Acquisition

5.2.1. Sampling frequency. The sampling frequency must be chosen according to the Nyquist criterion as well as to the possible presence of aliasing [6]. To our purposes, for low thermal conductivity materials, frequencies involved are low and a sampling rate of 1Hz is typically adopted.

5.2.2. Observation time. The observation time must assure a good resolution at low frequencies for thick materials and it must be long enough to observe the complete cooling transient during which a temperature difference between the defect and the sound zones is measurable.

By means of the regression program, the used criterion is

$$t_{oss} = 1.25\, t_{end} \tag{11}$$

where t_{end} is the time when a temperature difference between defected and sound zones is no longer detectable for the type of flaws that has to be investigated.

6. Experimental results

The adopted technique shows good results when applied on the 2 cm thick specimen. The depth measurement error, defined as

$$err = \frac{z_{exp} - z}{z} \tag{12}$$

is extremely low (< 0.3%) for the considered defects. The rule of thumb that the defect diameter should be equal to or greater than its depth for appropriate detection and identification, is partly confirmed, i.e., the defect D4z7 is not detectable.

Table 3. Experimental results, 2 cm plate

D [mm]	z [mm]	z_{exp} [mm]	err %
8	2,5	3,17	0,27
8	4	5,22	0,31
8	6	6,18	0,03
8	8	7,98	-0,00
6	6	6,18	0,03
4	3	3,83	0,28
4	5	5,22	0,04
4	7	x	x
10	10	9,77	-0,02

The 4 mm thick plate shows a few peculiar issues. All defects besides D2z3 are visible, but the most superficial defects are not clearly identifiable, due to the high phase noise at higher frequencies.

The technique shows instead very good results for 2 mm and 3 mm deep defects.

Table 4. Experimental results, 4mm plate

D[mm]	z[mm]	z_{exp}[mm]	err	Note
6	2	2.04	20	good characterization
6	3	3.03	10	visible only in phase images
4	1	n.a.	n.a.	visible but not identifiable
4	2	2.08	40	
4	3	3.03	10	
2	1	n.a.	n.a.	visible but not identifiable
2	2	n.a.	n.a.	visible but not identifiable
2	3	n.a.	n.a.	not visible

7. Conclusions

A numerical and experimental analysis have been carried out to better understand the pulse phase thermography mechanism. In particular, our analysis have been focused on the limitations of the use of conventional lamps instead of flashes.

A numerical model has been developed to predict the behavior of the defect set as well as to estimate the trial parameters and the correlation constant *C1* between the defect depth and the thermal diffusion length.

The technique adopted at the ThermALab at Politecnico di Milano has been presented, and in particular the phasegram correlation criterion to identify the frequency associated with the defect depth has been discussed.

Finally, it has been shown that the deepest defects, i.e., up to 10 mm, in low-conductivity materials are precisely identifiable, while defects just below the surface, i.e., 1 mm deep are visible but not yet identifiable at this time.

References

[1] Marinetti S, Plotnikov Y A, Winfree W P and Braggiotti A 1999 *proc. of SPIE* **3586** 230–38
[2] Ibarra Castanedo C 2005 *Quantitative subsurface defect evaluation by pulsed phase thermography: depth retrieval with phase*, PhD Thesis, Université Laval
[3] Ishikawa M, Hatta H, Habuka Y, Fukui R and Utsunomiya S 2013 *IR Phys. & Tech.* **57** 42-49
[4] Bertolotti M, Liakhou G L, Li Voti R, Paoloni S and Sibilla C 1999 *J Appl Phys*, **85**(7)
[5] Almond D P and Pickering S G 2012 *J. Appl. Phys.* **111**, 093510
[6] Gutiérrez Fajardo A 2005 *An Analytical and experimental study on defect recognition methodologies with Pulse Phase Thermography*, MS Thesis, Politecnico di Milano
[7] Lau S K, Almond D P and Milne J M 1991 NDT&E Int. **24** 040195-08
[8] Shepard S M, Lhota J R and Ahmed T 2009 *Proc. of SPIE* **72990**T 1-7

Journal of Physics: Conference Series **501** (2014) 012009 doi:10.1088/1742-6596/501/1/012009

Performances of infrared emitters applied to the porous thin materials drying

P Di Marco, S Filippeschi[1] and L Pietrasanta
Department of Energy, System, Land and Construction Engineering (DESTEC), University of Pisa
Largo Lucio Lazzarino 1, 56122 PISA - Italy

E-mail: p.dimarco@ing.unipi.it ; s.filippeschi@ing.unipi.it;

Abstract. Drying of solids is one of the oldest and most common unit operations found in diverse processes. In this paper the drying of hygroscopic textile materials is discussed. The authors have previously investigated the drying kinetic of different fabrics dried by a hot air jet. In this paper a comparison between the convective and electric IR drying is made. In particular two fabrics with fibers which show a different hygroscopic behaviour are analysed: wool and cellulose/cotton. Unlike the convective drying, IR drying is weakly affected by the radiation properties and by the hygroscopic behaviour of the two fabrics. This is likely due to a better diffusion of the heat flux, which is constant over the entire drying surface in the case of IR heating, and produces unexpected results on the nondimensional kinetic parameter (characteristic curve). Wool shows a complete different characteristic curve if dried with IR or with convective flow. The better performances have been reached with MW emitter, but it has been observed that this advantage decreases with the distance of the source from the surface to be dried.

1. Introduction

Drying of solids is one of the oldest and most common unit operations found in diverse processes such as those used in the agricultural, ceramic, chemical, food, pharmaceutical, pulp and paper, mineral, polymer, and textile industries. It is also one of the most complex and least understood operations because of the difficulties and deficiencies in mathematical descriptions of the phenomena of simultaneous—and often coupled and multiphase—transport of heat, mass, and momentum in solid media [1].

During the drying processes the heat may be supplied by convection (direct dryers), by conduction (contact or indirect dryers), radiation or volumetrically by placing the wet material in a microwave or RF electromagnetic field. Over 85% of industrial dryers are of the convective type, with hot air or direct combustion gases as the drying medium. In the case of multiple impinging gas jets, both round and slot, Mujumdar [1] claims that entrained ambient air may cause unpredictable effects and the heat transfer performance may worsen by 20–50%. Recently the present authors have analysed the hot air drying of thin textile materials [2-4], by focusing on round nozzle jets. The prediction errors of the heat and mass transfer correlations can be very high, depending on the hygroscopic behaviour of the

---

[1] Corresponding author

porous medium [4] even if during the first stage of drying, where only the external condition should influence the phenomenon.
Convective drying shows therefore several problems: the energy saving, the contamination (including food processing like baking, roasting, blanching and surface pasteurization), the irregular dehydration of thin sheet as paper, wood, textile and paints due to irregular mass flow rate impingent on the surface. The need for product quality puts heavy constraints on the chosen dryer and dryer operation. Quality depends on the end use of the product. For many bulk chemicals, handling considerations determine moisture content requirements. For foodstuffs, flavor retention, palatability, and rehydration properties are important. Timber must retain its strength and decorative properties after drying.
One of the increasingly popular methods to performing dehydration or a superficial treatment is infrared (IR) radiation. Some of the IR thermal treatment advantage are energy savings, simple coupling with convective heating, fast response and easy control of the device.
In the last decades IR radiations has been investigate for drying of several porous and nonporous materials as: rough rice by Abe and Afzal [5], paddy by Das et al. [6], apple slices by Nowak and Lewicki [6], banana by Pan and Shih [7], thin coated films by Navarri and Andrieu [8]-[9] and for heating of thermoplastic sheet by Shmidt and Maoult [10].
Two types of IR heaters dominate these applications: Gas Fired IR (GF-IR) emitter and Electrical IR (EL-IR) emitter. GF-IR heaters have achieved some attention in the past years with papers on experimental studies and modelling work but EL-IR are more versatile and efficient even if the literature is scarce. Studies on optimization of the EL-IR heater for the different drying applications are very fragmentary.
In this paper IR drying given by different EL-IR emitters has been experimentally compared with a convective one by drying different thin hygroscopic textile medium. The kinetic parameter of different fabrics has been evaluated, and the basic drying phenomenon has been detected for two different commercial EL-IR emitters.

2. Infrared emitters analysis

As mentioned before the IR Emitters which are installed in the most of industrial applications are Gas Fired or Electric emitters. In these types of emitters, the IR radiation is obtained by passing an electric current through a resistance, which raises its temperature. The emitters are therefore characterised by the maximum temperature of the emitting source and by the wavelength range associated with this temperature.

Table 1. Typical parameters of commercial IR emitters.

Band	Characteristics
SW	Wavelength range 0.85-1.6 μm, maximum emission at about 1.2 μm, power density 60-600 kW/m2, maximum running temperature 2800 K, response 1 s
MW	Wavelength range 2.1-2.75 μm, maximum emission at about 2.6 μm, power density 30-40 kW/m2, maximum running temperature 1800 K, response 30-90 s
LW	Wavelength range 3.5-6 μm, maximum emission at about 4 μm, power density 30-40 kW/m2, maximum running temperature 900 K, response 3-10 min

Infrared emitters can be made of various materials such as quartz glass, ceramic and metal. Generally speaking, EL-IR emitters can be classified in Long Wave (LW), Medium Wave (MW) and Short Wave (SW). Table 1 show typical EL-IR emitter bands and power capability. Because of the high

temperature and high energy density reached, SW emitters are not often used in drying and superficial heating of porous media. Excessive drying is wasteful; not only more heat, i.e. expense, is involved than necessary, but often overdrying results in a degraded product, as in the case of paper and timber.
The MW emitters are generally made of a filament, inserted into a quartz tube differently shaped (single, twin, row, etc). The backside of the lamp is usually coated with gold to ensure a high reflectivity. The chromium alloy filament is suspended in a quartz or metal sheath. Due to the material used and the relatively low temperatures, the filament can operate in open air. There are some papers which have attempted to model this kind of lamp if applied to paper pulp [12, 13].
The MW emitter which has been used in this experimental activity is shown in figure 1.a. It is made of a row of 7 quartz tubes, among which the central one is without any heating element. This element is supplied with 1kWel at maximum, and has a rectangular shape of 62x248 mm. Its peak radiation intensity is at about 2.5 μm.

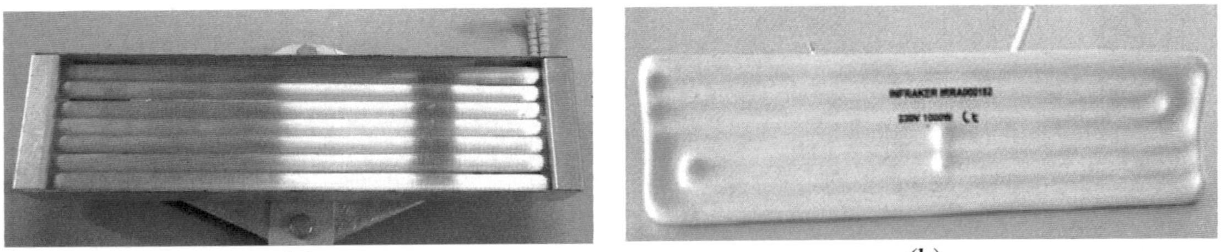

(a) (b)

Figure 1. MW IR emitter composed of 7 quartz tubes and Ni-Cr alloy filament (a), LW IR, ceramic surface (b).

The LW emitter consists of a ceramic matrix heated by an iron-chrome aluminium resistance wire. The temperature reached on the surface is lower than the other types. The maximum temperature is about 870 K, corresponding to a peak wavelength of 3.3 μm.
Usually this heater have a long start up time (7-10 minutes) but a more uniform temperature distribution, and therefore a diffuse radiative emission. The LW emitter which has been used in this experimental activity is shown in figure 1.b. This element is supplied with 1kWel at maximum, and has a rectangular shape of 60x245 mm.
Both the heaters can be tuned with a solid state power controller (GEFRAN GFX4 POWER) with a configurable proportional cycle time, which allows the power to be digitally controlled from 20% up to 100%.
The design and modeling of any radiative process always requires knowledge of the optical properties of these materials. Specifically for IR drying, this fact may be the clue to accomplish a safe and efficient process, because radiation properties of both the radiator and the material to be dried must be matched in order to obtain most efficient results. Emissivity, absorptivity, reflectivity, and transmissivity are the key radiation properties. Their relative magnitudes depend not only on the material, its thickness, and its surface finish, but also on the wavelength of the radiation. Furthermore the material soaked with water shows a different behaviour, and changes its radiative properties with respect to the dried material.
Generally, many fully wet materials have their minimum absorptivity at those wavelengths where water has its maximum transmissivity, pointing out the important role that water plays in radiation absorption. As drying proceeds, the material that is dried undergoes a change in its radiation properties, increasing its reflectivity, and consequently lowering its absorptivity at low water contents.
Water is transparent to the IR radiation (2.5 – 10 μm) with a transmittance of 0.85 at least, except for two windows (2.5 – 2.95 μm) and (5 – 7.14 μm) where the absorptivity is about 0.8. In this paper we study IR drying of textile materials. Several parameters influence the radiative properties of textile materials: areal density, fiber type and water content. The absorptivity in the NIR spectrum is however low and it is not affected by areal density. For wavelength higher than 2.5 μm, the weight of the fabric

largely affects the radiative properties. Transmittance decreases as the areal density increases, even if transmittance is never zero (20%) even for polyester fiber weighting 882 g/m^2 [14]. In the range 2.5 – 6 μm the most of fabrics are influenced by the weight and the transmittance can be 0.3 for the low density textile materials. For higher wavelengths the transmittance rapidly decreases approaching zero [14]. Fiber and water content largely influences the spectra behaviour of fabric. The spectral absorptivity at different water content is reported in figure 2 for wool and cotton, respectively.

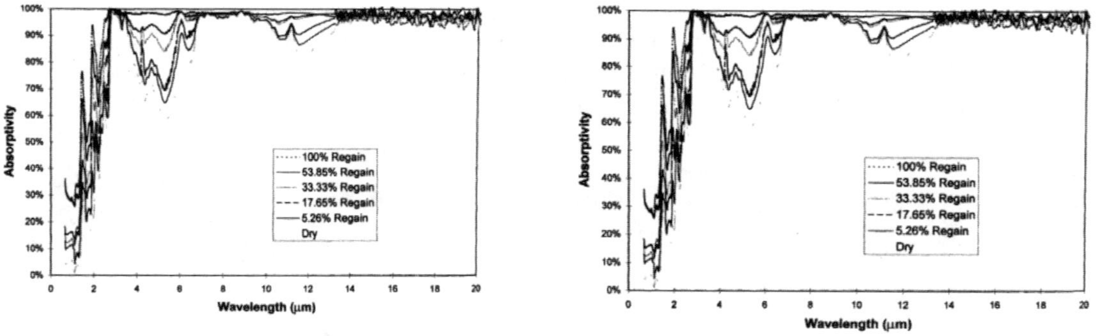

(a) (b)

Figure 2. Spectral IR Absorptivity for wool (173 g/m^2) at different water content (a), Spectral IR Absorptivity for cotton (193 g/m^2) at different water content (b) [14]

Figure 2 shows as the presence of the water inside the fiber increases the absorptivity for both the textile materials considered. The most of the influence is in the range (2.5 – 5 μm) typical of MW emitters.

3. The kinetic of thin porous IR fabric drying

The hygroscopic-porous media are characterised by a clearly recognizable pore space and a large amount of physically bound liquid. The drying process of a porous material can be therefore divided into four main stages [1]. In a first time period, the temperature of the media is lower than the wet bulb temperature, so that the temperature increases over time jointly with the drying rate N_w, while the water content X decreases. During the second time period (Stage I) the drying rate and the temperature of the hygroscopic material stay constant up to the third period (Stage II) where the drying rate decreases and the temperature increases. In this stage dry spots appear on the heated surface, and a part of bound moisture is removed. This stage is dominated by the diffusion and the capillarity of the medium. During the last drying stage (III) only the bound water is present inside the material, the drying rate decreases with a lower slope over the time as the equilibrium moisture is reached.

During the first stage of drying the phenomenon is dominated by the local heat and mass transfer and the characteristic of material should not affect the drying rate.

When the drying rate curves are determined over a range of conditions for a given solid, the curves appear to be geometrically similar and are simply a function of the extent to which drying has occurred. If these curves were normalized with respect to the initial drying rate and average moisture content, then all the curves could often be approximated to a single curve, ''characteristic'' of a particular substance, which does not depend on the way of drying. This is the characteristic drying curve. The normalized variables, the characteristic drying rate f and the characteristic moisture content ϕ are defined as follows:

$$f = \frac{N_W}{N_c} \tag{1}$$

$$\phi = \frac{\bar{X} - X^*}{X_{cr} - X^*} \quad (2)$$

where N_W is the rate of drying per unit surface, N_c is the constant rate during the Stage II, X is the average moisture content in the body (expressed as the moisture by weight of bone-dry material in the solid), X_{CR} is the critical moisture content, defined as the moisture content at the end of the first stage drying. Finally, the equilibrium moisture content X^* is the moisture content of the bound water, which depends on the humidity of the air. The saturation moisture content X_{sat} is defined as the maximum moisture content which the fabric can hold without dripping.

In [5] 4 different fabrics (large pore and small pore cotton, wool, cotton and cellulose) have been experimentally characterized by measuring the drying kinetic parameters for hygroscopic materials.

In [3] the authors experimentally evaluated these parameters for a wool and cotton and cellulose fabric with an accuracy of 10% by drying the media with a hot air convective flow. The drying kinetic curves obtained with a convective air drying are shown in figure 3 for wool and cellulose/cotton and the kinetic parameters are reported in table 2.

Table 2. Kinetic drying parameters for the tested samples.

Fabric type	Dry weight [g]	Areal density [g/m^2]	X_{crit} [g_{water}/g_{solid}]	X_{sat} [g_{water}/g_{solid}]	X^* [g_{water}/g_{solid}]	Experimental drying rate N_c [g/s m^2]
Cotton / cellulose	13.3	217.6	3.2	14.9	0.042	1.74
Wool	11.7	191.5	7.89	9.5	0.099	1.61

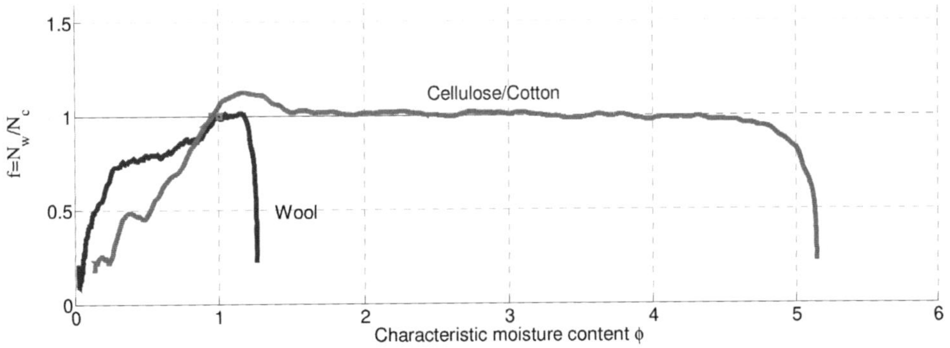

Figure 3. Characteristic drying curves for the wool and cellulose/cotton evaluated with convective drying.

4. Experimental Activity

A scheme of the experimental facility used during the tests for the IR drying analysis is shown in figure 4. It consists of a support, the heating element, the moistened fabric and of the control and acquisition devices.

The heating element and the power controller systems have been previously described in a dedicated section. The support allows the distance between the heating element and the moistened cloth to be changed with continuity in the range 5-50 cm. The electronic scale has 0.01 g of resolution and a maximum sampling rate of 10 s. The temperature spatial distribution over the moistened cloth has been observed with an infrared thermocamera, which has a focal plane array of two-dimensional micro-bolometric sensors, characterized by a digital resolution of 320x240 pixels, sensitive to wave lengths of 8÷14 μm. The minimum thermal resolution is 0.15 K. The optical system consists of a 35 mm lens, with a field of view of 25.8°x19.5°. The distance between the hot surface and the lens is 0.74 m, resulting in a pixel size of 1.04x1.04 mm^2.

For each power level, the drying rate was measured and each test was repeated 5 times. Four different fabrics were tested. Each fabric was previously completely dried in an oven, and immersed in demineralised water for 20 minutes. As the fabric was completely soaked of water, it was put on a metallic net hung to an electronic scale, to mechanically remove the water in excess by gravity. During this stage the weight of the patch linearly decreased with time first, then remained approximately constant. At that time the hygroscopic material was considered completely saturated of water, with all the excess water removed. The patch was then exposed to the IR radiation, and data were acquired every 5 seconds up to complete drying. When the moistened patch is exposed to the IR radiation, the heater had just reached a steady state operative regime.

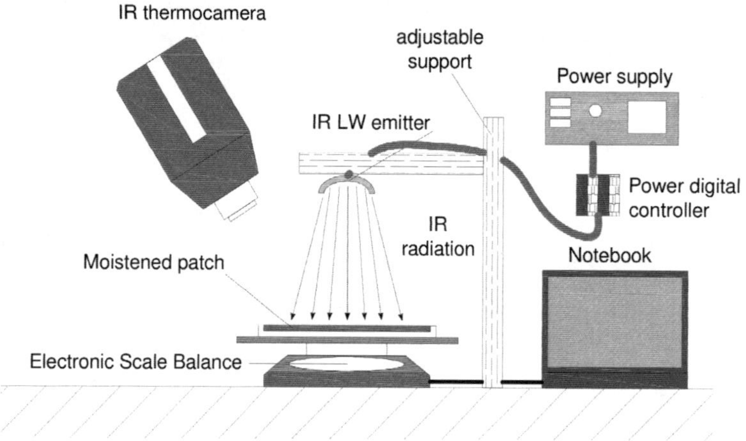

Figure 4. Scheme of the experimental facility and the acquisition systems

5. Results

This paper reports a preliminary experimental activity on the analysis of the drying of hygroscopic media in general, and fabrics in particular. The IR drying depends on the wet material which must be dried and by the emitters. The effect of the spectral radiation of two different emitters (Long Wave e Medium Wave) on two different fabrics is then investigated. The fabrics have been selected according to the hygroscopic behaviour of the materials, shown in figure 3.

If dried with a convective jet, wool shows the typical behaviour of the low hygroscopic media: no zone at a constant drying rate is observed, while for highly hygroscopic media (cellulose) the drying rate is constant for large moisture content inside the patch. These two materials show very similar density area and similar radiation properties. This experimental activity shows drying rate measured for both the emitters and the fabrics for two different distances between the patch and emitters (200 and 300 mm). For each distance the electric power of the two emitters has been tuned (from 60 to 100%) in order to change the radiation intensity and the temperature of the heating panel. The drying rate has been evaluated by measuring the mass of the fabric over time under a constant IR radiation.

An example of the weight decrease over time measured during the test is shown in figure 5. It is referred to the cellulose/cotton fabric with a distance of 200 mm. After an initial transient, the weight linearly decreases with time. All the data of this period were linearly fitted by least squares method, and the slope of this curve is the drying rate.

By knowing the drying rate over time, it is possible to evaluate also critical moisture content X_{cr} and therefore the characteristic curve for the wool and the cellulose. The characteristic curve depends only on the material and it does not depend of: drying power, emitters, distance between the emitters and the cloth. The characteristic curves evaluated for wool and cellulose are shown in figure 6.

The comparison between figures 3 and 6 shows which that in the case of wool the characteristic curves have a different shape if obtained with IR or convective drying. The characteristic curve should show the same shape at low characteristic moisture content, at least. The starting point can be different because it depends on the starting moisture content.

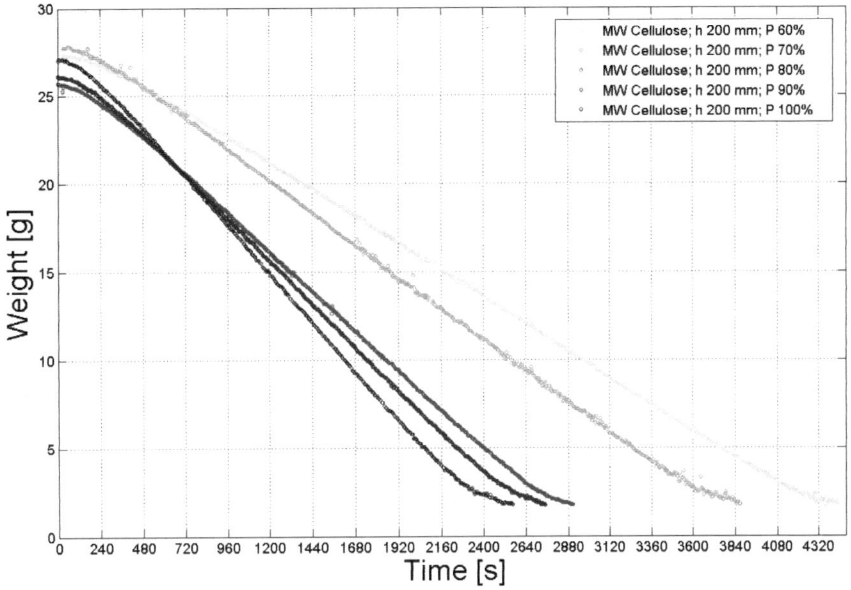

Figure 5. Weight of the wet cellulose/cotton fabric obtained with a LW emitter placed at 200 mm from the moistened fabric.

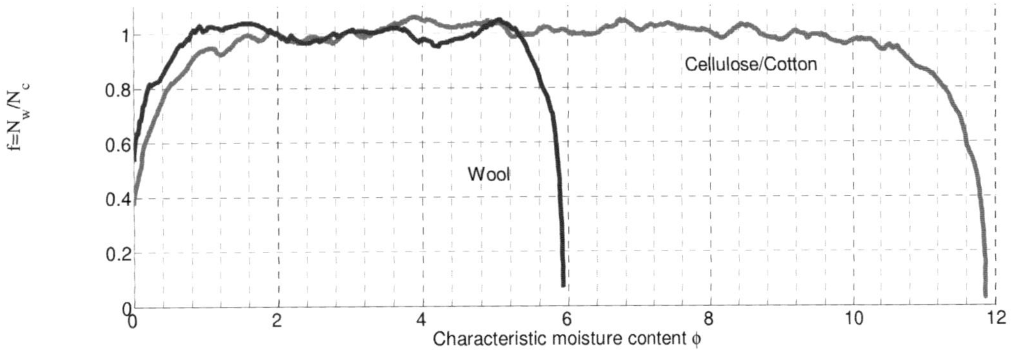

Figure 6. Characteristic drying curves for the wool and cellulose/cotton evaluated with IR drying.

This result is therefore unexpected. As the drying rate is constant, the mass flow rate of the water inside the medium is equal to the mass flow rate which evaporates from the surface. The experimental evaluation of the drying rate is however an integral method even if the local drying rate is not completely uniform. In the case of convective air jet, both in the case of slot and nozzle diffusers, the local distribution of temperature is not spatially uniform [4]. The IR drying made with the two IR panels used in this work, show a very diffuse heat flux and the spatial temperature distribution is very constant over the entire surface for both the materials, as shown in figure 7.

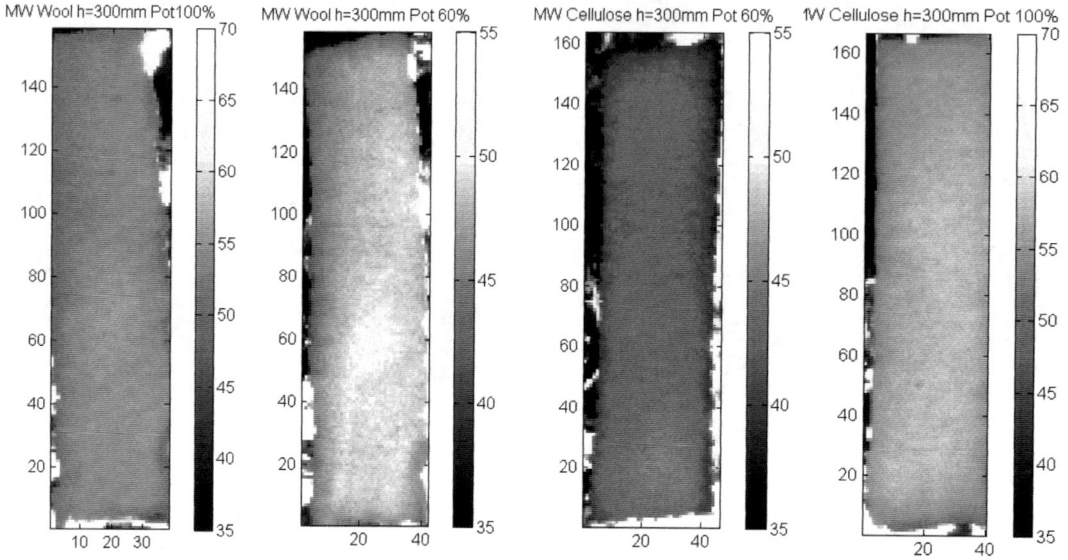

Figure 7. Temperature distribution of cellulose/cotton and wool at 600 and 1000 W at a distance of 300 mm by the MW emitter.

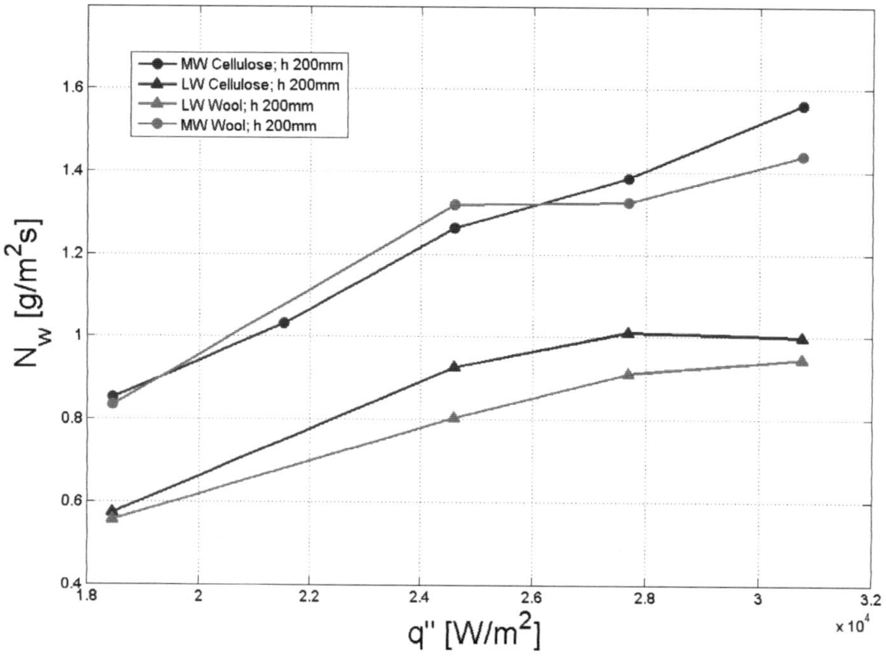

Figure 8. Drying rate of cellulose/cotton and wool at a distance of 200 mm under MW and LW emitters at different power areal density.

The mass flow rate in perpendicular direction with respect the heat flux is negligible if the heat flux is uniformly diffused, but it could be significant if the heat flux is non-uniform, as in the case of

convective drying. In the latter configuration the porous properties of the materials influence the local drying rate. In IR diffuse radiation, different hygroscopic materials show analogous quantitative behaviours (figure 6).
The drying rates obtained with an IR drying have been plotted versus the power areal density *q"*, which is here defined as half of the electric power of the emitters divided by the surface to be dried. Figures 8 and 9 show the drying rate plotted versus the power density for LW and MW emitters, at a distance of 200 and 300 mm, respectively.
Different hygroscopic materials present the same quantitative drying rate at different power areal densities both for LW and MW at different distances and, therefore, heat flux intensities. MW emitters at the same power areal density show higher drying rates than LW emitters, both at 200 mm and 300 mm. As the distance increases, MW emitters show better drying performances. At 200 mm and the maximum power, the drying rate in the case of MW emitters is about 45% more than the LW emitters, and this rate increases as the distance rises. At 300 mm the MW drying rate is about 60% more than the LW drying rate at the maximum power.

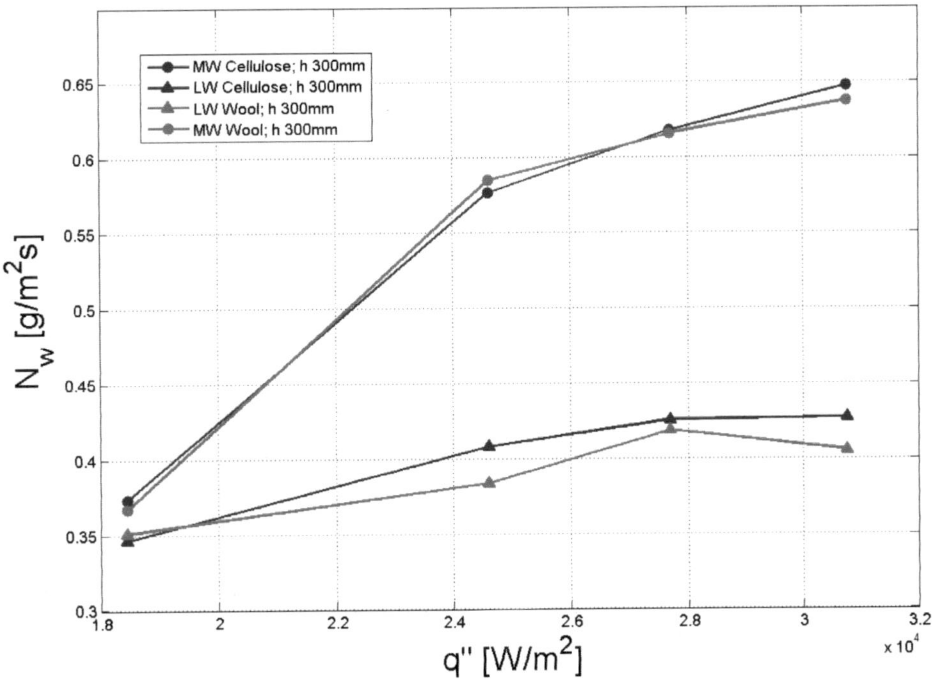

Figure 9. Drying rate of cellulose/cotton and wool at a distance of 300 mm by MW and LW emitter at different power areal density.

6. Conclusions

In this paper IR drying given by different EL-IR emitters has been experimentally compared with a convective one, by drying different thin hygroscopic textile media. In particular two fabrics with fibers which show a different hygroscopic behaviour have been analyzed: wool and cellulose/cotton. The areal densities of these two media are 217 g/m^2 for cellulose/cotton and 191 g/m^2 for wool.
In this preliminary paper, the performances of IR drying, with two different emitters, (MW and LW) has been qualitatively compared with that of a convective drying with similar power area density.
Unlike convective drying, IR drying is scarcely affected by the radiation properties and by the hygroscopic behaviour of the two fabrics. This is probably due to a better diffusion of the heat flux, which is uniform over the drying surface in the case of IR heating, and produces unexpected results on

the non-dimensional kinetic parameter (characteristic curve). In particular, wool shows a completely different characteristic curve if dried with IR or with convective flow.
The better performance has been reached with MW emitter, although it has been observed that this advantage decreases with the distance. This preliminary work does not allow for clearly detecting the influence of the distance on the drying rate for both the emitters. In the future, a wider experimental activity will be performed to clarify this aspect.

References

[1] Mujumdar A 2006 *Handbook of Industrial Drying*, 3rd edition, Taylor & Francis Group, LLC
[2] Di Marco P, Filippeschi S and Pieve M 2010 Experimental study on the drying power of a hot jet, *Proc. ASME-ATI-UIT Conference on Thermal and Environmental Issues in Energy Systems*, Sorrento, Italy
[3] Di Marco P, Filippeschi S and Pieve M 2011, Experimental study on drying of different kinds of fabric by an air hot jet, *Proc. XXIX UIT National Conference*, 20-22 June, Turin Italy
[4] Di Marco P, Filippeschi S 2012 Experimental investigation on influence of porous material properties on drying process by a hot air jet, *J. Phys.: Conf. Ser.*, **395**, 012139
[5] Abe T and Afzal T M 1997 Thin-layer infrared radiation drying of rough rice, *Journal of Agricultural Engineering Research*, **67**(4) 289-297
[6] Das I, Das S K and Bal S 2009 Drying kinetics of high moisture paddy undergoing vibration-assisted infrared (IR) drying, *J. Food Engineering* **95** 166–171
[7] Nowak D and Lewicki P P 2004 Infrared drying of apple slices, *Innovative Food Science and Emerging Technologies* **5** 353– 360
[8] Pan Z, Shih C, McHugh T H and Hirschberg E 2008 Study of banana dehydration using sequential infrared radiation heating and freeze-drying, *LWT - Food Science and Technology* **41** 1944-1951
[9] Navarri P and Andrieu J 1993 High-intensity infrared drying study. Part I: case of capillary-porous material, *Chemical Engineering and Processing* **32** 311-318
[10] Navarri P and Andrieu J 1993 High-intensity infrared drying study. Part II: case of thin coated films, *Chem. Eng. and Processing* **32** 319-325
[11] Schmidt F M, Le Maoult Y and Monteix S 2003 Modelling of infrared heating of thermoplastic sheet used in thermoforming process, *J. Materials Processing Technology* **143–144** 225–231
[12] Petterson M and Stenstrom S 2000 Modelling of an IR heater at transient and steady state conditions. Part I: model and validation, *Int. J. Heat Mass Trans.* **43** 1209-1222
[13] Petterson M and Stenstrom S 2000 Modelling of an IR heater at transient and steady state conditions. Part II: modelling a paper dryer, *Int. J. Heat Mass Trans.* **43** 1223-1232
[14] Pastore C M, Kiekens P 2001 *Surface characteristics of fiber and textiles*, CRC Press

Pore-scale simulation of laminar flow through porous media

M. Piller[1], **D. Casagrande**[2], **G. Schena**[1] **and M. Santini**[2]

[1] Dipartimento di Ingegneria e Architettura, Università degli Studi di Trieste, via A. Valerio, 6/1 - 34127 Trieste, Italy

[2] Dipartimento di Ingegneria, Università degli Studi di Bergamo, viale Marconi, 5 - 24044 Dalmine (BG), Italy

E-mail: [1] piller@units.it

Abstract. The experimental investigation of flow through porous media is inherently difficult due to the lack of optical access. The recent developments in the fields of X-ray micro-tomography (micro-CT hereafter), digital sample reconstruction by image-processing techniques and fluid-dynamics simulation, together with the increasing power of super-computers, allow to carry out pore-scale simulations through digitally-reconstructed porous samples. The scientific relevance of pore-scale simulations lies in the possibility of *upscaling* the pore-level data, yielding volume-averaged quantities useful for practical purposes. One of the best-known examples of upscaling is the calculation of absolute and relative permeability of reservoir rocks. This contribution presents a complete work-flow for setting up pore-scale simulations, starting from the micro-CT of a (in general small) porous sample. Relevant applications are discussed in order to reveal the potential of the proposed methodology.

1. Introduction

The experimental investigation of flow through porous media is inherently difficult due to the lack of optical access. High resolution computed tomography allows to construct a digital binary model representing the complex pore space geometry. This three-dimensional synthetic model can be used for pore-scale simulations of phenomena that previously were often approached with macro-scale lumped models or micro-scale simplified models. Fluid flow through porous media is relevant in a range of industrial applications. In reservoir evaluation, absolute and relative permeabilities and capillary pressure are probably the most important flow properties for the set-up of reservoir simulations and for devising exploitation scenarios [1, 2]. The numerical calculation of these quantities mimics the protocols that are routinely used for experimental measurements.

Open-cell metal foams provide another example of porous media of practical relevance. They are characterised by a large specific surface area and, because of that, are used in many practical applications, as enhanced heat transfer devices [3, 4] and compact heat exchangers [5], catalytic reactors [6, 7], rotors of centrifugal compressors [8]. Open-cell metal foams have been investigated experimentally, analytically and numerically.

Pore-scale simulations of porous media are relatively novel. Mostaghimi et al. [9] simulate low-Reynolds-number flow and transport through digital samples of rock cores at the scale of pores, using a finite-difference solver. The absolute hydraulic permeability and the longitudinal dispersion coefficient are calculated. The same method is used to compute the permeability of a range of consolidated and unconsolidated porous rocks [10]. Boomsma et al. [11] carry out numerical simulations of laminar flow through a cluster of two unit cells, each consisting of eight idealized pores of an open cell metal foam. A finite-volume method (FVM hereafter) is used to solve the incompressible, steady Navier-Stokes equations through the cell complex, using tetrahedral, unstructured meshes. Krishnan et al. [12] carry out pore-scale numerical simulations of both flow and heat transfer through an idealized, cubic unit cell, yielding good agreement with experimental data for foams with porosity larger than 94%. Gerbaux et al. [13] reconstruct digital samples of three open-cell metallic foams from X-ray tomography. Various unstructured, tetrahedral or cubic meshes are constructed. The microscopic Stokes flow through the foams is solved by two node-based FVM algorithms and one D3Q19 Lattice Boltzmann Method. The numerically computed and experimentally measured hydraulic permeabilities are in good agreement with each other and with available literature data. Bai et al. [4] carry out pore-scale numerical simulations for predicting the hydraulic head loss caused by a metal foam. They claim that the repetitive structure of metal foams allows for consideration of a single, idealized, representative unit cell, obtained by subtracting a sphere from a tetrakaidecahedron (sphere-centred Kelvin cell).

This study presents in detail a possible approach to the pore-scale flow simulation in digital porous systems reconstructed by X-ray computed tomography of real porous samples. Pore-scale numerical simulations are applied in three different studies, reported in the paper:

- Calculation of the absolute permeability tensor of samples of reservoir rocks.
- Calculation of both the relative permeability and the capillary pressure for laminar oil/water flow through a sample of sandstone rock.
- Investigation of the laminar flow through a sample of open-cell metal foam produced by infiltration casting [14, 15], within both the Darcy and the Darcy-Forchheimer regimes.

2. Materials and methods

2.1. X-ray computed tomography

Porous media are opaque to visible light which limits the measurements of their complex morphology often made up of stubby pores connected by narrower throats. The opacity prevents also the direct visualization of all the important phenomena related to the flow of multiphase fluids in the complex porous morphology. Among the tomographic imagining techniques, the X-ray computed tomography is the one with highest spatial resolution, good penetrating capability and thus it is the most suited for the morphological reconstruction of porous media samples with micrometric throats. Today the micrometric resolution obtainable at the tomographic beam-lines of synchrotron facilities is possible with micro- and nano-CT laboratory tomographic machines based on open-type cone-beam tubes. These tubes can focus the X-ray beam to less than 1 micron spot size and allow the production of a Bremsstrahlung (i.e. polychromatic and incoherent) X-ray beam with energies suitable for micrometric spatial resolution tomography in a millimetric size sample. A typical X-ray tomography station is made up with a planar digital detector (either CCD arrays or flat panels are used to capture intensity data after X-ray to visible light conversion on a scintillator) and a high precision rotating stage. The object is placed on the stage and rotated by small angular steps, a radiographic operation is repeated at each step. The collection of radiographies is made up of several hundreds of projections depending on the diameter of the object. The radiographic principle is conceptually simple: X-rays traverse a cross section of the sample along straight lines, the intensity of the attenuated X-ray signal emerging is converted by a scintillator into light and recorded as a digital image. The X-ray attenuation obeys the Beer-Lambert law that states that there is a logarithmic dependence between the transmission of a ray through an object and the distance it travels :

$$\ln\left(\frac{I_o}{I_i}\right) = \int_L \mu\left(l, E\right) \mathrm{d}l \qquad (1)$$

where μ is the linear attenuation coefficient of the material, I_o is the intensity of the incident impinging X-ray beam, I_i is the intensity emerging from the object and L is the material thickness that is the distance travelled. The linear attenuation coefficient is a function of the photon energy E. The incident intensity is read on the so called flat-field that is the radiography taken without sample before starting an acquisition. The logarithm of the intensity ratio is also referred to as projection. It contains an integral (through the length travelled in the material) information about the attenuation properties encountered by the ray. This information should be de-convoluted to extract information about the local internal (morphology) of the material.

2.1.1. Filtered back projection and segmentation The mathematical principle of the CT reconstruction for parallel rays was developed early in the last century by Radon et al. [16]. The Radon inverse recreates the map of the linear attenuation coefficients of the object cross-section from the collection of projections taken around it (i.e., from the *sinogram*). The process can be seen as a filtration of the projection in the Fourier domain and a back projection into the pixels of the image grid. The whole operation is often referred to as filtered back projection (FBP). The algorithm generalizing in 3D the Radon inverse for diverging cone-beams is due to Feldkamp et al. [17] and is also cast around the original filtered back projection technique.

2.1.2. Example of a home-made X-ray tomography devices The X-ray micro-tomographic facility available to the authors at University of Bergamo is based on an open type Hamamatsu X-ray source 160 kVp. An Aerotech rotary air-bearing drive stage provides superior angular positioning and velocity stability. The X-ray projections are collected by a Dexela-Perkin Elmer 1512 flat panel Complementary Metal-Oxide-Semiconductor (CMOS) X-ray detector. The detector has a pixel size of 75 micron and allows binning. The frame rate is 30fps at full resolution. The cabinet has 5 mm of lead shielding. To mitigate the effect of the beam hardening on imaging a thin aluminium filter plate is used to filter low energy X-rays exiting from the source. The commercial software VGStudio MAX from Volume Graphics GmbH [18] that includes beam-hardening correction is used for slice reconstruction and artefact reduction.

2.2. From the binary representation to a mesh

Two fundamental strategies can be followed to construct a mesh for a digital sample of a complex, porous matrix. The simplest approach, by which the present results have been obtained, consists in generating an unstructured mesh of cubes, where the computational cells coincide with the void voxels obtained from the X-ray tomography. Figure 1 shows a picture of a sample of metal foam, a digital volume extracted from the sample and a particular of the voxel-based mesh. The voxel size is 81μm.

(a) (b) (c)

Figure 1. (a) Picture of a sample of metal foam; (b) digital volume extracted from the sample ($21.789 \times 14.175 \times 12.798$mm); (c) particular of the voxel-based mesh.

In the second approach, an unstructured, tetrahedral or hybrid mesh is automatically generated once an STL representation of the pore-grain interface is available. Several commercial tools are available, to this end. Some experience has been gained with the +FE module by Simpleware [19], which provides utilities for generating both surface (triangular/quadrilateral) and volume (tetrahedral/hexahedral) meshes.

The voxel-based mesh generation approach presents some distinctive positive features, which are briefly outlined below.

- The fluid-solid interface reconstructed by X-ray tomography is inherently rough and any smoothing operation is arbitrary. The stair-step pattern of the interface represents the real surface roughness with length scale comparable to the voxel size.
- Global and local (by Octree techniques [20]) refinements can be easily achieved by splitting the cubic cells, without affecting the quality of the mesh.
- The mesh generation process is fast and automatic [13].

2.3. Image processing and morphological analysis

Sub-volumes of interest of size in excess to 1000^3 voxels are extracted from the stack of slices. The voxel volumes are treated with non-linear filters to mitigate the noise. The basic median filter with a $3 \times 3 \times 3$ kernel is a good candidate for removing noise whilst preserving edges.

The segmentation of the gray level voxel volume into a boolean tridimensional matrix where the pore space and solid material are represented by 0 and 1 respectively is based on the histogram of the luminosity value of all the voxels. The peaks and valleys in the histogram are used to discriminate between solid and void. The segmentation threshold is placed in the valley. The binary volume obtained is used for pore-space morphological measurements.

Figure 2(a) is a three-dimensional visualization of a sub-volume (120^3 voxels) of a sample of sandstone reservoir rock. The iso-surface representing the pore space walls is coloured in red. The material is shown as transparent grey. The pores are left void. The pore space is often sub-divided into pore-bodies that are relatively stubby and the narrow ducts or throats that connect these pores. The individual pores are segmented with a proprietary code that exploits a distance-transform-based (DT hereafter) watershed algorithm. The standard 3D watershed-transform is a parameter-free technique used to resolve pores in an objective manner. It is well known that the standard watershed suffers of severe over-segmentation. Here the pathology is successfully treated with a H-minima transform that suppresses all the local minima regions whose depth is less than a set scalar threshold [21]. In order to accurately calculate three-dimensional gradients, used by the watershed method to identify local minima, the resolution of the original volume is magnified by a factor 5. The segmented pores are visualized in figure 2(b) and their size distribution is shown in figure 2(c).

The medial axis is a path centred with respect to the boundary walls and its length is useful for pore space morphological characterization. As an example, the tortuosity is a morphological indicator of the easiness to pass through a porous media for a size-less particle. According to the most common definition, tortuosity is estimated as the ratio of the mean effective path length through a porous medium to the thickness of the medium or to the linear distance in the media between the start and the end of the particle path. Since one can connect an entrance point to an exit point with many paths, herewith the shortest and largest medial path is considered for tortuosity estimation. Figure 3 shows the path of a particle released at the top side of a tomographic volume with the objective to reach a point at the opposite side. The paths were extracted with a methodology based on the weighted geodesic distance transform [22, 23].

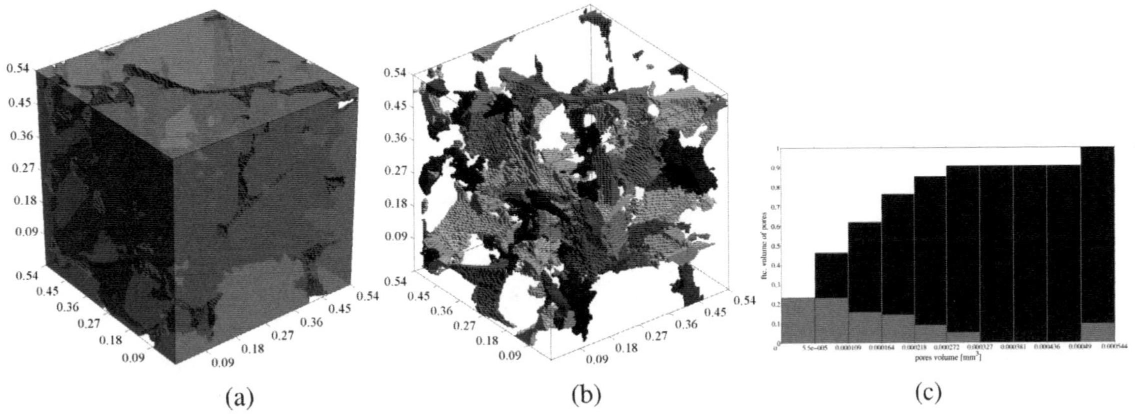

Figure 2. (a) Reservoir sample. (b) Segmented pore space. (c) Pore-size distribution (red) and cumulative pore-size distribution (blue). Lengths are expressed in millimetres.

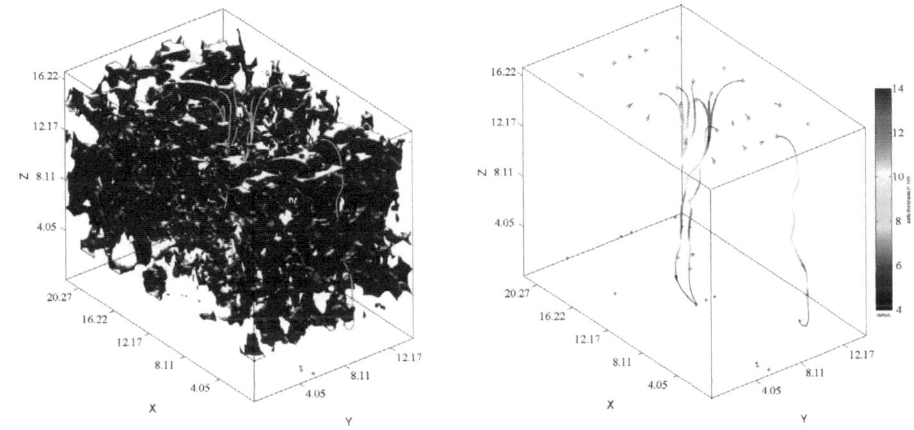

Figure 3. Path of a particle released at the top side of a tomographic volume with the objective to reach a point at the opposite side.

2.4. *Pore-scale fluid dynamics simulation*

In the following, the pore-network model [24] for single-phase flow and the Finite Volume method [25] for both single- and two-phase flow are briefly outlined.

2.4.1. Single-phase, pore-network models When a porous medium is synthetically represented as a set of *pores*, the laminar flow through it can be modelled as a linear, hydraulic network. Schena et al. [24] suggest to improve the pore-network modelling approach by taking into account the real shape of the cross-section of the branches, to calculate more reliable hydraulic resistance coefficients. Each branch is represented as a series arrangement of sub-branches with uniform cross-section. Poiseuille flow is assumed in each sub-branch. The specific hydraulic conductance of a sub-branch, K', is defined by the following equation

$$Q = K' \frac{\Delta p}{\rho g} \tag{2}$$

where Q denotes the volumetric flow rate, and the hydraulic conductance of the whole branch, K, is calculated as:

$$K = \left[\sum_j \frac{L_j}{K'_j} \right]^{-1} L_{\text{tot}}; \quad L_{\text{tot}} \equiv \sum_j L_j \tag{3}$$

where the index j denotes a specific sub-branch.

Specific hydraulic conductance and absolute permeability are related as follows:

$$K' = \frac{k \, \rho \, g \, A}{\mu \, L} \tag{4}$$

The absolute permeability of three digital samples of sandstone rock, identified as H, M, L, is compared against experimental results in table 1. Laboratory experiments for absolute permeability measurements were carried out in the petro-physical laboratory of an oil company using a high pressure Hassler sleeve core holder. The protocol is detailed in [26].

Table 1. Experimental and calculated values of absolute permeability for three samples of sandstone rock.

Sample	k [$m^2 \times 10^{-13}$] experimental	k [$m^2 \times 10^{-13}$] calculated
H	9.5	20.7
M	1.9	4.2
L	0.39	0.14

2.4.2. Single-phase FVM As we are only concerned with laminar flow of liquids at very low pore-Reynolds number, the well-known incompressible, steady-state Navier-Stokes equations are solved. Three different types of boundary conditions are enforced on the boundary of the computational domain. No-slip boundary conditions are enforced on the solid-fluid interface, while on the fluid-fluid open portions of the boundary either pressure-inlet or pressure-outlet conditions are enforced. Symmetry conditions are applied on the lateral faces of the sample. The choice of boundary conditions is of course rather arbitrary. Nevertheless, the point here is not to reproduce a flow subject to specific conditions, but rather to compute the transfer function of the porous sample with respect to applied pressure gradients. In this perspective, the enforcement of pressure boundary conditions seems appropriate.

Both the commercial software ANSYS Fluent® [27] and the open-source software OpenFoam® [28] are used to solve the Navier-Stokes equations. Both models adopt a cell-based, co-located discretization approach. The accuracy of the available spatial discretization algorithms ranges between first and fourth-order. In the present, reported applications, second-order upwind schemes for the advective term are preferred for robustness, while the viscous term is discretized by the least-squares or the Gauss methods.

2.4.3. Multiphase FVM Here the finite-volume/volume-of-fluid (FV-VOF) commercial solver ANSYS Fluent® [27] is used for the simulation of the oil/water flow in the pore-space of a sandstone rock that has been reconstructed using high resolution X-ray computed tomography. The VOF is a technique widely used to tackle problems in which two or more fluids are present and the dynamics of the interface is of interest. It was first introduced by Nicols et al. [29]. Various improvements have been developed subsequently and a non-exhaustive list is reported by Tang et al. [30]. Here VOF is used for tracking the shape of the water/oil interface. The phase volume fraction is defined at each computational cell of the fixed regular and uniform computational grid adopted. The volume fraction function is discontinuous passing through the interface. The piece-wise linear interpolation reconstruction (PLIC) algorithm is used in the present work for interface tracking. The method is coupled with the continuum surface force (CSF) algorithm for surface force modelling that adds a source term in the momentum equation, proportional to the interface curvature [31]:

$$\boldsymbol{F}_{\mathrm{CSF}} = \frac{2\,\rho}{\rho_w + \rho_o}\,\sigma_{w\,o}\,\kappa\,\nabla\,\alpha_w \tag{5}$$

where κ denotes the curvature of the oil-water interface, $\sigma_{w\,o}$ the water-oil surface tension, α_w the water volume-fraction and ρ is the volume-averaged mixture density:

$$\rho \equiv \alpha_w\,\rho_w + \alpha_o\,\rho_o \tag{6}$$

Wall adhesion is modelled by assuming a *contact angle* between the fluid-fluid interface and the solid wall. In the *dynamic* boundary condition implemented in ANSYS Fluent® [27], specification of the contact angle is used to adjust the interface curvature near the wall. Pressure-velocity coupling is realized via the pressure-implicit with splitting of operators (PISO) scheme [25]. A single pressure-correction step has been preferred in the present calculations for computational efficiency.

3. Results

3.1. Calculation of absolute permeability through metal foams and reservoir rocks

3.1.1. Metal foams An open-cell aluminium foam produced by infiltration casting was examined by X-ray CT [32]. A tomographic scan encompassing $269 \times 175 \times 158$ voxels is used, with resolution 81μm. The porosity of the sample and the mean pore diameter are calculated by image-processing techniques and result in $\epsilon = 79.4\%$ and $d_p = 1.72$mm, respectively. Further details and a through morphological characterization of this material can be found in Boschetto et al. [14, 15], Piller et al. [32]. Figure 4 shows the segmented pores of the aluminium foam of figure 1, coloured with a pallet associated to the pore size. Here the aluminium material is not shown. Avizo Fire® from FEI Visualization Sciences Group [33] is used to separate individual pores with a watershed driven by the DT map.

A set of simulations are performed for the sample of metal foam shown in figure 1. The same nature of the infiltration casting process [14] suggests that the inner structure of the considered foam is far from isotropic. Therefore, the fluid intrusion direction is arbitrarily selected as corresponding to the mean flow direction of molten aluminium during casting. The range of pore-scale Reynolds number, $\mathrm{Re}_p \equiv U_D\,d_p/\nu_f$, is selected in order to investigate the Darcy and the Darcy-Forchheimer flow regimes, see table 2.

A common correction to the Darcy model, aimed at extending its range of validity to the laminar inertial regime $\mathrm{Re}_p \lesssim 100$, adds a quadratic term, yielding the Darcy-Forchheimer model [34, 35] in non-dimensional form:

$$\mathrm{C_D} = \frac{1}{\mathrm{Re}_p\,\mathrm{Da}} + \frac{C_E}{\sqrt{\mathrm{Da}}} \tag{7}$$

The Darcy number Da and the drag coefficient $\mathrm{C_D}$ are defined as follows

$$\mathrm{Da} = \frac{k}{(d_p)_L^2}; \qquad \mathrm{C_D} = -\frac{\Delta p_f\,(d_p)_L}{\rho_f U_D^2\,L_x} \tag{8}$$

The Darcy number and the Ergun coefficient for the present sample are calculated by linear regression with equation (7), yielding:

$$\mathrm{Da} \;=\; 2.757 \times 10^{-2} \pm 4.2 \times 10^{-4}; \qquad C_E \;=\; 1.566 \times 10^{-1} \pm 4.1 \times 10^{-3} \tag{9}$$

Figure 4. The segmented pores coloured with a pallet associated to the pore size. Here the aluminium material is not shown.

where the 95% confidence bounds are also reported. The absolute permeability of the sample can be directly derived from the Darcy number and the mean pore diameter:

$$k = 8.2 \times 10^{-8}\ \mathrm{m}^2 \tag{10}$$

Table 2. Drag coefficient calculated at different pore-scale Reynolds numbers.

Case	A	B	C	D	E	F	G	H
Re_p	0.61	1.22	3.04	6.08	12.17	30.42	45.00	60.83
$(\mathrm{C_D})$	67.95	34.18	14.12	7.57	4.33	2.39	1.97	1.75

3.1.2. Reservoir rocks The full tensor of absolute hydraulic permeability is estimated for a sample of sandstone reservoir rock from a deep oil exploration and compared with laboratory experimental measurements. The rock porosity is estimated about 11% by image-processing techniques. Further morphological details can be found in Piller et al. [36]. The spatial resolution of the tomographic scan is $4.5\,\mu$m. The numerical simulations are carried out on a sub-volume encompassing 600^3 voxels. Comparison of morphological parameters against a $1.5\,\mu$m tomographic model of the same sample, scanned at the ID19 line of ESRF synchrotron at Grenoble (France), does not show significant discrepancies. The proposed methodology is validated by reproducing literature results on mono-disperse periodic arrays of spheres.

The hydraulic permeability tensor is positive-definite and symmetric [35]. In this study the full permeability tensor is estimated, by applying three sets of pressure gradients to the porous sample and computing the resulting specific discharge $\boldsymbol{q}$. With three experiments, nine equations are derived:

$$-\frac{1}{\mu}\sum_{j=1}^{3} k_{i\,j}\,\frac{\partial p^{(k)}}{\partial x_j} = q_i^{(k)} \tag{11}$$

where the superscript denotes a particular experiment. As in each experiment only a single component of the pressure gradient is non-zero, the following system of equations is obtained:

$$-\frac{1}{\mu}\begin{pmatrix} \frac{\partial p^{(1)}}{\partial x} & 0 & 0 & 0 & 0 & 0 \\ 0 & \frac{\partial p^{(1)}}{\partial x} & 0 & 0 & 0 & 0 \\ 0 & 0 & \frac{\partial p^{(1)}}{\partial x} & 0 & 0 & 0 \\ 0 & \frac{\partial p^{(2)}}{\partial y} & 0 & 0 & 0 & 0 \\ 0 & 0 & 0 & \frac{\partial p^{(2)}}{\partial y} & 0 & 0 \\ 0 & 0 & 0 & 0 & \frac{\partial p^{(2)}}{\partial y} & 0 \\ 0 & 0 & \frac{\partial p^{(3)}}{\partial z} & 0 & 0 & 0 \\ 0 & 0 & 0 & 0 & \frac{\partial p^{(3)}}{\partial z} & 0 \\ 0 & 0 & 0 & 0 & 0 & \frac{\partial p^{(3)}}{\partial z} \end{pmatrix} \begin{pmatrix} k_{11} \\ k_{12} \\ k_{13} \\ k_{22} \\ k_{23} \\ k_{33} \end{pmatrix} = \begin{pmatrix} q_1^{(1)} \\ q_2^{(1)} \\ q_3^{(1)} \\ q_1^{(2)} \\ q_2^{(2)} \\ q_3^{(2)} \\ q_1^{(3)} \\ q_2^{(3)} \\ q_3^{(3)} \end{pmatrix} \quad (12)$$

This means that in the first experiment the west face of the computational domain is a pressure inlet, while all other open boundary surfaces are classified as pressure outlets, with analogous arrangements for the second and third experiments.

The computed principal permeabilities are compared against the experimental measurement, the Kozeny-Carman correlation and the hydraulic network result [37] in table 3. Both the computed results and the Kozeny-Carman correlation overestimates the experimental measurement. As for the numerical results, this difference could be ascribed either to macroscopic inhomogeneities in the sample used in the experiments, or to insufficient resolution in the tomographic reconstruction, or to the inadequacy of the computational domain as a *representative elementary volume* (REV) [35].

Table 3. Computed principal components of the hydraulic permeability tensor, in $m^2 \times 10^{-13}$, for the real porous sample [36].

k_{I}		2.4
k_{II}		2.7
k_{III}		3.0
k_{exp}		1.9
$k_{\text{Kozeny-Carman}}$	$(M_s\ \mathrm{cm}^2/\mathrm{cm}^3)$	7.0 (207)
k_{netw}		4.2

3.2. *Calculation of relative permeability through reservoir rocks*

The major factors affecting relative permeability are pore structure, flow conditions, saturation history and wettability. However, for a given medium, the relative permeability is commonly assumed to be saturation-dependent only. Here the interest is to reproduce the steady-state method numerically, characterizing via relative permeability curves the dynamics of a two-phase (oil/water) flow within the digital sample of sandstone rock reconstructed by micro-CT, shown in figure 2(a). The pore-scale simulations are carried out using a 200^3 sub-volume extracted from the whole tomographic scan. The considered fluid properties are:

$$\rho_w = 998.2\ \mathrm{kg/m^3} \quad ; \quad \mu_w = 0.001\ \mathrm{kg/m\,s}$$
$$\rho_o = 800.0\ \mathrm{kg/m^3} \quad ; \quad \mu_o = 0.003\ \mathrm{kg/m\,s}$$
$$\sigma_{w,o} = 0.05\ \mathrm{N/m} \quad ; \quad \theta_{w,o} = 30^\circ$$

where ρ and μ denote density and dynamic viscosity of the fluids, $\sigma_{w,0}$ is the surface tension and $\theta_{w,o}$ is the equilibrium contact angle.

Once the digital domain is obtained the flow simulation can be carried out. The mixture velocity is imposed at the sample's inlet while the static pressure is enforced at the outlet. In turn, the latter condition implies zero capillary pressure at outlet, a boundary condition that is commonly used in special core analysis simulations. Oil and water volume ratio at the inlet section are fixed. The water layer covers the pore surface, while the oil fills the core of the pore body. Thus, at the fluid inlet section of the meshed pore-space, specialized image-processing tools are used to identify the boundaries of the pores and to adjust the thickness of the water layer according to the prescribed water/oil volumetric ratio. The zero-velocity (*no-slip*) condition is applied at the pore-fluid interface. The applied boundary conditions are summarized in table 4. The sample is initially filled with water. Once reached the stationary conditions, in terms of oil saturation S_o and pressure, the relative permeabilities for both oil, $k_{r,o}$ and water, $k_{r,w}$, are obtained using their definitions:

$$\boldsymbol{q}_o = -\frac{k\, k_{r,o}}{\mu_o} \nabla p_o \qquad \boldsymbol{q}_w = -\frac{k\, k_{r,w}}{\mu_w} \nabla p_w \tag{13}$$

The water and oil saturations are defined as:

$$S_o \equiv \frac{\text{volume of oil into pore space}}{\text{volume of pore space}} \qquad S_w \equiv 1 - S_o \tag{14}$$

Each test provides a couple of points in the relative oil and water permeability curves. The complete curves can be derived by changing oil/water injection ratio and measuring the microscopic variables when fluid saturation reaches equilibrium. The simulated data are fitted with the empirical relation proposed by Brooks and Corey and further compared to the experimental data obtained from a larger sample of the same rock using laboratory waterflood data [26]. Figure 5(a) shows two sets of data concerning oil (red) and water (blue). The circles represent relative permeabilities calculated with the numerical simulation, the dashed lines the fitting curves, the triangles the experimental data. Figure 5(b) shows a snapshot of the displacement process within the porous domain, where the oil phase is represented in red while the water phase appears in transparent blue.

Analysing the diagram and comparing the curves, it is possible to conclude that the simulated data are in good agreement with the experimental data, referring to the water; on the other hand, a discrepancy between the two sets of data for the oil is undeniable. The reasons of this difference are ascribable to the different dimensions of the samples used for the characterization: a one inch in diameter core plug for the laboratory experiment and a 200^3 voxels cube of side 1 mm for the numerical simulations. The simulation is carried out with fluid velocities lower than those recorded in the laboratory experiment in order to grant pore-Re numbers typical of a laminar flux.

Table 4. Boundary conditions used in the two-phase flow simulations.

Boundary	Condition
Inlet	Velocity inlet; α_o assigned
Outlet	Pressure outlet; either $\alpha_o = 1$ (imbibition) or $\alpha_w = 1$ (drainage) assigned to recirculating fluid
Pore/rock interface	No-slip condition; wall adhesion specified through static contact angle.

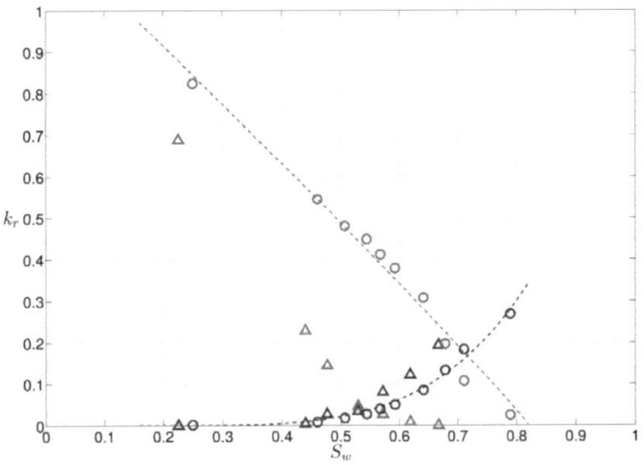

(a)

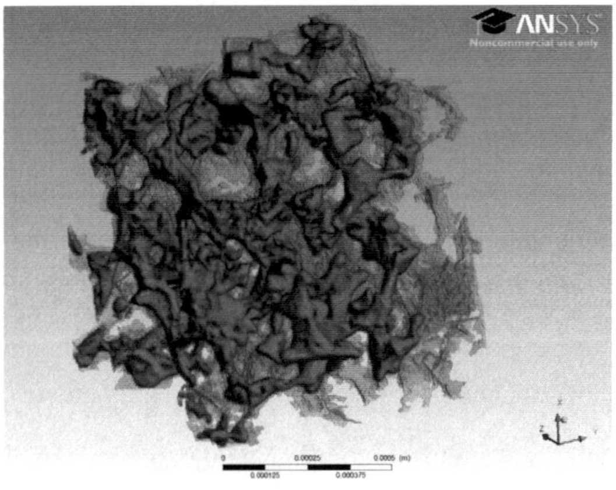

(b)

Figure 5. (a) Two sets of data concerning oil (red) and water (blue). The circles represent relative permeabilities calculated with the numerical simulation, the dashed lines the fitting curves, the triangles the experimental data. (b) Instantaneous visualization of oil phase (red) distribution within the pore space, initially filled with water (transparent blue).

4. Concluding remarks

The combined use of micro-CT systems, image analysis tools, volume meshing codes and CFD software allows reconstructing the complex morphology of the pore space of porous media and investigating their flow properties. This paper gives a thorough presentation of the work-flow to be used for predicting flow properties in both single- and multi-phase flows by pore-scale numerical simulation. The proposed methodology is used to evaluate the single-phase flow properties of a metallic foam, the full tensor of hydraulic permeability and the pressure induced and capillary driven multiphase flow of a sample of reservoir rock. Synthetic indicators such as absolute permeability, relative permeability and residual irreducible saturation can be computed on small samples rather than with costly and time consuming laboratory experiments.

Acknowledgments

The financial support from *Università degli Studi di Trieste - Finanziamento di Ateneo per Progetti di Ricerca Scientifica 2013 (FRA 2012)* is acknowledged.

References

[1] Loomis A G and Crowell D C 1962 *USBM Bulletin* **599**
[2] Akin S 2001 *J Petrol Sci Eng* **30**(1) 1

[3] Kho J C Y and Stevens R L 1975 *J Heat Transf* **97** 309
[4] Bai M and Chung J N 2011 *Int J Therm Sci* **50**(6) 869
[5] DeGroot C T, Straatman A G and Betchen L J 2009 *J Electron Packag* **131**(2) 021001
[6] Zong Y, Zhou B and Sobiesiak A 2006 *J Power Sources* **161** 143
[7] Faubion J M 2007 Ph.D. thesis Louisiana State University
[8] Benedik G, Širok B, Markič I and Eberlinc M 2011 *Heat Mass Transfer* **47**(6) 679
[9] Mostaghimi P, Bijeljic B and Blunt M J 2012 *SPE Journal* (**17**)(4) 1131
[10] Mostaghimi P, Blunt M J and Bijeljic B 2013 *Math Geosci* (**45**)(1) 103
[11] Boomsma K, Poulikakos D and Ventikos Y 2003 *Int J Heat Fluid Fl* **24** 825
[12] Krishnan S, Murthy J Y and Garimella S V 2006 *J Heat Transf* **128** 793
[13] Gerbaux O, Buyens F, Mourzenko V V, Memponteil A, Vabre A, Thovert J F and Adler P M 2010 *J Colloid Interf Sci* **342** 155
[14] Boschetto A, Costanza G, Quadrini F and Tata M E 2007 *Mater Lett* **61**(14-15) 2969
[15] Boschetto A, Campana F, Giordano V and Pilone D 2012 *Computational Modelling of Objects Represented in Images: Fundamentals, Methods and Applications III - Proceedings of the International Symposium, CompIMAGE* (*Rome, Italy*)
[16] Radon J and Parks P C 1986 *IEEE T Med Imaging* **5**(4) 170
[17] Feldkamp L A, Davis L C and Kreiss J W 1984 *J Opt Soc Am A* **1** 612
[18] VGStudio MAX from Volume Graphics GmbH, `http://www.volumegraphics.com/en/`
[19] +FE module by Simpleware, `http://www.simpleware.com/`
[20] Samet H 2006 *Foundations of multidimensional and metric data structures* Morgan Kauffman Publishers (San Francisco, USA)
[21] Soille P 1999 *Morphological Image Analysis: Principles and Applications* Springer-Verlag (Berlin)
[22] Deschamps T and Cohen L D 2001 *Med Image Anal* (**5**)(4) 281
[23] Peyré G and Cohen L 2005 *Heuristically driven front propagation for geodesic paths extraction. Lecture Notes in Computer Science* Springer-Verlag GmbH (Berlin)
[24] Schena G, Radaelli F, Piller M, Favretto S and Rossi E 2008 *Society of Petroleum Engineers - 70th European Association of Geoscientists and Engineers Conference and Exhibition* (*Rome, Italy*)
[25] Ferziger J H and Perić M 1972 *Computational Methods for Fluid Dynamics. 3rd ed.* Springer (Berlin, Germany)
[26] Tiab D and Donaldson E C 2004 *Petrophysics 2nd ed. - Theory and Practice of Measuring Reservoir Rock and Fluid Transport Properties* Elsevier (New York, USA)
[27] ANSYS Inc. 2010 *ANSYS FLUENT Theory Guide*
[28] OpenFoam website, `http://www.openfoam.com/`
[29] Nichols B D and Hirt C W 1981 *J Comp Phys* (**39**)(1) 201
[30] Tang H, Wrobel L C and Fan Z 2004 *Comput Mater Science* (**29**) 103
[31] Brackbill J U, Kothe D B and Zemach C 1992 *J Comp Phys* (**100**) 335
[32] Piller M, Boschetto A, Stalio E, Schena G and Errico O 2013 *J Porous Media* (**16**)
[33] Avizo Fire from FEI Visualization Sciences Group, `http://www.vsg3d.com/`
[34] Scheidegger A E 1960 *The physics of flow through porous media* University of Toronto Press (Toronto, CA)
[35] Bear J 1972 *Dynamics of Fluids in Porous Media* Dover Publications (New York, USA)

[36] Piller M, Schena G, Nolich M, Favretto S, Radaelli F and Rossi E 2009 *Transport Porous Med* (**80**)(1) 57

[37] Schena G and Favretto S 2007 *Transport Porous Med* (**70**)(2) 181

Journal of Physics: Conference Series **501** (2014) 012011 doi:10.1088/1742-6596/501/1/012011

A new Spool Piece for horizontal two-phase flow measurement

G Monni, M De Salve and B Panella

Politecnico di Torino, C.so Duca degli Abruzzi 24, 10129 Torino, Italy

E-mail: grazia.monni@polito.it

Abstract. This works presents the characterization of a Spool Piece (SP), made up of a Classical Venturi and a Wire Mesh Sensor (WMS), that are installed in a horizontal test section, in which an air-water mixture flows. The test section consists of a horizontal Plexiglas pipe of internal diameter 19.5 mm and total length of about 7 m. The flow quality ranges from 0 to 0.73 and the superficial velocity ranges from 0.14 to 32 m/s for air and from 0.019 to 2.62 m/s for water; the pressure ranges from atmospheric pressure to 4 bar depending on the experimental conditions. The observed flow patterns are stratified-bubbly-slug/plug-annular. The instruments response is analyzed and discussed. From the signal analysis the mass flow rate of each phase is obtained. The developed model allows the evaluation of the mass flow rate with an accuracy higher than 20% in the 84% of the cases and with an accuracy higher than 10% in the 73.3% of the cases. Finally the introduction of the estimated parameters in the SP model is considered and discussed.

1. Introduction

Within the framework of an Italian R&D program on Nuclear Fission, managed by ENEA and supported by the Ministry of Economic Development, the SPES3 experimental facility [1], able to simulate the innovative small and medium size PWR nuclear reactors, is being built and will be operated at SIET Company laboratories. In such facility some design and beyond design basis accidents, like LOCAs, with and without the emergency heat removal systems, will be simulated. An accurate accident analysis requires the measurement of the mixture mass flow rate and therefore instruments and methodologies to evaluate different two-phase flow parameters need to be developed.

Typically a set of instruments (Spool Piece - SP) must be installed in order to evaluate the mass flow rate of the phases in a large range of flow patterns, pressures and temperatures. Each instrument of the SP has to be sensitive to the different properties of the flow, like momentum, velocity, density, void fraction; moreover the selection of the instruments strongly depends on the experimental conditions: pressure, temperature and phases velocities. A different number of instruments can be coupled in a SP; a low number of instruments allows the reduction of costs, complexity and space requirements, but on the other hand, it requires the introduction of theoretical or experimental models able to compensate for the lacking of information. As the response of a meter in a two-phase flow tends to be highly sensitive to the flow pattern, to the upstream configuration and flow history, one of such instruments has to be able to measure the void fraction. Wire Mesh Sensors (WMS), based on the measurement of the local instantaneous conductivity of the two-phase mixture, are used for the evaluation of local void fraction, bubble size distributions and gas velocity distributions. The WMS has been used, in different geometry and for different configurations, to study mean cross-section void fraction and gas profile evolution [2]. In this paper a SP made up of a Venturi Flow Meter (VFM) with a Wire Mesh Sensor (WMS) is proposed. The calculation of fluid flow rate by reading the pressure drop across a pipe restriction is perhaps the most commonly used flow measurement technique in industrial applications. Compared with single-phase flow, the pressure drops in two-phase flow increase due to the interaction between the gas and the liquid phase; for this reason the value of the effective two-phase mass flow rate is over-predicted and must to be corrected using available correlations derived from experimental data or semi-empirical models.

In the present paper, the results of the experimental work carried out to analyze the response of a SP made up of a VFM and a WMS are shown. The SP has been tested in a horizontal test section for air-water flows in a large range of phases mass flow rates. The methodology for the signals interpretation, the developed model of the SP for the phases mass flow rates estimation and the results are presented and discussed.

2. Experimental Facility and Test Matrix

The experimental facility consists of the feed water loop (tap water with conductivity of about 620 μS is used), the feed air loop and the test section. The liquid flow rate is measured by means of an electromagnetic flow meter in a range of 0.9-36 m^3/h with a ±0.5% r.v. accuracy value. Lower values of the water flow rate are measured by means of rotameters in the ranges of 0-100 l/h and 100-400 l/h with an accuracy value of ±2% r.v.. The air flow rate is measured by means of different rotameters for the different ranges (100-500 Nl/h, 1000-5000 Nl/h and 6300-63000 Nl/h) with a ±2% f.s.v. accuracy value. The test section [3] consists of a 19.5 mm diameter and 7 m long pipe. The SP is installed at L/D = 192 from the entrance; the WMS is installed between two Plexiglas pipes having a length of 600 mm, while the VFM is installed between two Plexiglas pipes having a length of 500 mm and 490 mm upstream and downstream respectively. The inlet pressure ranges from atmospheric pressure to 4 bar depending on the experimental conditions. The flow is discharged at atmospheric pressure. The experimental test section is equipped with two quick closing valves (QCV) that allow the measurement of the volumetric void fraction in a length of 1300 mm.

The pressure drop in the VFM is measured by means of two differential pressure transducers having two different working range: -40÷40 mbar and 0÷1250 mbar. The instruments signals are acquired in a LabView® environment by the National Instruments DAQ (Digital AcQuisition) system (NI USB-6218). The WMS signals are instead acquired by means of the sensor software, synchronizing the measures through a trigger. The signals have been acquired using a frequency f = 1250 Hz for all the instruments for a total acquisition time of 20 s. Instruments characterization in single-phase flow of air and water has been performed. Two-phase experiments have been carried out at water temperature of about 20 °C. The flow quality ranges from 0 to 0.73 and the superficial velocity from 0.14 to 32 m/s for air and from 0.019 to 2.62 m/s for water. The typical observed flow patterns were stratified flow, intermittent flow (slug and plug) and annular flow. The population of samples ranges over different flow patterns and it is more representative of intermittent flow regime and annular wavy flow regime. From visual observation and maps results (figure 1) it also clear that some runs have been performed in transition regimes.

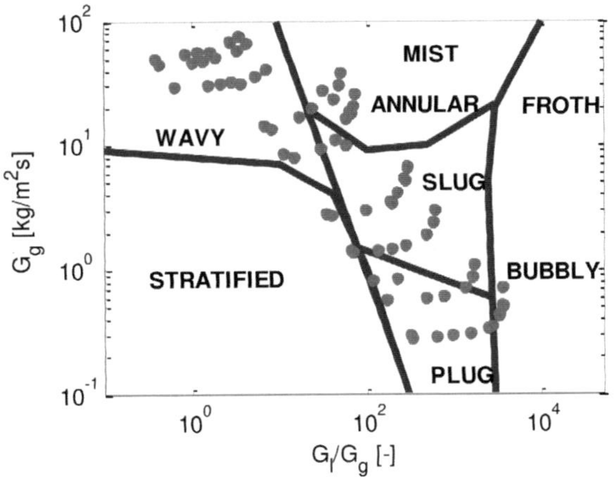

Figure 1. Baker's Map [6] prediction of Flow Patterns

3. Spool Piece

3.1 Wire Mesh Sensor Geometry and Signal Processing

The basic working principle of the sensor is the measurement of the conductivity of the fluid. Because air and water have different electrical properties (tap water is high conductive while air is very low conductive) the measurement of the conductance can be analyzed to detect the presence of each phase inside the pipe.

The sensor used in the present work has been constructed by Teletronic Rossendorf GmbH. The WMS consists of two planes of parallel wire grids (16x16) that are placed across the channel at a short distance from each other (1.5 mm) and span over the measuring cross section. The wires of both planes cross under an angle of 90°. The sensor has been designed to cover the cross section of a pipe having an internal diameter equal to 19.5 mm: the wires have a diameter D_{wire} of 70 μm and a pitch p equal to 1.3 mm, so that only the 5.4% of the pipe section is occupied by the sensor. The measuring grid allows to obtain a spatial resolution of the order of the pitch length (1.3 mm) and a very high time resolution. The WMS sensor signals are acquired by means of WMS200 electronics and processed in Matlab® environment. The output is a 3-D matrix $V(i,j,k)$ proportional to the local fluid conductivity. The indexes i and j refer to transmitting wires and to receiving wires respectively, while k is the time index. First of all the signal is normalized taken into account the single-phase reference matrix. The signal normalization can be considered as an approximation of the local void fraction value, if a linear relationship between conductivity and void fraction and a reference area equal to the square of the wire pitch p are assumed.

$$V^*(i,j,k)=\frac{V(i,j,k)-V_l(i,j)}{V_g(i,j)-V_l(i,j)} \qquad (1)$$

where V_l and V_g are the time averaged values of the signals with the pipe filled with water or air at the beginning of the test. From the normalized signal the mean cross-section instantaneous void fraction is derived taking into account the location of the wires respect to the pipe. The points of the grid, that are located near the wall, are analyzed taken into account the wall influence, while the points, that are located outside the cross section of the pipe are excluded from the analysis; so the local instantaneous void fraction is calculated as:

$$\alpha(i,j,k)=\frac{V^*(i,j,k)-1}{a(i,j)}+1 \qquad (2)$$

where $a(i,j)$ is the geometrical weight factor.

The local instantaneous void fraction α has been obtained with an acquisition frequency f_{acq} equal to 1250 Hz for a total observation time T_T equal to 20 s, so that the value of k ranges from 1 to k_T, where k_T is the total number of measured frames, that can be expressed by the eq.

$$k_T = f_{acq}\cdot T_T \qquad (3)$$

The averaged time cross-sectional void fraction is obtained by averaging over the observation time and over the measuring section:

$$\langle\alpha\rangle=\frac{1}{k_T}\sum_{k=1}^{k_T}\frac{1}{A}\sum_{i=1}^{16}\sum_{j=1}^{16}\alpha(i,j,k) \qquad (4)$$

where A is the pipe cross section.

3.2 WMS Measurement Accuracy

Errors of the wire-mesh sensor measurements for the gas fraction are mainly caused by the pitch of the wires which is 1.3 x 1.3 mm and the distance of the wire planes that is 1.5 mm.

Comparative measurements between the wire-mesh sensor and other research methods supplied information on the accuracy of the measurement technique and the evaluation algorithms for the experimental determination of these flow parameters. An air/water flow with gas volume fraction ranging from 0 to 100% shows that the deviations between wire-mesh sensor and gamma measurements are limited to ± 5% [2]. Comparative measurements between wire-mesh sensor and an X-ray tomography were also performed for air/water flow [4]. It was found that, the accuracy of the gas volume fraction averaged over the flow cross-section depends on the two-phase flow regime.

Differences in the absolute void fraction were determined for bubbly flows in the range of ± 1%, while a systematic underestimation of - 4 % was observed for slug flows. In the present work, the experimental void fraction is compared with the volumetric void fraction measured with the quick closing valves method. The volumetric void fraction

$$\langle \alpha_V \rangle = \frac{1}{V} \int_V \alpha(x, y, z) dV \qquad (5)$$

is measured in the 1300 mm long first part of the test section, with a precision of 3%.

Due to the intermittence of the flow, the volumetric void fraction is measured more than one time for each tested flow rates combinations in order to derive a representative mean value. In figure 2 the void fraction measured using the QCV technique and by means of the WMS are shown as a function of the flow quality. Compared to the volumetric void fraction, the void fraction measured with the WMS shows a higher dispersion in the range of intermittent flows and it's in good agreement at very high void fraction when the flow becomes annular. Moreover the WMS gives lower values of void fraction for the flow characterized by flow qualities higher than 0.01 and gives higher values at flow qualities lower than 0.001. Because the volumetric void is measured at the entrance of the test section, the flow is not completely developed, and an L/D effect can be present. A series of measurement in the volume where the WMS is installed will give more accurate information.

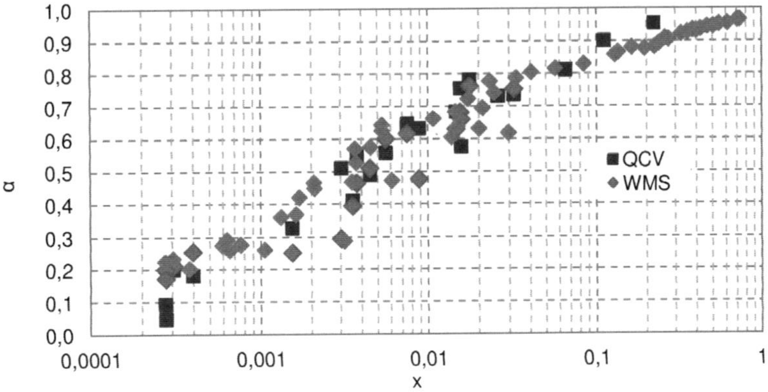

Figure 2. Mean void fraction measured using the QCV method and the WMS

3.3 Venturi Flow Meter

The characteristics of the Venturi flow meter are shown in table 1. Because in the experimented range the Re number is lower than 10^5 and because the sensor geometrical dimensions are not inside the range considered in the EN ISO 5167-1 [5], an in-situ calibration has been performed to obtain the single-phase discharge coefficient for air and water:

Journal of Physics: Conference Series **501** (2014) 012011 doi:10.1088/1742-6596/501/1/012011

$$C_{l,g} = \left[\frac{W}{F_a \cdot Y \cdot A_2} \cdot \left(\frac{\left(1-\beta^4\right)}{2\rho \cdot \Delta p} \right)^{0.5} \right]_{l,g} \quad (6)$$

The experimental discharge coefficient has been found equal to 0.94 in water flow and 0.97 in air flow, at Reynolds numbers higher than 15000.

Table 1. VFM Data

	Type Herschel (Fluid: Water)	
D_1	26	mm
D_2	10.251	mm
β	0.394	-
Δp	68.6 (700)	mbar (mmH_2O)
W_{water}	300	g/s
$L_{upstream}$	500	mm
$L_{downstream}$	490	mm

Compared to the single-phase flow measurements, in two-phase flow, the VFM pressure drops are higher due to phases interaction, and a model for the signal interpretation must to be used.

Baker [6] describes the different approaches used by the different researchers: estimation of the phases mass flow rates based on the estimation and/or calculation of the two-phase flow density and two-phase flow discharge coefficient and estimation of the phases mass flow rates based on the two-phase flow multipliers, relating single-phase and two-phase flow pressures drops. In 1949, Lockhart and Martinelli [7] researched the pressure drop and the liquid holdup of two-phase flow across the pipe, which became an important basis for the differential pressure method in two-phase flow measurement research. Later, many researchers studied the relationship between the mass flow rate of gas-liquid flow and the differential pressure across the flow meter. Flow rate measurement correlations were reported in [8] and [9], and recently in [10] and [11]. The developed correlations have been derived in a very limited operating range (pressure, temperature, phases velocities, fluids, etc..) and are strongly dependent on the VFM geometry (β, D_1, D_2, L, etc..). Moreover the most part of the available correlations are valid for very high ($x>0.95$) or very low ($x<0.1$) flow quality values.

The general equation of the VFM in a two-phase flow can be written as:

$$W = K \cdot \sqrt{2\Delta p_{TP} \rho_{TP}} \quad (7)$$

where

$$K = \frac{C_{TP} \cdot F_a \cdot Y}{\sqrt{1-\beta^4}} A_2 \quad (8)$$

Considering the value of Y and F_a equal to one, the two parameters, that have to be defined, are the two-phase discharge coefficient and the two-phase density.

In two-phase flow, due to the presence of the slip between the phases, different types of density can be defined, depending on the flow model hypothesis (Homogeneous, Separated flow, etc..). In the present work the density defined on the momentum balance is used to analyze the VFM response:

$$\rho_{TP} = \left[\frac{x^2}{\alpha \rho_g} + \frac{(1-x)^2}{(1-\alpha)\rho_l} \right]^{-1} \quad (9)$$

In order to estimate the density, the void fraction value must to be known; then the output of the WMS becomes an input for the VFM signal interpretation. The two-phase flow discharge coefficient is then evaluated, based on the experimental data (figure 3 (a)). The dependence of the parameter C_{TP} on the two-phase Re number has been modeled as (figure 3 (b)):

$$\begin{cases} C_{TP} = a + b\cdot\sqrt{\dfrac{10^c}{\mathrm{Re}_{TP}{}^d}} & \mathrm{Re}_{TP} < 1.5\cdot 10^4 \\ C_{TP} = 0.91 & \mathrm{Re}_{TP} > 1.5\cdot 10^4 \end{cases} \tag{10}$$

with the parameters evaluated on the basis of the experimental data: *a=1.1, b=0.6, c=3.5* e *d=1*.

The Reynolds number has been defined as:

$$\mathrm{Re}_{TP} = \frac{G_{TP} D_{pipe}}{\mu_{TP}} \tag{11}$$

and the dynamic viscosity μ, in accordance with the two-phase density (Eq. 9), has been defined as:

$$\mu_{TP} = \left[\frac{x^2}{\alpha\mu_g} + \frac{(1-x)^2}{(1-\alpha)\mu_l}\right]^{-1} \tag{12}$$

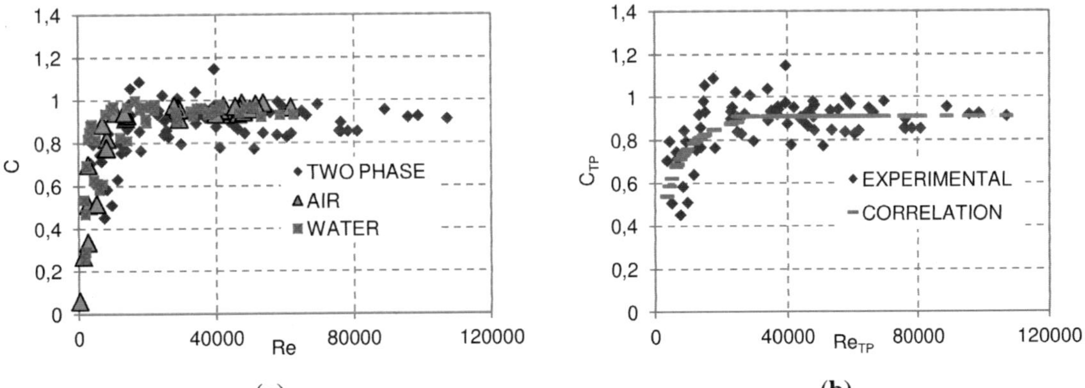

Figure 3. VFM discharge coefficient. (a) experimental single and TP flow; (b) TP flow correlation

The experimental pressure drops of the VFM have been analyzed in order to highlight the dependence on the superficial velocities of the phases. The Δp value increase is shown in figure 4 as a function of the two-phases superficial velocities. The regular behavior of the increasing curves suggests that a model of the instrument, depending on the phases velocities, can be developed in order to avoid the use of the discharge coefficient correlations, for the mass flow estimation. The lack of data between $J_g = 3$ m/s and $J_g = 6$ m/s is due to the non complete superimpositions ranges of the air mass flow meters.

In figure 5, the pressure drops in the VFM are expressed in terms of two-phase flow multipliers. The multipliers $\Phi^2_{l,g}$ has been evaluated as the ratio between the two-phase flow pressure drop and the single-phase pressure drop at the same phase mass flow rate:

$$\Phi_l^2 = \frac{\Delta p_{TP}}{\Delta p_l} \tag{13}$$

respect to the liquid single-phase flow, and

$$\Phi_g^2 = \frac{\Delta p_{TP}}{\Delta p_g} \tag{14}$$

respect to the air single-phase flow.

In the x-axis is reported the experimental Lockart-Martinelli parameter, that is evaluated as the ratio between liquid and gas VFM pressure drops:

$$\chi_{tt} = \frac{\Delta p_l}{\Delta p_g} \tag{15}$$

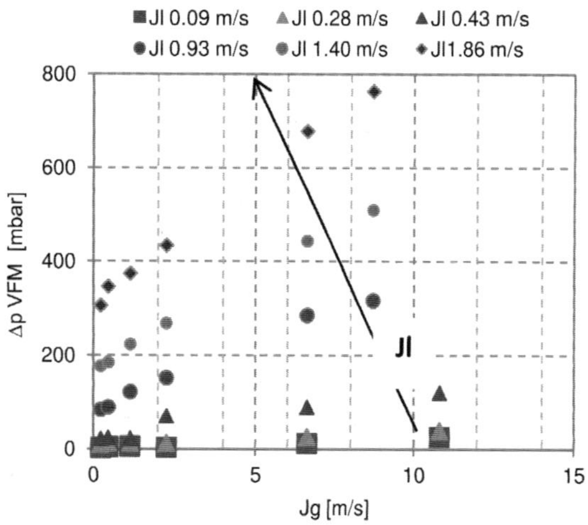

Figure 4. Experimental VFM pressure drops as a function of phases superficial velocities

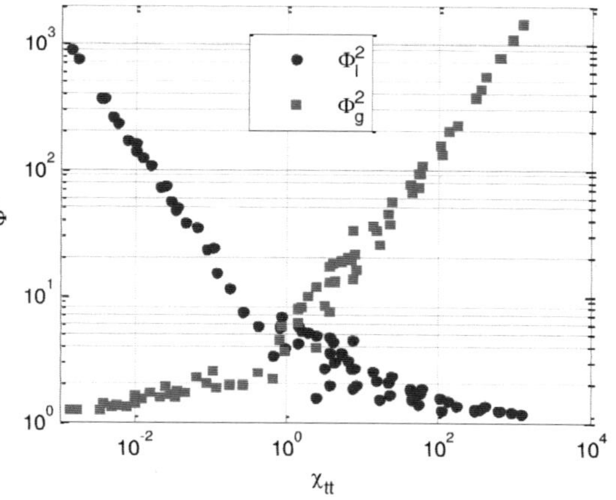

Figure 5. VFM pressure drops two-phase multipliers

3.4 Spool Piece Model

The SP model consists of a set of equations that, based on the measurement of the VFM pressure drop, and on the measurement of the void fraction in the WMS, together with the measurement of pressure and temperature of the fluid, allows the estimation of the mass flow rate of the phases in the test section:

$$\begin{cases} x = f_1(\alpha) \\ J_l = f_2(\alpha, \Delta p_{VFM}, p) \\ J_g = f_3(\alpha, \Delta p_{VFM}, p) \end{cases} \tag{16}$$

In the present work, an iterative approach is adopted for the estimation of the two mass flow rate of the phases. The model input are the void fraction, the pressure drop in the VFM, and the properties of air and water that are evaluated at the working pressure and temperature. An initial guess value for the flow quality is introduced. Using a literature x-α correlation the void fraction is calculated and the flow quality value is iterated till the error between the void fraction calculated and the void fraction measured by the WMS is lower than a threshold value (10^{-4}). When the required accuracy is reached, the two-phase flow Reynolds number and the discharge coefficient are calculated in a new iterative loop, where for the first iteration $C_{TP}=C_l$ is assumed. When the convergence is got ($W_{TP,i}-W_{TP,i-1}<10^{-4}$) the mass flow rate are calculated for the two-phases. For the *x-α* correlation used in the calculation, the Lockhart-Martinelli correlation (17) has been modified, based on the experimental data:

$$\alpha = \left[1+0.28\cdot\left(\frac{1-x}{x}\right)^{0.6}\cdot\left(\frac{\rho_g}{\rho_l}\right)^{0.36}\cdot\left(\frac{\mu_l}{\mu_g}\right)^{0.07}\right]^{-1} \tag{17}$$

$$\begin{cases} \alpha = \left[1+0.2\cdot\left(\frac{1-x}{x}\right)^{0.6}\cdot\left(\frac{\rho_g}{\rho_l}\right)^{0.36}\cdot\left(\frac{\mu_l}{\mu_g}\right)^{0.07}\right]^{-1} & x<1\cdot10^{-3} \\ \alpha = \left[1+0.28\cdot\left(\frac{1-x}{x}\right)^{0.6}\cdot\left(\frac{\rho_g}{\rho_l}\right)^{0.36}\cdot\left(\frac{\mu_l}{\mu_g}\right)^{0.07}\right]^{-1} & 10^{-3}\le x\le 6\cdot10^{-2} \\ \alpha = \left[1+0.45\cdot\left(\frac{1-x}{x}\right)^{0.6}\cdot\left(\frac{\rho_g}{\rho_l}\right)^{0.36}\cdot\left(\frac{\mu_l}{\mu_g}\right)^{0.07}\right]^{-1} & x>6\cdot10^{-2} \end{cases} \tag{18}$$

The comparison between the original correlation, the modified correlation and the void fraction value obtained from the WMS signal analysis is shown in figure 6.

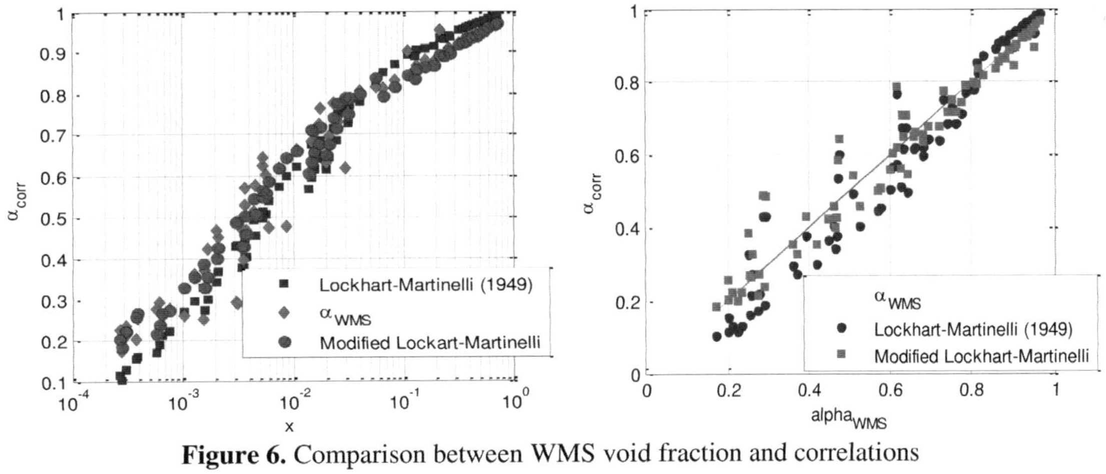

Figure 6. Comparison between WMS void fraction and correlations

4. Results

The developed SP Model has been applied to the experimental data in order to derive the mass flow rate of the phases. In figure 7, the flow quality, that is estimated by the SP model, is compared with the experimental value obtained by the phases flow rate measurements. The accuracy of the model has been evaluated as the degree of closeness of the estimated quantity to the experimental value; and the estimation error is considered the distance between the estimated quantity and the experimental value. The flow quality is estimated with an error lower than 20% for the flows in which the quality was higher than 0.1, and the error is lower than 10% for the 72% of the runs. For flow qualities lower than 0.1 the estimation error increases and is lower than the 50% for the 90% of the cases. The estimated mass flow rates are shown in figure 8 (a) (water flow rate), and figure 8 (b) (air flow rate). The total mass flow rate is estimated with an error that in the 73.3% of the cases is lower than 10% and in the 84% of the cases is lower than 20%. The highest estimation errors concern the air flow rate; in fact it is more affected by the errors in the quality estimation and in the total mass flow rate estimation due to the small values of this parameter. In this case the error is lower than 30% for the 73.3% of the cases. The highest errors have been obtained for the flows characterized by an intermittent behavior (slug/plug flow patterns). Moreover the intermittent flow has been realized in a region where the VFM discharge coefficient is strongly dependent on the Re number; and an error on the estimation of C_{TP} produces obviously an additional error on the mass flow rate estimation.

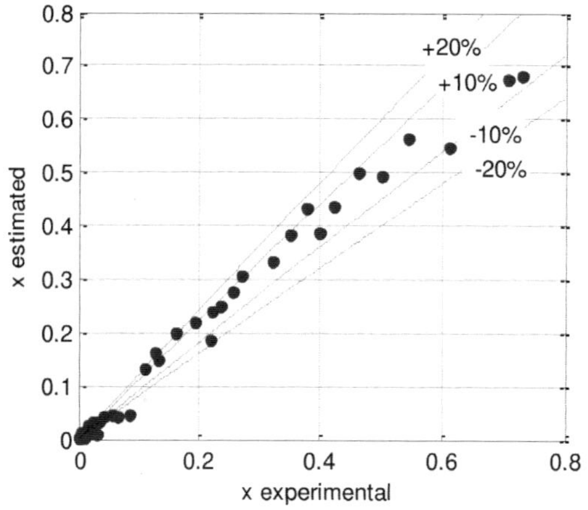

Figure 7. Experimental flow quality vs. SP flow quality

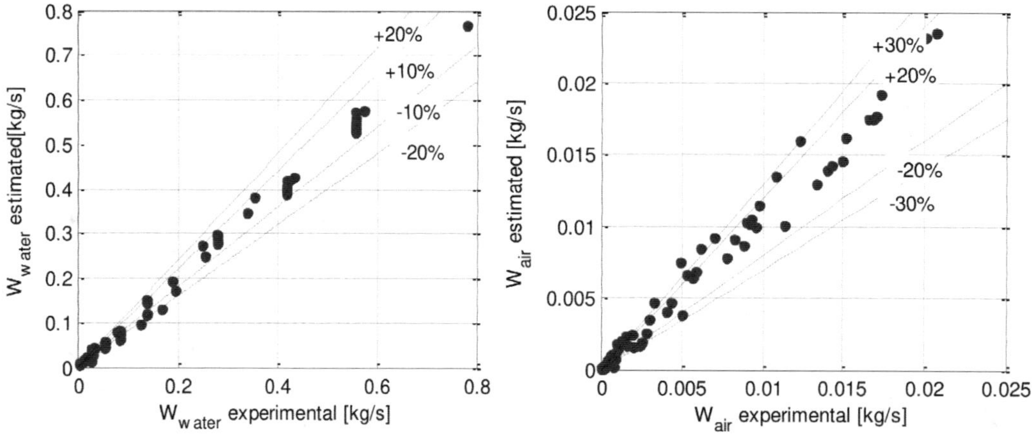

Figure 8. Experimental mass flow rate vs. SP mass flow rate for water phase (a) and air phase (b)

5. Conclusions

The analysis of the signals and the developed model of the SP, made up of VFM and WMS, allow us to estimate the mass flow rate of the phases in air-water horizontal flows with a good accuracy in a large range of the flow quality. The mass flow rate has been estimated with an error lower than 10% in the 73.3% of the cases. Moreover the estimation error is considerably lower for the flow characterized by high values of quality and void fraction. In this range, with reference to facility SPES3 test conditions, the error in the mass flow rate estimation is always lower than 10% and 20% for water and air respectively.

Nomenclature

A	area (m^2)	*Special characters*	
C	VFM discharge coefficient	α	void fraction
D	diameter (m)	β	D_2/D_1 VFM diameter ratio
F_a	VFM thermal expansion correction factor	μ	dynamic viscosity (Pa·s)
G	mass flux (kg/m^2s)	ρ	density (kg/m^3)
J	superficial velocity (m/s)	σ	surface tension (N/m)
Re	Reynolds number	Δp	differential pressure (bar)
SP	Spool Piece	*Subscripts*	
V*	WMS normalized signal	g	gas
W	mass flow rate (kg/s)	i	i-th vertical wire index
Y	VFM compressibility coefficient	j	j-th horizontal wire index
f	signal acquisition frequency (Hz)	k	WMS time index
f.s.v	full scale value	l	liquid
p	pressure (bar)	TP	Two-Phase
r.v	read value	1	pipe area subscript
x	flow quality	2	VFM constricted area subscript

Acknowledgement

The present research has been supported by ENEA and by the Ministry of Economic Development. The authors thank R. Costantino and G. Vannelli for the support through their work.

References

[1] Carelli M, Conway L, Dzodzo M et al. 2009 The SPES3 experimental facility design for the IRIS Reactor simulation *Sci. Technol. Nucl. Install.*
[2] Prasser H M, Bottger A and Zschau J 1998 A new electrode-mesh tomography for gas –liquid flows *Flow Meas. Instrum.* **9** 111
[3] De Salve M, Monni G and Panella B 2013 Horizontal Air-Water Flow Pattern Recognition *WIT Trans. Eng. Sci.* **79** (WIT Press 2013) doi:10.2495/MPF130381
[4] Prasser H M, Misawa M and Tiseanu I 2005 Comparison between WMS and ultra-fast X-ray tomography for an air-water flow in vertical pipe *Flow Meas. Instrum.* **16** 73
[5] ISO 5167-1:4 :2003 (E)
[6] Baker C R 2000 *Flow Measurement Handbook* (Cambridge University Press)
[7] Lockhart R W and Martinelli R C 1949 Proposed correlation of data for isothermal two- phase two-component flow in pipe *Chem. Eng. Prog.* **45** 39
[8] Murdock J W 1962 Two-phase flow measurement with orifices *J Basic Eng.* **84** 419
[9] Chisholm D and Rooney D H 1974 Research note: Two-phase flow through sharp-edged orifices *J Mech. Eng. Sci.* **16** 353

[10] De Leeuw R 1997 Liquid correction of venturi meter readings in wet gas flow In: *North sea workshop* paper 21
[11] Xu L, Xu J, Dong F and Zhang T 2003 On fluctuation of the dynamic differential pressure signal of Venturi meter for wet gas metering *Flow Meas. Instrum.* **14** 211

Journal of Physics: Conference Series **501** (2014) 012012 doi:10.1088/1742-6596/501/1/012012

Experimental and numerical investigation of a phase change energy storage system

G Casano and S Piva[1]

ENDIF- ENgineering Department In Ferrara, Università di Ferrara, Via Saragat 1, 44122 Ferrara, Italy

E-mail: stefano.piva@unife.it

Abstract. Latent heat storage systems are an effective way of storing thermal energy. Recently, phase change materials were considered also in the thermal control of compact electronic devices. In the present work a numerical and experimental investigation is presented for a solid-liquid phase change process dominated by heat conduction. In the experimental arrangement a plane slab of PCM is heated from above with an on-off thermal power simulating the behaviour of an electronic device. A two-dimensional finite volume code is used for the solution of the corresponding mathematical model. The comparison between numerical predictions and experimental data shows a good agreement. Finally, in order to characterize this thermal energy storage system, the time distribution of latent and sensible heat is analyzed.

1. Introduction

Heat storage systems using phase change materials (PCMs in the following) are an effective way of storing thermal energy due to the high energy storage density and the isothermal nature of the storage process. Latent heat storage systems have been widely used in building envelopes, residential heating and cooling, solar engineering, and spacecraft thermal control applications [1]. In recent years the utilization of PCMs has been also considered in the thermal control of compact electronic devices.

The heat generated by an electronic circuitry must be dissipated to prevent immediate failure and assure long term reliability. For high specific power and/or compactness, the limited capability of the traditional air cooling techniques requires the use of new technologies. In this area PCM based cooling is a very attractive technique of thermal control, considering the advantages of the PCMs such as: high specific heat, high latent heat, small volume change during phase change, availability of PCMs at convenient melting temperatures, non-toxicity, inertness and non-corrosiveness. A PCM energy storage system can be useful also to delay the heat release so to reduce the need of heat transfer surface, in particular for situations where the heat dissipation is of periodic nature or a sudden transient.

As the base case of thermal control unit (TCU in the following), a layer of PCM encapsulated in a hermetically sealed container can be considered. Referring to this solution, Alawadhi and Amon [2] investigated the thermal energy management issues associated with portable electronic equipment. The performance of a PCM based TCU was analysed for both constant and variable power operations. Tan and Tso [3], experimentally studied the cooling of a mobile electronic device using a PCM based TCU inside the device. They observed that the effectiveness of the TCU depends on the amount of PCM

[1] To whom any correspondence should be addressed.

used, as expected. Kandasamy *et al.* [4], investigated experimentally the effect of orientation of PCM based heat sinks and concluded that orientation has a limited effect on the thermal performance.

In applications [2-4] the low thermal conductivity of the PCMs made the charging and discharging slow during phase transitions. Different solutions were then developed to enhance the heat transfer in these PCM based TCUs, all related to the insertion of conducting paths or materials in the heat storage volume: discrete elements such as pins and fins, metal matrices or foams, nano and micro sized metal and metal oxide fillers, carbon nano tubes or fibers, graphite, exfoliated graphite or graphene [5,6]. Such a solution is referred to as a thermal conductivity enhancer (TCE in the following).

Limiting the analysis to TCUs where fins of different geometry are inserted in the PCM layer, Lamberg et al. [7] studied a system designed to store thermal energy when peaks of temperature are encountered in the operating conditions of a portable electronic device. Kandasamy *et al.* [8] analyzed experimentally and numerically the performance of a finned heat sink, where the throats were filled with a PCM. This energy storage system showed its potential as TCU in the transient electronic cooling applications. Wang *et al.* [9] carried out a parametric study on the performance of a finned heat sink filled with PCM. The parameters considered in the analysis were the aspect ratio, the filling of PCM, the level of under-cooling at the boundary. Fok *et al.* [10] focused the analysis on a finned heat sink filled of PCM for portable electronic devices. They concluded that this solution can increase the heat transfer rate during the charging stage, but did not seem to have any significant effect during the discharging stage. Baby and Balaji [11] investigated experimentally the performance of a finned TCE for the thermal management of portable electronic devices. Two different fins were compared. The experiments were carried for constant power dissipation. The time of melting and the final temperature were compared.

Finned surfaces partially filled with PCM, called hybrid systems, are somewhat more difficult to realize. However, these solutions show a lower thermal resistance than the finned TCUs. Gauché and Xu [12] analyzed numerically these hybrid PCM heat sinks providing evidence to the benefits, like a reduced size and the prevention of reliability issues. Krishnan *et al.* [13] investigated the ability of a hybrid PCM heat sink to operate continuously under time-varying cooling conditions. Stupar et al. [14] presented a hybrid PCM heat sink for temperature control of an electronic device. This was obtained by means of the combined action of a fan and a PCM-based energy storage system during peaks of power.

These last applications which combine both passive (PCM) and active (fins and fans) cooling solutions, seem to be of interest in high power amplifiers characterized by different levels of power dissipation. This is the case of the telecom base station power amplifiers, where the power is proportional to the traffic load.

In the present work some preliminary results of a numerical and experimental investigation are presented. The final goal of the research is the development of a hybrid PCM heat sink for power electronics applications, consisting of a parallel plate heat sink with part of the fins immersed in a suitable PCM. As a first step, a TCU consisting of a plane slab of PCM is analyzed for an on/off heating condition. For an easy confinement of the liquid phase, the cylinder geometry is chosen. For this energy storage system in a previous paper we analyzed the solid-liquid phase change process [15] for a constant heating. A two-dimensional finite volume code, validated for comparison with experimental data, was used for the solution of the corresponding mathematical model. In [16] we analyzed the solid-liquid phase change process for a sinusoidal heating. The study provided useful information for some applications, as the improvement of the thermal inertia of building walls.

The available experimental equipment [15-16] is now used to investigate the behaviour of a TCU subjected to an on/off heating. In the experiments, the power and the heat transfer surface are comparable to that of a telecom power amplifier. The specimen is placed horizontally and the heat flows from the top to the bottom. The on/off heating periods are equal. Energy parameters, as the amount of melting during the tests and the min/max temperature difference, are discussed. At this step of the research the phase change process is considered to be dominated by heat conduction. Finally, the time distribution of latent and sensible heat is analyzed.

Journal of Physics: Conference Series **501** (2014) 012012 doi:10.1088/1742-6596/501/1/012012

2. Experimental equipment and test procedure

The experimental equipment is described in full details in [15-16]; here just a short description is given. A schematic diagram of the equipment is shown in figure 1. The test cell consists of a hollow cylinder of polycarbonate (150 mm o.d., 140 mm i.d. and 210 mm height). A refrigeration system keeps this base at a temperature lower than the melting point of the test material. The uniform contact between the base and the cooler is enhanced by a thin layer of thermal grease. To reach easily this low temperature during the experiment, the cooling system uses Peltier cells. A uniform distribution of temperature over the whole surface of the bottom wall is obtained by means of nine thermoelectric modules (40 x 40 x 4 mm each). They are sandwiched between two square aluminium plates, 2 mm thick. One of these is finned (fins 40 mm high). The contact between the Peltier cells and the two aluminium plates is again enhanced by thin layers of thermal grease. To improve the heat release by the thermoelectric cooler, the fins are immersed in a constant temperature bath where the water is continuously stirred. The thermoelectric modules are supplied by DC power.

To reduce the heat transfer to the environment, the wall of the cylinder is insulated with a thick layer of expanded polystyrene foam (mean thickness 100 mm).

The test volume (height 51.1 mm) is closed on the top by the heater. The heat flow is obtained by dissipating via the Joule effect an assigned power in a resistor (450 W of nominal power at 220 V). Such a resistor is placed inside a copper disk soldered at the end of a copper pipe (135 mm o.d., 133 mm i.d. and 112 mm height). This copper pipe is placed coaxially inside the polycarbonate container. The gap between the two pipes is used as an auxiliary volume of expansion.

The temperature inside the specimen is measured by copper–constantan thermocouples at different heights. The positions of the thermocouples are shown in Table 1. The first and last thermocouples are in contact, respectively, with the cold and hot walls. The hot junctions of the thermocouples are supported by polycarbonate stands. These prismatic supports (3 mm of side width) guarantee the positions of the thermocouples and avoid the dragging of the solid during the freezing. The wires leave the rigid supports through holes and grooves machined in the bottom wall of the test cell.

A digital multimeter Agilent 3458A is used to measure the electromotive force for each thermocouple. The scanning of the thermocouples during the acquisition is carried out via a switch control unit Agilent 3488A. The data acquisition system is managed through a personal computer by means of an IEEE488 standard interface. The reference joint of the thermocouples is connected to an ice-point reference KAYE model K170. The time is measured with the inner clock of the personal computer.

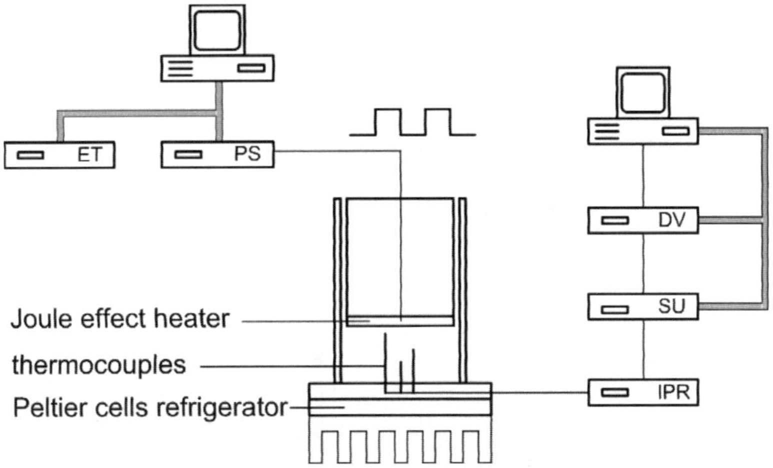

Figure 1 - Schematic diagram of the experimental equipment: ET, external trigger; PS, power supply; IPR, ice point reference; SU, switch control unit; DV, digital voltmeter.

Table 1 - Positions of the thermocouples (The origin is placed on the bottom of the specimen).

TC no.	1	2	3	4	5	6	7	8
z (mm)	0	6.2	16.2	25.4	30.3	36.7	44.6	51.1

The measured e.m.f. is converted into temperature by using a third degree polynomial, the interpolation coefficients being computed on the basis of the ASTM temperature–e.m.f. tables for standardized thermocouples [17].

The PCM test material used in the experiments is 99% pure n-octadecane($C_{18}H_{38}$), because it is non-corrosive and non-toxic, chemically inert and stable, with low vapour pressure, small volume changes during melting and large enthalpy release during the phase transition. The n-octadecane does not undergo any hysteresis cycle and its density decreases regularly as the temperature increases. Furthermore, its thermophysical properties, listed in table 2, are well known from the literature [18]. As can be seen, all the thermophysical properties except the density are considered constant with temperature and phase dependent.

While in [16] the heating was constant and in [15] sinusoidal in time, in this set of experiments we heated the sample with an on-off power. This power is obtained by varying the voltage via a power supply Agilent 6032A.

The uncertainty of the data gathered is estimated at the 95% confidence level, following the simplified procedure proposed by Moffat [19]. If values of fixed and random errors are available, the overall uncertainty assigned to the measured variable x is given by:

$$\varepsilon_x = \left[\left(B_{inst}\right)^2 + \left(2\sigma\right)^2 \right]^{1/2} \tag{1}$$

Otherwise, an overall uncertainty has been estimated on the basis of the manufacturers specifications.

The overall uncertainty assigned to the calculated parameter P is estimated using the root-sum-square propagation rule:

$$\varepsilon_P = \left[\sum_{i=1}^{N} \left(\frac{\partial P}{\partial x_i} \varepsilon_i \right)^2 \right]^{1/2} \tag{2}$$

The overall uncertainty of the temperature is 0.1 °C, whereas for the positions of the thermocouples is 0.3 mm. The measurement of the time is linked to the internal clock of the PC and it is accurate to 0.16 s.

3. Mathematical formulation and numerical solution

The physical system consists of a vertical cylinder of PCM heated from above and cooled from the bottom. The side walls are free to exchange heat with the environment.

Table 2 - Thermophysical properties of n-octadecane

Properties	Value	
Liquid density	814	kg/m^3
Solid density	814	kg/m^3
Liquid thermal conductivity	0.157	W/m K
Solid thermal conductivity	0.390	W/m K
Liquid specific heat	2200	J/kgK
Solid specific heat	1900	J/kgK
Latent heat of fusion	241360	J/kg
Melting temperature	28.18	°C

The geometry and the boundary conditions make the problem axisymmetric (excluding any angular dependence in the temperature distribution).

The mathematical model formulated to represent the physical system is based on the following assumptions:

- the PCM is homogeneous and isotropic;
- the thermophysical properties are constant in each phase;
- the phase-change occurs at a single temperature;
- the heat transfer is controlled only by conduction;
- the problem is two-dimensional;
- the difference of density between solid and liquid does not create appreciable local motion of the liquid.

The problem is then governed by the Fourier equation, to be solved for the two phases, solid and liquid (i = s or l for solid or liquid):

$$\frac{\partial}{\partial z}\left(\lambda_i \frac{\partial T_i}{\partial z}\right)+\frac{1}{r}\frac{\partial}{\partial r}\left(\lambda_i r \frac{\partial T}{\partial r}\right)=\frac{\partial\left(\rho_i c_i T_i\right)}{\partial t} \tag{3}$$

The initial condition is given by:

$$T(z,r,t=0)=T_0(z,r) \tag{4}$$

The boundary conditions are given by

- cold surface $\quad T=T_b(z=0,r,t) \quad 0 \le r \le R, \quad t>0$ (5)
- hot surface $\quad T=T_t(z=H,r,t) \quad 0 \le r \le R, \quad t>0$ (6)

On the side surface the thermal coupling with the environment is modelled through boundary condition of the third kind:

$$\left[-\lambda\frac{\partial T}{\partial r}\right]_{r=R}=h\left[T-T_e(t)\right] \qquad 0 \le z \le H,\ r=R,\ t>0 \tag{7}$$

At the phase-change interface two further conditions must be satisfied. At the solid-liquid interface the energy balance equation can be written in the following form, as proposed by Özisik [20]:

$$\left[1+\left(\frac{\partial Z}{\partial r}\right)^2\right]\left[\lambda_s\frac{\partial T_s}{\partial z}-\lambda_l\frac{\partial T_l}{\partial z}\right]_{z=Z(r,t)}=\rho_s r_f\frac{\partial Z(r,t)}{\partial t} \tag{8}$$

while continuity of temperature is given by:

$$T_s=T_l=T_f \tag{9}$$

Equations (8) and (9) are those typical of the Stefan problem. These consider the release or the storage of energy during the freezing or melting; generally the interface position is unknown a priori. For the numerical solution of the problem Finite Volume Method is used. The numerical procedure and some validation exercises are described in full details in [15]. Here it is relevant to underline that a simplified 2-D approach is followed. Inside each control volume affected by the change of phase, the interface is assumed to advance only in the z direction, neglecting the radial derivative of the interface position, equation (8). The resulting melting/solidification front assumes thence a stepwise shape.

4. Results and discussion

Experimental tests were carried out to obtain cyclic processes of melting and freezing in the PCM. An on/off heating was used to produce these cyclic processes. The tests always started by switching on the cooler in order to obtain a steady temperature distribution inside the specimen. After 1 h of data acquisition, a time enough to attain the steady state, the heating system was activated for 7 h, whereupon was disconnected. Successively, the test was continued for 2 periods of on/off heating 16 h long. The data acquisition was then 48 h long. On the bottom a temperature of about 15°C was maintained. The temperature was sampled every 4 min.

Even if strongly very low, the thermal coupling between cylinder and environment was taken into account through an overall heat transfer coefficient equal to 0.5 W/m^{2}K. The external temperature, recorded every two hours, was considered variable in the model and taken into account through a step-wise temperature-time function.

The measured and computed temperature distributions are compared in figure 2. In this figure the symbols TC1 - TC8 refer to the thermocouples of table 1. TC1 and TC8 are those used as the boundary conditions in the mathematical model. The continuous line is due to the high number of samples gathered (15 per hour in the 48 h of sampling). For reason of clarity, in figure 2 for TC2-TC7 just a limited numbers of samples are shown (1 per hour).

On the whole the agreement between experiment and prediction is very good. The presence of an interface that separates an upper zone, where the PCM is liquid, from a lower zone at the solid state, is clearly evident. After the activation of the phase change process, the liquid zone is characterized by faster variations of temperature, whereas in the solid zone this change is slower. At the third on/off cycle a steady periodic state seems to be attained. While on the heater the peaks of temperature and the switching off are exactly in phase, inside the specimen the peaks of temperature move forward in time due to the thermal inertia of the system.

In the liquid phase, the agreement between numerical results and experimental data is very good; the measured temperature is slightly different than that computed. The maximum difference is of 1.66 °C. In the solid phase the agreement seen in the liquid, is lost. The difference between measured and predicted temperatures increases, in particular for TC4. The maximum difference is of 3.23 °C. As already discussed in [16], for periodic heating the quantitative disagreement is less significant than for the case of constant heating [15].

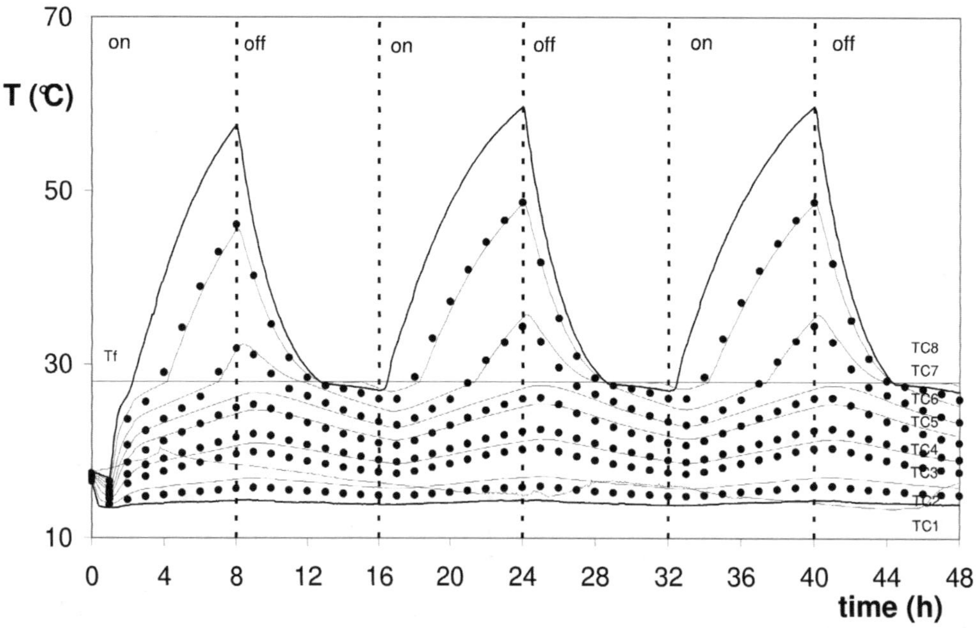

Figure 2 - Experimental and numerical comparison of the test (points: experimental data; continuous lines: calculated temperature; dotted line: ambient temperature).

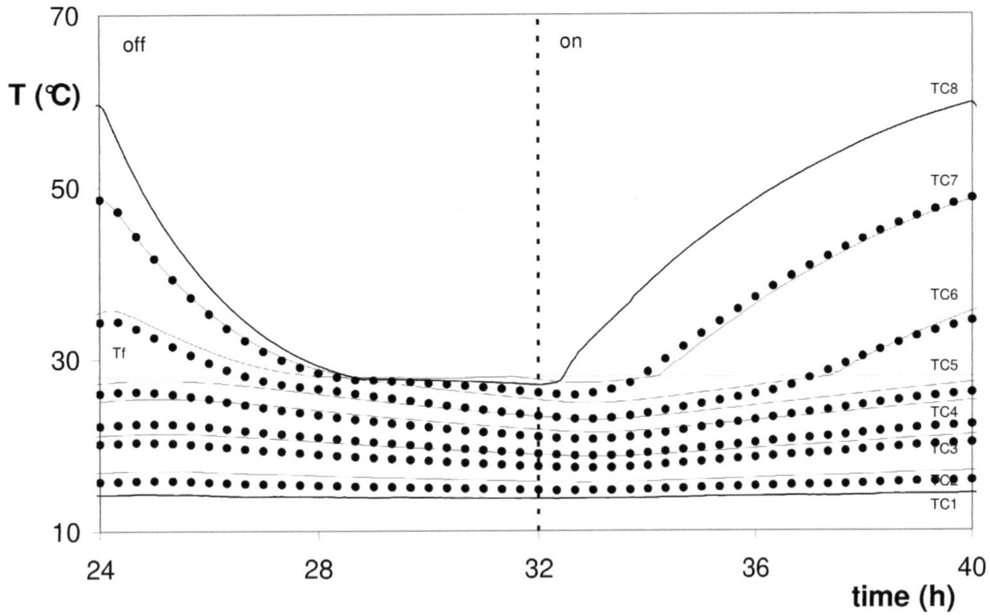

Figure 3 - Detail of the experimental and numerical comparison of the test (points: experimental data; continuous lines: calculated temperature; dotted line: ambient temperature).

In the numerical simulation the melting interface moves faster that in the experiment. This can be easily explained by a literature value of the latent heat of fusion [18] higher than expected. It is enough to reduce the latent heat of a 10% to improve the comparison.

As can be seen in figure 3, where a detail of the final period is shown, in the numerical simulation, during cooling, in particular when the specimen returns completely solid, TC7 becomes hotter than TC8, that used to impose the boundary condition. This is probably due to a real melting temperature lower than that proposed in the literature [18]. In figure 2 the experimental melting temperature seems to be about 27°C. It is enough to reduce this value in the model up to 27°C to improve the comparison.

The solid-liquid phase change interests only the first two thermocouples near the heater. This means that the latent heat storage allows to maintain two thirds of the sample under the melting point. This is because the great part of the heat flux introduced in the sample via the heater is used for the advancement of the interface and only a limited amount of this flux is still available for the heating of the remaining solid.

The coexistence of more phase change surfaces was not detected. This is due to the constant cooling of the bottom; the heat flux is never extracted from the top surface and the solid phase is significantly sub-cooled at the beginning of the experiment.

As experienced in [16], also for this steady-periodic heating the assumption to neglect the natural convection in the liquid phase gives rise to numerical results and experimental data in very good accordance, in the liquid region. Conversely, for a constant heating [22] the effect of natural convection is found to be predominant. As suggested by Hasan *et al.* [21] for a similar experimental arrangement but for a different heating scheme, the prevailing role played by the conduction is confirmed. It seems that the steady-periodic heating could be able to control and reduce the occurrence of natural convection in the liquid phase due to the thermal coupling with the environment through the side walls of the cylinder.

The distributions of latent, sensible and total heat storage per unit area, calculated with the mathematical model, are shown in figure 4. Here the two almost overlapped curves represent the total heat storage calculated both as the time integration of the difference between input and output heat fluxes, and as the time accumulation of internal energy and latent heat in each control volume.

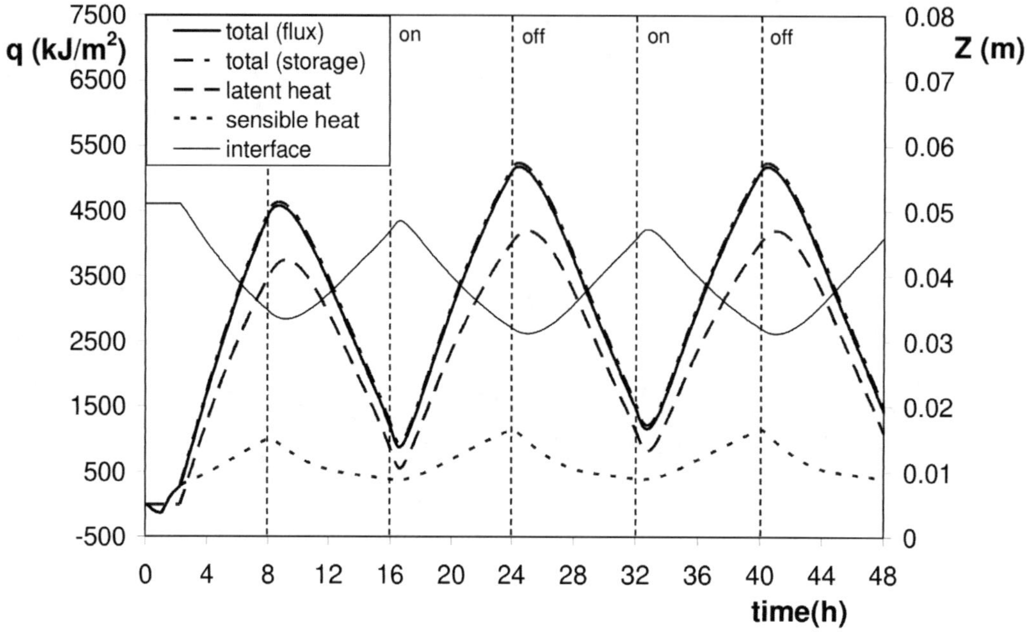

Figure 4 - Energy storage

The maximum amount of stored latent heat is about four times the sensible one. The maximum of sensible heat storage is in phase with the on/off heating. The maximum of latent heat storage is strictly correlated to the advancement of the solid-liquid interface. This interface moves also after the stop of the heating, due to the energy available above the melting interface.

In figure 4 the calculated position of the melting interface in r = 0.016 m is also shown. As already anticipated in the discussion of figure 2, the periodic movements of the interface produce the melting and freezing of 19.9 mm of PCM. This is due to the low thermal conductivity of the PCM. In the practical applications an enhancement of this parameter, for instance with fins, is mandatory.

Even if the specimen is heated with an on/off heat flux, in the model the boundary condition on the heater is of "assigned temperature". This choice is due to the high thermal inertia of the heater, obtained with a thick layer of copper, and to the heat dissipation through the copper walls of the heater. It was impossible to quantify the true amount of heating of the specimen and its time distribution. For this reason as the boundary condition the effect of the heating on the PCM, that is the temperature, was used.

5. Concluding remarks

A two-dimensional phase-change problem for which conduction is prevalent on the other heat transfer effects has been investigated experimentally and numerically. The process is characterized by an on/off heating.

The measured and calculated distributions of temperature are in good accordance. In this comparison only minor discrepancies arise. For this comparison, a very good knowledge of the thermal properties of the PCM is important. However, this is difficult to be obtained, because these properties are strongly dependent on the purity of the PCM, with large range of variations [22].

The analysis of the energy behaviour of the system shows that the energy stored oscillates in time with the on/off boundary condition.

In general it can be observed that PCM based TCUs are suitable to be used when high powers act in short times. After a peak of power a phase change can occur and the storage system has the possibility to release the stored energy with a freezing process.

Finally, the numerical code demonstrated to be an effective tool to analyse the energy behaviour of

Journal of Physics: Conference Series **501** (2014) 012012 doi:10.1088/1742-6596/501/1/012012

a storage system in those applications characterized by on-off heating.

Nomenclature

B	fixed error	
c	specific heat	J/(kgK)
h	heat transfer coefficient	W/(m^{2}K)
H	specimen height	m
P	calculated parameter	
q	heat per unit area	J/m^2
R	specimen radius	m
r	radial coordinate	m
r_f	latent heat of fusion	J/kg
t	time	h
T	temperature	°C
x	measured variable	
z	axial coordinate	m
Z	interface position	m

Greek symbols

ε	overall uncertainty	
λ	thermal conductivity	W/(m K)
ρ	density	kg/m^3
σ	random errors	

Subscripts

b	bottom
e	environment
f	melting
H	hot surface
inst	instrument
l	liquid
P	calculated parameter
r	radial
s	solid
t	top
x	measured variable
0	cold surface

References

[1] Sharma A, Tyagi V V , Chen C R and Buddhi D 2009 Review on thermal energy storage with phase change materials and applications *Renewable and Sustainable Energy Reviews* **13** 318-45

[2] Alawadhi E M and Amon C H 2003 PCM Thermal control unit for portable electronic devices: experimental and numerical studies *IEEE Trans. Components and Packaging Technologies* **26** 116-25

[3] Tan F L and Tso C P 2004 Cooling of mobile electronic devices using phase change materials, *Applied Thermal Engineering* **24** 159-69

[4] Kandasamy R, Wang X Q and Mujumdar A S 2007 Application of phase change materials in thermal management of electronics, *Applied Thermal Engineering* **27** 2822-32

[5] Fan L, Khodadadi J M 2011 Thermal conductivity enhancement of phase change materials for thermal energy storage: a review, *Renewable and Sustainable Energy Reviews* **15** 24-46

[6] Sarier N, Onder E 2012 Organic phase change materials and their textile applications: an overview, *Thermochimica Acta* **540** 7-60

[7] Lamberg P, Lehtiniemi R and Henel A M 2004 Numerical and experimental investigation of melting and freezing processes in phase change material storage *Int. J. Thermal Sciences* **43** 277-87

[8] Kandasamy R, Wang X Q and Mujumdar A S 2008 Transient cooling of electronics using phase change material (PCM)-based heat sinks, *Applied Thermal Engineering* **28** 1047-57

[9] Wang X Q, Yap C and Mujumdar A S 2008 A parametric study of phase change material (PCM)-based heat sinks *Int. J. Thermal Sciences* **47** 1055-68

[10] Fok S C, Shen W and Tan F L 2010 Cooling of portable hand-held electronic devices using phase change materials in finned heat sinks *Int. J. Thermal Sciences* **49** 109-17

[11] Baby R and Balaji C 2012 Experimental investigation on phase change material based finned heat sinks for electronic equipment cooling *Int. J. Heat and Mass Transfer* **55** 1642-49

[12] Gauché P, Xu W 2000 Modeling phase change material in electronics using CFD – A case study *Proc. Int. Conf. on High-Density Interconnect and System Packaging* 402-07

[13] Krishnan S, Garimella S V, Kang S S 2005 A novel hybrid heat sink using phase change materials for transient thermal management of electronics, *IEEE Trans. Components and*

Packaging Technologies **28** 281-89

[14] Stupar A, Drofenik U and Kolar J W 2010 Application of phase change materials for low duty cycle high peak load power supplies *Proc. Int. Conf. on Integrated Power Electronics Systems (Nuremberg, D, 16-18 March 2010)*

[15] Pinelli M, Casano G and Piva S 2000 Solid-liquid phase-change heat transfer in a vertical cylinder heated from above *Int. J. Heat and Technology* **18** 61-7

[16] Casano G and Piva S 2002 Experimental and numerical investigation of the steady periodic solid-liquid phase change heat transfer *Int. J. Heat and Mass Transfer* **45** 4181-90

[17] ASTM 1987, Standard E230-87 Temperature–electromotive force (EMF) tables for standardised thermocouples

[18] Chung J D, Lee J S, Yoo H 1997 Thermal instability during the melting process in an isothermally heated horizontal cylinder *Int. J. Heat Mass Transfer* **40** 3899–907

[19] Moffat R J 1994 Establishing the credibility of experimental work *Exp. Thermal Fluid Science* **8** inside back cover

[20] Ozisik M N 1993 *Heat conduction*, J. Wiley & S, New York

[21] Hasan M, Mujumdar A S and Weber M E 1991 Cyclic melting and freezing *Chemical Engeering Science* **46** 1573-87

[22] Pinelli M, Piva S 2003 Solid/Liquid phase change in presence of natural convection: A thermal energy storage case study *ASME J. Energy Resoucers Technology* **125** 190-98

Journal of Physics: Conference Series **501** (2014) 012013 doi:10.1088/1742-6596/501/1/012013

Experimental study on the flow patterns and the two-phase pressure drops in a horizontal impacting T-Junction

C Bertani, M Malandrone and B Panella
Dipartimento Energia, Politecnico di Torino, corso Duca degli Abruzzi 24, 10129 Torino, Italy

E-mail: bruno.panella@polito.it

Abstract. The present paper analyzes the experimental results concerning the flow patterns and pressure drops in two-phase flow through a horizontal impacting T-junction, whose outlet pipes are aligned and perpendicular to the inlet pipe. The test section consists of plexiglass pipes with inner diameter of 10 mm. A mixture of water and air at ambient temperature and pressures up to 2.4 bar flows through the T-junction, with different splitting of flow rates in the two outlet branches; superficial velocities of air and water in the inlet pipe have been varied up to a maximum of 35 m/s and 3.5 m/s respectively. The flow patterns occurring in the inlet and branch pipes are compared with the predictions of the Baker and Taitel – Dukler maps. The pressure drops along the branches have been measured relatively to different splitting of the flow rate through the two branches and the pressure loss coefficients in the junction have been evaluated. Friction pressure drops have allowed us to evaluate two-phase friction multipliers, which have then been compared to the predictions of Lockhart-Martinelli, and Friedel correlations. Local pressure drops have been extrapolated at the junction centre and analyzed; the two-phase multiplier has been evaluated and compared with the predictions of Chisholm correlation; the value of the empirical coefficient that minimizes the discrepancy has also been evaluated.

1. Introduction

In both conventional and nuclear power plants there are several T-junctions where two-phase flow may occur in normal and accidental conditions. Both configurations of dividing tees, i.e. the impacting tee and the branching tee, have been studied in literature since the 80's [1-6]. An outlet pipe is aligned with the inlet pipe in the case of a branching tee, while both outlet pipes are perpendicular to the inlet pipe in the impacting tee.

The mass qualities in the two outlet branches can be different, as the T-junction can induce an uneven phase separation. At different mass fluxes and different qualities and void fractions, the local pressure drops in the junction as well as the friction pressure drops in the branches are measured, and different flow regimes are observed; moreover the orientation and the inclination of the T-junction strongly affect two-phase flow phenomena.

Even though several experiments have already been carried out, the prediction of the phase separation in T-junctions is still difficult and further experimental work is needed.

In the present study experiments on the horizontal T-junction that has been described in [7] have been performed. A plexiglass test section with 10 mm inner diameter has been chosen, since few experimental data for small diameters are available in the literature. Superficial velocities of air and water in the inlet pipe are varied up to a maximum of 35 m/s and 3.5 m/s respectively, with different partitioning of the flow rate through the outlet pipes.

Flow patterns in the inlet and branch pipes, that have been previously [7] compared with the Mandhane map, have been further analysed by means of the Baker and Taitel–Dukler maps. Moreover, the measured pressures have allowed the evaluation of the friction pressure drops along the branches, of the singular pressure drops in the junction as well as of the two-phase multipliers.

The test data of the friction pressure drop have been compared with the Lockhart-Martinelli and Friedel multipliers; as regards singular pressure drops the Chisholm correlation has been used and an empirical coefficient that minimises the discrepancy from the experimental values has been determined.

2. Experimental facility

The experimental facility is schematically shown in figure 1. It mainly consists of the feeding lines for air and water, the mixer, the T-junction, the outlet pipes and an air-water separator having a volume of approximately 0.4 m^3. The two-phase mixture develops in a T-mixer that is located at the inlet of the test section.

The test section is shown in figure 2. It consists of three plexiglass pipes (10 mm in diameter, about 1 m long) named inlet, right and left. Four pressure taps are placed on each of them; these measurements allow us to determine the pressure behaviour along the pipes and to extrapolate the pressure losses between the inlet and the branches. The flow patterns have been visualised by a Panasonic video camera Model AG-DVC30E, with a shutter opening time of 1/8000 s. Instrumentation and its accuracy is described in [7]. Table 1 summarizes the test conditions.

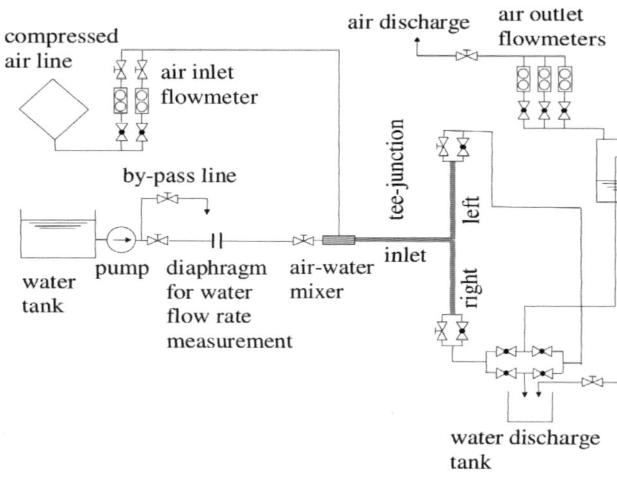

Figure 1. Experimental facility

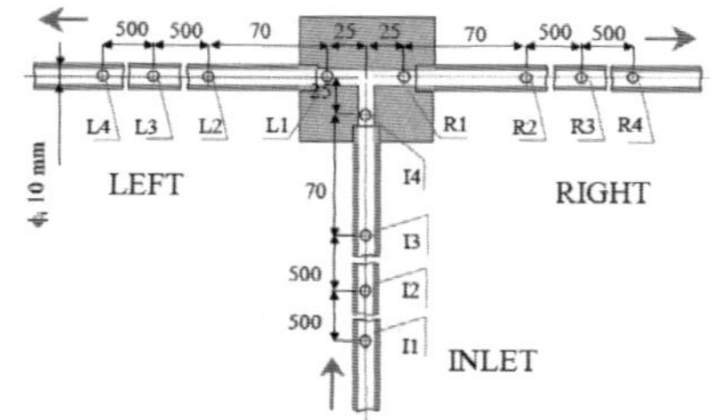

Figure 2. Test section

Table 1. Tests conditions.

Water inlet flow rate: six values from 80 g/s to 280 g/s
Air inlet flow rate: three values for each water flow rate (inlet pressure from 1.5 bar to 2.4 bar)
Extraction ratios: four values from 0 to 1 for each pair of water and air flow rates
Flow pattern in the inlet pipe: annular, intermittent and bubbly

3. Experimental results

Measurements of the two-phase flow have been carried out for six different values of water flow rate. At each water flow rate, the feed pressure has been varied in order to obtain three values of the air flow rate; several groups of three measurements have been carried out by varying the splitting of the phases and therefore the extraction ratio.

The mass velocity G in the inlet branch and the flow quality x in each pipe are reported in table 2, where for each run the minimum (min) and the maximum (max) measured values are reported. Table 2 also shows the values of the absolute pressure measured at pressure tap I1 at the test section inlet. The extraction ratios ER_R and ER_L of the right and left branch range from 0 to 0.5 and from 0.5 and 1 respectively.

4. Flow patterns

The flow patterns observed in all experimental tests of groups 1, 2, 3 and 4 are intermittent flow (plug – slug, I) and annular flow (A), as defined in [8].

More than one flow pattern has been observed in most cases: annular flow and intermittent flow are found alternatively, but the annular regime is prevailing (A-I); in other tests the intermittent regime prevails (I-A). Many flow patterns are therefore close to the transition conditions.

The flow patterns in the groups 5 and 6 have not been systematically analysed, as the rather high mass flux makes difficult the observation; anyway, in such tests bubble flow is prevailing in the inlet pipe. Table 3 reports the percentage of occurrence of the different flow patterns.

Table 2. Two-phase flow data

Group	p_{I1} [MPa]	G_I [kg/(sm^2)]	x_I [%]	x_L [%]	x_R [%]
1	0.157-0.241	1019-1117	1.45-5.19	1.33-6.05	1.46-5.36
2	0.159-0.240	1289-1394	0.78-4.85	0.19-5.23	0.77-4.59
3	0.160-0.236	1614-1760	0.36-3.01	0.13-3.10	0.34-2.97
4	0.156-0.232	2133-2345	0.13-1.52	0.00-1.59	0.13-1.57
5	0.154-0.227	2717-2984	0.08-0.77	0.0-0.77	0.11-0.84
6	0.174-0.223	3266-3528	0.06-0.43	0.00-0.41	0.07-0.46

Table 3. Percentage of occurrence of the flow pattern

Observed flow pattern	% of the total
annular	1.5
annular - intermittent	18.9
intermittent - annular	22.0
intermittent	57.6

As part of the experimental points are in disagreement with the Mandhane map [7], the comparison has been extended to the Baker [9] and Taitel - Dukler [9] maps. The Baker maps are represented in figures 3a, 3b, 3c and 3d. They show that the experimental points of the test groups 1 and 2 predominantly lie in the annular flow region of the map; on the other hand the experimental points of group 3 partly lie also in the slug flow region and those of group 4 predominantly lie in the slug flow region.

As it can be seen in table 2, mass velocities in the experiments of test group 1 are slightly higher than 1000 kg/(s m^2), while the mass velocities of the tests of the other groups are higher than 2000 kg/(s m^2) in test group 4.

Journal of Physics: Conference Series **501** (2014) 012013 doi:10.1088/1742-6596/501/1/012013

The comparison shows a poor agreement at low mass velocities, as annular flow is foreseen while intermittent flows are observed; a better agreement is evident at higher flow rates, as the experimental points lie mainly in the slug flow region of the map and the observed flow regime is actually predominantly intermittent.

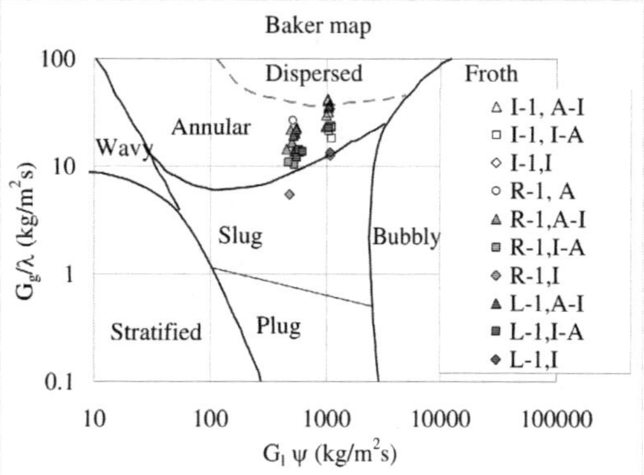

Figure 3a. Comparison of the group 1 results with the Baker Map; I (Inlet), R (Right), L (Left), 1 (Group 1)

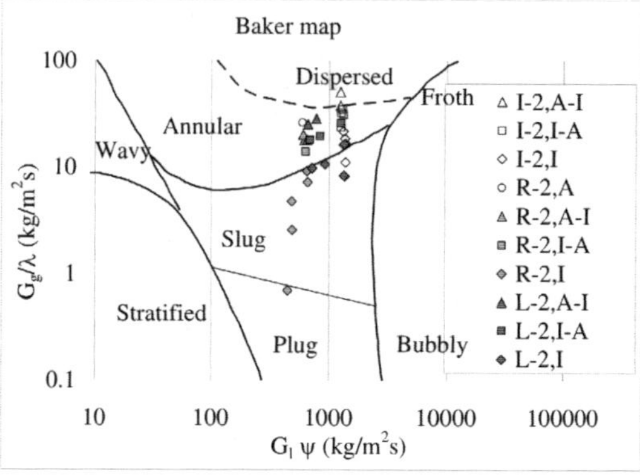

Figure 3b. Comparison of the group 2 results with the Baker Map; 2 (Group 2)

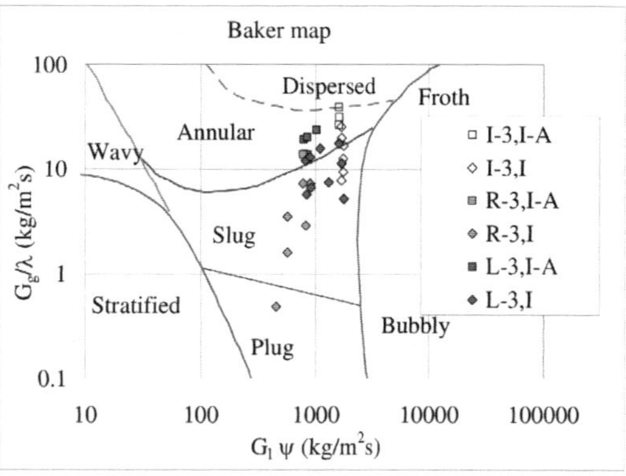

Figure 3c. Comparison of the group 3 results with the Baker Map; 3 (Group 3)

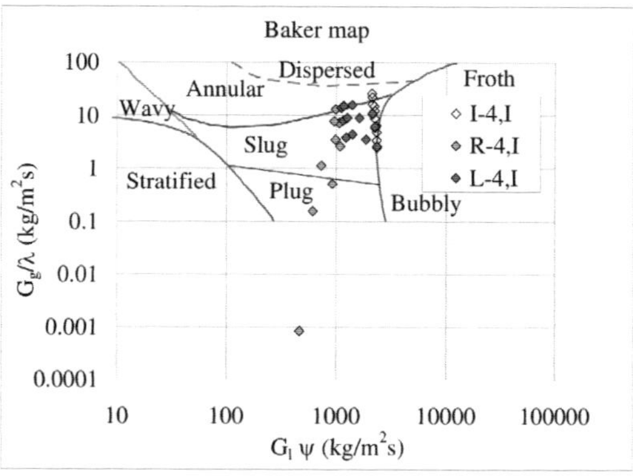

Figure 3d. Comparison of the group 4 results with the Baker Map; 4 (Group 4)

The comparison with Taitel–Dukler maps is shown in figures 4a- 4h, which reports the Froude number Fr and the non-dimensional parameter T versus the Martinelli parameter χ.
In most of the experimental tests of test groups 1, 2 and 3, the annular flow has been observed, but it alternated with intermittent flow. The Taitel-Duckler maps show that experimental points of test groups 1, 2 and 3 lie in the transition zone. Some points of test group 1 are in the intermittent flow region.

All experiments of test group 4 revealed an intermittent flow and their relative points lie in the intermittent zone of Taitel-Duckler map. Therefore we can state that the Taitel-Duckler maps give a good prediction of the flow pattern of all tests.

5. Friction pressure drops

Friction pressure gradients along the branches have been evaluated by using the differential pressure measured between pressure taps I1, I2, I3, L1, L2, L3, R1, R2, R3 and by neglecting the acceleration pressure drop, as it is lower than 1% of the total pressure difference. The experimental results have then been used in order to determine the two-phase friction multipliers.

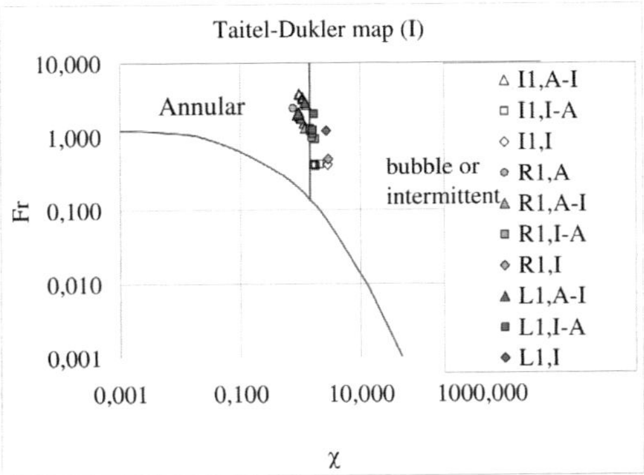

Figure 4a. Comparison of the group 1 results with the Taitel- Duckler maps; I (Inlet), R (Right), L (Left), 1 (Group 1)

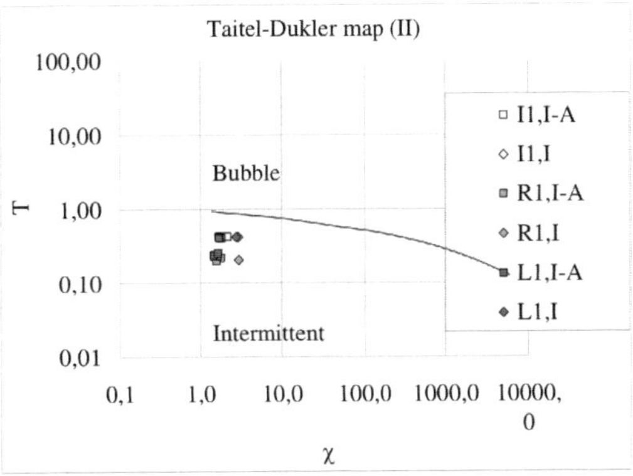

Figure 4b. Comparison of the group 1 results with the Taitel- Duckler maps

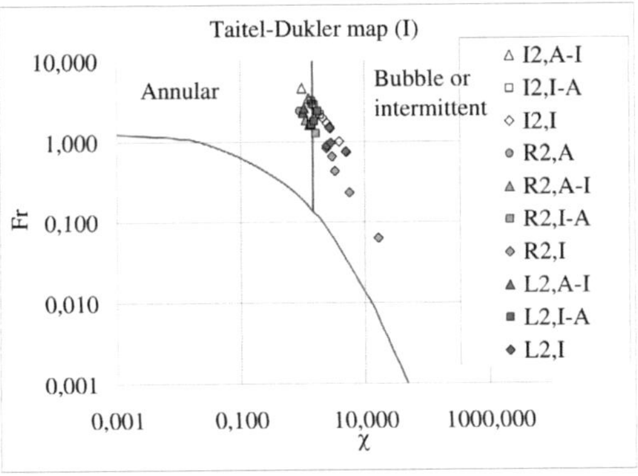

Figure 4c. Comparison of the group 2 results with the Taitel- Duckler maps

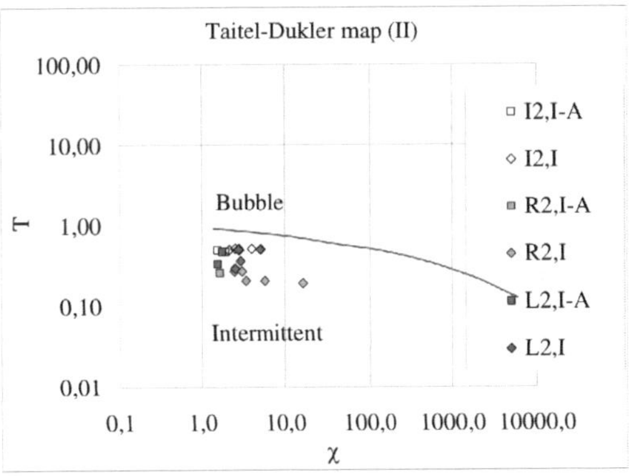

Figure 4d. Comparison of the group 2 results with the Taitel- Duckler maps

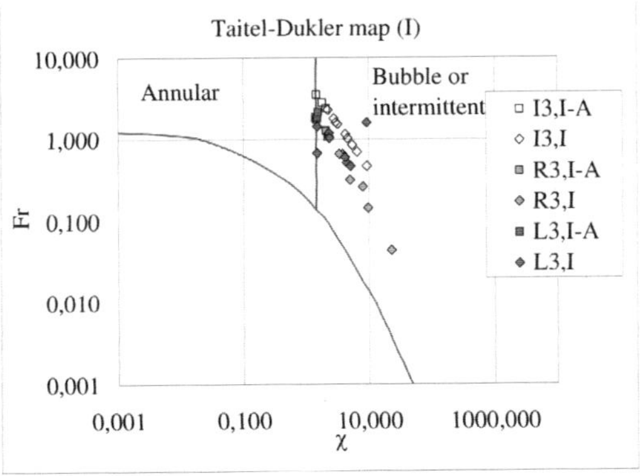

Figure 4e. Comparison of the group 3 results with the Taitel- Duckler maps

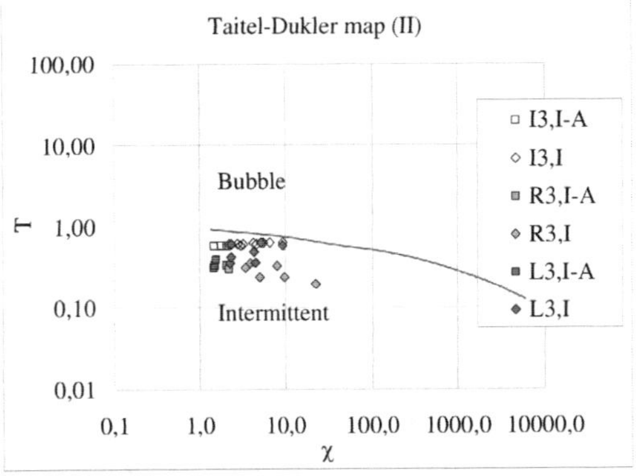

Figure 4f. Comparison of the group 3 results with the Taitel- Duckler maps

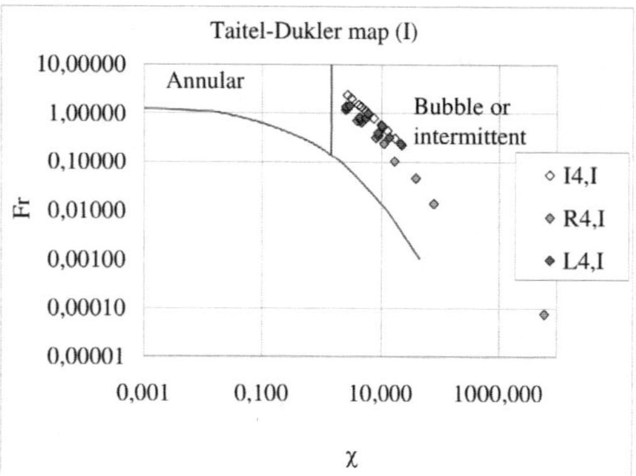

Figure 4g. Comparison of the group 4 results with the Taitel- Duckler maps

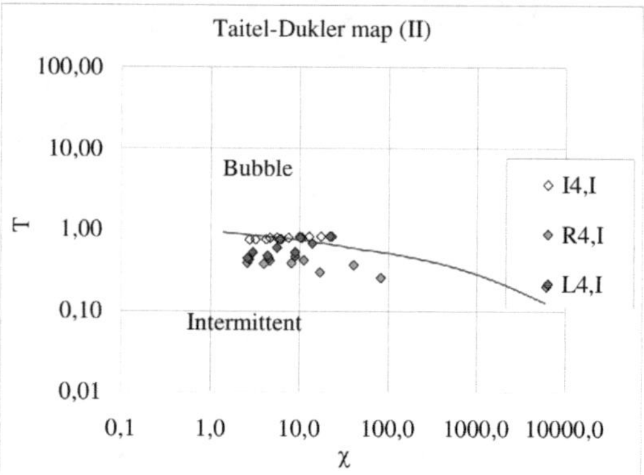

Figure 4h. Comparison of the group 4 results with the Taitel- Duckler maps

The two-phase friction pressure drop has been related to the single-phase pressure drop that is evaluated considering the friction factor deduced from the experiments that have been carried out with single-phase water. The two-phase multipliers ϕ_l^2 e ϕ_g^2 have been then evaluated: they are reported in figures 5a-5b for the test groups 1 and 2 and in figures 5c-5d for the test groups 3 and 4 versus Martinelli parameter χ. The figures also report the curves of the correlations that have been proposed by Chisholm [9] with C equal to 10 (turbulent flow of liquid, laminar flow of gas) and 20 (turbulent flow for both gas and liquid):

$$\phi_l^2 = 1 + \frac{C}{\chi} + \frac{1}{\chi^2} \qquad \phi_g^2 = 1 + C\chi + \chi^2 \qquad (1)$$

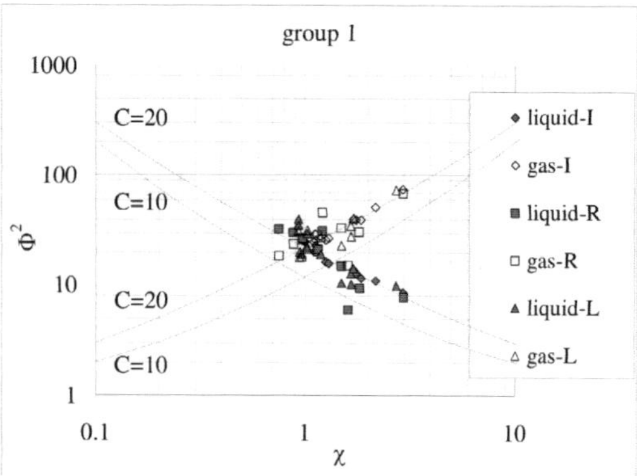

Figure 5a. Comparison of the two-phase multipliers of group 1 with the Lockhart-Martinelli correlation

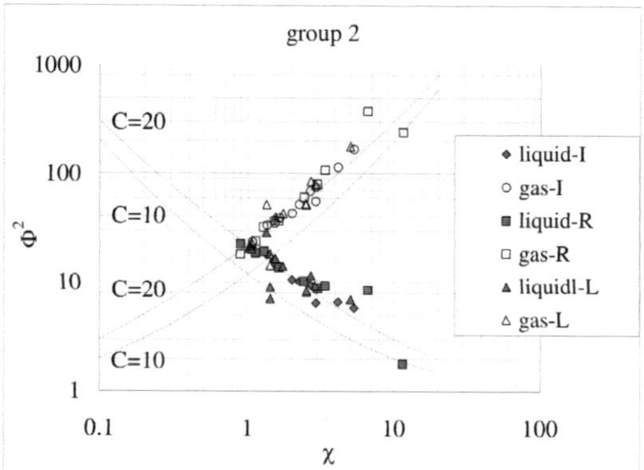

Figure 5b. Comparison of the two-phase multipliers of group 2 with the Lockhart-Martinelli correlation

The spread of the experimental points is higher for group 1 tests (lower mass velocities); anyway the experimental points are in reasonable agreement with the Lockhart–Martinelli correlation with C = 20; nevertheless the correlation tends to underestimate the majority of experimental results.

Journal of Physics: Conference Series **501** (2014) 012013 doi:10.1088/1742-6596/501/1/012013

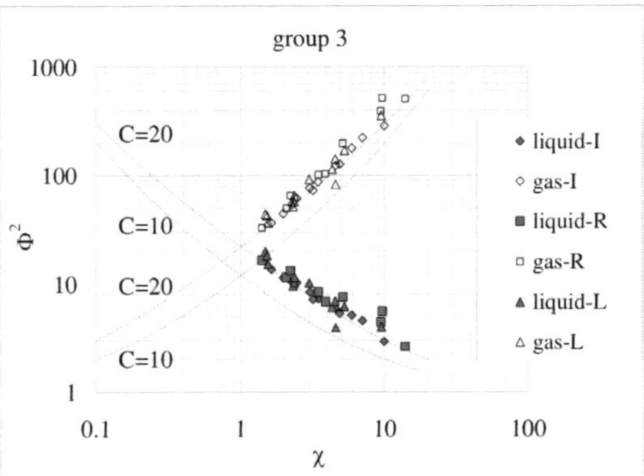

Figure 5c. Comparison of the two-phase multipliers of group 3 with the Lockhart-Martinelli correlation

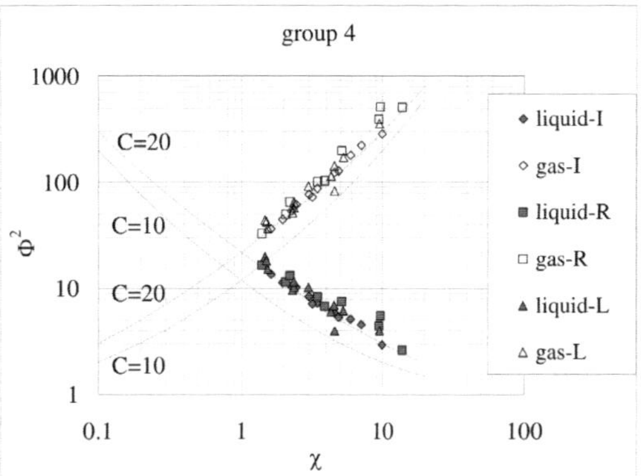

Figure 5d. Comparison of the two-phase multipliers of group 4 with the Lockhart-Martinelli correlation

The results have also been compared with the prediction of the Friedel correlation [10], that is valid also for horizontal pipes:

$$\phi_{lo}^2 = E + \frac{3.24\,F\,H}{Fr^{0.045} We^{0.035}} \qquad E = (1-x)^2 + x^2 \frac{\rho_l C_{f,go}}{\rho_g C_{f,lo}} \qquad F = x^{0.78}(1-x)^{0.224}$$

$$H = \left(\frac{\rho_l}{\rho_g}\right)^{0.91} \left(\frac{\mu_g}{\mu_l}\right)^{0.19} \left(1 - \frac{\mu_g}{\mu_l}\right)^{0.7} \qquad Fr = \frac{G^2}{g\,d\,\rho_h^2} \qquad We = \frac{G^2 d}{\sigma\,\rho_h} \qquad \rho_h = \left(\frac{x}{\rho_g} + \frac{1-x}{\rho_l}\right)^{-1} \tag{2}$$

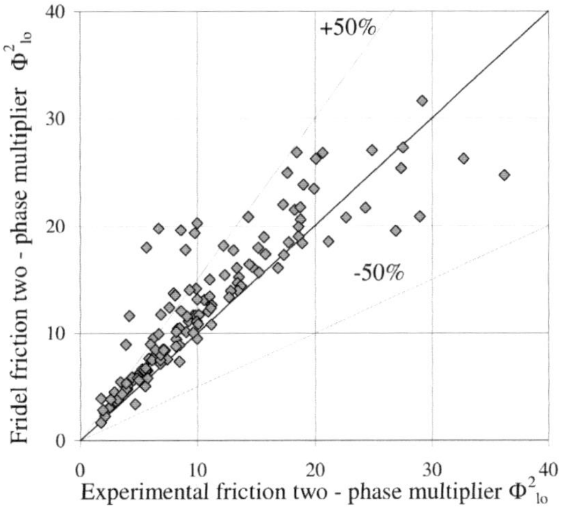

Figure 6. Comparison of the experimental two-phase multiplier with the Friedel multiplier

The experimental values and the Friedel correlation prediction are reported in figure 6, which shows a rather good agreement: it correlates the test data within 50% at a probability level of 86%, within 30% at a probability level of 73% and within 20% at a probability level of 46% . The discrepancies higher than 50% that can be observed in fig.6 (in a few tests they exceed 100%) occur

Journal of Physics: Conference Series **501** (2014) 012013 doi:10.1088/1742-6596/501/1/012013

predominantly when the pressure drops are low, as these tests are affected by a higher relative uncertainty. The error analysis gives an uncertainty of the two-phase multipliers of Lockhart-Martinelli and Friedel in the range 5-25%.

6. Local pressure drops

The local pressure drops have been evaluated using an extrapolated pressure at the junction determined by the methodology reported in [7]; in fact, the procedure that is generally followed in the literature is based on extrapolated pressures; these values will be different from the real ones measured at pressure taps I4, L1 and R1, but at least they are not affected by local effects occurring near the junction.

The two-phase local pressure drops were evaluated by the Separated Flow Model SFM proposed in [3] and [11] for the pressure drop between the inlet pipe and the perpendicular branch pipe of a branching T-junction. The application to the case on an impacting T gives:

$$\Delta p_{I,j} = \frac{\rho_{hj}}{2}\left(\frac{G_j^2}{\rho_j^2} - \frac{G_I^2}{\rho_I^2}\right) + K_{Ij}\frac{G_I^2}{2\rho_l}\phi_{lo}^2 \tag{3}$$

The subscript I refers to the inlet pipe and j to the right or left outlet branch; K_{Ij} is the local loss coefficient evaluated in single-phase flow; ρ_{hj} is the homogeneous density, while ρ_j and ρ_I are the energy-weighted densities in the branches and inlet pipe evaluated as follows [3]:

$$\rho = \left[\frac{(1-x)^3}{(1-\alpha)^2\rho^2} + \frac{x^3}{\alpha^2\rho_g^2}\right]^{-0.5} \tag{4}$$

The drift flux model for horizontal slug flow has been adopted to evaluate the void fraction.

$$\alpha = \frac{1}{1.2\left[1 + G_l\rho_g/(G_g\rho_l)\right]} \tag{5}$$

The two-phase multiplier has been evaluated by the Chisholm correlation that is valid for T-junctions [8] and, in general, for several types of local pressure drops [11], by using different values of the coefficient C_i (the Martinelli parameter is evaluated for turbulent flow in both liquid and gas):

$$\phi_{lo}^2 = (1-x_I)^2\left(1 + \frac{C_{Ij}}{\chi_{tt}} + \frac{1}{\chi_{tt}^2}\right) \qquad \chi_{tt} = \left[\left(\frac{x_I}{1-x_I}\right)\left(\frac{\rho_l}{\rho_g}\right)^{0.5}\right]^{-1}$$

$$C_{Ij} = \left[1 + (C_j - 1)\left(\frac{\rho_l - \rho_g}{\rho_l}\right)^{0.5}\right]\left[\left(\frac{\rho_l}{\rho_g}\right)^{0.5} + \left(\frac{\rho_g}{\rho_l}\right)^{0.5}\right] \tag{6}$$

The coefficient K_{Ij} has been expressed as a function of the extraction rate ER by the polynomial (7), which has been obtained by applying the minimum square root method to the single-phase tests with air only and water only:

$$K_{Ij} = -0.5723\ ER^2 + 0.9958\ ER + 0.5309 \tag{7}$$

In [8] a value of $C_j = 1.75$ is suggested for the pressure drop between inlet and branch pipes of a branching T-junction (no other values are suggested in [8] for the impacting T).

The extrapolated values of the differential pressure are used to evaluate the two-phase multiplier, that have then been compared with the prediction of Chisholm correlation. Using $C_i = 1.75$ the Chisholm two-phase multipliers are much higher than the experimental values (the discrepancy is lower than 50% only in 9% of the tests).

The minimum mean square error method has been used to determine the value of C_i (the value of 0.432 has been found); figure 7 compares the experimental two-phase multiplier with the Chisholm prediction, that correlates the test data within 50% at a probability level of 86%, within 30% at a probability level of 49%, and within 20% at a probability level of 32%. The error analysis gives an uncertainty of the two-phase multipliers for the local pressure drops in the range 6-20%.

The minimum mean square error method has also been used to evaluate C_{lj}, which has resulted to be 9.53; considering this value, the discrepancy between experiment and prediction is very similar to that one shown in figure 7.

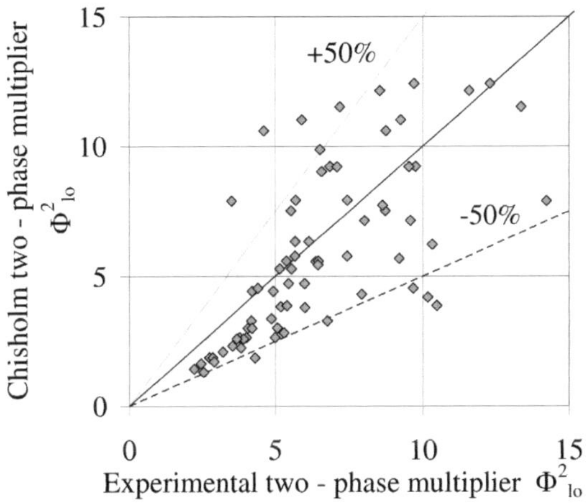

Figure 7. Comparison between the experimental two-phase multiplier and the Chisholm prediction with $C_j = 0.432$

7. Conclusions

The present paper further analyzes the results of tests with a two-phase mixture of air and water flowing through a horizontal impacting T- junction, that have been presented in [7].

The experimental flow patterns have been compared with the prediction of the Baker and Taitel–Dukler maps. The agreement to the Baker map is rather poor at low mass velocity, but it is satisfying at higher mass velocities. On the other hand, the Taitel – Dukler maps appropriately predicts flow regimes at every flow rate.

The friction pressure drops have also been investigated and the two-phase friction multipliers have been evaluated: on the average, they are higher than the values predicted by the Lockhart–Martinelli correlation. The Friedel correlation is in better agreement with the experimental results, even though it slightly overestimates the two-phase multiplier and can be applied along the branches up to the junction, with discrepancies that are typical of the two-phase flow correlations.

The pressure drops at the junction centre have been extrapolated to determine the two-phase multipliers of the irreversible local pressure drop in the junction; the experimental values have been compared with the predictions of Chisholm correlation. The optimum value of the coefficient C_i was determined with reference to the present experimental data.

Acknowledgment

The authors wish to thank M. Fogliacco for her valuable help.

Nomenclature

C_f	Fanning friction factor	T	$\left[\left\|(dp/dz)_l\right\|/g(\rho_l-\rho_g)\right]^{1/2}$	σ	surface tension (N/m)
d	inner diameter (m)	x	mass flow quality	$\phi^2_{l,g}$	$\left[(dp/dz)_{TP}/(dp/dz)_{l,g}\right]$
ER	extraction ratio (G_j/G_I)	z	axial coordinate (m)	ϕ^2_{lo}	$\left[(dp/dz)_{TP}/(dp/dz)_{lo}\right]$
Fr	$G_g/\left[\rho_g(\rho_l-\rho_g)d\,g\right]^{1/2}$	α	void fraction	χ	$\left[(dp/dz)_l/(dp/dz)g\right]^{0.5}$
g	gravity constant (m/s^2)	λ	$\left[(\rho_g/\rho_a)\cdot(\rho_l/\rho_w)\right]^{0.5}$	χ_{tt}	$1/[x_I-(1-x_I)(\rho_l/\rho_g)0.5]$
G	mass velocity (kg/m^2 s)	μ	dynamic viscosity (kg/m s)	ψ	$\sigma_w/\sigma\left[\mu_l/\mu_w(\rho_w/\rho_l)^2\right]^{1/3}$
p	pressure (Pa)	ρ	density (kg/m^3)		

Subscripts

A	air at ambient condition	l, lo	liquid, liquid only
g, go	gas, gas only	L	left
H	homogenous	R	right
j	right, left	TP	two - phase
I	inlet	w	water at ambient conditions

References

[1] Azzopardi B J and Whalley P B 1982 The effect of flow patterns on two-phase flow in a T junction *Int. J. Multiphase Flow* **8** 491

[2] Oranje L 1983 Handling two-phase condensate flow in offshore pipeline systems *Oil and Gas J.* **81** 128

[3] Buell J R, Soliman H Mand Sims G E 1994 Two-phase pressure drop and phase separation at a horizontal tee junction *Int. J. Multiphase Flow* **20** 819

[4] El-Shaboury A M F, Soliman H M and Sims G E 2007 Two-Phase flow in a horizontal equal-sided impacting tee junction *Int. J. Multiphase Flow* **33** 411

[5] Mohamed M A, Soliman H M and Sims G E 2011 Experimental investigation of two-phase flow splitting in an equal-sided impacting tee junction with inclined outlets *Exp.Thermal and Fluid Science* **35** 1193

[6] Elazhary A M and Soliman H M 2012 Two-phase flow in a horizontal mini-size impacting T-junction with a rectangular cross-section *Int. J. Multiphase Flow* **42** 104

[7] Bertani C, Malandrone M and Panella B 2012 Experimental study on two-phase flow in a horizontal impacting T-Junction, *Proc.30th UIT Nat. Heat Transfer Conf. Bologna* 25-27 June 2012 pp 157- 162

[8] Collier J G and Thome J R 1994 *Convective boiling and condensation* (Oxford: Clarendon Press)

[9] Whalley P B 1987 *Boiling, condensation and gas–liquid flow*, (Oxford: Clarendon Press)

[10] Friedel L 1979 Improved friction pressure drop correlation for horizontal and vertical two-phase flow *European Two-Phase Flow Group Meeting* Ispra, Italy

[11] Saba N, Lahey R T 1984 The analysis of phase separation phenomena in branching conduits *Int. J. Multiphase Flow* **10** 1

Journal of Physics: Conference Series **501** (2014) 012014 doi:10.1088/1742-6596/501/1/012014

Single and two-phase flow fluid dynamics in parallel helical coils

M De Salve, M Orio and B Panella[1]
Dipartimento Energia Politecnico di Torino
Corso Duca degli Abruzzi, 24, 10129 Torino, Italy

E-mail: bruno.panella@polito.it

Abstract. The design of helical coiled steam generators requires the knowledge of the single and two-phase fluid dynamics. The present work reports the results of an experimental campaign on single-phase and two phase pressure drops and void fraction in three parallel helicoidal pipes, in which the total water flow rate is splitted by means of a branch. With this test configuration the distribution of the water flow rate in the helicoidal pipes and the phenomena of the instability of the two-phase flow have been experimentally investigated.

1. Introduction

The helicoidal pipes have been used extensively in chemical engineering and nuclear power, like heat exchangers and steam generators. They are adopted in the nuclear field for their compactness and the heat transfer high efficiency; in order to perform a good design it is important the knowledge of pressure drops and heat transfer, both in single and in two phase flow conditions. A review on flow and heat transfer on curved tubes is due to Naphon and Wongwises [1]; in the framework of integral Small-medium Modular Reactors (SMRs) steam generators, Santini et al [2,3], Papini et al. [4], Bertani et al. [5] have studied several aspects of pressure drops and two-phase flow in helical pipes. Some preliminary results have been presented at XXX UIT Heat Transfer Conference [6]. The approach by means of the CFD codes has been used by Jayakumar et al. [7] to compare the CFD results with experimental friction factors.

The present work reports the results of an experimental campaign on single-phase and two phase pressure drops and void fraction in three parallel helical pipes, in which the total water flow rate is splitted by means of an inlet branch; with this test configuration the main objective is to experimentally investigate the distribution of the water flow rates in helical pipes and the phenomenon of the instability of the two-phase flow.

2. Experimental facility

The experimental facility at the Energy Department of the Polytechnic of Turin, consists of three helicoidal test sections, the water and air lines, an inlet branch and an outlet branch, the instrumentation for measuring the flow rates and the differential and absolute pressures, and the acquisition data system. The figure 1 shows the experimental facility schematic and a picture of it.

[1] To whom any correspondence should be addressed.

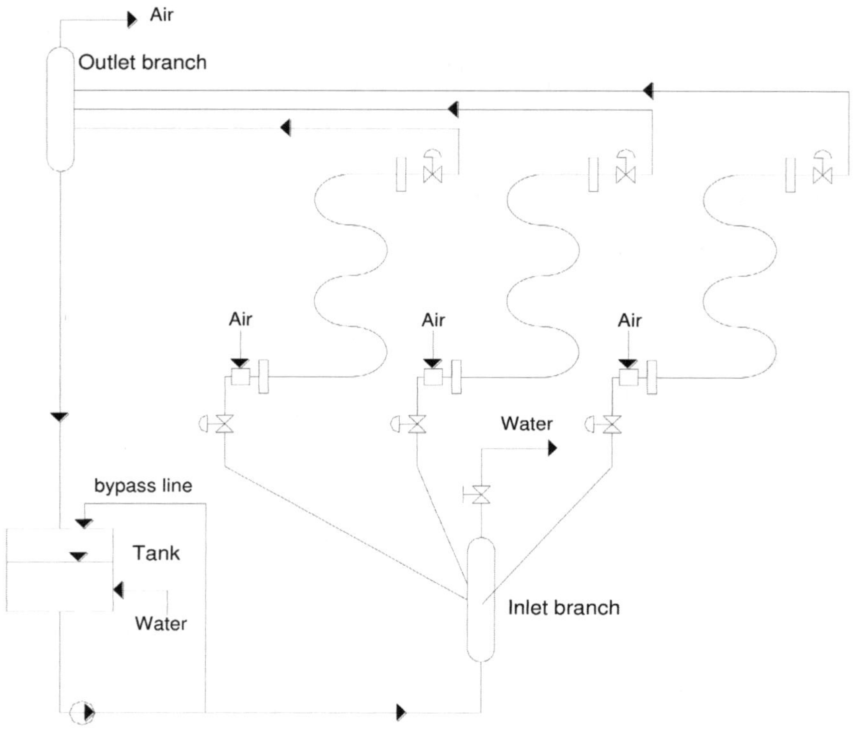

(a)

(b)

Figure 1. Experimental facility: schematic **(a)** and picture **(b)**

The three test sections have the helix diameter of 0.64 m, 1 m and 1.39 m and pitches of 0.485 m, 0.79 m 0.954 m respectively; they are made with tubes of polymethylmethacrylate whose inner diameter is equal to 12 mm; the geometry characteristics of the helical coils are shown in table 1.

Table 1. Geometry parameters

Test section pipe number	1	2	3
Helix diameter D [m]	0.64	1	1.39
Helix pitch p [m]	0.485	0.79	0.954
Inner diameter tube d [mm]	12	12	12
Helix length L [m]	10.85	10	13.9
Turns	5	3	3
Branch length [m]	1.2	1.2	1.2
Outlet straight pipe length [m]	2.82	0.71	5.52

A centrifugal pump sucks water from a tank, where the water flow in turn returns by means of an outlet branch. A bypass system that is placed after the pump provides for the regulation of the water flow rate. Downstream of the pump there is the inlet branch, from which the branches of pipes which feed the three test sections depart.

The feed branches of the three helical coils are placed symmetrically at 120 ° from each other and are located at half the height of the inlet branch. In each horizontal branch that is connected to each test section there is a flow meter of water. Upstream and downstream of each test section, there are two pneumatic quick closing valves for the rapid and simultaneous interception of the mixture inside the sections for measuring the volumetric void fraction. To get the two-phase flow conditions air is injected directly into the pipe. At the end of the three helicoidal test sections and downstream of the pneumatic valves, the mixture of water and air flows to an outlet manifold where the phases are separated.

The liquid flow rate ranging up to 800 kg /h is measured by means of three orifices with a 1% f.s.v accuracy value. The air flow rate ranging up to 500 Nl/h is measured by means of a rotameter with a 5% f.s.v. accuracy value. The Statham type pressure drops transducers accuracy is 0.1 % f.s.v.. The void fraction is measured by means of the quick closing valves technique with a 5 % accuracy.

The single phase tests were carried out by using only water and the total flow rate, sum of the three water flow rates in each test section, ranges between 70 g/s and 600 g/s. The pressure drops across each helicoidal test section are presented in figure 2.

The experimental results (figure 2) show the effects of the helix lengths. The effect of the pitch and the coils diameter on pressure drops is small. The Fanning friction factor f can be derived from the measured pressure drops by the equation:

$$f = \frac{\Delta p \cdot d}{2 \cdot \rho \cdot u^2 \cdot L} \tag{1}$$

The friction factor is shown in figure 3 versus the Reynolds number; it is also compared with the Ito correlation prediction [6], that is valid for curved tubes, and with the formulations for straight pipes for laminar and turbulent flow: the experimental results are well approximated by the correlation for straight pipes in turbulent flow.

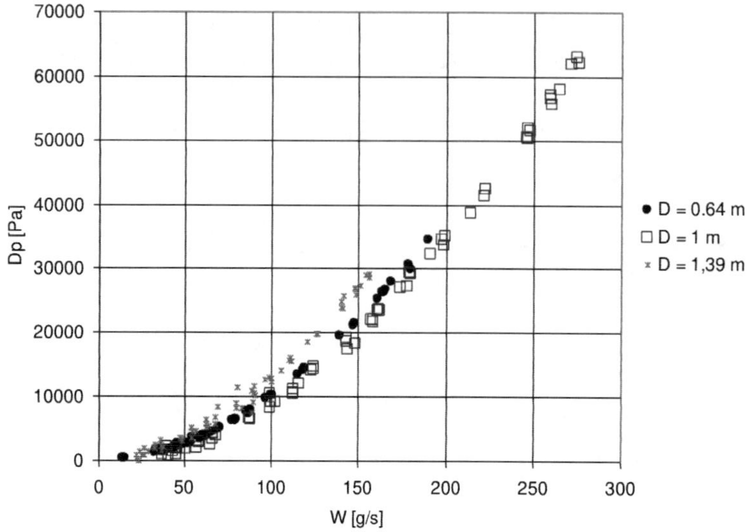

Figure 2. Measured pressure drop in single phase flow

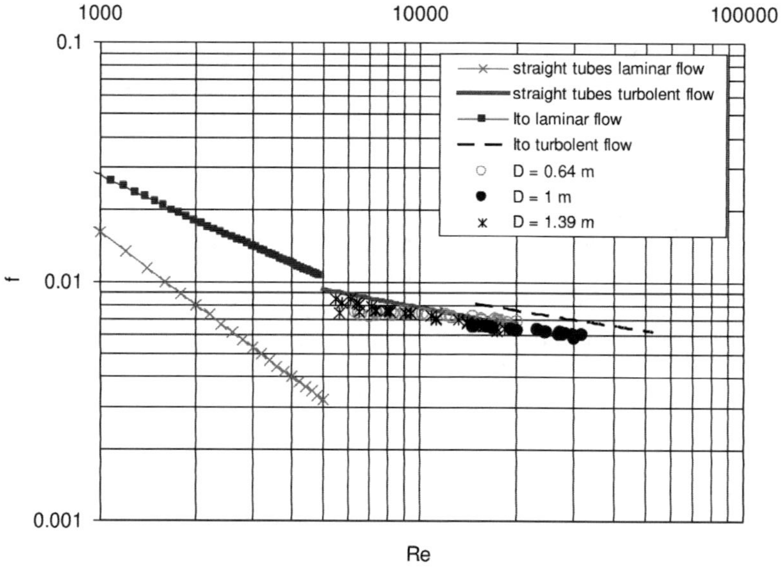

Figure 3. Friction factor versus Reynolds number

In the transition region the friction factor value is lower than that one estimated by the straight tubes turbulent flow correlations.

3. Experimental results for two phase flow

Two different experimental procedures have been adopted for two-phase flow conditions:

- two-phase flow in one section a time;
- two-phase flow in the three test sections.

In both cases the inlet water flow rate is measured for the three helical coils, while the air flow rate is measured in the two-phase test sections; the pressure drops across the test sections, as well as the volumetric void fraction and the average pressure are measured too.

The first set of runs was carried out by imposing the two-phase flow in a single helicoidal pipe and the single phase flow in the remaining two test sections; the water flow rate to the branch ranges from 290 g/s to 580 g/s, and for each water flow rate the air flow rate range is from 0.25 g/s to 0.85 g/s (table 2).

The second set of runs was carried out by imposing the two-phase flow in all three tubes simultaneously.

Table 2. Experimental matrix

	Water flow rate [g/s]	Air flow rate [g/s]
Two phase flow one section a time(test A)	290 – 530 – 560 – 580	0 – 0.25 – 0.58 – 0.85
Three test sections in two-phase flow (test B)	290 – 530 – 560 – 580	0 – 0.25 – 0.58 – 0.85
Three test sections in two-phase flow (test C)	112 – 240	1.45 – 1.65 – 1.9

The water flow rate to the branch and the air flow rate within the test sections, for the test B1, are the same as the first set of measurements, but two lower water flow rates and three higher air flow rates (test B2), have been tested too (see table 2). The total water flow rate is divided into the three test sections in a different way due to the different lengths of the three helical coils and of the outlet pipes. The figure 4 shows an example of the redistribution of the total flow rate (W_{tot} = 290 g/s) for different air flow rates in the three test sections for the test A.

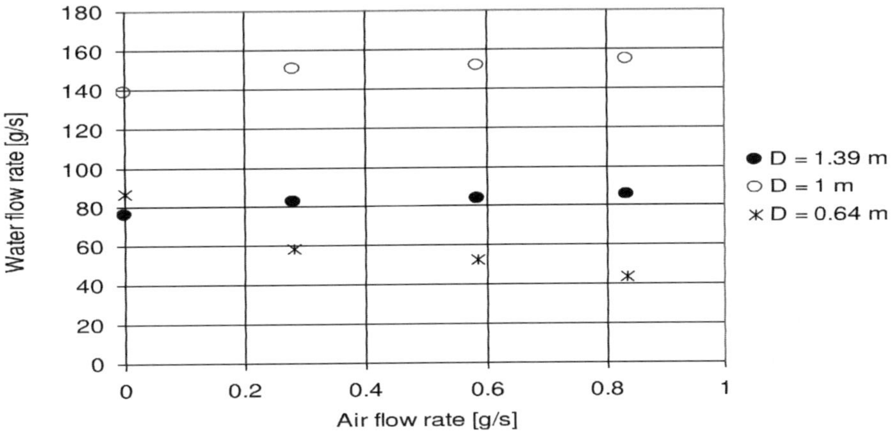

Figure 4. Example of total water flow rate distribution (Test A)

For the case of injection of the air flow in a single coil, at the same pump power, a reduction of total liquid flow rate occurs but the liquid flow rate in the small and medium coils increases while the liquid flow rate in the large coil decreases when air is injected in the coil number 1 that is characterised by the lowest helix diameter: the flow rate distribution in the coils is affected by the outlet straight pipe length that is lower for the coil number 2 and higher for the coil number 3 (see table 1).

The void fraction is measured by means of the quick closing valves technique and is presented in figure 5 as a function of the flow quality, for the test A; the flow quality is lower than 0.02.

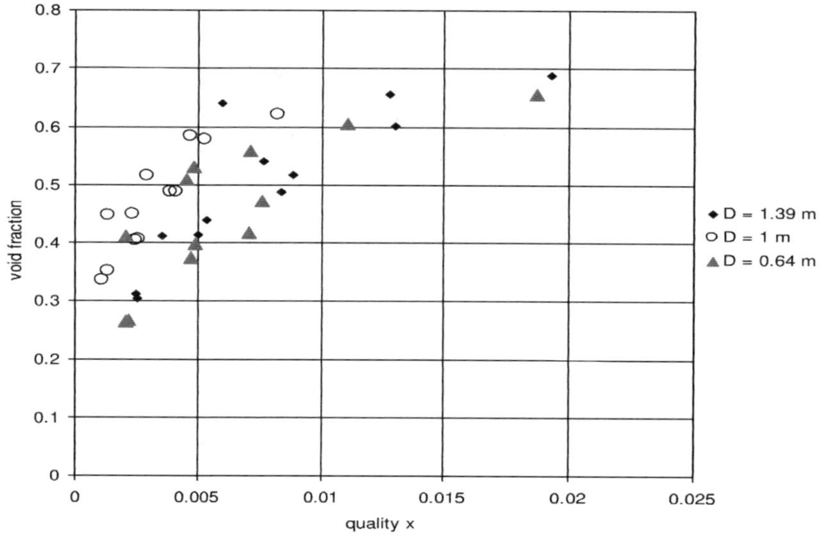

Figure 5. Void fraction versus flow quality (Test A)

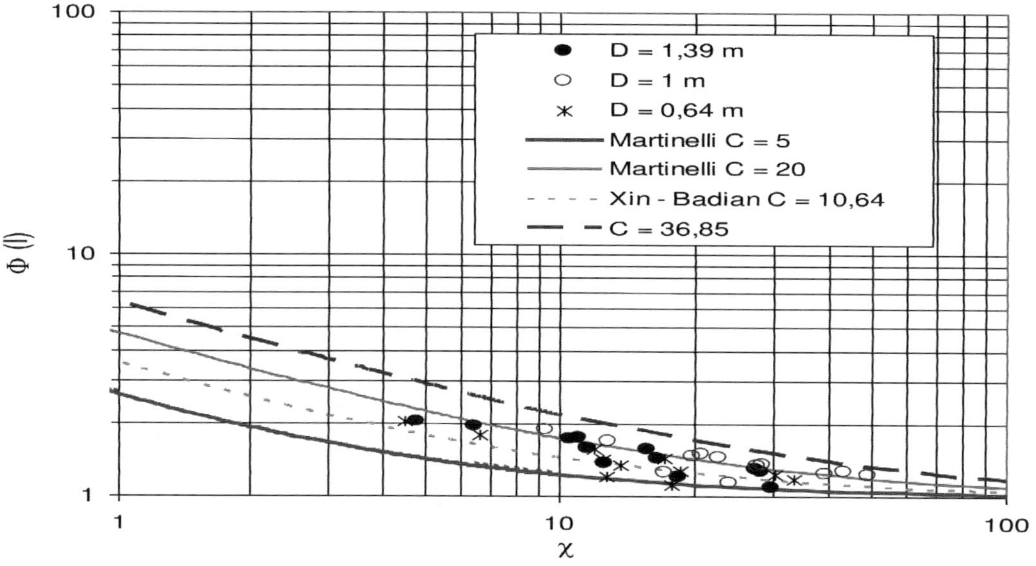

Figure 6. Experimental liquid two-phase multiplier and Lockhart-Martinelli prediction

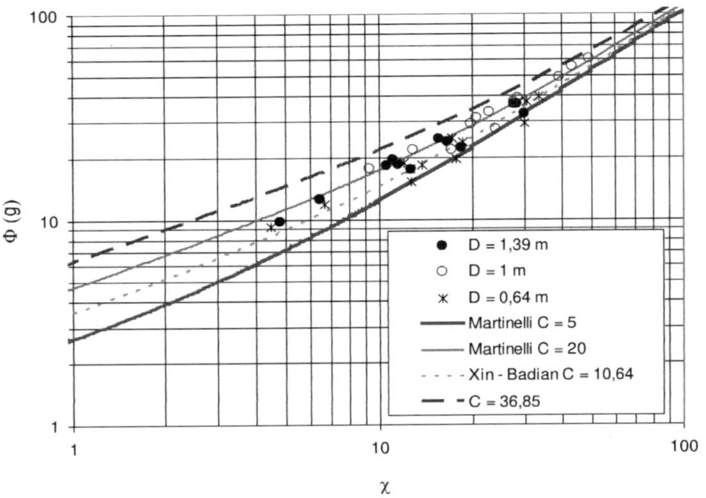

Figure 7. Experimental gas two-phase multiplier and Lockhart-Martinelli prediction

The experimental two-phase multiplier and void fraction have been compared with the Lockhart-Martinelli prediction only for the test A (figures 6 and 7).
The Martinelli parameter χ^2 and friction multipliers for two phase flow Φ_l and Φ_g, for the liquid and gas phases, are defined in eqs. (2), (3) and (4):

$$\chi^2 = \frac{\left(\frac{dp}{dz}\right)_{friction\_monophase\_water}}{\left(\frac{dp}{dz}\right)_{friction\_monophase\_air}} \tag{2}$$

$$\Phi_l^{\ 2} = 1 + \frac{C}{\chi} + \frac{1}{\chi^2} \tag{3}$$

$$\Phi_g^{\ 2} = 1 + C\chi + \chi^2 . \tag{4}$$

The parameter C, for straight tubes, according to the model of Lockhart-Martinelli, has a value between 5 and 20, depending on the laminar or turbulent flow for the liquid and gas phases. Xin et al. [9,10] have found a value of the coefficient C of 10.64, by applying the model of Lockhart-Martinelli to helicoidal pipes with an annular flow pattern. In the test campaigns, that was performed previously with the present test sections [5], the value of 36.85 was estimated for the parameter C. However for the tests that have been performed in the present experimental campaign a value of C that is adequate to estimate the friction two-phase multipliers was not found for each test condition, but the a value between 20 and 10 for C seems in agreement with the test data (figures 6 and 7). The experimental void fraction has been compared with the values predicted by the model of Lockhart: the prediction underestimates the measured values.

The flow rate oscillations in the helicoidal test sections have been analyzed too. Figures 8, 9 and 10 show some examples of the flow rate oscillations in the three helicoidal test sections for the test A.

The analysis of the results for total water flow rates lower than 250 g/s and for an air flow rate higher than 1.4 g/s shows that the oscillations were so large that was not straightforward to estimate an average water flow rate in the test sections.

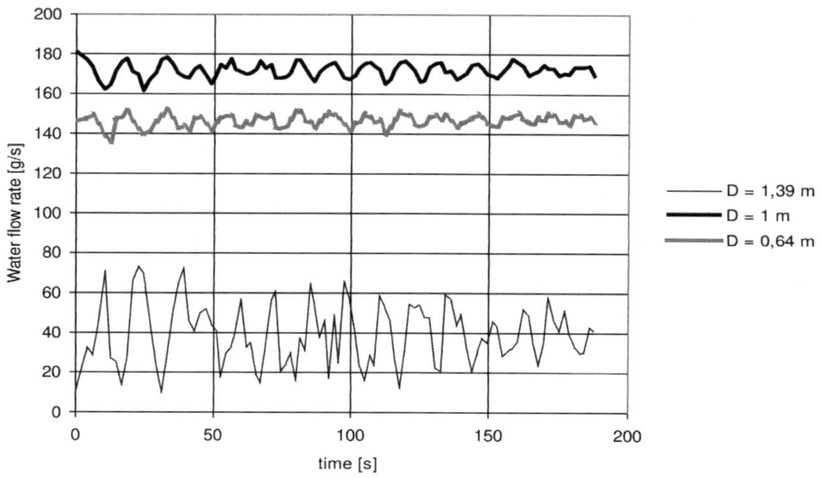

Figure 8. Flow rate oscillations in the helicoidal test sections for test A.

W_{tot} = 290 g/s, W_{air} in the three helical coils = 300 l/h

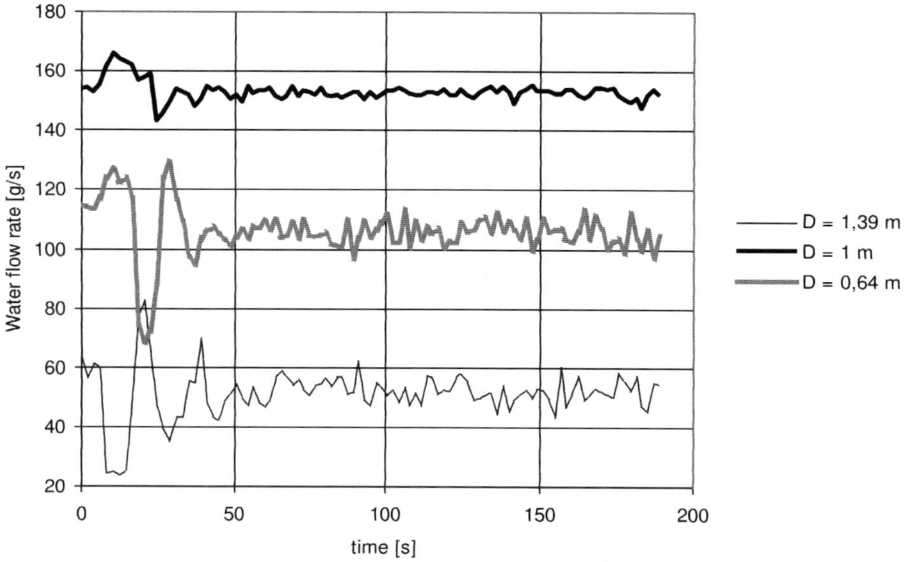

Figure 9. Flow rate oscillations in the helicoidal test sections for test A.

W_{tot} = 290 g/s, W_{air} in the three helical coils = 200 l/h

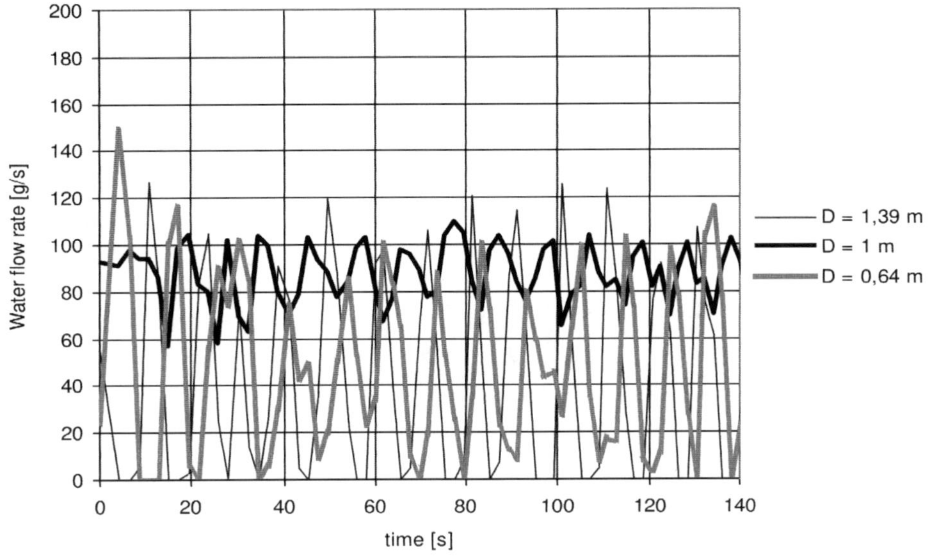

Figure 10. Flow rate oscillations in the helicoidal test sections for test B2.
W_{tot} = 240 g/s, W_{air} in the three helical coils = 500 l/h

4. Conclusions

The experimental results of single phase and two phase pressure drops and void fraction for three parallel helicoidal pipes have been presented.

In single phase flow the three helical coils have about the same pressure drop and it is possible to approximate the friction factor, derived from the experimental results, with the friction factor that is predicted by the correlation for straight pipes in turbulent flow.

In two-phase flow two different experimental procedures have been adopted: two-phase flow in one test section a time or two-phase flow in all the test sections. The model of Lockhart-Martinelli is not adequate to estimate the friction two-phase multipliers for each test condition and also the prediction of the experimental void fraction by means of the model of Lockhart is rather poor.

The flow rate oscillations in the helicoidal test sections are higher for total water flow rates lower than 250 g/s and for air flow rates higher than 1.4 g/s, and for these cases it is not straightforward to estimate an average water flow rate in the test sections, due to oscillatory behaviour.

In the next future CFD codes will be used to evaluate the main flow parameters and the prediction will be compared with the test data.

Acknowledgements

The present research has been supported by ENEA and by the Ministry of Economic Development.
The authors thank R. Costantino and G. Vannelli for their support during the experiments.

Nomenclature

A	tube cross section	m^2
d	inner diameter tube	mm
D	coil diameter	m
f	friction factor	
L	coil length	m
p	pressure	Pa
p_i	helical pitch	m
u	velocity	m/s
Re	Reynolds number	
W_i	channel flow rate	kg/s
W_{tot}	total flow rate	kg/s
x	flow quality	
α	void fraction	
Φ	two-phase multiplier	
χ^2	Martinelli parameter	

References

[1] Xin R C, Awwad A, Dong Z F and Ebadian M A 1996 An investigation and comparative study of the pressure drop in air-water two-phase flow in vertical helicoidal pipes *Int. J. Heat Mass Transfer* **39** 735

[2] Santini L, Cioncolini A, Lombardi C and Ricotti M E 2008 Two-phase pressure drops in a helically coiled steam generator *Int. J. Heat Mass Transfer* **51** 4926

[3] Santini L and Cioncolini A 2005 An experimental investigation regarding the laminar to turbulent flow transition in helically coiled pipes *Exp. Thermal and Fluid Science* **39** 735

[4] Papini D and Ricotti M E 2010 Prove sperimentali per canali in parallelo e relative zone di instabilità *Cerse-Polimi* 1134

[5] Bertani C, De Salve M, Malandrone M, Orio M and Panella B 2009 Fluidodinamica bifase in condotti elicoidali: influenza della geometria e delle portate *Proc. XXVII Congr. Nazionale UIT sulla Trasmissione del Calore* Reggio Emilia 22-24 Giugno 2009 pp 103- 112

[6] De Salve M, Giraudo M G and Panella B 2012 Experimental study of air-water two-phase flow in parallel helically coiled pipes, *Proc. 30th UIT Heat Transfer Conf* Bologna 25-27 June 2012 Publ. Esculapio pp 171-17.

[7] Jayakumar J S, Mahajani S M, Mandal J C, Iyer K N and Vijayan P K 2010 Thermal hydraulic characteristics of air-water two-phase flows in helical pipes *Chem. Eng. Res. and Des.* **88** 51

[8] Ito H 195 Friction factors for turbulent flow in curved pipes *Trans. Amer. Soc. Mech. Engrs.* **D81** 123

[9] Xin R C, Awwad A, Dong Z F and Ebadian A 1997 An experimental study of single phase and two-phase flow pressure drop in annular helicoidal pipes *Int. J. Heat and Fluid Flow* **18**, 5, 482

[10] Awwad A, Xin RC, Dong Z F, Ebadian M A and Soliman H M 1995 Measurement and correlation of pressure drop in air-water two-phase flow in horizontal helicoidal pipes *Int. J. Multiphase Flow*, **21**, 4, 607-619.

Measurement of the oil holdup for a two-phase oil-water flow through a sudden contraction in a horizontal pipe

L P M Colombo, M Guilizzoni, G M Sotgia, S Bortolotti and L Pavan

Politecnico di Milano – Department of Energy, Via Lambruschini 4, 20156 Milano, Italy

E-mail: luigi.colombo@polimi.it

Abstract. Oil-water two-phase flow experiments were conducted in a horizontal duct made of Plexiglas® to determine the holdup of oil by means of the quick closing valves technique, using mineral oil (viscosity: 0.838 Pa s at 20 °C; density: 890 kg m^{-3}) and tap water. The duct presents a sudden contraction, with contraction ratio of 0.64. About 200 tests were performed by varying the flow rates of the phases. Flow patterns were investigated for both the up- and downstream pipe. Due to the relatively high value of the contraction ratio, it was not observed any relevant variation of the flow patterns across the sudden contraction. Data were then compared with predictions of a specific correlation for oil-water flow and some correlations for gas-water flow. A drift-flux model was also applied to determine the distribution parameter. The results agree quite well with flow pattern visualization.

1. Introduction

This work is framed in the research activity on multiphase flows by now well-established at the Department of Energy, Politecnico di Milano [1-3]. The main subject concerns the measurement of the oil holdup by means of the quick closing valves technique for a two-phase oil-water flow downstream a sudden contraction in a horizontal pipe.

This matter has been deeply investigated in the literature especially for gas-liquid flows and a very comprehensive review can be found in [4]. Nevertheless, for liquid-liquid flows the information, mainly related to the pressure drop, is still lacking: concerning oil-water flows across a sudden contraction, the authors were able to find in the literature only two contributions, namely the work by Wang and Pal [5], dealing with emulsions, and the more recent one by Balakhrisna et al. [6], which considers both the phase distribution and the pressure drop for small diameter tubes (0.0254 m and 0.012 m). On the other hand, the applications of such kind of two-phase flow is very relevant in many technical fields. In particular, for petroleum engineering applications, the knowledge of the effect of singularities both in terms of flow pattern and pressure drop variation is important in the pipeline design.

The holdup, i.e. the "in situ" volume fraction of a phase, plays a considerable importance to understand the flow distribution and is used in mechanistic models to predict both the flow pattern and the pressure drop. There are three major techniques to measure the holdup: the shut-in method, usually involving quick closing valves [7], suitable for steady-state measurements on non-intermittent streams; the probe method, based on resistance or capacitance sensors [8] able to detect instantaneous fractions and to provide even local information, according to the features of the probe itself; the nuclear method (γ-ray or X-ray densitometer) [9]. Nevertheless, for oil-water flows difficulties arise with impedance probes because the difference in the permittivity of the two phases is not marked and, furthermore, if the probe is directly wetted by the mixture, the oil tends to adhere permanently, preventing a correct detection. On the other hand, nuclear techniques are very expensive and require special care for a safe operation. For these reasons, the shut-in method has been applied in this work due to its relative simplicity and reliability compared to the other techniques.

2. Experimental setup

The liquid-liquid flow facility available in the Multiphase Thermo-Fluid Dynamics Laboratory at the Department of Energy, Politecnico di Milano is sketched in figure 1. More details can be found in [1-3]. Oil (Milpar 220, μ_o=890 kg/m^3, μ_o=0.9 Pa·s at 20 °C) and water (tap water, μ_w=1.026×10^{-3} Pa·s at 20 °C) are pumped separately from their storage tanks. The water flow rate is measured by a magnetic flowmeter (accuracy ±0.5% of the reading), while a calibrated metering pump is used for the oil.

The two liquids pass through a coaxial mixer, where oil flows parallel to the pipe axis while water is injected through an annulus into the oil stream, then the mixture enters the test section. Plexiglas® pipes are used to allow flow visualization.

The test section consists in a 11 m long circular pipe: the sudden contraction is realized joining tubes, respectively 50 mm and 40 mm inner diameter: hence the contraction area ratio is 0.64. Downstream, at a distance of 2.5 m from the contraction, two manually operated valves are inserted enclosing a volume of about 1.4 dm^3. A drainage cock is mounted at the bottom.

Image recordings are taken by video/photo cameras.

After the test section, the mixture flows into a tank where effective separation of the two liquids is obtained due to gravity.

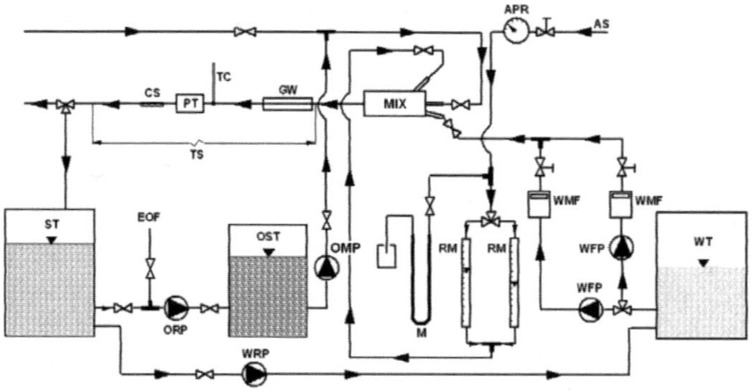

Figure 1. Schematic representation of the oil-water loop. APR air pressure regulation, AS air supplying line, CS capacitance sensor, EOF external oil feeding, GW glass window, M manometer, MIX phase inlet mixer, OMP oil metering pump, ORP oil recovering pump, OST oil supply tank (0.5 m^3), PT pressure transducer, RM rotameter, ST phase collector/separator tank (1.0 m^3), TC thermocouple (K type), TS test section, WFP water feeding pump, WMF water magnetic flow meter, WRP water recovering pump, WT water supply tank (5 m^3).

3. Governing parameters

The *superficial velocity or volumetric flux J* (m/s) is defined as the ratio between the volume flow rate of each single phase and the area of the pipe cross-section. The *mixture superficial velocity or total volumetric flux*, is defined as

$$J_{w-o} = J_w + J_o \tag{1}$$

The *oil input volume fraction* ε_o is defined as

$$\varepsilon_o = \frac{J_o}{J_{w-o}} \tag{2}$$

The *water input volume fraction* is simply given by $\varepsilon_w = 1 - \varepsilon_o$.

The *oil holdup* H_o is defined as

$$H_o = \frac{V_o}{V_{tot}} \tag{3}$$

where V_{tot} and V_o (m^3) represent respectively the volume enclosed within the two valves and the portion occupied by the oil. The *water holdup* is simply given by $H_w = 1 - H_o$.

The *oil slip velocity ratio* is the ratio between the oil velocity U_o and the water velocity U_w

$$s_o = \frac{U_o}{U_w} \tag{4}$$

where for the generic i-th phase $U_i = J_i / H_i$.

The experimental conditions, summarized in table 1, are set by varying the following quantities, referred to the downstream pipe, in their proper ranges.

Table 1. Summary of the experimental conditions.

J_o (ms^{-1})	J_w (ms^{-1})	ε_o	J_o (ms^{-1})	J_w (ms^{-1})	ε_o
0.43	0.34	0.56	0.83	0.38	0.69
	1.11	0.28		0.44	0.65
0.57	0.34	0.63		0.56	0.60
	1.11	0.34		0.67	0.55
	1.33	0.30		0.78	0.52
0.70	0.56	0.56		0.89	0.48
	0.67	0.51		1.00	0.45
	0.78	0.47		1.11	0.43
	0.89	0.44		1.22	0.40
	1.33	0.35		1.33	0.38

The reported conditions correspond to different flow regimes mainly classified as annular and dispersed. Stratified flow patterns with oil in contact to the wall have also been observed, but, being chaotic flow regimes, they have not been considered except for defining the bounds of the present investigation, which is devoted to flows with the water adjoining the wall. Furthermore, a finest distinction can be made for annular flow patterns according to either the eccentricity or the presence of drops at the oil-water interface. Examples covering the widest range of observations are reported in figure 2.

4. Experimental procedure

Tests were run by introducing in the test section the water starting from the maximum value of the superficial velocity $J_{w,max}$. Then, oil was added at the selected superficial velocity J_o. At each run J_w was decreased until its minimum value was reached. The value of J_o was then changed and the sequence was repeated. Concerning the measurement of the oil holdup, once steady-state conditions are achieved, both the valves are closed and simultaneously the pumps are switched off in order to avoid water hammer. After a few minutes the oil and the water trapped within the two valves are separated by gravity. Then, opening the drainage cock, the water is collected in a graduated tank previously calibrated to give the water holdup within 0.3% accuracy. At least ten measurements of the holdup have been repeated for each experimental condition.

	J_o (ms^{-1})	J_w (ms^{-1})	Classification
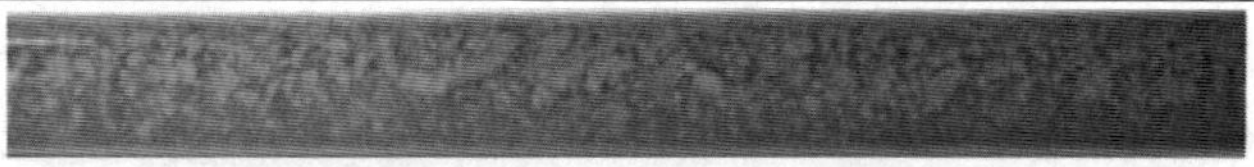	0.70	1.33	Dispersed (D)
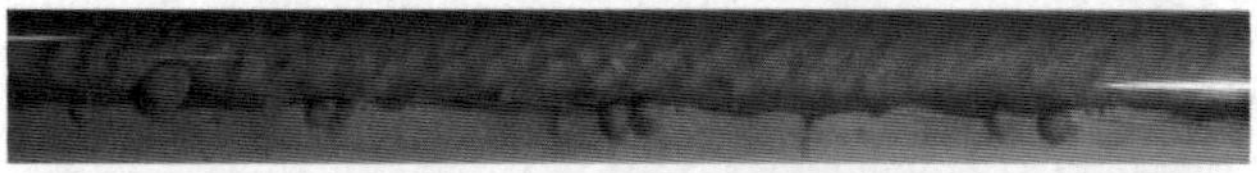	0.83	0.38	Eccentric annular with big drops (EA-D)
	0.83	0.44	Eccentric Annular (EA)
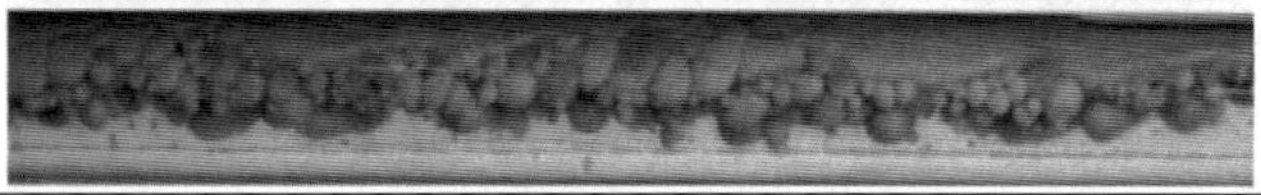	0.43	0.34	Stratified (oil at the wall) with big drops (S-D)

Figure 2. Example of the observed flow patterns.

5. Results and discussion

The effect of the sudden contraction on the flow regime has been investigated by comparing the flow pattern maps downstream and upstream the contraction. A significant selection of the data is shown in figure 3 where it is seen that the flow patterns do not change dramatically. Nevertheless, it is observed a tendency towards an increase of oil dispersion.

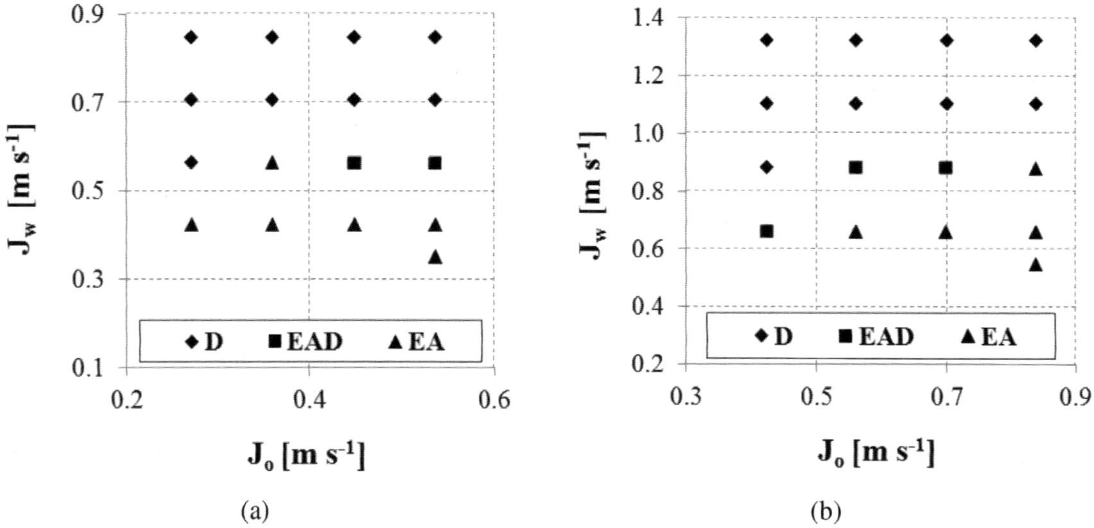

Figure 3. Flow pattern map for the upstream (a) and the downstream pipe (b). EA eccentric annular, EA-D eccentric annular with drops, D dispersed.

Figure 4 (a) shows the oil holdup versus the oil input volume fraction for the whole data set. The superficial velocity varies between 0.34 and 1.33 m s^{-1} for the water and from 0.43 to 0.83 m s^{-1} for the oil. A different marker has been used to identify each flow pattern. It can be seen at first that all the data but one fall below the bisector, i.e. the two phase flow cannot be considered as equivalent to a pseudo-homogeneous one. Besides, the oil holdup is lower than the oil input volume fraction suggesting that the effective average velocity is greater for oil than for water or, equivalently, the slip ratio is larger than unity. This result is not surprising for the conditions corresponding to the annular flow regime, where it is expected that the oil in the central core runs faster than the water in the annulus adjoining the pipe wall. Similar findings have been reported by Arney and Oliemans [4]. Actually, most of the data refer to such flow regime. On the other hand, the same behaviour is also shown by the data related to the dispersed flow regime where a closer similarity with the homogeneous flow might be expected. Nevertheless, it must be noticed that, according to the flow visualisations, the dispersed flows have an inhomogeneous appearance with most of the drops crowding in the upper part of the pipe while the water keeps wetting the pipe wall.

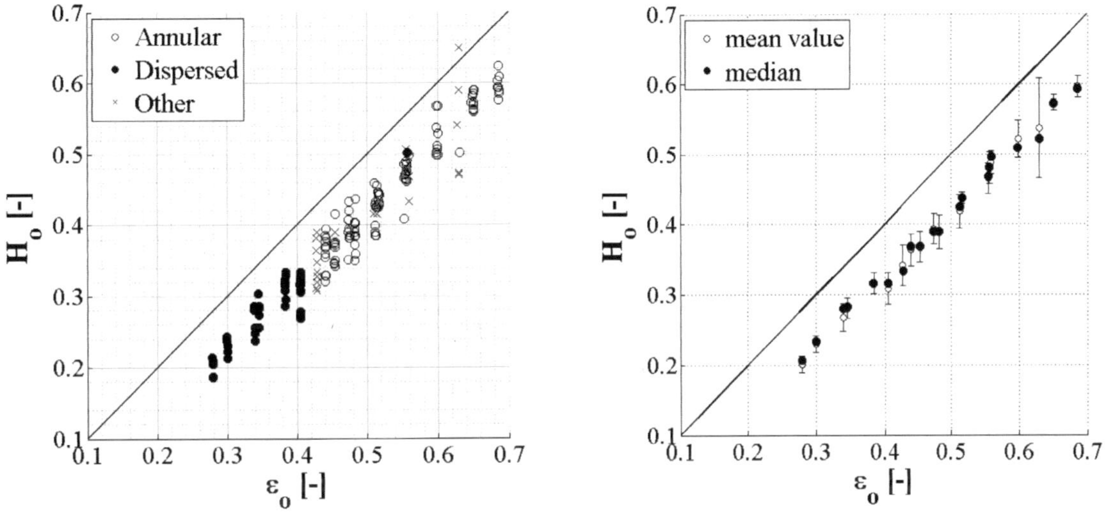

Figure 4. Oil holdup versus oil input volume fraction. Whole data (a) and statistics (b).

Figure 4 (b) shows the statistics for the whole data set. It can be seen that in all the cases but three the mean and the median have the same value. The three conditions for which the mean and the median are different are also characterised by the highest standard deviation. Still relying on visual observations, it can be inferred that the best repeatability has been attained for the conditions corresponding to well-defined flow regimes (either annular or dispersed), whereas the data dispersion increases for strongly eccentric annular flows, for annular flows with drops or near to a transition (annular-stratified, annular-dispersed). A deeper analysis of the measurements can be made according to the approach of Zuber and Findlay [10]. In particular, the experimental data are reported on the oil velocity (U_o) – volumetric flux ($J_{w\text{-}o}$) plane as depicted in figure 5. It is recalled that data falling on the bisector would represent a homogeneous flow with slip velocity ratio equal to unity. In this case, the measurements lie above the bisector, thus indicating a slip velocity ratio greater than unity, i.e. the oil velocity is higher than the water velocity as previously observed. Additional information arises from the linear regression of the data: the slope of the straight line is 1.38, meaning that the oil distribution is not uniform, being the wall always wetted by water. Actually, this is verified for the annular flow regime which is observed in most of the experimental conditions; on the other hand, it can be inferred that the dispersed flow regimes are not characterised by a homogeneous mixing of the two phases, and the pipe wall is mainly adjoined by water, in agreement with the visual observations. This is confirmed also when the data corresponding to dispersed flows alone are considered: in this case, the slope of the straight line is reduced to about 1.1, but is still greater than unity.

Journal of Physics: Conference Series **501** (2014) 012015 doi:10.1088/1742-6596/501/1/012015

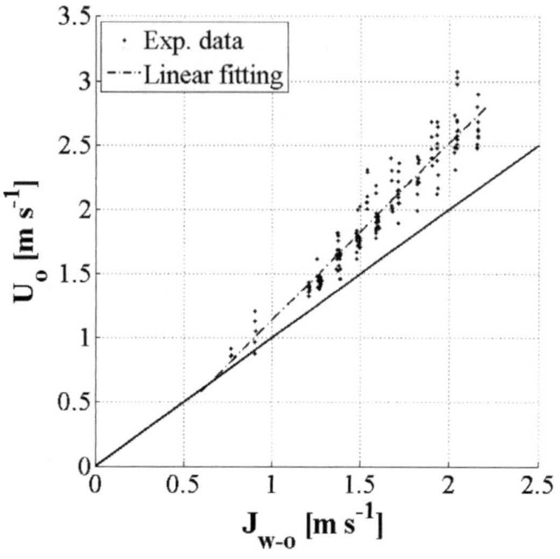

Figure 5. Mean velocity – flux density plane for the oil phase.

The measurements have been compared with the prediction of the empirical correlation developed by Arney et al. [11] who performed quite similar experiments for straight tubes using quick closing valves as in the present work. Nevertheless, it has to be remarked that the considered flow patterns were always annular. In the original formulation, the water hold up can be estimated as:

$$H_w = \varepsilon_w \left[1 + 0.35 \left(1 - \varepsilon_w \right) \right] \tag{4}$$

Hence, the oil holdup is computed as $H_o = 1 - H_w$. The result is reported in figure 6 where it is compared with the experimental data, showing a very good agreement; in particular, considering the maximum relative error between the prediction and the average measurements is 5.15%, whereas from the parity plot shown in figure 7 it is evident that most of the data fall within ±15% relative error. On the other hand, the deviation itself seems to be independent of the flow pattern, even though the correlation has been formulated with data from annular flows only.

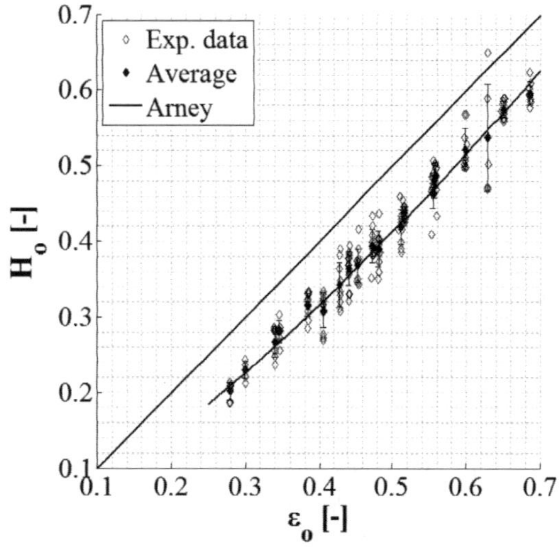

Figure 6. Comparison between the data and the prediction by Arney et al. [8]

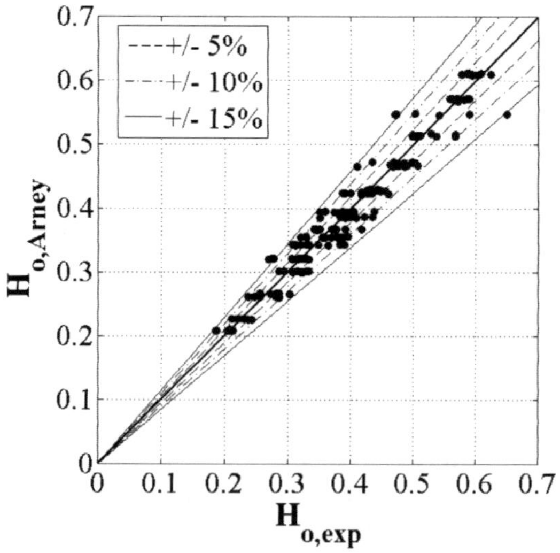

Figure 7. Parity plot for the correlation of Arney et al. [8]

Journal of Physics: Conference Series **501** (2014) 012015 doi:10.1088/1742-6596/501/1/012015

As observed in the Introduction, a large number of experimental correlations to predict the holdup in gas-liquid flows are available in the literature. Taking into account the wide survey presented in [1] a selection of these models has been considered to check the possibility to extend their application to oil-water flows. Considering that the experimental data refer to annular or dispersed flow regimes, which are also met in gas-liquid flows, with the oil phase playing the role of the gas phase, the rheological properties of the gas were simply replaced in the correlations with the ones of the oil. Five models, listed in table 2, returned a very good agreement with the data, showing average relative errors lower than 10 %.

Table 2. Relative errors for some gas-liquid prediction models.

Correlation	Average error (%)	Maximum Error (%)
Armand [12]	5.73	25.26
Rouhani [13]	6.52	29.18
Chisholm [14]	7.30	33.02
Dix [15]	9.91	37.44

6. Conclusions

The shut-in method was successfully applied to the measurement of the oil holdup downstream a sudden contraction with 0.64 area ratio. The operating conditions was set such that the majority of the flow regimes was eccentric annular since this flow pattern is the most convenient for pumping. On the other hand, dispersed flow regimes were observed at the highest values of the input water superficial velocity. Regardless of the flow pattern, the results can be predicted quite well by the correlation of Arney *et al.*, originally developed for annular oil-water flow in horizontal straight pipes. On the other hand, some correlations originally developed for gas-liquid flows are able to provide predictions within 10 % relative error. The flow visualisations show that the contraction does not modify significantly the flow pattern, even though it is observed a tendency towards an increase of oil dispersion. Thus, further investigation on contractions with more severe area ratios can be useful to clarify the influence on the flow pattern.

References

[1] Sotgia G M, Tartarini P and Stalio E 2008 *Int. J. Multiphase Flow* **34** 1161
[2] Poesio P, Strazza D, Sotgia G M 2009 *Chemical Engineering Science* **64**(6) 1136
[3] Colombo L P M, Guilizzoni M, Sotgia G M 2012 *Experiments in Fluids* **53**(5) 1617
[4] Woldesemayat M A and Ghajar A J 2007 *Int. J. Multiphase Flow* **33**(4) 347
[5] Hwang C J and Pal R 1997 *Chem. Eng. J.*, **68**(2-3) 157
[6] Balakhrisna T, Ghosh S, Das G and Das P K 2010 *Int. J. Multiphase Flow* **36**(1) 13
[7] Oddie G, Shi H, Durlofsky L J, Aziz K, Pfeffer B and Holmes J A 2003 *Int. J. Multiphase Flow* **29** 527
[8] Du M, Jin N D, Gao Z K, Wang Z Y and Zhai L S 2012 *Int. J. Multiphase Flow* **41** 91
[9] Rodriguez O M H and Oliemans R V A 2006 *Int. J. Multiphase Flow* **32** 323
[10] Zuber N and Findlay J A 1965 *J. Heat Transfer* 453
[11] Arney M S, Bai R, Guevara E, Joseph D D and Liu K 1993 *Int. J. Multiphase Flow* **19**(6) 1061
[12] Armand A A 1946 Izv Vse Tepl Inst. **1** 16
[13] Rouhani S Z and Axelsson E 1970 *Int. J. Heat Mass Transfer* **13** 383
[14] Chisholm D 1973 *Int. J. Heat Mass Transfer* **16** 347
[15] Coddington P Macian R 2002 *Nucl. Eng. Design* **215** 199

Prediction of two-phase choked-flow through safety valves

G Arnulfo, C Bertani and M De Salve
Dipartimento Energia, Politecnico di Torino, corso Duca degli Abruzzi 24, 10129 Torino, Italy

E-mail: cristina.bertani@polito.it

Abstract. Different models of two-phase choked flow through safety valves are applied in order to evaluate their capabilities of prediction in different thermal-hydraulic conditions. Experimental data available in the literature for two-phase fluid and subcooled liquid upstream the safety valve have been compared with the models predictions. Both flashing flows and non-flashing flows of liquid and incondensable gases have been considered. The present paper shows that for flashing flows good predictions are obtained by using the two-phase valve discharge coefficient defined by Lenzing and multiplying it by the critical flow rate in an ideal nozzle evaluated by either Omega Method or the Homogeneous Non-equilibrium Direct Integration. In case of non-flashing flows of water and air, Leung/Darby formulation of the two-phase valve discharge coefficient together with the Omega Method is more suitable to the prediction of flow rate.

1. Introduction

Safety relief valves (SRV) play a fundamental role in many industries, including power plants (both conventional and nuclear), aerospace, automotive, printing industries and multi-purpose plants for the production of fine chemical and pharmaceutical products, where recipes are frequently changed and various processes are carried out; they have to ensure that the operating pressure does not exceed unsafe limits.

The depressurisation rate of a pressure vessel or a plant system is limited by choked flow rates through the safety valve and the prediction of flow rates through it is a crucial issue in the design of both pressure vessels and downstream collecting and handling systems.

It might occur that, due to off-normal or accidental conditions or to a change of recipe in a multi-purpose plant, safety relief valves initially sized for saturated steam or for highly subcooled liquid may be required to discharge a mixture of liquid and vapour or subcooled liquid that flashes along the valve flow path.

Sizing methods and flow models for single-phase flow through safety valves are well established; however, this is not the case for two-phase flow, as complex phenomena occur between the two phases.

In fact, the two phases have different velocities and also their temperatures could differ; therefore momentum and energy exchanges are not merely due to mass transfer. Thermodynamic non-equilibrium phenomena could also occur and the presence of metastable liquid causes vaporisation delays. The sound velocity in two phase flow strongly depends on the fraction of steam and is strongly reduced as soon as vapour appears.

The knowledge about two-phase phenomena in safety valves is still insufficient and the development of a theoretical model able to accurately predict two-phase flow rate through safety valves is not an easy task even in steady state conditions.

In single-phase choked flow through safety valves (SRV), the flow rate is a function of upstream thermodynamic conditions (pressure, temperature), minimum valve section, and a discharge coefficient, which is defined as the ratio between the actual flow rate and the flow rate through an ideal nozzle operated at the same upstream conditions. During the choked flow the fluid velocity in the minimum section is equal to the sound speed in the fluid. The discharge coefficients for single-phase liquid flow (K_{dL}) and single-phase gas flow (K_{dG}) are determined experimentally and normally are provided by the valve manufacturers.

Most authors [1-5] suggest that two-phase choked flow through safety valves can be evaluated by simplified two-phase flow models whose approach is similar to the one in single-phase: they suggest to multiply the two-phase choked flow rate through an ideal convergent-divergent nozzle and a two-phase discharge coefficient. Even though normally manufacturers provide the discharge coefficient for the liquid and sometimes for gas flow, no such data are provided in two-phase flow.

Several models for the two-phase discharge coefficient have been developed; they evaluate the two-phase valve discharge coefficient as function of K_{dG}, K_{dL}, properties of the two phases and quality or void fraction of the two-phase mixture; each of them applies to a certain ideal nozzle model.

Both theoretical models for safety valves and experimental results have been analysed and some of them have been considered in the present work in order to compare models predictions with experimental results available in the literature and to highlight discrepancies. Experimental/modelling studies of flashing water through small valves have been carried out Boccardi et al. [6-9].

The authors compared the predictions of two-phase flow rate through ideal nozzle by the Homogeneous Equilibrium Model (HEM), the ω method (first developed by Leung [10]), Slip Equilibrium Models (SEM), Homogeneous Non Equilibrium models like the one developed by Henry and Fauske in 1970 (HNE) [11] and afterwards improved by Diener and Schimdt (HNE-DS) [1], Equilibrium Rate Model (ERM) [2], Homogeneous Frozen Models (HFM) [3] or by Homogeneous Direct Integration (HDI) and Homogeneous Non-equilibrium Direct Integration (HNDI) methods [4].

Afterwards the authors considered the models of two phase valve discharge coefficient developed by Lenzing [3], Darby [4], and Leung [5] and combined them with the different models of two-phase choked flow in ideal nozzles, in order to evaluate which couple of the two gives the best prediction of experimental data in different thermal-hydraulic conditions.

2. Two-phase flow through ideal nozzles.

Here the models used for the evaluation of the flow rate trough an ideal nozzle are mentioned.

The ω-method is a form of Homogeneous Equilibrium Model that can be applied to choking and non-choking flows, flashing and non-flashing systems and its applicability has been extended to initial subcooled conditions. Firstly developed by Leung [10], it has been revised and modified by other authors and is recommended by known standards, such as API 520 [12].

Different formulation have been proposed for flashing flow of initially saturated and subcooled fluid and for non-flashing flow.

The HNE model, developed by Henry and Fauske in 1970, considers an incompressible liquid phase and describes the vapour phase by a polytropic law with the Tangren's exponent; the model also considers flashing delays when the exit quality is less than 0.05.

The model, reported in [3], can be applied to saturated two-phase mixture at the nozzle inlet.

The HFM model supposes that the quality does not change and the vapour is considered an ideal gas undergoing an isoentropic process. Different formulation of this model are available: the Starkman formulation [13] and the Nastoll formulation [3] were used in the present work.

The ERM model [2] was originally developed by Fauske (1985), is valid for saturated and subcooled liquid at the nozzle inlet and cannot be applied to two-phase mixtures. It suppose that thermal equilibrium occurs if the pipe length is greater than 0.1 m. For shorter pipes it takes into account of thermodynamic non-equilibrium phenomena by introducing a non-equilibrium correction factor.

The HDI model [4] evaluates the mass flow rate by integration of Bernoulli's equation using a simple numerical approximation of the fluid properties of the two-phase mixture. It can be applied to saturated mixture and to non-flashing flows.

A modification of the model was developed in order to consider thermodynamic non-equilibrium (Homogeneous Non-Equilibrium Direct Integration Model, HNDIM [14]).

3. Two-phase valve discharge coefficient.

The models by Lenzing, Leung and Darby for the evaluation of the two-phase discharge coefficient are reported.

Lenzing formulation considers a homogeneous flow of the phases; each of them occupies a portion of the flow area and behaves as a single-phase fluid. Lenzing uses the void fraction in the nozzle mouth to evaluate the two-phase discharge coefficient [5]:

$$K_{dTP} = \alpha_e K_{dG} + (1-\alpha_e) K_{dL} \tag{1}$$

The void fraction α_e at safety valve entrance can be evaluated by the ω method, the HFM by Nastol, HNEM and others.

Leung formulation [5] was developed to be used with ω method, the HFM by Nastol and HNEM and is an extension of the Jobson and Bragg's work on compressible discharge through an orifice to the case of compressible two-phase discharge: it yields the two-phase discharge coefficient as a function of liquid discharge coefficient and the ω parameter. Diagrams are also available.

Darby [4] states that the discharge coefficient is strongly dependent on flow conditions. For choked flow the two-phase coefficient is equal to the gas discharge coefficient, while when the flow is not choked the liquid discharge coefficient is used. This formulation was developed to be used with HDI and HNDI models. The authors suggested the model to be used for the evaluation of flow rate through the ideal nozzles. These ones, together with the discrepancies with respect to the experimental results that they considered, are reported in table 1.

Table 1. Coupling models for choked flow through an ideal nozzle and formulations for the two-phase discharge coefficient

Author	**K_{dTP} formulation**	**Choked flow model**	**Discrepancy (%)**
Lenzing	$K_{dTP} = \alpha_e K_{dG} + (1-\alpha_e) K_{dL}$	Omega Method, HFM (Nastoll), HNE (Henry-Fauske)	±10% HNE ±40% HFM/Omega
Leung	Diagrams	Omega Method, HFM, HNE(Henry-Fauske)	-
Darby	Choked => $K_{dTP} = K_{dG}$ Not choked => $K_{dTP} = K_{dL}$	HDI, HNDI	-

Table 2. Valves characteristics.

Valve	K_{dG}	K_{dL}	Orifice diameter (mm)	Exit area (mm^2)
Crosby 1x2"E"(JLT/JBS).	0.962	0.729	13.5	143.2
ARI DN25/40	0.81	0.81	22.5	397.6
LESER DN25/40 (441)	0.77	0.51	23	415.5

Journal of Physics: Conference Series **501** (2014) 012016 doi:10.1088/1742-6596/501/1/012016

Table 3. Experimental data of mixtures of water and steam.

Valve	p_0(bar)	T_0(°C)	x	G_{exp} (kg/m^{2}s)
LESER DN25/40 (441)	5.4	Ts(P0)	0.012	3750
LESER DN25/40 (441)	5.4	Ts(P0)	0.008	4240
LESER DN25/40 (441)	5.4	Ts(P0)	0.0065	4280
LESER DN25/40 (441)	5.4	Ts(P0)	0.0051	4460
LESER DN25/40 (441)	8	Ts(P0)	0.022	4410
LESER DN25/40 (441)	8	Ts(P0)	0.015	4580
LESER DN25/40 (441)	8	Ts(P0)	0.012	5000
LESER DN25/40 (441)	8	Ts(P0)	0.0046	5960
LESER DN25/40 (441)	10.6	Ts(P0)	0.04	4200
LESER DN25/40 (441)	10.6	Ts(P0)	0.03	4830
LESER DN25/40 (441)	10.6	Ts(P0)	0.025	4900
LESER DN25/40 (441)	10.6	Ts(P0)	0.016	5500
LESER DN25/40 (441)	10.6	Ts(P0)	0.012	6000
LESER DN25/40 (441)	10.6	Ts(P0)	0.007	6700
LESER DN25/40 (441)	10.6	Ts(P0)	0.0028	7300
LESER DN25/40 (441)	10.6	Ts(P0)	0.0011	7900

Table 4. Experimental data of mixtures of water and air.

Valve	p_0(bar)	T_0(°C)	x	G_{exp} (kg/m^{2}s)
LESER DN25/40 (441)	5	25	0.106	2700
LESER DN25/40 (441)	5	25	0.04	4200
LESER DN25/40 (441)	5	25	0.03	5000
LESER DN25/40 (441)	5	25	0.016	6450
LESER DN25/40 (441)	5	25	0.012	7400
LESER DN25/40 (441)	5	25	0.0073	9000
Crosby 1x2"E"(JLT/JBS)	5	25	0.12	3100
Crosby 1x2"E"(JLT/JBS)	5	25	0.021	6500
Crosby 1x2"E"(JLT/JBS)	5	25	0.0078	9500
Crosby 1x2"E"(JLT/JBS)	5	25	0.006	11900
Crosby 1x2"E"(JLT/JBS)	5	25	0.001	19050
ARI DN25/40	5	25	0.15	3000
ARI DN25/40	5	25	0.04	5000
ARI DN25/40	5	25	0.0068	10400
ARI DN25/40	5	25	0.0059	11160
ARI DN25/40	5	25	0.0031	14000
LESER DN25/40 (441)	8	25	0.315	2600
LESER DN25/40 (441)	8	25	0.048	7000
LESER DN25/40 (441)	8	25	0.027	8450
LESER DN25/40 (441)	8	25	0.018	9700
LESER DN25/40 (441)	8	25	0.0092	12400
LESER DN25/40 (441)	8	25	0.006	14000
ARI DN25/40	8	25	0.23	3100
ARI DN25/40	8	25	0.095	5500
ARI DN25/40	8	25	0.028	9600
ARI DN25/40	8	25	0.015	12400
ARI DN25/40	8	25	0.0096	14100
ARI DN25/40	8	25	0.009	15000
ARI DN25/40	8	25	0.0048	18000

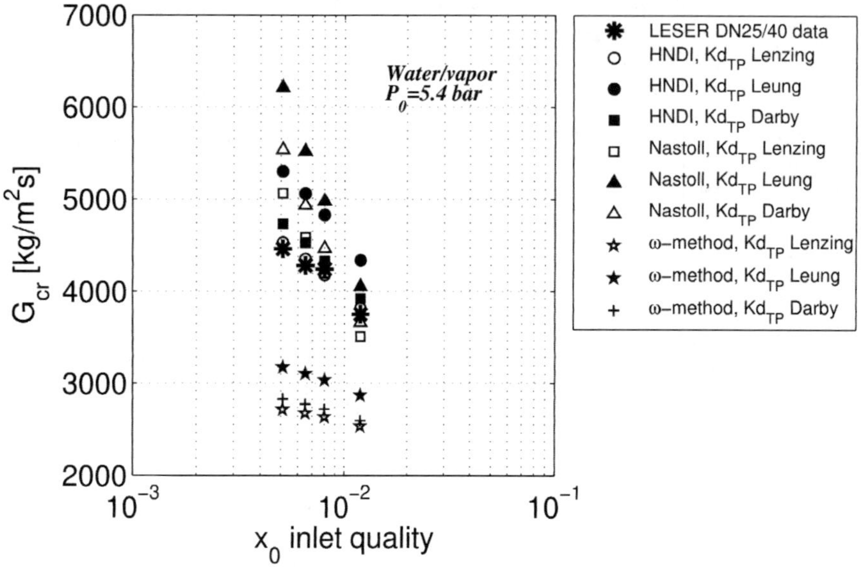

Figure 1. Comparison of the prediction of HNDIM, Nastol HFM and ω-method coupled with Lenzing's, Leung's and Darby's formulations of discharge coefficients with the flow rate of water/vapour discharged through a LESER DN25/40 valve at 5.4 bar.

4. Experimental data on safety valves.

Experimental data on spring safety valves have been taken from the scientific literature. Different valves have been used in the experimental studies available at present [4, 6-9].

As far as discharge of saturated mixture in the stagnation section is concerned, we considered experimental data obtained with the LESER DN25/40 (441) valve. Experimental data of non-flashing choked flow through three different valves (LESER DN25/40 (441), ARI DN25/40 and Crosby 1x2"E"(JLT/JBS)) are considered. Their characteristics [4] are reported in table 2.

Table 3 reports test conditions and the values of the mass flow rate obtained experimentally with mixture of water and steam. Table 4 refers to the experimental tests with water and air.

5. Models predictions.

We applied the formulations of discharge coefficient proposed by Lenzing, Leung and Darby to several choke models for ideal nozzle, in order to find out which couple of models best predicts the flow rate.

The HEM, ω-method and HDIM provided approximately the same predicted flow rate; therefore only the values predicted by the ω-method are reported here.

The flow rate predictions of two-phase mixture at the stagnation point are reported in figures 1, 2,3 relatively to stagnation pressures of 5.4 bar, 8 bar and 10.6 bar respectively.

The discrepancies, in terms of relative error, between predicted and experimental values are highlighted in table 5, which reports the minimum and the maximum percentage error (respectively in the first and second line of each cell of the table) for each couple of models.

The ω-method always underestimates the flow rate, while the HNE model strongly overestimates it. The best prediction is given by the HNDI model coupled with the Lenzing two-phase discharge coefficient.

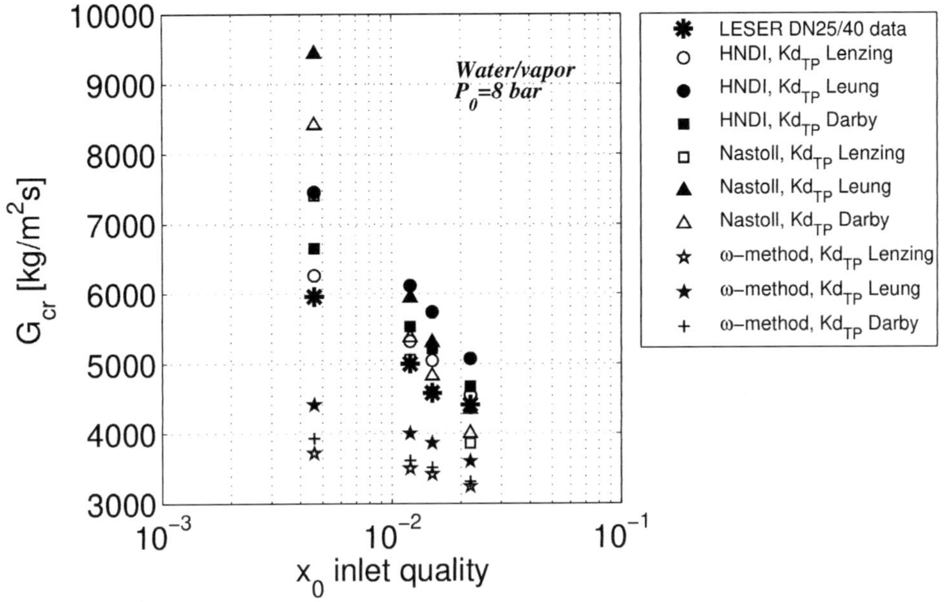

Figure 2. Comparison of the prediction of HNDIM, Nastol HFM and ω-method coupled with Lenzing's, Leung's and Darby's formulations of discharge coefficients with the flow rate of water/vapour discharged through a LESER DN25/40 valve at 8 bar.

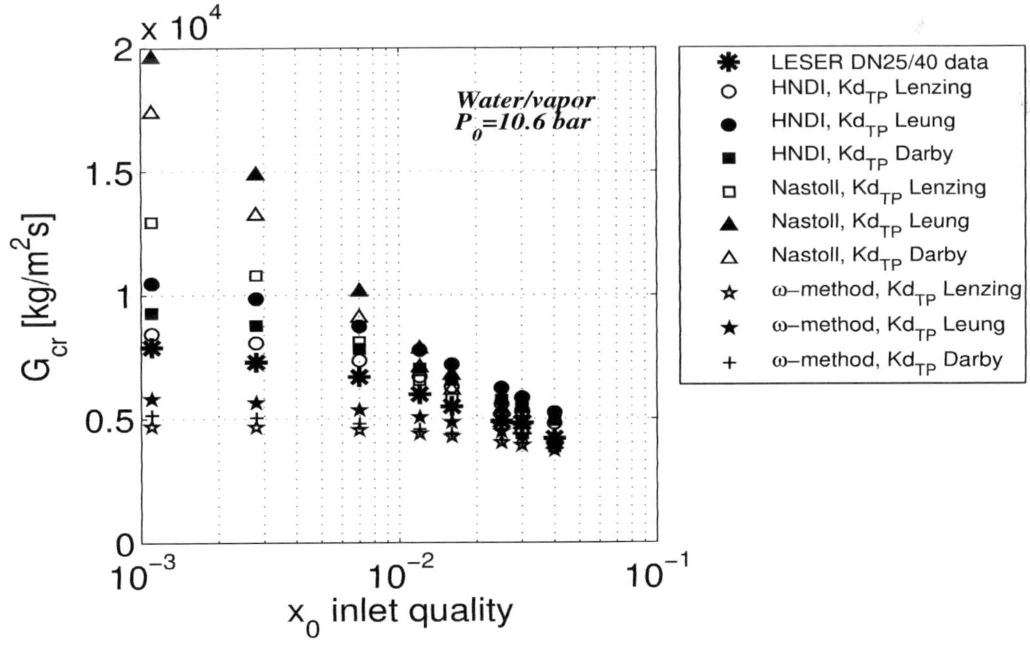

Figure 3. Comparison of the prediction of HNDIM, Nastol HFM and ω-method coupled with Lenzing's, Leung's and Darby's formulations of discharge coefficients with the flow rate of water/vapour discharged through a LESER DN25/40 valve at 10.6 bar.

Table 5. Discrepancies between predicted and experimental flow rate of water/vapour mixture through a LESER valve (expressed as % error).

Model		Lenzing	Leung	Darby	Lenzing	Leung	Darby	Lenzing	Leung	Darby
	p_0 (bar)	5.4	5.4	5.4	8	8	8	10.6	10.6	10.6
Omega	Error range	-32.5%	-23.5%	-30.8%	-25.3%	-15.6%	-23.2%	-11.6%	**-4.3%**	-10.3%
		-39.2%	-28.9%	-36.6%	-37.6%	-26.0%	-34.0%	-40.7%	**-26.4%**	-34.7%
HNDI	Error range	**-1.4%**	**13.9%**	**+2.1%**	**+3.1%**	**15.1%**	**+5.9%**	**+6.9%**	+21.3%	**12.6%**
		+1.7%	**18.9%**	**+6.0%**	**+10.2%**	**25.4%**	**14.0%**	**+14.7%**	+35.1%	**20.3%**
HFM Nastoll	Error range	-6.5%	+8.0%	-2.3%	-12.4%	-1.4%	-9.3%	-9.0%	+0.4%	-5.9%
		+13.6%	39.3%	24.2%	+24.3%	58.3%	41.2%	+64.1%	148.0%	+120%
HNE	Error range	+5.7%	+25%	13.1%	-0.2%	14.6%	+5.4%	+2.4%	+16.4%	+8.1%
		+16.7%	49.4%	33.2%	+20.0%	61.1%	43.7%	+37.9%	117.3%	92.7%
HFM Starkman	Error range	+29.6%	50.8%	36.4%	+25.4%	41.1%	29.8%	-20.5%	+20.2%	+6.6%
		+34.8%	65.4%	47.5%	+36.5%	62.8%	45.2%	+38.7%	+62.6%	47.1%

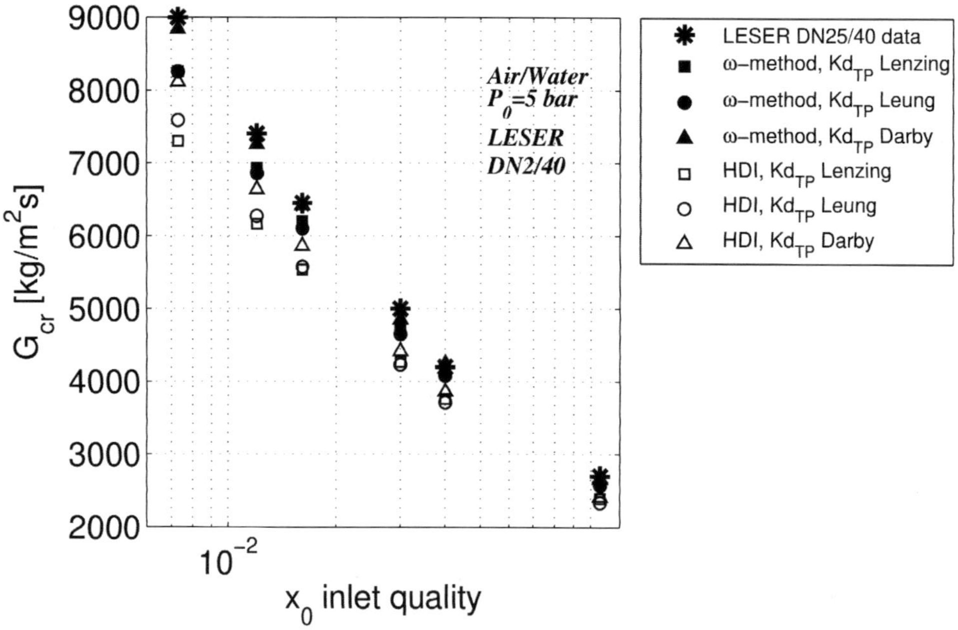

Figure 4. Comparison of the prediction of HNDIM, Nastol HFM and ω-method coupled with Lenzing's, Leung's and Darby's formulations of discharge coefficients with the flow rate of water/air at 5 bar through a LESER valve.

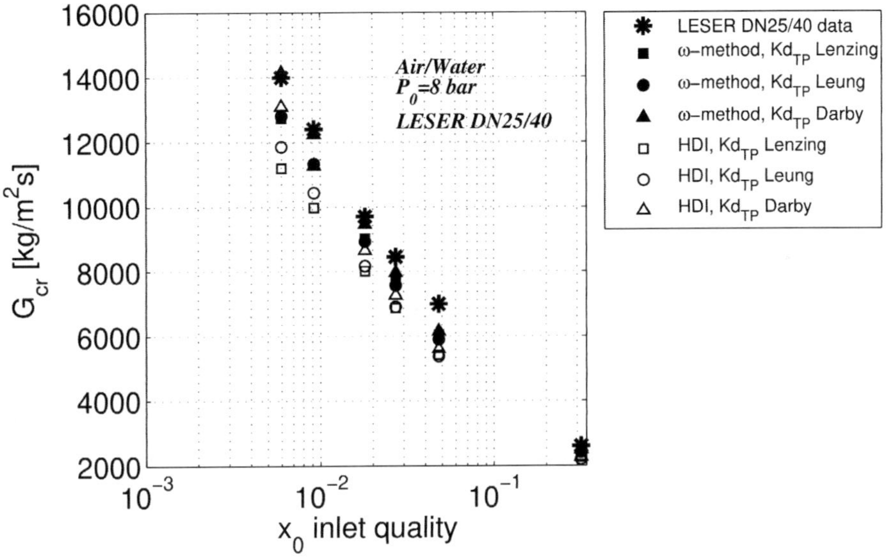

Figure 5. Comparison of the prediction of HNDIM, Nastol HFM and ω-method coupled with Lenzing's, Leung's and Darby's formulations of discharge coefficients with the flow rate of water/air at 8 bar through a LESER valve.

Table 6. Discrepancies between predicted and experimental flow rate of water/air mixture through the LESER DN25/40 (441) valve (expressed as % error).

Model		Lenzing	Leung	Darby	Lenzing	Leung	Darby
	p_0 (bar)	5	5	5	8	8	8
	Error	**-0,45%**	**-2.84%**	**-2.90%**	**-4.30%**	**-7.10%**	**-12.00%**
Omega	range	**-8.20%**	**-8.25%**	**1.09%**	**-13.80%**	**-15.70%**	**1.06%**
	Error	-10.47%	-11.65%	-8.10%	-13.40%	-15.31%	-6.44%
HDI	range	-18.90%	-15.73%	-11.70%	-22.60%	-23.30%	-19.94%
HFM	Error	-7.20%	-3.78%	3.00%	-30.14%	-25.50%	-17.67%
Starkman	range	52.80%	49.02%	54.11%	52.50%	48.30%	53.10%
HFM	Error	-34.68%	-34.08%	-29.42%	-31.00%	-30.50%	-23.20%
Nastoll	range	-37.69%	-38.11%	-35.42%	-42.81%	-43.50%	-41.00%

Boccardi et al. [6-9] compared experimental results on commercial valves with 10 mm orifice diameter with ω-method, HNE and found that the geometry of the valve strongly influences flow rate and therefore the accuracy of the prediction method. In particular it was found that the ω-method always underpredicts the flow rate of water and steam. The HNE model gave better results, but overpredicted the flow rate at low inlet quality in valves with slightly convergent-divergent nozzle.

The underprediction of the ω-method is due to the fact that it is not able to take into account thermodynamic non-equilibrium effects occurring through safety valves, which causes an increase of flow rate with respect to thermodynamic equilibrium conditions.

Journal of Physics: Conference Series **501** (2014) 012016 doi:10.1088/1742-6596/501/1/012016

The flow rate of water/air mixtures have been compared with the experimental values through the LESER valve, the Crosby valve and the ARI valves reported in table 2. Figures 4 and 5 show the comparison with experimental flow rates through a LESER valve at 5 and 8 bar respectively.

The discrepancies, in terms of relative error, between predicted and experimental values are highlighted in table 6,7 and 8, which report the minimum and the maximum percentage error (respectively in the first and second line of each cell of the table) for each couple of models applied to the LESER, Crosby and ARI valves.

It can be seen that all models tend to underestimate the flow rate. The best predictions are reported in the table in bold characters.

The model that globally best predicts the flow rate of non-flashing mixtures of water and air is the ω-method, coupled with the Darby's discharge coefficient.

The presence of an incondensable gas influences seems to influence the discharge capability of the safety valve and the discharge coefficient tends to the value for gas flow.

Table 7. Discrepancies between predicted and experimental flow rate of water/air mixture through the Crosby 1x2 E JLT/JBS valve (expressed as % error).

Model		Lenzing	Leung	Darby
	p_0 (bar)	5	5	5
		8.29%	4.95%	-0.19%
Omega	Error range	-10.18%	-9.02%	13.36%
		-3.12%	**-3.66%**	**4.06%**
HDI	Error range	**-18.62%**	**-15.55%**	**-8.39%**
		-64.76%	-61.90%	-56.70%
HFM Starkman	Error range	57.70%	46.18%	58.60%
		-24.40%	-25.00%	-16.10%
HFM Nastoll	Error range	-33.09%	-37.80%	-32.54%

Table 8. Discrepancies between predicted and experimental flow rate of water/air mixture through the ARI DN25/40 valve (expressed as % error).

Model		Lenzing	Leung	Darby	Lenzing	Leung	Darby
	p_0 (bar)	5	5	5	8	8	8
Omega	Error range	**-11.77%** **-21.50%**	**-11.18%** **-22.03%**	**-8.16%** **-21.27%**	**-1.51%** **-19.47%**	**-2.23%** **-17.20%**	**-1.27%** **-14.50%**
HDI	Error range	-20.10% -29.00%	-17.10% -29.30%	-15.60% -28.56%	-10.90% -28.28%	-11.30% -23.35%	-10.42% -22.40%
HFM Starkman	Error range	-42.35% 28.70%	-37.50% 29.21%	-33.40% 31.14%	43.63% -55.70%	-38.90% 54.80%	-34.70% 56.40%
HFM Nastoll	Error range	-39.21% -47.10%	-35.80% -47.40%	-31.50% -46.90%	-31.36% -43.72%	-31.74% -43.52%	-31.06% -42.08%

6. Conclusions

A thorough analysis of the models available in the literature has been carried out.

The authors compared the predictions of two-phase flow rate through ideal nozzle by several models (HEM, ω-method, SEMs, Henry-Fauske's Homogeneous Non Equilibrium, ERM, HFM, HDI, HNDI) and used the models developed by Lenzing, Darby, and Leung for the two phase valve discharge coefficient.

The results show that for flashing flows good predictions are obtained by multiplying the Lenzing's formulation of the two-phase valve discharge coefficient and the critical flow rate obtained by the Homogeneous Non-equilibrium Direct Integration.

On the other hand, Leung/Darby formulation of the two-phase valve discharge coefficient together with Omega Method is the most suitable to the prediction of flow rate in case of non-flashing flows of air and water.

References

[1] Diener R and Schmidt J 2004, Sizing of throttling device for gas/liquid two-phase flow Part 1: safety valves, *Process Safety progress* **23** 335-344

[2] Fauske H K 1985 Flashing flows: some practical guidelines for emergency releases, *Plant/operation Progress* **4**, n.3

[3] Lenzing T, Friedel L, Cremers J and Alhusein M 1998, Prediction of the maximum full lift safety valve two-phase flow capacity, *J. Loss Prevention in the Process Industries* **11** 307-321

[4] Darby R 2004, On two-phase frozen and flashing flows in safety relief valves: recommended calculation method and the proper use of the discharge coefficient, *J. Loss Prevention in the Process Industries* **17** 255-259

[5] Leung J 2004, A theory on the discharge coefficient for safety relief valves, *J. Loss Prevention in the Process Industries* **17** 301-313

[6] Boccardi G., Bubbico R, Celata G P, Cumo M and Di Tosto F 2004, Geometry influence on two-phase flow in safety valves, *5th Int. Conference On Multiphase Flow ICMF'04*, Jokohama, Japan

[7] Boccardi G., Bubbico R, Celata G P and Mazzarotta B 2005, Two-phase flow through pressure safety valves. experimental investigation and model prediction, *Chemical Engineering Science* **60** 5284-5293

[8] Boccardi G., Bubbico R, Celata G P and Di Tosto F 2008, Geometry influence on safety valves sizing in two-phase flow, *J. Loss Prevention in the Process Industries* **21** 66-73

[9] Boccardi G., Bubbico R, Celata G P, Di Tosto F and Trinchieri R 2010, Comparison among three prediction methods for safety valves design in two-phase flow in the case of a small valve, *Chemical Engineering Transactions* **10** 175-181

[10] Leung J 1994, The Omega Method for discharge rate evaluation, Update on Relief System Design, *AIChE Meeting*, Denver, Colorado, USA

[11]Henry R E and Fauske H K 1970, Two-phase critical flow at low qualities, *Nucl. Sci. Eng.* **41** 79-91

[12] API, Recommended practice 520: Sizing, selection and installation of pressure-relieving devices in refineries, 2000

[13] Whalley P B 1987, *Boiling, condensation and gas/liquid flows*, Claredon Press,Oxford

[14] Darby R, Freeman S E and Victor E H 2002, Properly size pressure-relief valves for two-phase flow: a number of factors must be kept in mind when using different models, *Chemical Engineering* 68-74

Experimental analysis of R134a flow boiling inside a 5 PPI copper foam

A Diani, S Mancin[1] and L Rossetto

Dipartimento di Ingegneria Industriale, Università di Padova, Padova 35131, Italia

E-mail: simone.mancin@unipd.it

Abstract. Heat dissipation is one of the most important issues for the reliability of electronic equipment. Boiling can be a very efficient heat transfer mechanism when used to face with the electronic technology needs of efficient and compact heat sinks. Recently, cellular structured materials both stochastic and periodic, particularly open cell metal foams, have been proposed as possible enhanced surfaces to lower the junction temperatures at high heat fluxes. Up today, most of the research on metal foams only regards single phase flow, whereas the two phase flow is still almost unexplored. This paper presents an experimental study on the heat transfer of R134a during flow boiling inside a 5 PPI (Pores Per linear Inch) copper foam, which is 5 mm high, 10 mm wide and 200 mm long, and it is brazed on a 10 mm thick copper plate. The experimental measurements were carried out by imposing three different heat fluxes (50, 75, and 100 kW m^{-2}) and by varying the refrigerant mass velocity between 50 and 200 kg m^{-2} s^{-1} and the vapour quality from 0.2 to 0.90, at constant saturation temperature (30°C). The effects of the refrigerant mass flow rate, heat flux and vapour quality on the heat transfer coefficient, dry out phenomenon, and pressure drop are studied.

1. Introduction

Boiling is the heat transfer mechanism with the highest heat transfer coefficients, thus it can be used to spread high fluxes and to maintain the wall temperature at low values in compact heat sinks. Moreover, new surfaces with microporous coatings, Carbon Nano Tubes (CNTs) coatings, or microstuctured surfaces are now available to enhance the boiling phenomenon. Nevertheless, a lot of work is still needed to deeply understand the boiling mechanism in such surfaces, where a huge number of variables (heat flux, saturation temperature, flow pattern, gravity, subcooling, wall surface, and others) are linked together and play important and crucial roles.

Metal foams are interesting new surfaces, that can be used in compact heat exchangers, to reject high heat fluxes. Metal foams are a class of cellular structured materials with open cells randomly oriented and mostly homogeneous in size and shape. Most of the work regarding metal foams is focused on single phase flow, and only few works involve the two phase process. Mancin et al. [1] collected experimental data during air flow through 21 aluminum foams, in order to highlight the effect of the geometrical parameters (pore density, porosity, and core foam height) on heat transfer and pressure drop. They also proposed two empirical correlations to estimate the heat transfer coefficients and pressure gradients. Boomsma et al. [2] performed experiments with water flowing through different compressed open-cell aluminum foam heat exchangers. They found that the metal foam heat

[1] To whom any correspondence should be addressed.

exchangers decreased the thermal resistance by nearly half when compared to conventional heat exchangers designed for the same applications.

Zhao et al. [3] experimentally studied R134a flow boiling in foam filled tubes. They found that the heat transfer coefficient is almost doubled when increasing the linear porosity (i.e. PPI) from 20 to 40 PPI, thanks to more surface area and flow mixing. The effects of mass velocity, vapour quality, heat flux, and operative pressure were also analysed.

Kim et al. [4] explored the possible viable use of metal foams inserted in a channel as porous medium for a cold plate. They considered three different copper foam configurations: 95% porosity and 10 PPI, 95% porosity and 20 PPI, and 92% porosity and 20 PPI. They used water and FC-72 as working fluids. The 10 PPI copper foam was found to be the best choice in terms of heat transfer coefficient.

Li and Leong [5] investigated the water and FC-72 flow boiling in aluminium foams. The heat transfer process prior to the onset of nucleate boiling and the hysteresis effect were investigated. They also studied the critical heat flux for FC-72. They found that hysteresis occurred for water but not for FC-72. Numerical simulations were performed for both single and two phase water heat transfer through 40 PPI foams.

Another work about water flow boiling in metal foams is that of Ji and Xu [6]. They conducted experiments to study pressure drop in copper foams embedded in a rectangular copper channel. They used water as working fluid with mass velocity between 30 and 200 kg m^{-2} s^{-1}. Three different copper foams were tested, each with a porosity of 0.88 but with different linear porosity (30, 60 and 90 PPI). Effects of mass flux, vapour quality, and average pore diameter of metal foams on boiling were analysed: the pressure drops are increased when vapour quality, mass flux, and PPI values increase. In addition, an empirical correlation to estimate the two phase pressure drop was proposed.

This paper is one of the first works on the flow boiling through metal foams and it presents some preliminary results of R134a flow boiling inside a copper foam, electrically heated from the bottom. The tested metal foam has 5 PPI and a porosity, defined as the ratio of total void volume to the total volume occupied by the solid matrix and void volumes, of 0.93. The experimental measurements have been carried out by imposing three different heat fluxes of 50, 75 and 100 kW m^{-2}, at constant saturation temperature of 30°C; the refrigerant mass velocity has been varied between 50 and 200 kg m^{-2} s^{-1} and the vapour quality from 0.2 to 0.90. The results are presented in terms of heat transfer coefficients and pressure drops; furthermore, two-phase flow visualizations are proposed to give a direct evidence of the effect of the foam's structure on the heat transfer process.

2. Experimental set up

The experimental set up is located at the Heat Transfer in Micro-Geometries Labs (HTMg-Labs) of the Dipartimento di Ingegneria Industriale of the University of Padova. As shown in figure 1, the experimental facility consists of three loops: the refrigerant, the cooling water and the hot water loop. The rig was designed for heat transfer and pressure drop measurements and flow visualization during either vaporization or condensation of pure refrigerants and refrigerants mixtures inside structured micro-geometries. The facility has a maximum working pressure of 3 MPa, while refrigerant mass fluxes can be varied up to 400 kg m^{-2} s^{-1} in a section of 50 mm^2.

In the first loop the refrigerant is pumped through the circuit by means of a magnetically coupled gear pump, it is vaporized and superheated in a brazed plate heat exchanger fed with the hot water. Superheated vapour then partially condenses in a pre-condenser fed with the cold water to achieve the set quality at the inlet of the test section. The refrigerant enters the test section at a known mass velocity and vapour quality and then it is vaporized by means of a calibrated Ni-Cr wire resistance. The fluid leaves the test section and passes through a post-condenser, a brazed plate heat exchanger, where it is fully condensed and subcooled. Then, the subcooled liquid is sent back to the boiler by a pump.

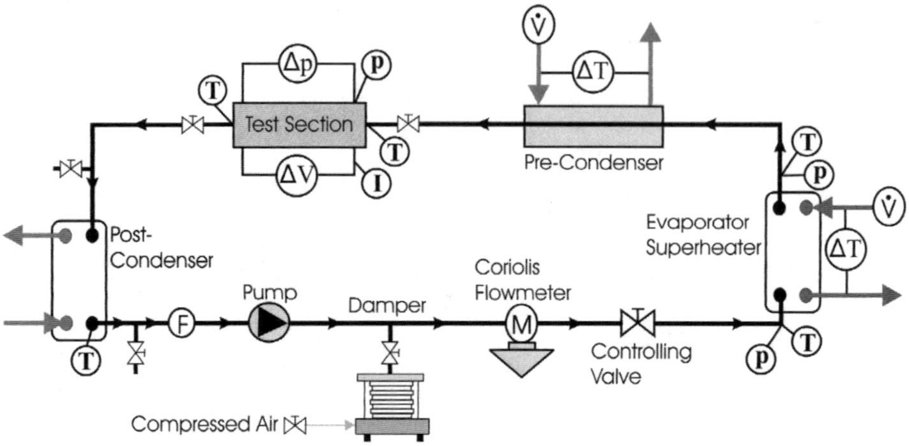

Figure 1. Schematic of the experimental apparatus.

A damper connected to the compressed air line operates as pressure regulator to control the saturation condition in the refrigerant loop. As shown in figure 1, the refrigerant pressure and temperature are measured in several locations throughout the circuit to know the refrigerant properties at the inlet and outlet of each heat exchanger.

The refrigerant flow can be independently controlled by the gear pump and it is measured by means of a Coriolis effect flowmeter. The inlet vapor quality to the test section is determined by the heat extracted in the precondenser, which can be controlled by varying water temperature and flow rate. The cold water loop consists of a stabilized chiller connected to the precondenser. The hot water circuit consists of a pump, an electrical heater and a controlling valve; it permits to set both the water flow rate and the inlet water temperature. Water flow rates in the precondenser and boiler sections are measured by means of magnetic type flow meters, while the water temperature differences are measured using 4-junction T-type thermopiles. Table 1 lists the values of accuracy of the instruments implemented in the experimental facility.

Figure 2 reports a photo of the designed test section, it consists of a 440x130x50 mm of low conductivity MISOGLASS block. This material has been selected by virtue of its interesting properties: high working temperature (greater than 200°C), low thermal conductivity (around 0.35 $W\ m^{-1}\ K^{-1}$). The block has been machined to obtain two plenums where the refrigerant enters and exits and where both the refrigerant temperature and pressure are measured by means of calibrated T-type thermocouples and high accuracy pressure transducers, respectively.

Table 1. Values of instruments accuracy.

Transducer	Accuracy
T-type thermocouples	± 0.05 K
T-type thermopiles	± 0.03 K
Electric power	± 0.13% of the reading
Coriolis mass flowmeter (refrigerant loop)	± 0.10% of the reading
Magnetic volumetric flowmeters	± 0.25% of the reading
Differential pressure transducer (test section)	± 25 Pa
Absolute pressure transducers	± 1950 Pa

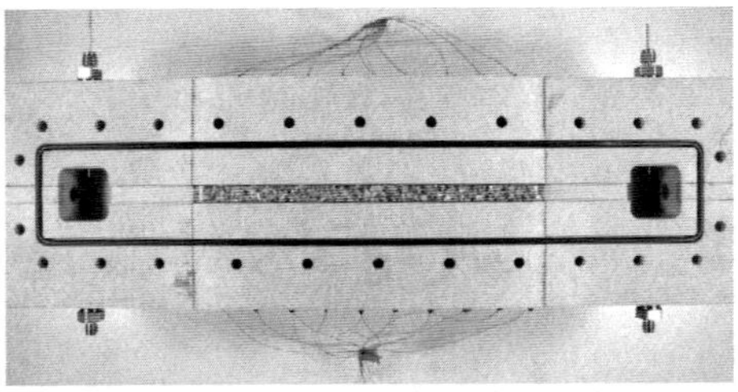

Figure 2. Photo of the test section.

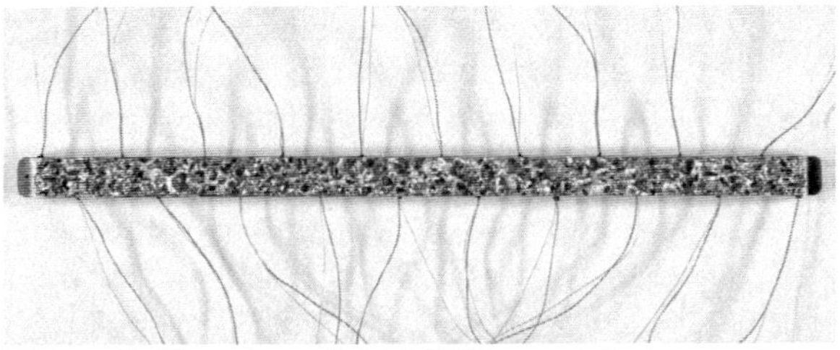

Figure 3. Top view of the 5 PPI copper foam sample.

A guide is milled in the center of the block to locate the heater and the foam sample. A photo of the tested foam sample is reported in figure 3, while table 2 lists its main geometrical characteristics; it is 10 mm wide and 190 mm long, it has 5 PPI and a porosity ε of 0.93; it has been brazed over a 200x10x10 mm copper plate. As shown in figure 3, 20 holes (5 mm deep) have been drilled just 1 mm under the foam-wall brazing surface to monitor the wall temperature distribution by locating as many calibrated T-type thermocouples. The copper foam sample is heated from the bottom using a heater, which is a 7 mm thick copper plate where a 2 mm deep guide is milled to locate a calibrated Ni-Cr wire resistance. The heat is supplied by means of stabilized DC power supplier, which is able to supply up to 900 W. The top wall consists of a 19 mm high tempered glass positioned over the test section, the sealing is accomplished using an EPDM o-ring located in a guide previously milled on the top of the section block. Two stainless steel plates are located on the top and on the bottom of the assembled test section and they are bolted together.

Table 2. Major geometrical characteristics of the copper foam sample.

Type	**PPI** [in^{-1}]	ε [-]	**Fiber thickness** [mm]	**Fiber lenght** [mm]	a [m^2 m^{-3}]
Cu-5-7.0	5	0.93	0.588	2.049	310

3. Data reduction

The subcooled liquid is pumped to the boiler where it is vaporized and superheated; the refrigerant temperature and pressure are measured at both inlet and outlet of the heat exchanger. The vapor quality at the inlet of test section depends on the refrigerant conditions at the inlet of the precondenser and on the heat flow rate exchanged in the tube-in-tube heat exchanger and obtained from a thermal balance on the cooling water side, as given by:

$$q_{pc} = \dot{m}_{w,pc} \cdot c_{p,w} \cdot \left(t_{w,pc,out} - t_{w,pc,in}\right) = \dot{m}_{ref}\left(h_{vs} - h_{TS,in}\right) \tag{1}$$

where $\dot{m}_{w,pc}$ is the water mass flow rate at the precondenser, $c_{p,w}$ the water specific heat at constant pressure, $t_{w,pc,out}$ and $t_{w,pc,in}$ the water temperatures at the outlet and inlet of the precondenser, respectively. Considering the right-hand side of (1), $\dot{m}_{ref}$ the refrigerant mass flow rate, while h_{vs} is the enthalpy of the superheated gas at the inlet of the precondenser, and $h_{TS,in}$ the enthalpy of the refrigerant at the inlet of the test section. The vapor quality at the inlet of the test section (x_{in}) can be calculated from the heat balance, as:

$$x_{in} = \frac{h_{TS,in} - h_L}{h_V - h_L} \tag{2}$$

where h_L and h_V are the specific enthalpies of the saturated liquid and vapor, respectively, evaluated at the saturation pressure of the refrigerant measured at the inlet of the test section. The electrical power supplied to the sample is indirectly measured by means of a calibrated reference resistance (shunt) and by the measurement of the effective EDP (Electrical Difference Potential) of the resistance wire inserted in the copper heater. The current can be calculated from the Ohm's law.

Preliminary heat transfer measurements permitted to estimate the heat losses (q_{loss}) due to conduction through the test section as a function of the mean wall temperature. The tests were run under vacuum conditions by supplying the power needed to maintain the mean wall temperature at a fixed value. The measurements were carried out by varying the mean wall temperature from around 30°C to more than 60°C. The results of these calibration tests are shown in figure 4, where the heat lost through the test section is plotted against the mean wall temperature. As it clearly appears the relationship between the heat lost and the wall temperature is linear; in this way, the actual value of heat supplied to the sample can be evaluated.

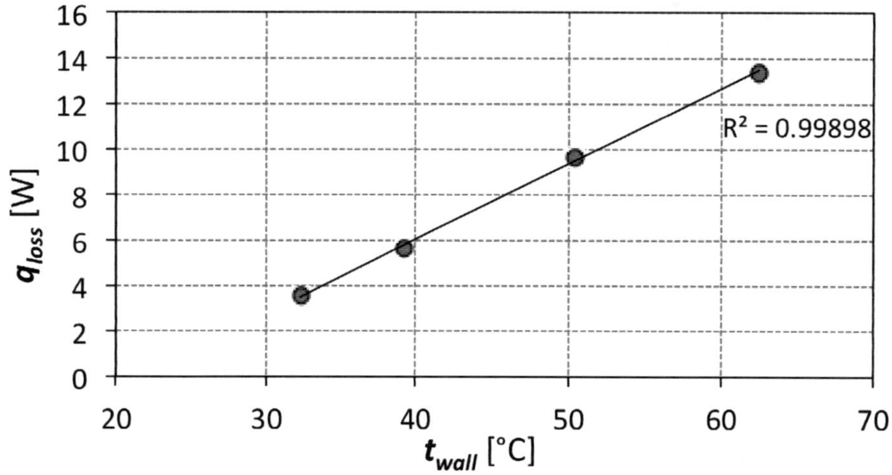

Figure 4: Estimated values of heat losses through the test section.

The heat losses through the test section are given by:

$$q_{loss}\,[\mathrm{W}] = 0.3311 \cdot \bar{t}_{wall}[°\mathrm{C}] - 7.128 \tag{3}$$

where $\bar{t}_{wall}$ is the mean wall temperature; thus, the actual heat flow rate supplied to the foam is given by:

$$q_{TS} = P_{EL} - q_{loss} = \Delta V \cdot I - q_{loss} \tag{4}$$

where P_{EL} is the electrical power supplied, ΔV is the electric potential, and I is the current. It was estimated that the heat lost was always less than 3%. The specific enthalpy at the outlet of the test section can be calculated from the thermal balance applied to the test section:

$$h_{TS,out} = h_{TS,in} + \frac{q_{TS}}{\dot{m}_{ref}} \tag{5}$$

Then, vapor quality x_{out} is given by:

$$x_{out} = \frac{h_{TS,out} - h_L}{h_V - h_L} \tag{6}$$

In this case the thermophysical properties of the refrigerant are evaluated at the outlet saturation pressure. The two phase heat transfer coefficient *HTC*, referred to the bare base area A_{base}, can now be defined as:

$$HTC = \frac{q_{TS}}{A_{base} \cdot (\bar{t}_{wall} - \bar{t}_{sat})} \tag{7}$$

where $\bar{t}_{sat}$ is the average value of the saturation temperatures obtained from the measured values of the pressure, as:

$$\bar{t}_{wall} = \frac{1}{20}\sum_{i=1}^{20} t_{wall,i} \qquad \bar{t}_{sat} = \frac{t_{sat,in}(p_{sat,in}) + t_{sat,out}(p_{sat,out})}{2} \tag{8}$$

where $t_{wall,i}$ is the temperature measured by the *i*-th thermocouple, and $p_{sat,in}$ and $p_{sat,out}$ are the saturation pressure at the inlet and outlet of the test section, respectively. Thermodynamic and transport properties of the refrigerant are estimated from RefProp 8.0 [7]. From the error propagation analysis, it has been estimated that the uncertainty on the two phase heat transfer coefficient shows a mean value of ±2.5% and a maximum value of ±3.8%, while the uncertainty on the vapor quality is ±0.035. Two phase experimental runs have shown a fine agreement between saturation temperatures calculated by Refprop v8.0 [7] from saturation pressure and the ones measured by thermocouples at the inlet and outlet of the test section, with a maximum difference lower than 0.1 K.

4. Experimental results

The experimental measurements were carried out at constant saturation temperature of 30°C, which can be considered suitable for the case of electronic cooling. The tests investigated the effects of different parameters: refrigerant mass velocity, vapour quality, and heat flux, on the vaporization of R134a inside a 5 PPI copper foam.

Figure 5 shows the effects of the imposed heat flux on the two-phase heat transfer coefficient as a function of the mean vapour quality, at constant refrigerant mass velocity of 100 kg m^{-2} s^{-1}. At first sight, we can state that the heat transfer coefficient increases as the imposed heat flux increases. For instance, at x_{mean}=0.5, the heat transfer coefficient is equal to around 19000, 20500, and 22900 W m^{-2} K^{-1} at 50, 75 and 100 kW m^{-2}, respectively. Moreover, at all the investigated heat fluxes, the heat transfer coefficient also increases when increasing the mean vapour quality, then it suddenly decreases because dryout occurs. The apexes of the heat transfer coefficient's curves highlight the vapour quality values at the onset of dryout; since no experiments were run at vapour qualities greater than 0.9, it is not possible to determine the end of the dryout out phenomenon. In the case of electronic cooling application, it is fundamental the knowledge of the vapour quality at the onset of dryout because it must be avoided during operation. As shown in figure. 5, the onset vapour quality decreases as the imposed heat flux increases varying from 0.78 to 0.7, when the heat flux goes from 50 to 100 kW m^{-2}.

The effects of the refrigerant mass velocity are highlighted in figure 6 where the two-phase heat transfer coefficient is plotted against the vapour quality at constant heat flux of 100 kW m^{-2}. As previously described also in this case, at all the investigated mass velocities, the flow boiling heat transfer coefficient increases as the vapour quality increases up to around x_{mean}=0.66-0.72, then it suddenly decreases because the dryout phenomenon occurs. It is worthy to underline that 50 and 100 kg m^{-2} s^{-1} exhibit similar heat transfer coefficient values. The effect of the mass flow rate is evident when mass velocity varies from 100 kg m^{-2} s^{-1} to 200 kg m^{-2} s^{-1}: the heat transfer coefficient increases by about 30-35%. Moreover, at vapour quality greater than 0.75-0.77 with developed dryout phenomenon, all the mass fluxes are characterized by similar *HTC* values which exhibit the same slope of trend profiles. Single phase liquid heat transfer measurements were also carried out, the heat transfer coefficient is equal to 4000, 5000, and 7000 W m^{-2} K^{-1} at 50, 100, and 200 kg m^{-2} s^{-1}, respectively; this results highlight the interesting heat transfer augmentation capabilities of metal foams during the phase change process.

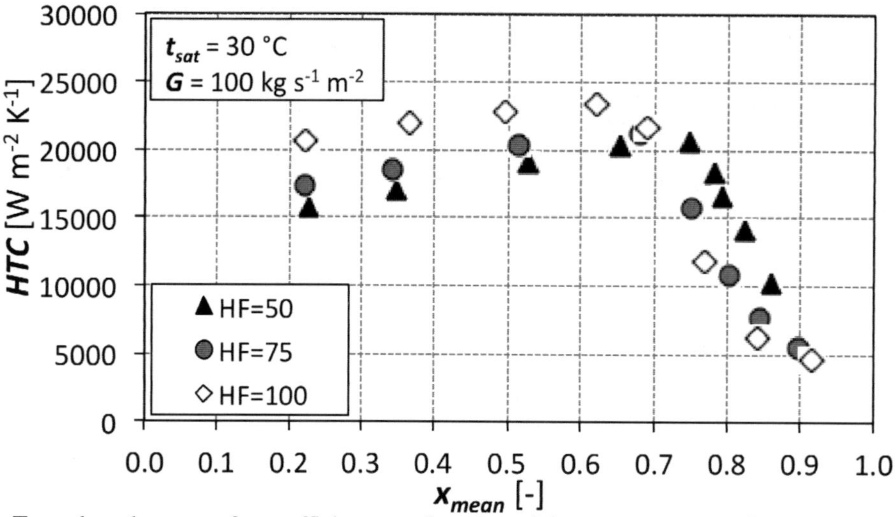

Figure 5. Two-phase heat transfer coefficient as a function of the mean vapor quality at constant refrigerant mass velocity. *HF* is expressed in [kW m^{-2}].

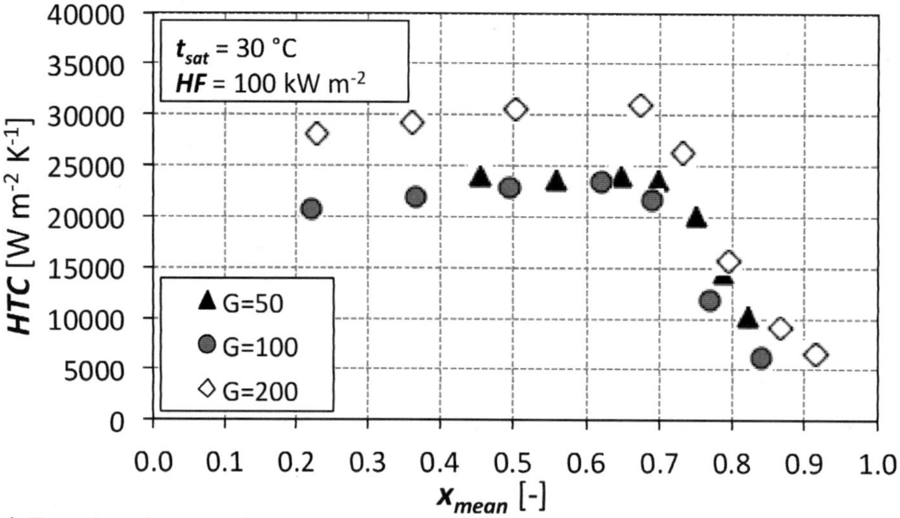

Figure 6. Two phase heat transfer coefficient as a function of the mean vapor quality at constant heat flux. *G* is expressed in [kg m^{-2} s^{-1}].

G=50 kg m^{-2} s^{-1}; x_{in}=0.5; HF=100 kW m^{-2}

G=100 kg m^{-2} s^{-1}; x_{in} =0.5; HF=100 kW m^{-2}

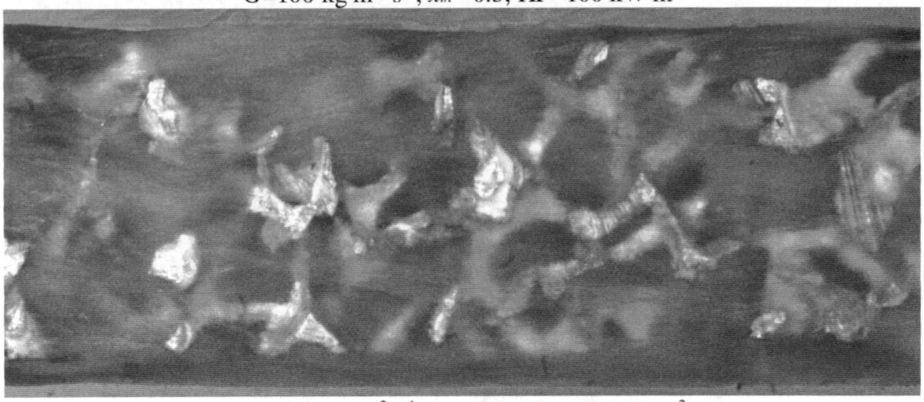

G=200 kg m^{-2} s^{-1}; x_{in} =0.5; HF=100 kW m^{-2}

Figure 7. Effect of the refrigerant mass velocity on the two-phase flow at constant inlet vapour quality and heat flux. Saturation temperature 30°C.

This behaviour can be explained using the photos reported in figure 7; the two-phase flow was recorded at different refrigerant mass velocities, keeping constant the heat flux at 100 kW m^{-2}, by means of a high speed B/W camera, Phantom v9.1 positioned on the top of the test section. The images show the first part of the foam sample, the focus is around 2 cm far from the inlet section and they are taken at constant inlet vapour quality of x_{in}=0.5. They permit to clarify the effects of the foam's structure on the phase change process; starting from the top photo, it shows the phase change process at 50 kg m^{-2} s^{-1}, in these operative conditions, the refrigerant does not fill the channel, it flows on the bottom in a stratified like flow, the nucleate boiling controls the vaporization process and the foam structure only helps in mixing the bubbles nucleated on the surface avoiding any coalescence process. When increasing the refrigerant mass velocity to 100 kg m^{-2} s^{-1}, even if the channel is not completely filled by the refrigerant, some liquid entrainment caused by the vapour shear through the foam fibers starts to be visible but, it is not sufficient to overtake the nucleate boiling and the heat transfer coefficients remain the same measured at 50 kg m^{-2} s^{-1} (Middle photo).

The mixing properties of the foam structure associated to the high shear stress are clearly highlighted in the bottom photo taken at 200 kg m^{-2} s^{-1}. Liquid and vapour are highly mixed and the fluid streams are deviated by the foam's fibers through a tortuous path; in this way, the convective contribute is extremely enhanced and seems to dominate the phase change process.

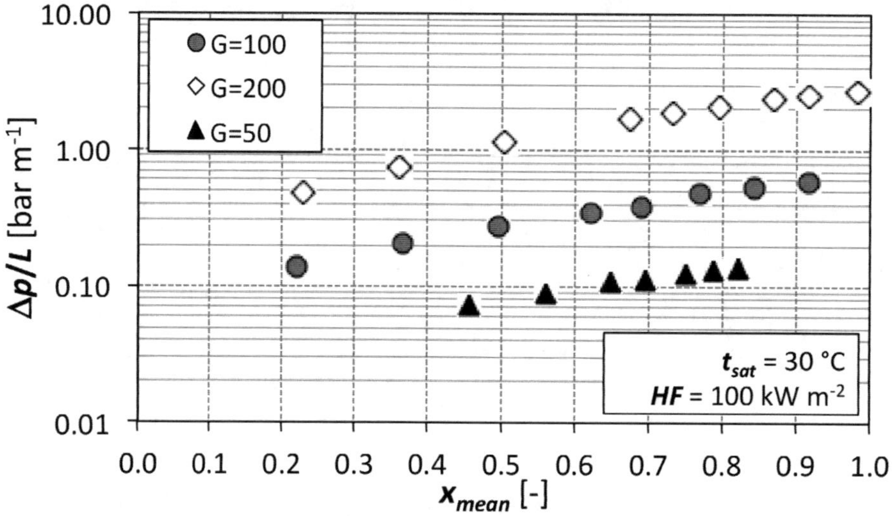

Figure 8. Two-phase pressure gradient plotted against the vapour quality at constant heat flux. G is expressed in [kg m^{-2} s^{-1}].

Figure 8 shows the total two phase pressure gradient plotted against the mean vapor quality as a function of mass velocity; as expected, at constant mass velocity, pressure drop increases as vapor quality increases; furthermore, keeping constant the vapor quality, it increases when increasing mass velocity.

5. Conclusions

This paper presents some experimental results relative to R134a flow boiling heat transfer and pressure drop inside an open cell metal foam with 5 PPI and porosity of 0.93. The measurements were carried out at constant saturation temperature of 30°C, by varying heat flux, mass velocity and vapour quality. The experimental measurements highlight the effects of these parameters on the flow boiling process; furthermore, using a high speed camera it was possible to give direct visual explanations of what deducted from the measured profiles of the two-phase heat transfer coefficients. At 100 kg m^{-2} s^{-1}, the heat transfer coefficients show to increase when increasing the heat flux; moreover, the higher the heat flux, the earlier the onset of dryout. Considering the effects of the refrigerant mass velocity at constant heat flux of 100 kW m^{-2}, 50 and 100 kg m^{-2} s^{-1} exhibit similar heat transfer coefficients. Then, passing from 100 to 200 kg m^{-2} s^{-1}, the heat transfer coefficient is enhanced of around 30-35%, highlighting the two phase forced convection effects. Furthermore, the heat transfer measurements give important information of the onset of the dryout phenomenon, which, for the investigated mass velocities, occurs at around x_{mean}=0.66-0.78. The two phase total pressure drops increase with both vapour quality and mass velocity. Finally, the results here presented demonstrate the interesting heat transfer capabilities of metal foams during the flow boiling heat transfer; more experimental work is needed to deeply understand the two phase heat transfer process inside these new enhanced surfaces.

Acknowledgement
The support of the MIUR through the PRIN Project 2009TSYPM7_003 is gratefully acknowledged. Collaboration of Alessandro Scardoni during his MS thesis work is gratefully acknowledged.

References

[1] Mancin S, Zilio C, Diani A and Rossetto L 2013 Air forced convection through metal foams: Experimental results and modeling *Int. J. Heat Mass Transf.* **62** 112-123
[2] Boomsma K, Poulikakos D and Zwick F 2003 Metal foams as compact high performance heat exchangers *Mech. Mater.* **35** 1161-1176
[3] Zhao C Y, Lu W and Tassou S A 2009 Flow Boiling Heat Transfer in Horizontal Metal-Foam Tubes *J. Heat Transf.* **131** 121002-1-8
[4] Kim D W, Bar-Cohen A and Han B 2008 Forced convection and flow boiling of a dielectric liquid in a foam-filled channel *Proc. 11th Intersoc. C. Thermal T. ITHERM* **11** 86-94
[5] Li H Y and Leong K C 2001 Experimental and numerical study of single and two-phase flow and heat transfer in aluminum foams *Int. J. Heat Mass Transf.* **54** 4904-4912
[6] Ji X and Xu J 2012 Experimental study of the two-phase pressure drop in copper foams *Int. J. Heat Mass Transf.* **55** 153-164
[7] NIST National Institute of Standard and Technology 2007 RefProp Version 8.0 Boulder Colorado USA

Journal of Physics: Conference Series **501** (2014) 012018 doi:10.1088/1742-6596/501/1/012018

Experimental sizing and assessment of two-phase pressure drop correlations for a capillary tube with transcritical and subcritical carbon dioxide flow

R Trinchieri[1], G Boccardi[1], N Calabrese[1], GP Celata[2] and G Zummo[1]
[1] ENEA, Laboratory of Thermal Fluid-Dinamics Applied to Energy Systems, Italy
[2] ENEA, Division of Advanced Technologies for Energy and Industry, Italy

E-mail: raniero.trinchieri@enea.it

Abstract. In the last years, CO_2 was proposed as an alternative refrigerant for different refrigeration applications (automotive air conditioning, heat pumps, refrigerant plants, etc.) In the case of low power refrigeration applications, as a household refrigerator, the use of too expensive components is not economically sustainable; therefore, even if the use of CO_2 as the refrigerant is desired, it is preferable to use conventional components as much as possible. For these reasons, the capillary tube is frequently proposed as expansion system. Then, it is necessary to characterize the capillary in terms of knowledge of the evolving mass flow rate and the associate pressure drop under all possible operative conditions.
For this aim, an experimental campaign has been carried out on the ENEA test loop "CADORE" to measure the performance of three capillary tubes having same inner diameter (0.55 mm) but different lengths (4, 6 and 8 meters). The test range of inlet pressure is between about 60 and 110 bar, whereas external temperatures are between about 20 to 42 °C.
The two-phase pressure drop through the capillary tube is detected and experimental values are compared with the predictions obtained with the more widely used correlations available in the literature. Correlations have been tested over a wide range of variation of inlet flow conditions, as a function of different inlet parameters.

1. Introduction

Carbon dioxide has been proposed as an alternative to HFCs in refrigeration systems for its almost optimal environmental features. For this, several possible CO_2 fields of application (automotive air conditioning, residential heating and cooling, water heating, commercial refrigeration, etc.) have been studied and a number of plants have been built. In many of these applications, in particular in the case of medium-large powers, the target to obtain high COPs led to the achievements of complex and expensive engineering solutions. Instead, in the case of lower power applications, the simplicity of installation and the containment of plant costs induce the adoption of technical solutions which are simpler and with lower economic impact.

One possible CO_2 small size application is related to the household refrigeration: this application requires obtaining of rather low evaporation temperatures (T_{ev}), typically from -25 to -30 °C, with cooling capacities (Q_{ev}) of the order of 60-180 W. If we do not want to use complex regulation systems for the execution of the expansion phase in the refrigeration CO_2 cycle, it is still possible using a capillary tube. This cheaper option can be pursued as, in the case of transcritical cycle, the capillary tube demonstrates an intrinsic ability to adjust the higher pressure of CO_2 to a value close to the optimal one to vary the outlet temperature of gas-cooler [1].

The aim of this work is the characterization and the experimental testing of a capillary tube to be used in a prototypic household refrigerator. For this purpose, an experimental campaign was carried out on the test loop "CADORE" to measure the performance of three capillary tubes having the same inner diameter (0.55 mm) but different lengths (4, 6 and 8 meters). In particular, we tried to reach the working conditions of a CO_2 household refrigerator using the three capillary tubes; in this way, the influence of the choice of capillary tube dimensions on the performances was investigated. The test range of inlet pressure was between about 60 and 110 bar, whereas external temperatures were between about 20 to 42 °C.

The two-phase pressure drop through the capillary tube is detected and experimental values compared with the predictions obtained with the more widely used correlations available in the literature. Correlations have been tested over a wide range of variation of inlet flow conditions, as a function of different inlet parameters.

Moreover, the influence of the values of the subcooling, mass flow-rate, pressure and temperature at the capillary inlet on the pressure drop is evaluated.

2. Experiment

The tests have been carried out on the ENEA test loop CADORE (Carbon Dioxide Refrigeration) [2].

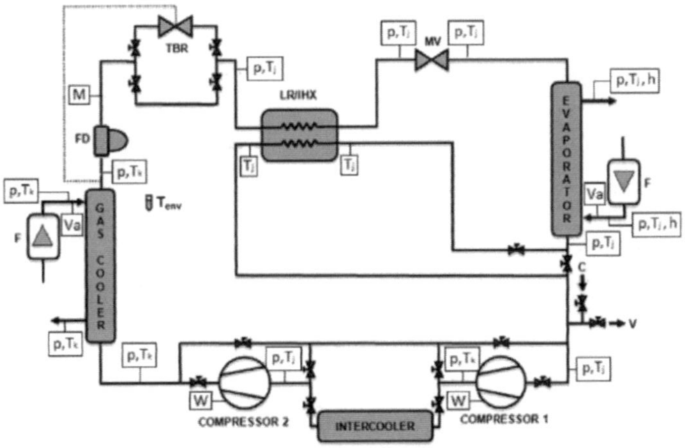

p	Pressure gauge	TBR	Back-pressure valve
T_j	J-type thermocouple	MV	Metering valve
T_k	K-type thermocouple	LR/IHX	Liquid receiver / Internal heat exchanger
M	Mass flow meter	FD	Filter dryer
W	Wattmeter	F	Fan
Va	Hot wire anemometer	C	Charge system
h	Humidity transmitter	V	Vacuum system

Figure 1. "CADORE" CO_2 circuit layout

Figure 1 shows a schematic of the experimental plant, which is composed of a carbon dioxide refrigeration loop, two auxiliary systems devoted to control the secondary fluid and a measurement and data acquisition system. The refrigeration loop is basically equipped with two hermetic single-stage reciprocating compressors, three finned coil heat exchangers (evaporator, gas cooler and intercooler), a double-stage expansion system and a filter dryer. Bypass circuits could allow the testing of different plant configurations.

Two auxiliary PID-controlled closed loop systems are employed to regulate and control the heat transfer rate in both the evaporator and the gas cooler, by setting the desired temperature, mass flow-rate and air humidity value at the heat exchangers inlet.

Temperatures at the inlet and the outlet of each component of the refrigeration loop are measured using Cr/Al K-type and Fe/Co J-type thermocouples. In correspondence of each thermocouple, except for the internal heat exchanger inlet and outlet, pressure transmitters with an accuracy of ±0.08% FS is used. Absorbed power and mass flow-rate are measured by a digital wattmeter with an accuracy of ±0.5% of reading and a Coriolis mass flow meter placed downstream the gas cooler and the filter dryer, respectively. Secondary fluid conditions are measured both for the evaporator and the gas cooler loops with several thermocouples placed alongside the air circuit, absolute and differential pressure transducers, hot wire anemometers and humidity active transmitters.

The output signals from the transducers and thermocouples are recorded in real time through a computerized data acquisition system and elaborated by a software developed in LabVIEW environment.

From the measured data, we calculated the thermodynamic and thermophysical properties (enthalpy, specific heat, etc.) using the computer code REFPROP version 8.0 of the NIST.

An important feature of CO_2 is the great variability of these properties around the critical point, as shown in figure 2 and 3 for the isobaric specific heat and for dynamic viscosity [3].

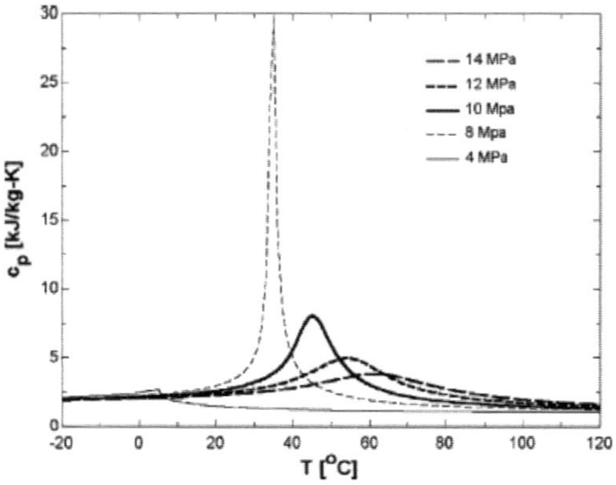

Figure 2. Isobaric specific heat of CO_2 vs temperature. Parameter pressure.

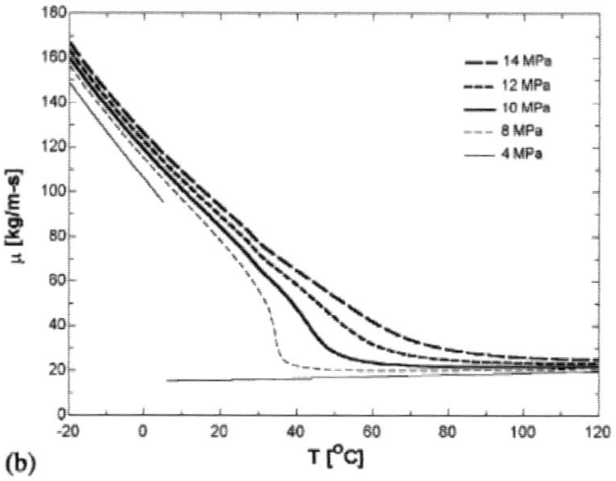

(b)

Figure 3. Dynamic Viscosity of CO_2 vs temperature. Parameter pressure.

The facility, as said before, can work in different configurations; in this research, it was used in the basic configuration of a vapour compression refrigerating machine. Three experimental campaigns (for a total of 129 tests) with capillary tubes of the same inlet diameter (0.55 mm) and different lengths (4, 6 and 8 m) were carried out.

Another important feature of the CO_2 transcritical cycle is the existence of an optimum value for the gas cooler pressure p_{GC}, which leads to obtain the maximum COP, once the other operating conditions have been fixed [4]. Figure 4 shows that the COP theoretically obtainable is a function of p_{GC} for every CO_2 gas cooler outlet temperature ($T_{CO2outGC}$). A numerical simulation performed using the data reported in table 1, allows to obtain the COP trend as a function of gas cooler pressure (figure 5).

Table 1. Data for performance evaluation of CO_2 vapour compression refrigerating cycles

- **Temperature of isobaric evaporation T_{ev}** = -25°C
- **Isentropic efficiency of adiabatic compression** η_{is}= 0.7
- **Vapour superheat at compressor inlet ΔT_{surr}** = 5°C
- **Isenthalpic transformation in the capillary tube**

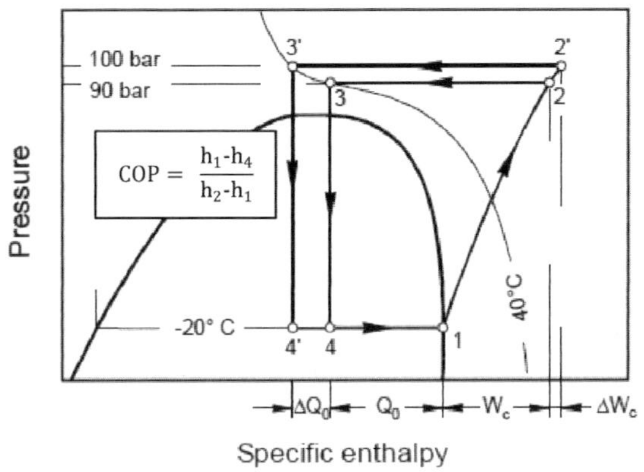

Figure 4. Transcritical cycle with different value of gas cooler pressure [4]

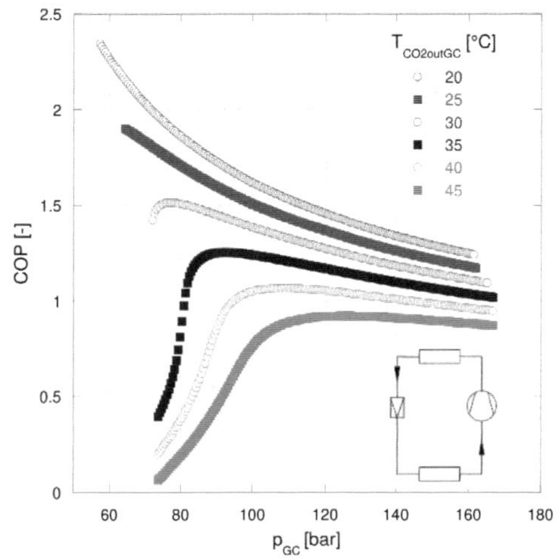

Figure 5. Theoretical performance evaluation of CO_2 vapour compression refrigerating cycle

This figure depicts the importance of working with pressures close to that corresponding to the highest COP, especially for $T_{CO2outGC}$ higher than 30°C. This behavior confirms the importance of the throttling system and the expansion devices.

The compressor used in these experimental campaigns is a prototype of 500 W nominal cooling capacity and the heat exchangers (gas cooler and evaporator) are those planned for a CO_2 household refrigerator prototype. In particular, the gas cooler works in all tests with efficiency very high, so the CO_2 temperature at the inlet of capillary tube is always very similar to the temperature of the environment external to the household refrigerator.

Furthermore, operating conditions of the heat exchangers and the refrigerant charge are regulated to obtain the required different circuit operating conditions (in table 2, the operative conditions at inlet and outlet of the capillary tubes are indicated).

Table 2. Test Matrix

Capillary length m	Number of test	P_{in} bar		T_{in} °C		P_{out} bar		G kg/h		ΔT_{sub} °C		charge g	
		min	max	min	max	min	max	min	max	min	max	min	max
4	41	55	105.5	17.7	42.4	12.8	20.2	3.4	6	1.2	13	500	700
6	63	54.2	107	16	41	10.8	18.2	2.3	5	0.6	15	500	585
8	25	55.2	103.5	16	40.5	10.8	18	2.1	4.7	0.7	15	500	700

Assuming an isoenthalpic expansion from the inlet conditions we can obtain the saturation pressure and therefore the pressure drops in the single phase and in the two-phase. In the first zone, the fluid can be considered incompressible and the single phase length (Δl in equation 1) can be calculated by the equation [5]:

$$\left(\frac{\Delta p}{\Delta l}\right)_{SP} = \frac{f_{SP} G_S^2}{2d\rho} \tag{1}$$

The single phase friction factor f_{SP} can be evaluated with the explicit equation proposed by Swamee and Jain [6] and used by Madsen et al. [5] for CO_2:

$$f_{SP} = 0.25 \left[log\left(\frac{e/d}{3.7} + \frac{5.74}{Re^{0.9}} \right) \right]^{-2} \quad (2)$$

where "e" is the surface roughness and "d" the diameter of capillary tube. This equation shows a maximum deviation of 1% from the Colebrook equation for Re values between 5000 and 10^8 and "e/d" values between 10^{-6} and 0.01: all experimental values of Re and e/d are in these ranges.

To improve calculation precision in (1), Δp between the initial state and saturation state is divided into 10 parts: the total length of tube required for the single phase flow is therefore calculated and consequently also the length available for two-phase flow. The experimental two-phase length in this way obtained is divided in 10 parts and for each section the two-phase pressure drop is calculated using the main correlations proposed in literature.

In two-phase conditions, the pressure drop in horizontal tubes is the sum of the momentum pressure drop (acceleration) and the frictional pressure drop:

$$\Delta p_{TP} = \Delta p_a + \Delta p_{fr} \quad (3)$$

The momentum pressure drop can be evaluated, for each section, with the equation:

$$\Delta p_a = G_s^2 (v_{out} - v_{in}) \quad (4)$$

To calculate this term it is necessary to know the thermodynamic outlet conditions, which depend on the total pressure drop. For their evaluation an iterative method has been used: the values of the momentum pressure drop are, for our tests, between 2 and 4 bar.
Kim e Mudawar [7] provided an exhaustive list of main prediction methods for the frictional pressure drop calculation. Some methods are semi-empirical, whereas others proceed, as the famous model of Lockhart-Martinelli, to the definition of a two-phase multiplier in order to obtain the Δp_{fr} using an equation like that:

$$\left(\frac{dp}{dz}\right)_{fr} = \left(\frac{dp}{dz}\right)_l \phi_l^2 \quad (5)$$

Among the best-known models, we tested the semi-empirical correlation of Friedel [8], a homogeneous equilibrium model and the method of Sun and Mishima [9], which, starting from the Lockhart-Martinelli model, changes the two-phase multiplier using data obtained by experimental tests on the expansion of various refrigerants (including CO_2) in tubes of diameter between 0.5 and 12mm.

The correlation of Friedel, among other, was proposed by Madsen [5] for the capillary tube sizing in a transcritical refrigeration system and it is considered by Cecchinato et el. [1,10] as one of the most accurate methods for sizing of capillary tubes with CO_2.

3. Results and discussion

The thermodynamic capillary inlet conditions can be very different as the boundary conditions change (e.g. ambient temperature). Figure 6 shows the CO_2 capillary inlet conditions in the performed tests; in the experimental test loop "CADORE", with a refrigerant charge of about 500 g and varying the outside air temperature from about 20 to 40 °C, we can observe that the upper pressure, depending on gas cooler outlet temperature, increases from about 60 to 100 bar, moving from subcritical to transcritical (or supercritical) conditions. We know that CO_2 is in supercritical conditions when the pressure and the temperature are both higher than the critical pressure (p_c= 7377 kPa) and critical temperature (T_c= 30.98°C), in transcritical conditions when the pressure is higher than the p_c but the temperature is lower than the T_c and in subcooled state when the pressure is lower than the critical value and temperature is between T_c and the evaporation temperature.

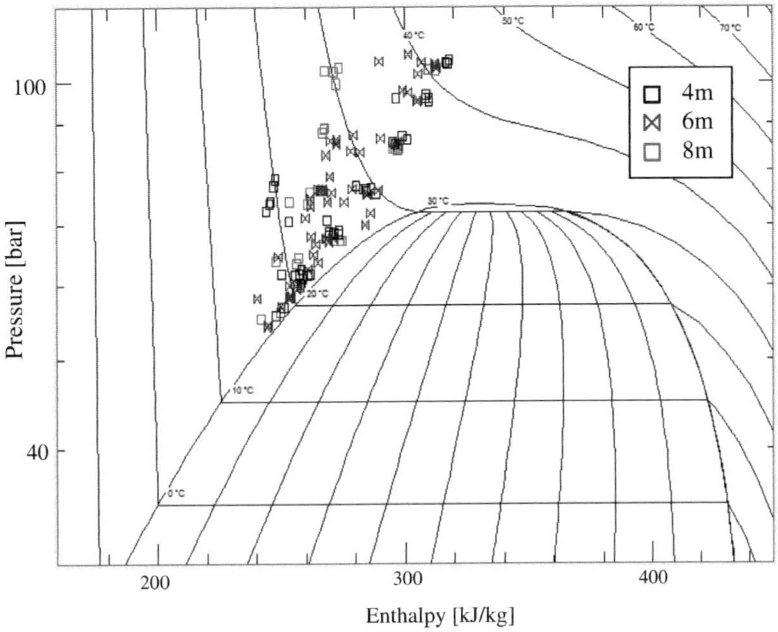

Figure 6. Experimental capillary inlet points on the CO_2 p-h diagram

In the next sections we present:

- the performance obtainable (cooling capacity, Q_{ev}, evaporation temperature, T_{ev}) with the three capillaries in our refrigerant circuit;
- the analysis of some process variables, with particular regard to the relation between pressure drops and other parameters;
- the prediction capacity of pressure drop of some correlations available in the literature.

3.1 Comparison of performance obtained with the three capillary tubes

The total pressure drop (Δp_{cap}) in the capillary tube depends mainly on the CO_2 inlet thermodynamic conditions and on the refrigerant charge: in particular, it is noted that the pressure drop increases approximately linearly with the inlet capillary temperature, T_{inCAP} (figure 7, where all performed tests are plotted). We can also observe that, for the same CO_2 inlet temperatures and capillary length (figure 8, in the case of 8 m length), a considerable increase in the pressure drop may be obtained raising the refrigerant charge.

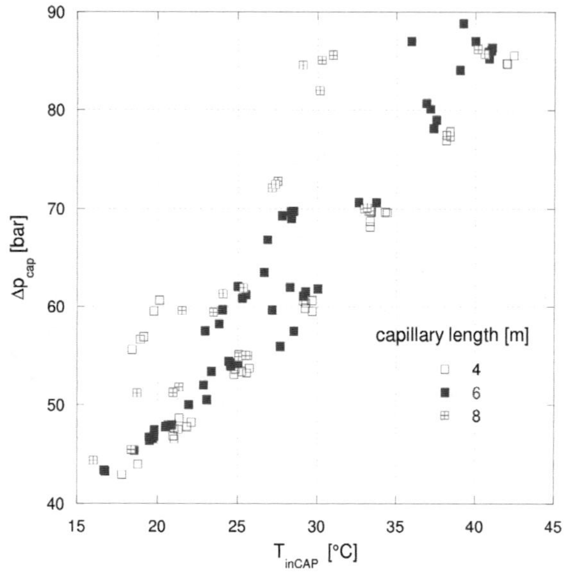

Figure 7. Total pressure drop in the capillary tubes vsT_{inCAP}; parameter capillary length

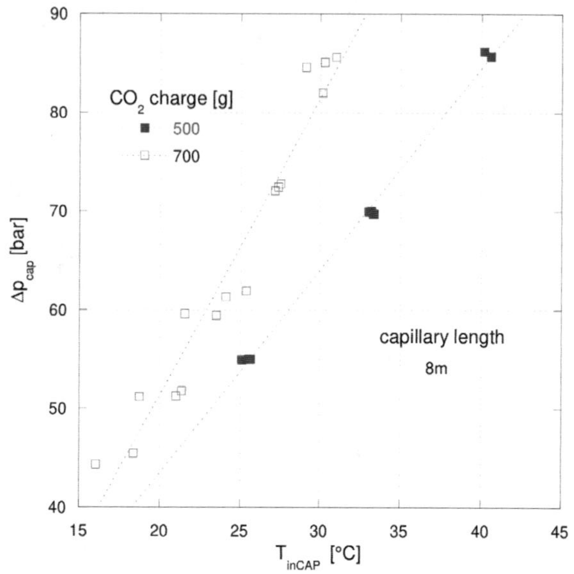

Figure 8. Total pressure drop in the 8 m capillary tubes as a function of T_{inCAP}

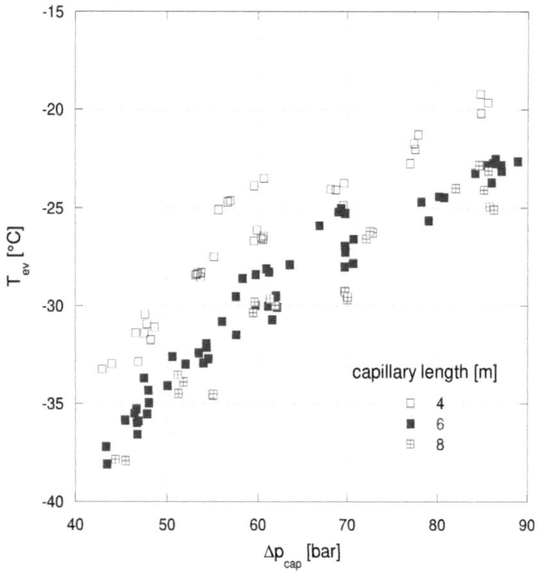

Figure 9. T_{ev} as a function of Δp_{cap}; parameter capillary length

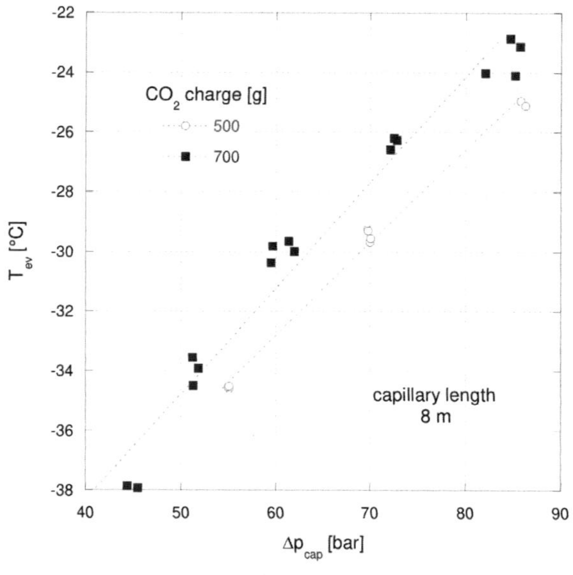

Figure 10. T_{ev} as a function of Δp_{cap} for 8 m capillary tube

Figure 9 shows the evaporation temperatures obtainable with the three capillary tubes versus the pressure drop for all the performed tests. The evaporation temperatures seem to show a dependence on the refrigerant charge (figure 10, where, for clarity, only the tests with capillary length of 8m are presented), whose mechanisms will be investigated in a future work; figure 9 also shows that, for capillary pressure drops larger than 80 bar, corresponding to inlet temperatures above 35 °C, see figure 7, evaporation temperatures equal to or lower than the target of -25°C are obtainable only with capillaries longer than 6 m.

Figure 11 shows the cooling capacity obtained with the three capillary tubes as a function of the pressure drop for all the performed tests. For the cooling capacity, the dependence on the charge appears to be more evident (figure 12, where only data obtained with 8 m capillary length are shown). Figure 11 shows how the target of 150 W is achieved in almost all cases (except for few tests with the capillaries of 6 and 8 m and low pressure).

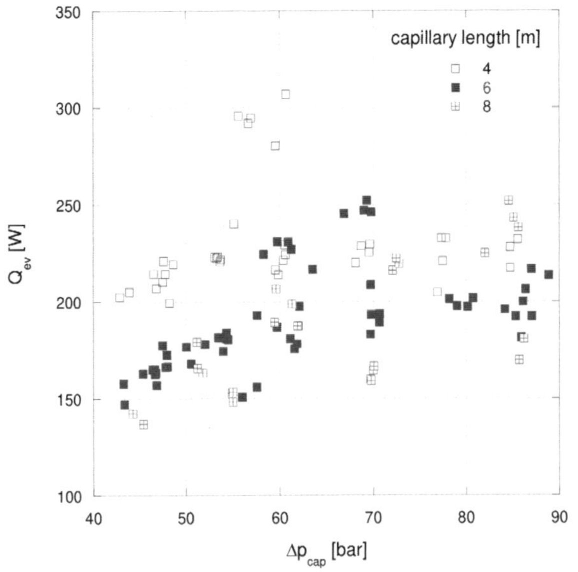

Figure 11. Q_{ev} as a function of Δp_{cap}; parameter capillary length

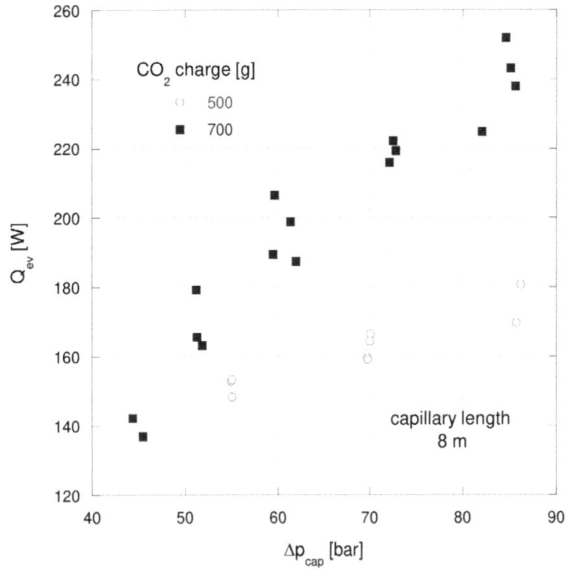

Figure 12. Q_{ev} as a function of Δp_{cap} in 8 m capillary tube

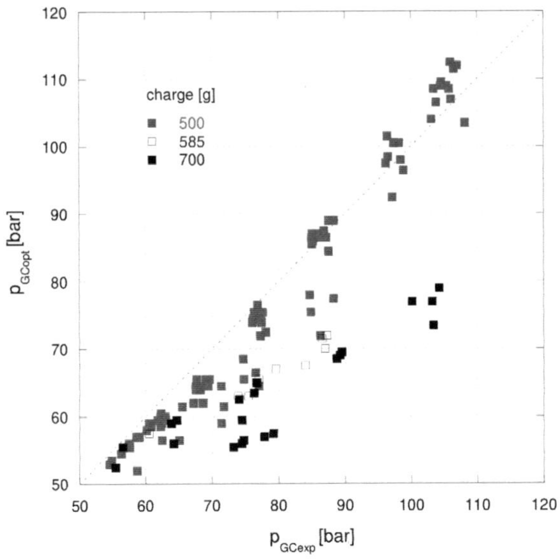

Figure 13. p_{GCopt} vs p_{GCexp} for all data; parameter refrigerant charge

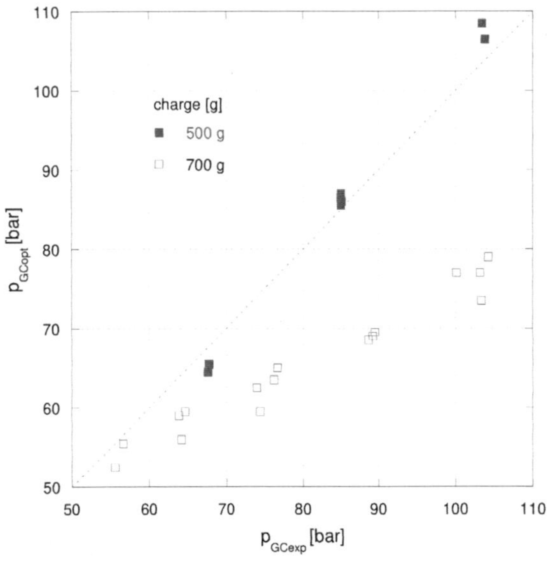

Figure 14. p_{GCopt} vs p_{GCexp} for 8m capillary tube; parameter refrigerant charge

Regarding to capillary tube capability to adjust the gas cooler pressure p_{GC} towards optimal value, the figure 13 shows that, for all tests with a charge of 500g, the experimental "upper pressure" p_{GCexp} is always very close to optimum value p_{GCopt} (calculated assuming the experimental values of T_{ev}, ΔT_{surr} and $T_{CO2outGC}$, isoenthalpic expansion and an isentropic efficiency η_{is} of compressor of 0.7).

This data would seem to confirm that, by setting a suitable refrigerant charge, it is possible to work with upper pressures close to the optimum value even using a capillary tube as expansion device, for a large range of external conditions. In the case of our prototype, the best value of the charge seems to be 500g. For each tube length, increasing the charge, p_{GC} increases with respect to optimal value (figure 14, for the 8 m capillary tube). This leads to greater COP penalizations when the $T_{CO2outGC}$ is of the order of 20-30 °C (figure 5).

Quantitatively, the maximum deviation of the experimental pressure from the optimal one is 5% with a charge of 500g and can reach the 30% with a charge of 700g.

3.2 Evaluation of pressure drop trough the capillary tubes

The upper pressure varies from 54 to 107 bar, the inlet temperature ranges from 0.6 (with use of an internal heat exchanger) to 41°C while the outlet pressure is between 10 bar and 20.2 bar. These test conditions are chosen to check the typical operating conditions of household refrigerators, which usually adopt capillary tubes as throttling devices. Figure 15 shows that, in our tests, the choking condition is never achieved. In fact, chocking pressure, p_{ch}, calculated using the correlation proposed by Zhang e Ding [11], ranges from 4 to 12 bar, values always lower than the achieved evaporation pressures p_{ev}.

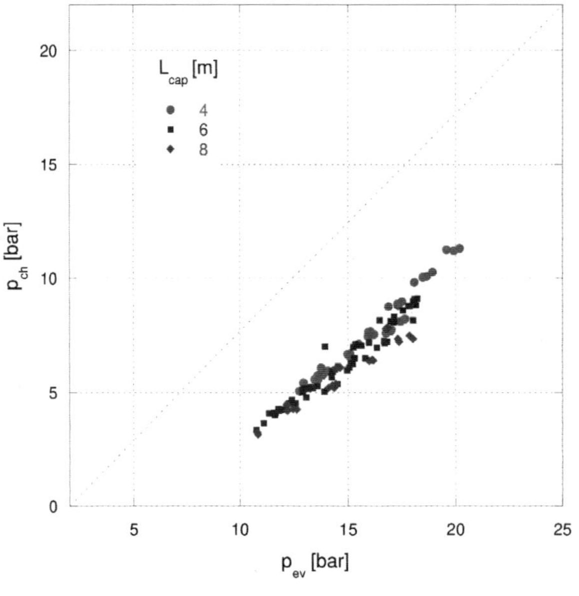

Figure 15. Evaluation of the choked flow: p_{ch} vs p_{ev}

The performance of the tested correlations can be evaluated using the statistical parameters, standard deviation σ and root mean square RMS:

$$\sigma = \sqrt{\frac{\Sigma(\Delta_i - \Delta_{med})^2}{N-1}} \tag{6}$$

$$RMS = \sqrt{\frac{\Sigma(\Delta_i)^2}{N-1}} \tag{7}$$

where Δ_i indicates the percentage deviation from the measured value for a single test, Δ_{med} is the average of Δ_i, and N is the number of tests performed.

Table 3 shows the performance of the main tested methods, implemented as described in [7]. In the following, we will analyze in particular the best three correlations: the homogeneous equilibrium model, Friedel and Sun-Mishima.

Table 3. Statistical reproductive accuracy of the tested methods

Correlation	Δ_{i_min} (%)	Δ_{i_max} (%)	σ (%)	RMS (%)
HOMOGENEOUS MODEL	-59.1	+22.3	14.1	30.4
FRIEDEL	-50.5	+26.9	13.6	23.1
MULLER-STEINHAGEN	-92.0	+134.3	33.1	54.4
SUN-MISHIMA	-51.1	+89.5	23.9	24.2
BEATTYE_WHALLEY	-46.1	+212.9	48.5	58.6

Figure 16 shows the comparison between the frictional pressure drops calculated with the equilibrium homogeneous model and the measured ones. This method has considerable underestimation for low Δp and a few overestimations when the Δp approaches 60 bar. The high value of RMS compared to the standard deviation indicates a high dispersion of the predictions.

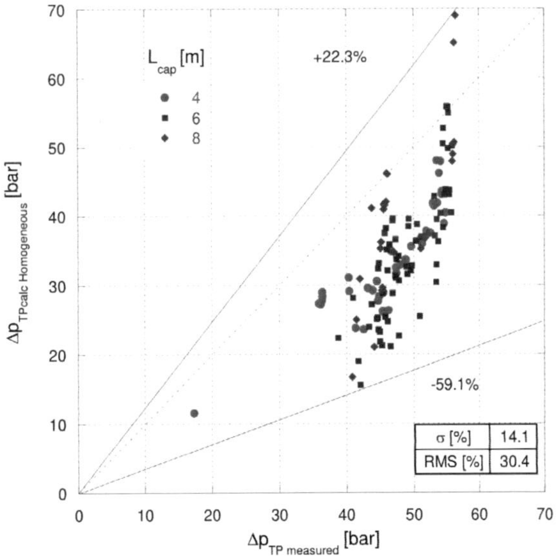

Figure 16. Δp_{fr} calculated with the omogeneous model vs measured Δp_{fr}

The predictions of the Friedel method have a trend similar to those of the homogeneous model, but are more centered around the measured value (figure 17). In fact, the RMS is lower, indicating a lower dispersion of predictions.

Figure 18 shows the quality of the prediction of the Friedel correlation as a function of saturation pressure. The capacity of prediction has a trend fairly stable up to pressures of the order of 70 bar with a tendency to underestimation of 20% - 40%. Only when the pressure is above 70 bar the method presents some overestimation, which, apart from one single experimental point, remains lower than 10%. We can not note any particular trend with respect to Reynolds number.

Journal of Physics: Conference Series **501** (2014) 012018 doi:10.1088/1742-6596/501/1/012018

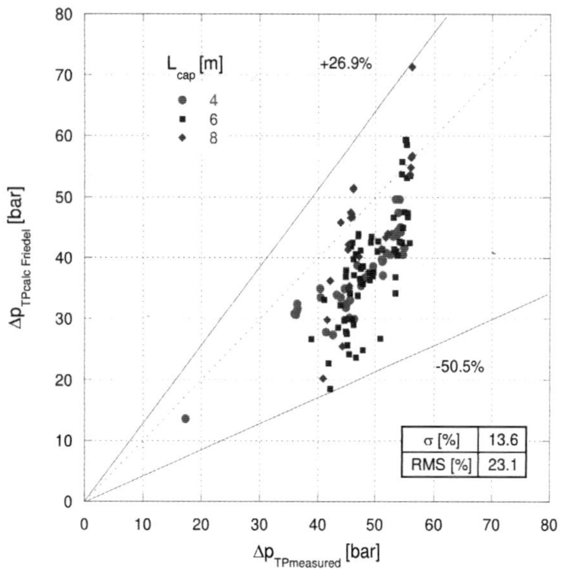

Figure 17. Δp_{fr} calculated with the Friedel correlation vs measured Δp_{fr}

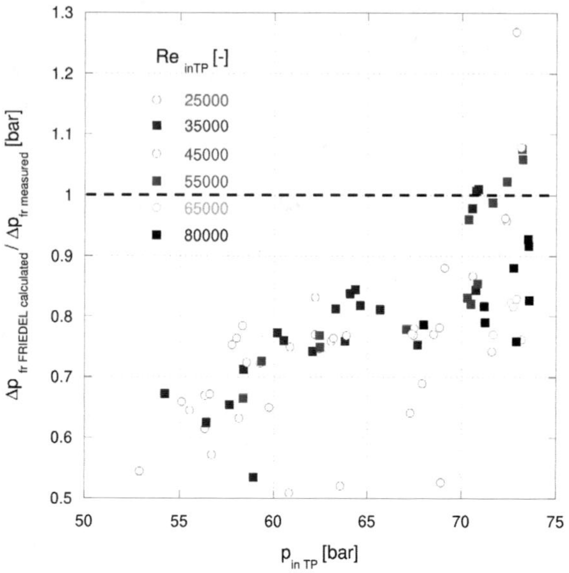

Figure 18. Friedel correlation Δp_{fr} predictions vs isoenthalpic saturation pressure; parameter: Re at starting of two-phase flow

Figure 19 shows the predictions of the Sun and Mishima correlation. This correlation adapts the correlation of Lockhart-Martinelli, in particular the two-phase multiplier, also using experimental data obtained with CO_2 and pipes of diameter between 0.5 and 12mm. The correlation, implemented as reported in [7], provides dispersed predictions with high overestimation values, up to 90% for high saturation pressures (figure 20). On the other hand, for low saturation pressures the underestimation can even reach 50%.

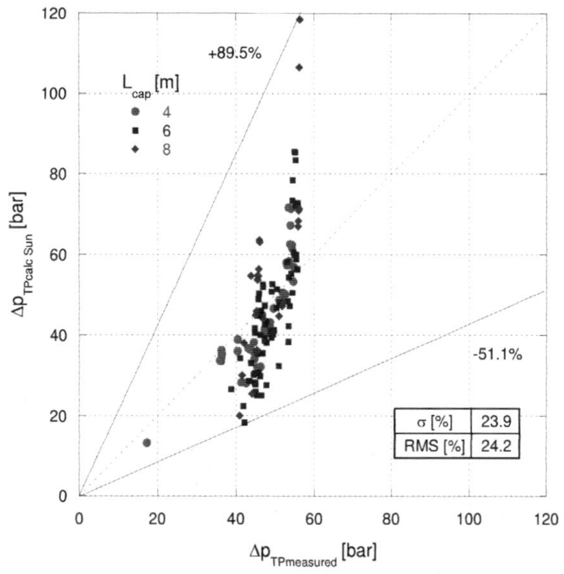

Figure 19. Δp_{fr} calculated with the Sun-Mishima correlation vs measured Δp_{fr}

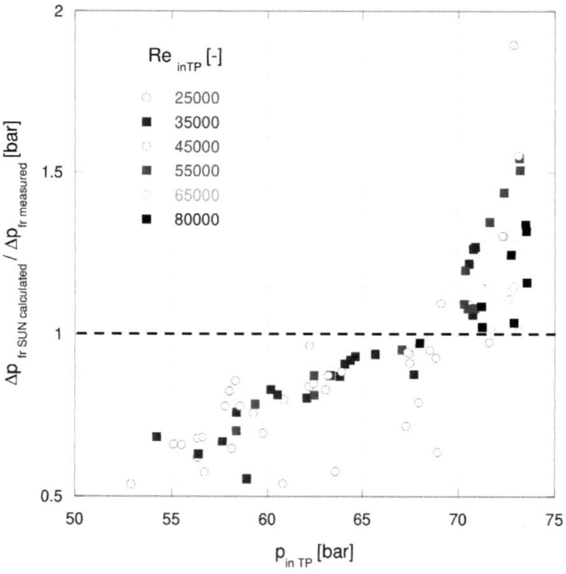

Figure 20. Sun-Mishima correlation Δp_{fr} predictions vs isoenthalpic saturation pressure; parameter: Re at starting of two-phase flow

Figure 20 shows the quality of the prediction of the Sun and Mishima correlation as a function of the saturation pressure. Prediction capability tends to grow linearly with the saturation pressure up to pressures of the order of 70 bar and then rapidly increase beyond the optimal values for higher pressures. Even in this case, we can not note any particular trend with respect to Re number.

4. Conclusions

This work shows the results of experimental tests, carried out on 3 capillary tubes of different lengths, to characterize and size a capillary tube that ensures the achievement of certain cooling performance within definite reference conditions. This capillary is a part of a prototype of a household refrigerator that uses CO_2 as refrigerant.

Regarding to the refrigeration performance, tests have shown that:

- with unfavourable environmental conditions (i.e. with outside temperatures higher than 32-35 °C) and with capillary lengths of 4 and 6 m, the desired evaporation temperature (almost -25 °C) can not be reached;
- the cooling capacity obtained is often larger or equal to the desired one.

Regarding to operating conditions, experimental data seem to confirm that, by setting a suitable refrigerant charge, it is possible to work with upper pressures close to the optimum values even using a capillary tube as expansion device, for a large range of external conditions.

The evaluation of the pressure drop during the expansion is important, among other, for the definition of the characteristic curve of the capillary tube, which, coupled with that of the compressor, defines the operating points and the performance of the refrigeration circuit.

Among the tested methods for the calculation of two-phase pressure drop, only the Friedel and homogeneous methods provide satisfactory predictions. All models have a noticeable worsening of the predictions when the saturation pressure approaches the critical one. This may be due to the assumption of calculation values for thermodynamic and thermophysical properties very different from the actual measurements. Indeed, these properties are computed as a function of some measured parameters (for example, isobaric specific heat and dynamic viscosity are a function of temperature, figure 2 and 3). Then a small error in the measurement, possible despite the use of high-precision instruments, can lead to large errors in the calculation of the thermodynamic and thermophysical properties.

References

[1] Cecchinato L, Corradi M, Fornasieri E, Schiocet G and Zilio C 2009 Assessment on the use of common correlations to predict the mass-flow rate of carbon dioxide through capillary tubes in transcritical cycles, *Int J Refrig* **32** 1041-1048

[2] Boccardi G, Calabrese N, Celata GP, Mastrullo R, Mauro AW, Perrone A and Trinchieri R 2013 Experimental performance evaluation for a carbon dioxide light commercial cooling application under transcritical and subcritical conditions, *Appl Therm Eng* **54** 528-535

[3] Kim MH, Pettersen J and Bullard CW 2004, Fundamental process and system design issues in CO_2 vapour compression systems *Prog Energ Combust* **30** 119-174

[4] Corradi M 2004 Analisi teorica e sperimentale di sistemi frigoriferi operanti con CO_2, *dottorato di ricerca in fisica tecnica; Università degli Studi di Padova*

[5] Madsen KB, Poulsen CS and Wiesenfarth M 2005 Study of capillary tubes in a transcritical CO_2 refrigeration system *Int J Refrig* 28 1212-1218

[6] Swamee PK and Jain AK 1976 Explicit equations for pipe-flow problems *J HydrEng Div-ASCE* **102** 5 657-664

[7] Kim SM and Mudawar I 2013 Universal approach to predicting two-phase frictional pressure drop for mini/micro-channel saturated flow boiling *Int J Heat Mass Tran* **58** 718–734

[8] Friedel L 1979 Improved friction pressure drop correlations for horizontal and vertical two phase pipe flow *European Two-Phase Group Meeting, Ispra (Va), Paper E2*

[9] Sun L and Mishima K 2009 Evaluation analysis of prediction methods for two-phase flow pressure drop in mini-channels, *Int. J. Multiphas Flow* **35** 47-54

[10] Cecchinato L, Corradi M, Schiocet G and Zilio C 2007 Analisi teorica e sperimentale del moto adiabatico nella laminazione di anidride carbonica in tubi capillari *Proceedings of 62° Congresso Nazionale ATI* 32-37

[11] Zhang C and Ding G 2004 Approximate analytic solutions of adiabatic capillary tube *Int J Refrig* **27** 17-24

Journal of Physics: Conference Series **501** (2014) 012019 doi:10.1088/1742-6596/501/1/012019

Experimental investigation of free surface vortices and definition of gas entrainment occurrence maps

G Caruso, L Cristofano, M Nobili and D Vitale Di Maio

"Sapienza" Università di Roma
Dip. Ingegneria Astronautica, Elettrica ed Energetica (DIAEE) – Area Nucleare
C.so Vittorio Emanuele II, 244 – 00186 Roma (Italy)

E-mail: gianfranco.caruso@uniroma1.it, damiano.vitaledimaio@uniroma1.it, luca.cristofano@uniroma1.it, matteo.nobili@uniroma1.it

Abstract. For the future development of Generation IV nuclear reactors, both safety and economic targets have to be achieved. In order to increase, at the same time, the power density generation and the safety features, a huge R&D effort is still required. Referring especially to Liquid Metal Cooled Fast Reactors, much attention is placed on Gas Entrainment (GE) phenomena, which could cause unlikely positive reactivity insertion accident. The GETS experimental facility (Gas Entrainment Test Section), especially aimed at studying the free surface vortices occurrence, has been built in the thermal-hydraulics laboratory of the DIAEE. The main purpose of this facility is to identify the most important parameters affecting the whirlpools formation and evolution. Experimental tests and preliminary observations have been performed. Different vortex behaviours related to different experimental conditions have been identified and presented in the present paper. 2D occurrence maps as function of different dimensionless groups (Reynolds, Froude and Weber numbers and $H^* = H/d$ ratio) have been defined. In the present paper, the results of a first experimental campaign, carried out with tap water, are discussed.

1. Introduction

The free surface vortices occurrence is an important issue in many hydraulic intakes (i.e. pumps and turbines) since strong free surface vortices introduce swirl flow at the intake causing entrainment of floating matters and gas (air or other gases). These conditions are considered more or less harmful for both the intake itself and the downstream devices. The free surface vortices formation has been studied in the last fifty years due to the important issues related to this phenomenon. In the hydraulic field, many studies have been performed to prevent the free surface vortex formation while, only in the last years, the GE phenomenon was considered an important issue in the nuclear industry. The GE, due to free surface vortex formation, can strongly affect the reactor safety performance especially in liquid metal cooled reactors. A volume of slightly pressurized inert gas is always maintained above the liquid metal free surface in order to avoid chemical reaction issues both due to the liquid metal or to the air. On the other hand, the presence of the inert gas, above the free surface, can cause gas bubbles entrainment into the primary coolant. Because of the GEN IV nuclear reactors have to be economically competitive with LWRs, it was necessary to foresee a reactor vessel size reduction, an increase in the reactor density power and hence in the coolant velocity within the primary circuit. This means increasing in the upper plenum coolant velocity causing instability in the free surface and enhancing the risk of free surface vortex formation as well as other GE phenomena occurrence, if any special solution will not be foreseen (e.g. vortex breaker).

The presence of any gas bubble entrained into the reactor coolant, and hence the passage through the reactor core, can affect the reactor safety leading to abnormal conditions up to accidents. In particular, GE occurrence can cause:

- reactor instability (i.e. reactivity fluctuation);
- thermal crisis due to sudden reduction of the local heat transfer coefficient;
- thermal stresses, especially to those structures in contact with both liquid metal and cover gas;
- cover gas activation, which can reduce the safety system performance;
- increasing in the instrumentation noise.

Due to these important safety issues, the free surface vortex phenomenon has to be deeply investigated, in order to understand the formation mechanism and to prevent the occurrence. Because of its own physical structure (e.g. gas-liquid interface, instability, etc.), the free surface vortex phenomenon results to be very difficult to study and its dynamics is not yet well understood.

Several paths have been followed to study the free vortex formation and evolution, and they can be grouped in analytical, numerical and experimental approaches, respectively: definition of the velocity fields from the Navier-Stokes equations, CFD simulations and development of empirical correlations.

The first theoretical vortex model, aimed at defining the tangential flow velocity, was proposed in the mid '800 by Rankine [1]. Following this first approach, the work of Burgers [2] was focused at the solution of the Navier-Stokes equations, considering the axial velocity linearly dependent from the axial position. The first complex models, based on experimental observations, were proposed by Chen et al. [3]. Odgaard [4] proposed correlations to calculate the depth of incipient air entrainment critical submergence and investigated on the most important factors affecting the phenomenon. A more consistent theory, focused on vortices within a rotating vessel, was proposed by Lundgren [5] and Andersen et al. [6] added later the effects of the surface tension and of the viscosity in the Ekman boundary layer near the outlet orifice. Later, Stepanyants et al. [7] extended this model to a non-rotating vessel. An analytical expression of the free surface vortex velocity was derived by Hite et al. [8]. Several dimensional analyses and experimental tests, aimed at identifying dimensionless groups able to describe the phenomenon, were performed by many researcher groups [8]-[13]. In order to make the phenomenon more stable during the time, a special geometrical configuration (i.e. imposition of a tangential inlet) was selected by Moriya [14] and Monji et al. [15] to study the velocity distribution around the vortex. Some experimental tests were carried out by Kimura et al. [16]-[18] to study the free vortex occurrence under specific conditions (e.g. in the Japanese Sodium Fast Reactor - JSFR): the authors developed a scaled down partial model of the JSFR to understand the main factors and mechanisms related to the GE within this geometry. This scaled model was aimed at investigating how the fluid physical properties affect the GE phenomena, and to evaluate the size distribution of entrained bubbles from free surface vortices. Some studies based on the numerical approach, especially those of Sakai et al. [19] and Ito et al. [20]-[21], were focused on the development of a high precision methodology based on Volume Of Fluid – VOF - algorithm to be used in numerical simulations. The above mentioned most important research works guaranteed steps forward in the understanding of vortices formation but the free surface vortex formation and its evolution mechanisms are not yet well understood.

In this work experimental tests, carried out to analyse the vortex formation and the GE phenomena, and their occurrence maps based on qualitative observations of the vortex, are presented. Dimensionless groups have been investigated to evaluate the influence of the outlet tube diameter, surface level and flow rate.

2. Experiments

To carry out the experimental tests, the GETS facility (Gas Entrainment Test Section), with water at the ambient temperature as working fluid, was used. The experimental facility includes a parallelepiped-shaped tank (made of PMMA), a circulation pump, valves and a flow meter. It was designed and built mainly to study the GE phenomena, due to the free surface vortex, in a no vortices promoting geometry (figure 1). In order to investigate how the drain diameter affects the GE phenomenon, a special flanged connection has been studied; it allows to easily change the diameter of the draining pipe without any other changes in the facility layout. Moreover, the most important parameters that can be changed are the mass flow rate and the free surface level. From this experimental campaign, different occurrence maps, focused on the GE onset condition, have been obtained.

2.1. Experimental setup
In figure 1 a flow diagram and a 3D view of the GETS facility are shown. The facility is realized to allow an easy filling and draining of the tank, to control finely the mass flow rate and to change the inlet flow configuration (i.e. passing from two symmetrical inlet flows up to an asymmetrical inlet). To perform experimental tests both in steady state and transient conditions (blow-down) the GETS facility can operate as a closed or open loop.

The draining and refilling operations are carried out by an auxiliary line, which is connected through a shut-off valve to the pump upstream line. The water flow rate can be controlled both by the regulation valve (coarse regulation) and by the inverter installed on the pump motor (fine regulation). Concerning the inlet flow condition, the steady state with symmetrical inlets (valves fully opened) has been considered in the present work.

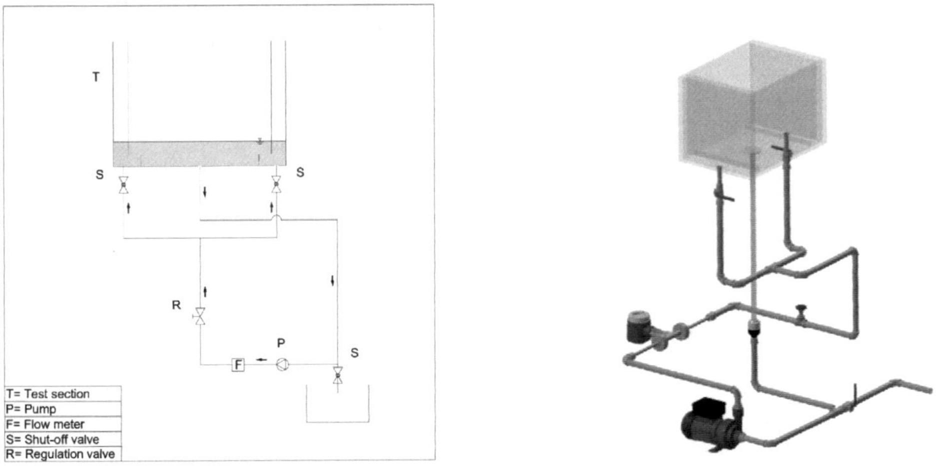

Figure 1. A schematic and a 3D views of the GETS test section

The GETS layout foresees that water is drawn from the tank through the circular orifice to the draining pipe up to reaching the circulating pump. Downstream the circulating pump, a flow meter with its upstream and downstream free lengths, to avoid any effect on the flow rate measurements, is foreseen. A vortex flowmeter, Prowirl 73F produced by Endress-Hauser, has been used to measure flow rates from 85 l/h to 3400 l/h with an accuracy of 0.75% of the measured value.

Before reaching the tank again, the flow passes through the regulating valve and, after a tee joint, it reaches the two shutoff valves upstream the two inlet nozzles. Finally, the flow, through two completely symmetrical inlets, enters the active volume of the tank (external dimensions 700 mm × 500 mm × 500 mm - height×length×width - with walls 20 mm thick), where the free surface vortex observations can be performed. In order to overcome the constraints of a free surface vortex induced by the facility geometry, as in the most of the previous research works, the GETS facility has been designed to allow completely symmetrical inlet flows, which do not generate a rotating flow within the tank. In order to avoid (or limit) any free surface perturbation, vertical baffles are foreseen between the inlet and outlet nozzles. Thanks to these baffles, the upward inlet flows can reach horizontally the active zone of the tank, near the outlet orifice, limiting the free surface perturbations to the lateral zones (not included in the active volume of the tank).

The outlet orifice, in the centre of the tank bottom wall, has been equipped with three different draining pipes (i.e. d = 14 mm, 26 mm and 38 mm inner diameter) thanks to the flanged connecting system; the connection flange has a diameter of 100 mm with four screws to assure a fixed position. The connected draining pipe is also made of PMMA in order to make easier the observation of the outlet flow for several diameter under the tank.

In figure 2 the GETS facility, during current experiments aimed at measuring the velocity fields by the PIV (Particle Image Velocimetry) technique, is shown.

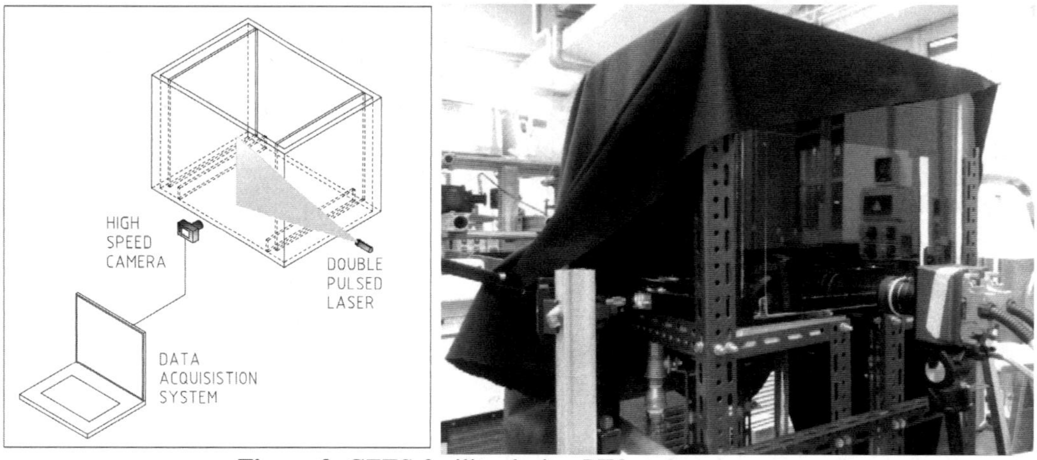

Figure 2. GETS facility during PIV measurements

2.2. *Experimental procedure*

The present study is focused on the development of occurrence maps, based on qualitative classification criteria, aimed at defining the onset condition for GE under different operating conditions. To obtain comparable observations, the following specific procedure has been developed and followed during the experimental tests:

- initial condition: from the previous test, tank filling or level changing;
- waiting period: a waiting time of 10 minutes is required to reduce free surface oscillations and to prevent vortex induction from the starting transient, until a steady condition within the tank is reached (the coarse regulation of the pump flow rate is performed before this waiting period);
- fine regulation: by means of the regulating valve and the pump inverter, the fine regulation of the flow rate can be completed;
- observation: the tank active zone is observed for a period of 10 minutes, recording the mass flow rate, the temperature (to evaluate correct values of the thermodynamic water properties) and the flow qualitative condition (see Table 1);
- conclusion: after 10 minutes of observation, the regulating valve and the free surface level are regulated for the next test.

Overall, each test lasted 1200 seconds. This procedure allows to compare data from different experimental tests preventing any possibility of vortex induction by the initial flow condition. The qualitative observation criteria, used in the present work, is derived from Hecker's [22] method and it is shown in Table 1.

Table 1. Qualitative classification criteria

Stage	Name	Description
1	Surface swirl and very small dimple	A constant visible swirl and a very small deformation on the free surface. Curvature observable only by means of the refraction phenomenon.
2	Well-developed dimple	A deformation of at least a few millimetres in the centre of the swirl.
3	Bubble entraining core	Vortex pulls air bubbles into the intake.
4	Full air core	Vortex pulls constant stream of air into the intake.

2.3. Experimental conditions

The experimental tests have been performed with a quite wide range of the main parameters affecting the free surface vortex phenomenon. The minimum and maximum values of the main parameters are shown in Table 2.

Table 2. Main parameter ranges.

Parameter	-	Minimum value	Maximum value
Pipe inner diameter	[mm]	14	38
Mass flow rate	[kg / s]	0.025	0.625
Free surface level	[mm]	50	225

In this study, the following dimensionless groups have been considered:

- the Reynolds number $$Re = \frac{\rho(T) U d}{\mu(T)}$$
- the Froude number $$Fr = \frac{U}{\sqrt{g H}}$$
- the Weber number $$We = \frac{\rho(T) U^2 d}{\sigma(T)}$$
- the dimensionless depth $$H^* = \frac{H}{d}$$

where d is the inner diameter of the pipe, m; g, gravity acceleration, 9.806 m s^{-2}; H, water level, m; T, water temperature, K; U, average outlet velocity, m s^{-1}; μ, dynamic viscosity, Pa s; ρ, water density, kg m^{-3}; σ, gas-liquid surface tension, N m^{-1}. Water temperature varied from 18.6° C to 32.5 °C. In this range, the water density and dynamic viscosity varied, respectively, from 998.48 to 994.9 kg/m^3, and from 1036.92 to 758.26 μPa·s, Surface tension varied from 0.07295 to 0.07082 N/m. Reference has been made to IAPWS- 95 (Int. Ass. for the Properties of Water and Steam, Formulation 1995 for the Thermodynamic Properties of Ordinary Water Substance for General and Scientific Use).

3. Results and discussion

3.1. Experimental observations

During the experimental tests all the four vortex stages have been observed and occurrence maps have been generated merging experimental data obtained with the different outlet tube diameters. A not perfectly stable condition has been identified in all the performed tests. The vortex position moved during the time, driven by the instantaneous conditions of the flow. Moreover, the vortices strength changed during the tests, resulting in a continuous variation of the vortex depth. It is not clear if this instability is a peculiar characteristic of the physical phenomenon, or if it only caused by the tank geometry and flow perturbations introduced by the pump. Anyway, each experimental condition has been classified considering the worst vortex stage observed during the test.

Figure 3. Pictures of vortex different stages: a) Stage1, b) Stage 2, c) Stage 3, d) Stage 4.

Figure 4 shows the four different stages considered in this analysis. Mirror images visible in the pictures are due to the reflection of vortex on the water free surface. This effect can't be avoided due to the position of the camera below the free level surface.

The vortex *Stage 1* (figure 4-a), which is the initial stage for the formation of a vortex, is characterized by a constant swirl and a very small deformation of the free surface, which can be only observed through the light refraction. Thanks to tracer particles, it has been observed that, even in this stage, the rotational flow is not confined only to the free surface but it affects the entire zone near the outlet orifice.

The *Stage 2* includes all vortices that have an appreciable free surface deformation (more than 1 mm). As can be seen in figure 4-b, in this stage the vortex acquires an axial dimension and a gaseous core extends toward the outlet orifice; in this stage there is not bubbles detachment from the tip. As the vortex depth increases, the tip becomes thinner, and the surface tension has a greater importance in the shape definition of the free surface. When the depression of the gas-liquid interface is small the flow in the vortex core can be considered axi-symmetric while, when the vortex is well-developed, this is not true anymore. In these cases, the axis of the vortex is inclined because of the local conditions of the flow; probably because well-developed vortices occur at higher velocities and their thin structure is more sensitive to flow disturbances.

The *Stage 3* is reached when, from the vortex, some bubbles are pinched off from the tip by the downward flow and are dragged into the outlet orifice (figure 4-c); this stage, therefore, involves a gas entrainment into the experimental loop, and it is particularly unstable, rapidly evolving to a worst condition (*Stage 4*) or comes back to *Stage 2*. Figure 4-c shows that the size of the entrained bubbles is comparable with the thickness of the tip of the gas core.

The *Stage 4* conditions correspond to a full developed vortex whose tip reaches the outlet orifice (figure 4-d) causing a continuous gas entrainment into the experimental loop. When too much air is entrained, pump cavitation occurs and the flow rate goes down; thereby, the downward flow decreases and the vortex goes back to the *Stage 2*, passing through the unstable *Stage 3*. When no more air is entrained, the pump recovers its efficiency and the flow rate returns to the previous level, recreating the occurrence conditions for *Stage 4* vortices.

Empirical correlations of the different transition zone boundaries have been obtained in [23], to reformulate the transition criteria in a dimensionless form.

3.2. Occurrence maps

The experimental results have been used to build 2D occurrence maps. In each map, the *Stage 3* and the *Stage 4* cases are gathered together in a single group (*GE*), because both conditions lead to gas entrainment phenomena. An evaluation of the different vortex stage zone boundaries has been done in the different occurrence maps; however, this must be considered only a preliminary and rough estimation of the transition boundaries, because of the qualitative classification criteria adopted in this work. The figure 5 shows the *H*-Re* occurrence map of the experimental tests carried out with the smaller tube diameter (*d* = 14 mm). It can be seen that the *No vortices-Stage 1* and *Stage 1-Stage 2* transition zones are well defined, while the *Stage 2-GE* transition zone is characterized by some mismatches; this is probably due to the greater importance that other dimensionless groups (*Fr* and *We)*, that are not considered in this map, have on the physical dynamics of the gas entrainment phenomenon. Moreover, the high flow rate in some cases, which correspond to high *Re* numbers, introduced oscillations of the free surface and tiny waves that seem to have an inhibitory effect on the vortex intensification and the gas entrainment occurrence.

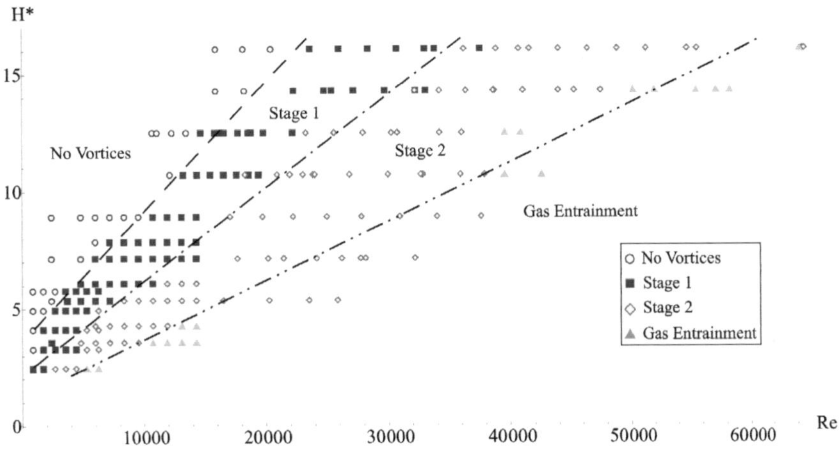

Figure 4. *H*-Re* Occurrence map (*d*=14 mm)

Other two 2D occurrence maps have been drawn plotting the experimental data, obtained with the smaller outlet tube diameter, in *H*-We* (figure 6) and *H*-Fr* (figure 7) graphs. Even in these two maps, the transition zones are quite well defined, and the most evident mismatches are localized in correspondence of the *Stage 2-GE* transition. Moreover, while the boundaries of the *H*-Re* map have an almost linear trend, the functional dependence of *H** from *Fr* and *We* is clearly non-linear.

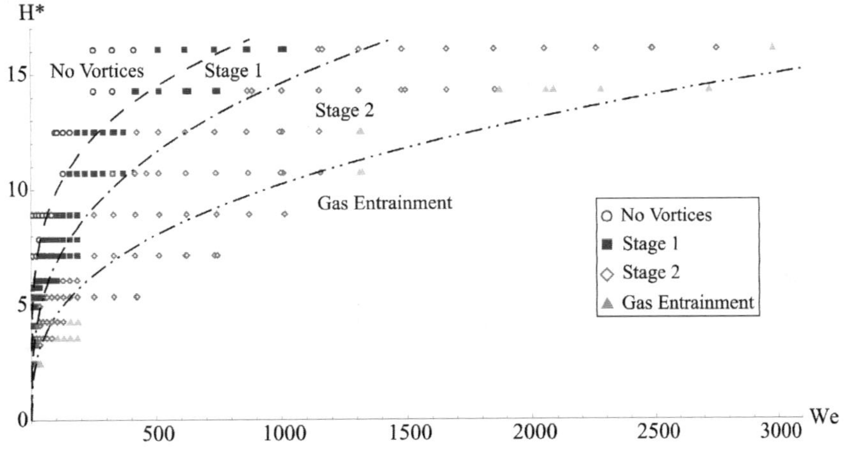

Figure 5. *H*-We* Occurrence map (*d*=14 mm)

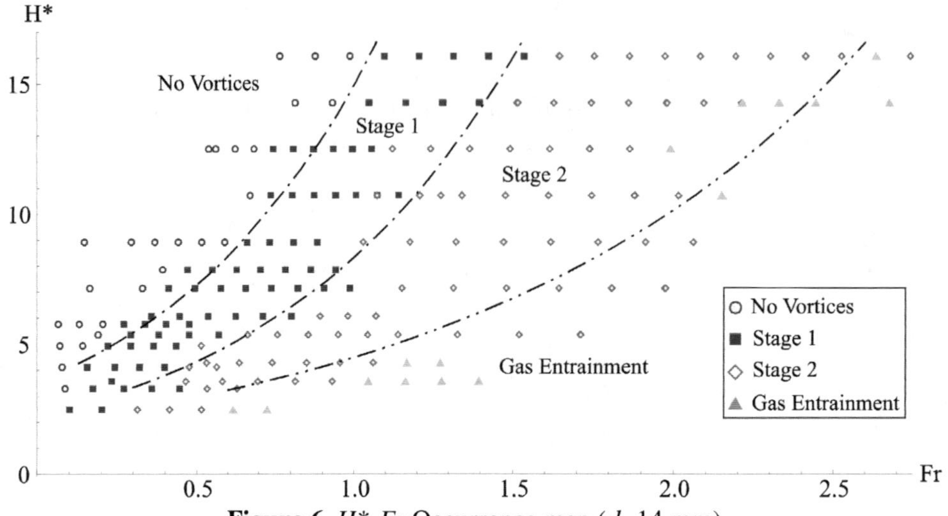

Figure 6. *H*-Fr* Occurrence map (*d*=14 mm)

The occurrence maps of the experimental results obtained with the two other outlet pipe diameters have similar trends to those shown in figure 5, figure 6 and figure 7.

In figure 8 and figure 9 two overall occurrence maps, obtained merging together the experimental results obtained with the three outlet pipe diameters, are presented. The data series characterized by different outlet diameters are not in perfect agreement with each other, and this means that the parameter *d* has a non-negligible effect on the formation and evolution of free surface vortices. However, the transition boundaries of the different vortex stage zones have been preliminary identified. Concerning the *H*-Re* map (figure 8) and the *H*-We* map (figure 9), a general agreement of the experimental results can still be noticed, at least for trends of the different transition zones.

Otherwise, in the *H*-Fr* map (gathering together all the experimental data) no vortex stage zone boundaries have been identified. In this case, the series of experimental data obtained with different outlet diameters do not match well together. This fact could be explained considering that the *Fr* number, as defined in this paper, does not explicitly depend on outlet pipe diameter, unlike the *Re* and *We* numbers. In any case, it is evident that, concerning the *Fr* number influence on vortex formation and GE occurrence, the outlet pipe diameter has non-negligible effects.

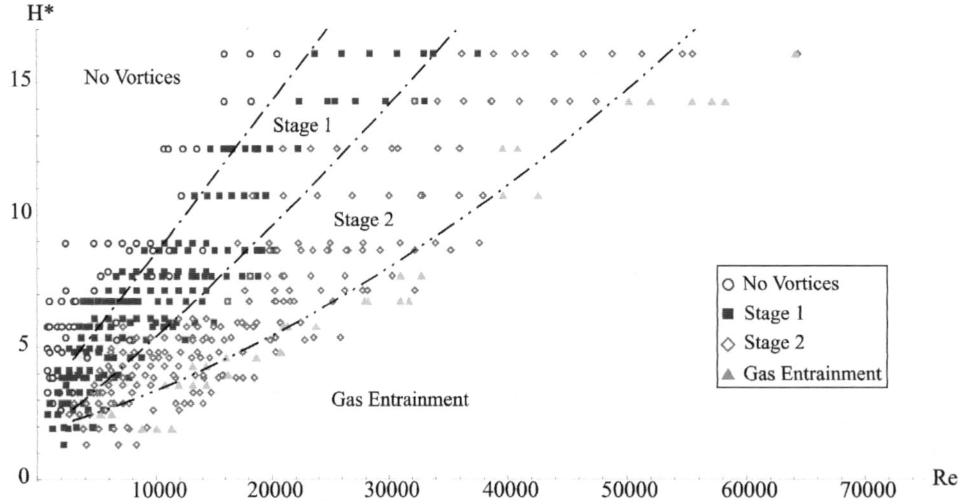

Figure 7. *H*-Re* Occurrence map

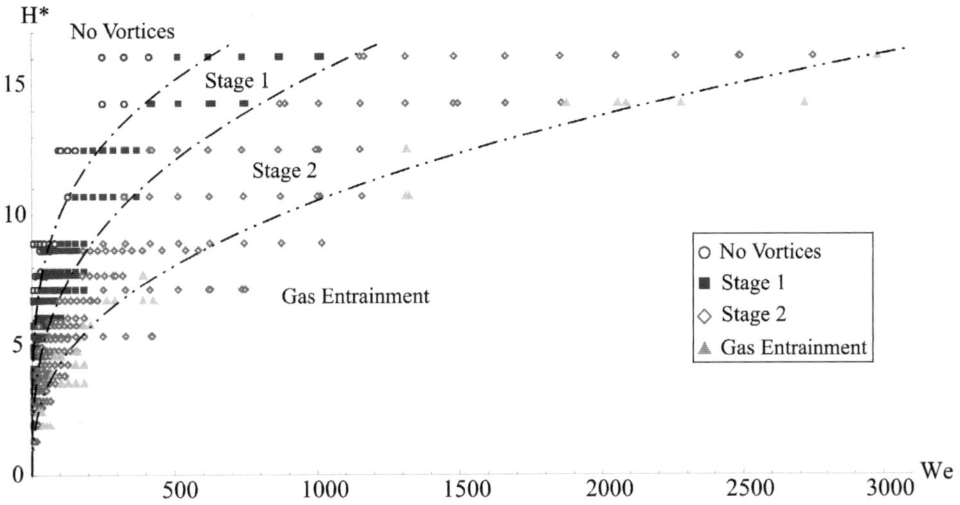

Figure 8. *H*-We* Occurrence map

4. Conclusions

The GETS facility has been built with the aim of experimentally study the formation and evolution of free surface vortices and the GE occurrence, and a first experimental campaign has been carried out. The observed phenomena have been classified according to qualitative criteria, and different 2D occurrence maps have been obtained.

The occurrence maps, related to a single outlet pipe diameter, show transition zone boundaries quite well defined, with the main mismatches located at the *Stage 2-GE* transition zone. This probably means that the *Stage 2-GE* transition is influenced, in a non-negligible way, by all the dimensionless parameters, that cannot be considered together in 2D maps. Furthermore, strong perturbations of the free surface, introduced by the high flow rates that characterize this transition zone, could result in experimental anomalies.

Concerning the overall occurrence maps, instead, quite good accordance has been obtained with the *H*-Re* and *H*-We* maps, and preliminary transition zone boundaries have been identified. Otherwise, a poor agreement of the experimental results, obtained with different outlet pipe diameters, has been observed in the *H*-Fr* map.

The GETS experimental facility is very useful to study of free surface vortices since it allows to observe the physical phenomenon and to analyse the influence of different parameters on it. However, further quantitative studies are needed to have a better comprehension of the free surface vortex formation, with the aim of defining some empirical correlations. With this purpose, an experimental campaign is being carried out in order to characterize the vortex velocity fields by the PIV (Particle Image Velocimetry) technique. Furthermore it will be important to evaluate the behaviour of different fluids and to determinate the dominating phenomena in each stage [23].

5. References

[1] Rankine W J M 1858 *Manual of Applied Mechanics* (London:Griffin)
[2] Burgers J M 1948 A mathematical model illustrating the theory of turbulence *Adv. In Appl. Mech* 171-99 (New York:Academic Press)
[3] Chen Y I, Wu C, Ye M and Ju X 2007 Hydraulic characteristics of vertical vortex at hydraulic intakes *J. Hydrodynamics* **19-2** 143-49
[4] Odgaard A J 1986 Free-surface air core vortex *J. Hydr. Engrg.* **112** 610-20
[5] Lundgren T S 1985 The vortical flow above the drain-hole in a rotating vessel *J. Fluid Mech* **155** 381–412

[6] Andersen A, Bohr T, Stenum B, Jull Rasmussen J and Lautrup B 2003 Anatomy of a bathtub vortex *Phys. Rev. Lett.* **91(10)**
[7] Stepanyants Y A and Yeoh G H 2008 Stationary bathtub vortices and a critical regime of liquid discharge *J. Fluid Mech.* **604** 77-98
[8] Gulliver J S and Rindels A J 1985 Vortices at vertical intakes *Proc. ASCE Hydraulic Div. Specialty Conference*, 973-78
[9] Hite E J and Mih W C 1994 Velocity of vertical vortices at hydraulic intakes *J.Hydr Engrg.* **120** 284-97
[10] Anwar H O Weller J A and Amphlett M B 1978 Similarity of free-vortex at horizontal intake *J. Hydr. Research* **16** 95-105
[11] Jain A K, Ranga Raju K G and Garde R J 1978 Vortex formation at vertical pipe intakes *J. Hydr. Div.* **104** 1429-48
[12] Baum M R and Cook M E 1975 Gas entrainment at the free surface of a liquid: entrainment inception at a vortex with an unstable gas core, *Nucl. Eng. Des.* **32-2** 239-245
[13] Padmanabhan M and Hecker G E 1984 Scale effects in pump sump models *J.Hydr Engrg.* **110** 1540-56
[14] Moriya S 1998 Estimation of hydraulic characteristics of free surface vortices extension vortex, theory and fine model test measurements *Central Research Institute of Electric Power Industry Report*, No. U93004
[15] Monji H, Akimoto T, Miwa D and Kamide H 2004 Unsteady behavior of gas entraining vortex on free surface in cylindrical vessel *Proc. of Fourth Japan-Korea Symposium on Nuclear Thermal Hydraulics and Safety* 190-94.
[16] Ezure T, Kimura N, Miyakoshi H and Kamide H 2011 Experimental investigation on bubble characteristics entrained by surface vortex *Nucl. Eng. Des.* **241** 11
[17] Kimura N, Ezure T, Miyakoshi H and Kamide H 2010 Experimental study on gas entrainment due to nonstationary vortex in a sodium cooled fast reactor-comparison of onset conditions between sodium and water *J. Eng. for Gas Turbines and Power* **132** 10
[18] Kimura N, Ezure T, Tobita A and Kamide H 2008 Experimental study on gas entrainment at free surface in reactor vessel of a compact sodium-cooled fast reactor *J. Nucl. Sci. and Tech.* **45** 1053-62.
[19] Sakai T, Eguchi Y, Monji H, Ito K and Ohshima H 2008 Proposal of design criteria for gas entrainment from vortex dimples based on a computational fluid dynamics method *Heat Transfer Engineering* **29** 731-39
[20] Ito K, Kunugi T, Ohshima H and Kawamura T 2009 Formulations and validations of high-precision volume-of fluid algorithm on nonorthogonal meshes for numerical simulations of gas entrainment phenomena *J. Nucl. Sci. and Tech.* **46** 336-73
[21] Ito K, Kunugi T, Ohshima H and Kawamura T 2011 Two-phase flow simulation of gas entrainment phenomena in large-scale experimental model of sodium-cooled fast reactor *Prog. Nucl. Sci. and Tech.* **2** 114-19
[22] Hecker G E 1981 Model-Prototype Comparison of Free-Surface Vortices *J. Hydr. Div.* **107** 1243-59
[23] Cristofano L, Nobili M and Caruso G 2013 Experimental study on unstable free surface vortices and gas entrainment onset conditions *Exp. Therm. Fluid Sci.* doi: 10.1016/j.expthermflusci.2013.09.015

Journal of Physics: Conference Series **501** (2014) 012020 doi:10.1088/1742-6596/501/1/012020

A dynamic mixed subgrid-scale model for large eddy simulation on unstructured grids: application to turbulent pipe flows

P Lampitella, E Colombo and F Inzoli

Department of Energy, Politecnico di Milano, Via Lambruschini 4, 20156 Milano

E-mail: paolo.lampitella@mail.polimi.it

Abstract. The paper presents a consistent large eddy simulation (LES) framework which is particularly suited for implicitly filtered LES with unstructured finite volume (FV) codes. From the analysis of the subgrid-scale (SGS) stress tensor arising in this new LES formulation, a novel form of scale-similar SGS model is proposed and combined with a classical eddy viscosity term. The constants in the resulting mixed model are then computed trough a new, cheaper, dynamic procedure based on a consistent redefinition of the Germano identity within the new LES framework. The dynamic mixed model is implemented in a commercial, unstructured, finite volume solver and numerical tests are performed on the turbulent pipe flow at $\mathrm{Re}_\tau = 320-1142$, showing the flexibility and improvements of the approach over classical modeling strategies. Some limitations of the proposed implementation are also highlighted.

1. Introduction

In the last two decades, LES has emerged as the working horse of turbulence modeling approaches for several classes of flows and increasingly tends to be used for complex multi-physics applications [1]. However, while the numerical techniques and practices have strongly evolved in the years, recent LES approaches still tend to be strongly tied to the earliest formalization of the field [2]. This apparently innocuous trend can, nonetheless, become a limit when the so called implicit filtering approach is adopted. In implicitly filtered LES the separation between large, resolved scales and small, unresolved ones is implicitly determined by the grid and the numerical method in use; as a consequence, the features of the resolved flow field and the requirements for the necessary SGS model are directly affected.

A well known example of this comes from the implicit filtering approach with the FV method, where the flow equations are affected by the top-hat filtering implied by the integral formulation and a closure problem arises at the flux level. Nonetheless, this is not usually considered and most of the LES formulations for FV usually rely on the classical SGS closures originally developed in a continuous differential framework. Hence, as the SGS model is the only one responsible for representing the filter effects, a basic contradiction arises for implicitly filtered LES with FV: not only the discretization error is neglected but, more importantly, the form of the underlying implicit filter is assumed as completely different from the real one [3].

The aim of the present work is to introduce an LES framework which is a generalization of previous works [4, 5] and allows a clear, consistent identification of filter effects which is independent from the specific numerical method adopted. The framework is then applied to the case of an implicitly filtered FV discretization and a Taylor series analysis of the resulting SGS stress tensor is used to derive a form of scale-similar model which is fully consistent with the numerical formulation. Finally, the scale-similar term is combined with a classical eddy viscosity closure in a new form of dynamic procedure which is substantially cheaper than the classical one. In extension of preliminary results obtained for the turbulent plane channel flow [6], the formulation is tested for the first time with unstructured hexaedral grids in the simulation of the turbulent pipe flow at $\mathrm{Re}_\tau = 320 - 1142$ [7, 8] using the commercial FV solver Fluent.

2. LES formulation

As in classical LES, the filtering operator G^i is introduced:

$$\bar{\phi}\left(\vec{x},t,\Delta^i\right) = G^i * \phi\left(\vec{x},t\right) = \int_\Omega G\left[\vec{x}-\vec{\xi},\Delta^i\left(\vec{x}\right)\right]\phi\left(\vec{\xi},t\right)\mathrm{d}\vec{\xi} \tag{1}$$

however, differently from classical LES, here the fluid domain Ω is considered finite and the filter kernel G^i has a spatially varying width $\Delta^i(\vec{x})$. Notice that this is not a convolution filter, hence commutation with spatial derivative operators does not hold; however, the interaction with the domain boundaries can be avoided (as assumed here) by a proper reduction of the filter width. For future reference, we also introduce the conventions:

$$\begin{gathered}\bar{\phi}^n = G^n * \bar{\phi}^{n-1} = G^n * G^{n-1} * \ldots * G^1 * G^0 * \bar{\phi}^0 \\ G^0 * \bar{\phi}^0 = \bar{\phi}^0 = \phi\,; \quad G^1 * \bar{\phi}^0 = \bar{\phi}\,; \quad \tilde{\phi}^n = \overline{\rho\phi}^n/\bar{\rho}^n\end{gathered} \tag{2}$$

where the ith filter kernel G^i has an associated filter width $\Delta^i \geq \Delta^{i-1}$. To make the approach as general as possible, the classical Favre filtering is also introduced, ρ being the density of the fluid. With this notation, it is possible to express the Navier-Stokes equations at a generic filter level $n \geq 0$; limiting the discussion to the momentum equations, we get:

$$\begin{aligned}\frac{\partial\left(\bar{\rho}^n\tilde{u}_i^n\right)}{\partial t} + \overline{\frac{\partial}{\partial x_j}\left(\bar{\rho}^n\tilde{u}_i^n\tilde{u}_j^n + \tilde{\sigma}_{ij}^n\right)}^n &= \overline{\frac{\partial}{\partial x_j}\left(\tau_{ij}^{n-0}\right)}^n \\ \tilde{\sigma}_{ij}^n &= \bar{p}^n\delta_{ij} - 2\mu\left(\tilde{S}_{ij}^n - \tfrac{1}{3}\tilde{S}_{gg}^n\delta_{ij}\right) \\ \tilde{S}_{ij}^n &= \tfrac{1}{2}\left(\frac{\partial\tilde{u}_i^n}{\partial x_j} + \frac{\partial\tilde{u}_j^n}{\partial x_i}\right) \\ \tau_{ij}^{n-m} &= \left(\bar{\rho}^n\tilde{u}_i^n\tilde{u}_j^n - \bar{\rho}^m\tilde{u}_i^m\tilde{u}_j^m\right) + \left(\tilde{\sigma}_{ij}^n - \tilde{\sigma}_{ij}^m\right)\end{aligned} \tag{3}$$

where τ_{ij}^{n-0} is the SGS stress tensor arising in this new formulation. With respect to classical LES formulations, we notice that no commutation with spatial derivatives is ever required and, as a consequence, the resulting SGS stress tensor is completely different as it now also includes pressure and viscous terms. A classical criticism with the proposed formulation is the lack of Galilean invariance; however, this is a property of any quantity which is non-uniformly filtered in space, hence it is natural that equations based on such quantities inherit this property too[9].

3. Finite volume based LES

The proposed formulation has a clear advantage in the fact that the filter (1) can be readily interpreted, without any additional assumption, as a FV discretization of the equations:

$$\bar{\phi}\left(\vec{x},t,V\right) = \frac{1}{\left|V\left(\vec{x}\right)\right|}\int_{V(\vec{x})}\phi\left(\vec{\xi},t\right)\mathrm{d}\vec{\xi} \tag{4}$$

with $|V(\vec{x})| = \left(\Delta^1(\vec{x})\right)^3$ being the volume of the generic grid cell. As the resulting equations now exactly fit those usually employed in general FV codes, two important consequences follow: a) FV codes are inherently based on fully consistent (i.e., without commutation errors) LES equations based on a top-hat filter and do not require specific modifications besides SGS modeling; b) the SGS stress tensor arising in the FV based LES approach is substantially different from the classical SGS stress tensor and, in theory, a dedicated modeling is required.

With respect to point (a), notice that it holds not only for the present density-based formulation, used just for the sake of generality, but also for pressure-based ones; indeed, it is a matter of simple manipulations to show that the following exact pressure equation holds:

$$\overline{\frac{\partial}{\partial x_i}\frac{\partial}{\partial x_j}\left(\bar{p}^n\delta_{ij}\right)}^n = \frac{\partial^2\bar{\rho}^n}{\partial t^2} - \overline{\frac{\partial}{\partial x_i}\frac{\partial}{\partial x_j}\left[\bar{\rho}^n\tilde{u}_i^n\tilde{u}_j^n - 2\mu\left(\tilde{S}_{ij}^n - \frac{1}{3}\tilde{S}_{gg}^n\delta_{ij}\right) - \tau_{ij}^{n-0}\right]}^n \tag{5}$$

which, once interpreted according to (4), exactly represents the usual FV discretization of the pressure equation (possibly assuming a constant density).

With respect to point (b), which is fully developed in the next sections, here we want to remark its importance in the implicitly filtered approach, as the SGS model becomes the only one representing the underlying filter operator and the resulting LES approach as a whole. However, even if this is not considered and a classical SGS model is used, an important difference still arises in the Germano Identity (GI) which holds in the present formulation and the resulting dynamic procedure [10]. Indeed, for any (n, m, k) combination of filter levels, the following generalized GI holds among the underlying SGS stresses:

$$\tau_{ij}^{m-k} - \tau_{ij}^{n-k} = \tau_{ij}^{m-n} \tag{6}$$

which presents several differences with respect to the classical one usually adopted. First, it does not involve any test filtered tensor, which means that the resulting dynamic procedure does not require the constants to be arbitrarily extracted from the test filter operation; also, this implies a substantial saving in computational costs as only the single variables need to be test filtered (including pressure and, possibly the density), in contrast to the classical dynamic procedure which requires filtering products of variables and additional non-linear terms as present in the adopted SGS model. As an example, for incompressible flows, the computation of the constant in a dynamic Smagorinsky model would only require 3 applications of the test filter with the present method against the 15 of the classical dynamic procedure (the 3 velocity components plus the 6 independent components of two additional tensors). Second, due to the peculiarity of the SGS stresses, the present GI also involves the pressure and viscous terms. The saving in the computational costs of this new GI will be exploited by a dynamic, two parameters, mixed model whose scale-similar part is developed in the next section.

4. SGS tensor analysis

While the appearance of the SGS stress tensor in a different form would generally require a different form of SGS model, this does not seem to be a relevant issue when an eddy viscosity parameterization is used. Indeed, as long as the required role of the model is an added dissipation, its specific implementation can be considered of secondary significance with respect to the main modeling assumption. However, this reasoning loses validity when a structural modeling approach is adopted, like with scale-similarity models [11]. Indeed, structural SGS models make no assumptions on the nature of the interaction between resolved and non resolved scales but try to recover the full structure of the SGS stress tensor (i.e., the eigenvectors) from information in the resolved scales; this, in turn, allows the possibility for backscatter and anisotropy which, for example, are particularly important in wall-bounded turbulent flows. Among the structural SGS models, scale-similar ones are peculiar in the fact that they extract the required information from the smallest resolved scales trough repeated explicit filtering, as these scales are the most involved in the interaction with the sub-grid ones; also, according to a scale-similarity hypothesis, the SGS stress tensor at the basic filter level is approximated with an equivalent tensor based on the smallest resolved scales. Hence, in order for scale-similar models to be representative of the actual SGS stress tensor, a correct definition is required for the true SGS stress tensor and the underlying filter determining it.

In order to leverage the several benefits offered by structural models with respect to simple eddy viscosity closures, in the present work we want to introduce a scale-similar SGS model which is suitable for the present FV based LES formulation, and take advantage of the reduced computational costs of the new GI by using it in a two parameters, dynamic mixed SGS model. In order to understand what is the proper form of a candidate scale-similar term, we analyze the Taylor series development for τ_{ij}^{m-n} assuming that, for $m > n$, the following hold [13]:

$$\bar{\phi}^m = \bar{\phi}^n + \Delta_m^2 M_k \frac{\partial^2 \bar{\phi}^n}{\partial x_k^2} + O\left(\Delta_m^4\right) \quad \Delta_m^2 = \sum_{i=0}^{m-n} \left(\Delta^{m-i}\right)^2 \tag{7}$$

where M_k are filter dependent coefficients and Δ_m is the filter width resulting from the consecutive application of the filters G^i for $i = n, n+1, \ldots, m$. Under this circumstance, which is especially relevant for a FV discretization, the substitution of the expansion (7) in the definition for τ_{ij}^{m-n} in (3) and additional manipulations (omitted for the sake of conciseness) lead to the following estimates:

$$\bar{\rho}^m \tilde{u}_i^m \tilde{u}_j^m - \bar{\rho}^n \tilde{u}_i^n \tilde{u}_j^n = \Delta_m^2 M_k \frac{\partial^2 \bar{\rho}^n \tilde{u}_i^n \tilde{u}_j^n}{\partial x_k^2} - 2\Delta_m^2 M_k \bar{\rho}^n \frac{\partial \tilde{u}_i^n}{\partial x_k}\frac{\partial \tilde{u}_j^n}{\partial x_k} + O\left(\Delta_m^4\right) \tag{8}$$

$$\bar{p}^m - \bar{p}^n = \Delta_m^2 M_k \frac{\partial^2 \bar{p}^n}{\partial x_k^2} + O\left(\Delta_m^4\right) \tag{9}$$

$$\tilde{S}_{ij}^m - \tilde{S}_{ij}^n = \frac{1}{2}\left[\frac{\partial}{\partial x_j}\left(\Delta_m^2 M_k \frac{\partial^2 \tilde{u}_i^n}{\partial x_k^2}\right) + \frac{\partial}{\partial x_i}\left(\Delta_m^2 M_k \frac{\partial^2 \tilde{u}_j^n}{\partial x_k^2}\right)\right] \tag{10}$$
$$+\frac{1}{2}\left[\frac{\partial}{\partial x_j}\left(2\frac{\Delta_m^2 M_k}{\bar{\rho}^n}\frac{\partial \bar{\rho}^n}{\partial x_k}\frac{\partial \tilde{u}_i^n}{\partial x_k}\right) + \frac{\partial}{\partial x_i}\left(2\frac{\Delta_m^2 M_k}{\bar{\rho}^n}\frac{\partial \bar{\rho}^n}{\partial x_k}\frac{\partial \tilde{u}_j^n}{\partial x_k}\right)\right] + O\left(\Delta_m^4\right)$$

The resulting approximation for τ_{ij}^{m-n} can then be summarized as follows:

$$\tau_{ij}^{m-n} = \Delta_m^2 f\left(\bar{\rho}^n, \tilde{u}_i^n, \tilde{u}_j^n\right) + \frac{\partial \Delta_m^2}{\partial x_j} h\left(\bar{\rho}^n, \tilde{u}_i^n\right) + \frac{\partial \Delta_m^2}{\partial x_i} h\left(\bar{\rho}^n, \tilde{u}_j^n\right) + O\left(\Delta_m^4\right) \tag{11}$$

where, following the product differentiation rule, f and h are functional forms due, respectively, to the constant and variable Δ_m^2 parts in the estimates above (h being due just to the viscous part in (10)). However, when the same analysis is performed for the true SGS stress tensor τ_{ij}^{n-0} for sufficiently smooth variations of Δ_n, the fact that $\bar{\phi}^m = \bar{\phi}^n + O\left(\Delta_m^2\right)$ allows (within a $O\left(\Delta_m^4\right)$ accuracy order) a substitution of the n-level variables with the corresponding m-level variables in the right hand side of equations (8-10); hence, the resulting estimate for τ_{ij}^{n-0} is:

$$\tau_{ij}^{n-0} = \Delta_n^2 f\left(\bar{\rho}^n, \tilde{u}_i^n, \tilde{u}_j^n\right) + \frac{\partial \Delta_n^2}{\partial x_j} h\left(\bar{\rho}^n, \tilde{u}_i^n\right) + \frac{\partial \Delta_n^2}{\partial x_i} h\left(\bar{\rho}^n, \tilde{u}_j^n\right) + O\left(\Delta_n^4\right) \tag{12}$$

The main conclusion to be drawn from this analysis is that a scale-similar model based on τ_{ij}^{m-n}, as originally proposed in [4], has a second order difference with respect to the true SGS stress tensor τ_{ij}^{n-0}. As a consequence, such a model would be unable to represent any of the filter effects, including those at the lowest order represented by the $O\left(\Delta_n^2\right)$ term. In order to overcome such deficiency, we propose to scale the model according to the following estimate:

$$\frac{\Delta_n^2}{\Delta_m^2}\tau_{ij}^{m-n} - \tau_{ij}^{n-0} = \Delta_n^2\left[h\left(\bar{\rho}^n, \tilde{u}_j^n\right)\beta_i + h\left(\bar{\rho}^n, \tilde{u}_i^n\right)\beta_j\right] + O\left(\Delta_m^2\Delta_n^2\right) + O\left(\Delta_n^4\right) \quad (13)$$

$$\beta_i = \frac{1}{\Delta_m^2}\frac{\partial \Delta_m^2}{\partial x_i} - \frac{1}{\Delta_n^2}\frac{\partial \Delta_n^2}{\partial x_i} = \frac{\partial}{\partial x_i}\left[\log\left(\frac{\Delta_m^2}{\Delta_n^2}\right)\right]$$

From which it follows that a scale-similar tensor of the form $\left(\Delta_n^2/\Delta_m^2\right)\tau_{ij}^{m-n}$ approximates the convective/pressure part of the tensor τ_{ij}^{n-0} with $O\left(\Delta_n^4\right)$ accuracy; if, in addition, the ratio Δ_n^2/Δ_m^2 is constant in space, the accuracy is restored also for the diffusive part. This form of scale-similar term is thus a better candidate to represent the implicit filtering effects due to the FV formulation. However, it is worth noting that numerical error effects, due to the discretization of the fluxes on the faces of the FV cells, are not considered here; also, the relation (13) says nothing about its applicability and the relative importance of the neglected higher order terms. From the point of view of the practical implementation in general FV solvers, the nth filter level quantities in the model are just those already available at the basic implicit filter level determined by the FV discretization (i.e., $n = 1$ and $\Delta_n = \Delta^1$). In contrast, the mth filter level quantities require explicit test filtering according to the assumption (7); while this form of differential filters is relatively easy to implement in unstructured solvers, both in explicit or implicit form, in the present work we adopt a simpler volume average filter over an extended neighbor stencil (see figure 1) in order to save computational time. While the present model is strictly dependent on the selected test filter, this choice can be justified by the fact that the model only considers the implicit filtering effects due to the FV formulation and not those due to the discretization, hence an additional approximation would still be present; also, as the model is implemented in a dynamic version, it is reasonable to expect a partial compensation effect for these errors due to the dynamic procedure. The limitations arising from this choice will be partially assessed in the following while more consistent options for the test filter will be the object of future works.

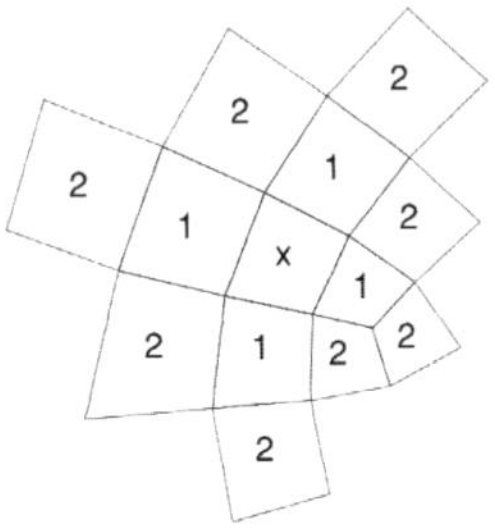

Figure 1. Test filter stencil for the cell x with nearest (1) and extended (2) neighbor cells.

5. Dynamic mixed SGS modeling

Taking advantage of the reduced costs of the dynamic procedure within the FV framework, the new scale-similar term is combined with an eddy viscosity closure in a two parameter, dynamic mixed model. Introducing the test filter levels m and r ($\Delta_r > \Delta_m > \Delta_n$), the proposed parameterization for the SGS stresses at the basic (n) and test (m) filter levels reads:

Journal of Physics: Conference Series **501** (2014) 012020 doi:10.1088/1742-6596/501/1/012020

$$\begin{aligned} \tau_{ij}^{n-0} &= C_{ev} 2\bar{\rho}^n \left|\tilde{S}^n\right|^\theta \left(k_{SGS}^{n-m}\right)^{\frac{1-\theta}{2}} \Delta_n^{1+\theta} \left(\tilde{S}_{ij}^n - \tfrac{1}{3}\tilde{S}_{gg}^n \delta_{ij}\right) + C_{ss} \left(\tfrac{\Delta_n}{\Delta_m}\right)^2 \tau_{ij}^{m-n} \\ \tau_{ij}^{m-0} &= C_{ev} 2\bar{\rho}^m \left|\tilde{S}^m\right|^\theta \left(k_{SGS}^{m-r}\right)^{\frac{1-\theta}{2}} \Delta_m^{1+\theta} \left(\tilde{S}_{ij}^m - \tfrac{1}{3}\tilde{S}_{gg}^m \delta_{ij}\right) + C_{ss} \left(\tfrac{\Delta_m}{\Delta_r}\right)^2 \tau_{ij}^{r-m} \end{aligned} \quad (14)$$

where C_{ev} and C_{ss} are the dynamic constants to be computed and:

$$\begin{aligned} k_{SGS}^{n-m} &= \tfrac{1}{2}\left(\tilde{u}_j^n - \tilde{u}_j^m\right)\left(\tilde{u}_j^n - \tilde{u}_j^m\right) \\ \tilde{S}^n &= \sqrt{2\tilde{S}_{ij}^n \tilde{S}_{ij}^n} \end{aligned} \quad (15)$$

If the relations (14) are expressed in compact form as:

$$\begin{aligned} \tau_{ij}^{n-0} &= C_{ev} A_{ij}^n + C_{ss} B_{ij}^n \\ \tau_{ij}^{m-0} &= C_{ev} A_{ij}^m + C_{ss} B_{ij}^m \end{aligned} \quad (16)$$

and inserted in the new GI (6), the constants in the model can be computed by a least-squares error minimization [10, 11], leading to:

$$\begin{aligned} C_{ss} &= \frac{A_{ij}A_{ij}L_{rs}B_{rs} - L_{ij}A_{ij}A_{rs}B_{rs}}{A_{ij}A_{ij}B_{rs}B_{rs} - A_{ij}B_{ij}A_{rs}B_{rs}} \\ C_{ev} &= \frac{L_{ij}A_{ij} - C_{ss}A_{ij}B_{ij}}{A_{ij}A_{ij}} \end{aligned} \quad (17)$$

where $A_{ij} = A_{ij}^n - A_{ij}^m$, $B_{ij} = B_{ij}^n - B_{ij}^m$ and $L_{ij} = \tau_{ij}^{n-m}$. According to the considerations made in section 4, a standard mixed-scale model [14] is selected as eddy-viscosity model, the parameter θ being free to vary in order to obtain different closures. When compared to classical, two parameters, dynamic mixed SGS models [11], the present approach still provides a substantial saving in computational costs as, for the incompressible case, only 6 scalar fields need to be test filtered (3 velocity components per each test filter level involved, the isotropic pressure term being usually excluded from the modeling) against the 24 of a classical approach with a scale-similar model (3 velocity components per test filter level, plus the 6 components of the Leonard tensor per test filter level and the 6 components of the SGS model tensor at the basic filter level).

6. Flow description and computational setup

The test case considered is represented by the fully developed, turbulent, incompressible, isothermal flow in a straight circular pipe of radius R. The mean flow direction is along the z axis, while the r and θ directions define, respectively, the radial and tangential coordinates in planes normal to the z direction. The velocities in the three coordinate directions are, respectively, u_z, u_r and u_θ. Two distinct cases at different Reynolds numbers are considered, respectively $Re_D = 10,000$ and $Re_D = 44,000$ based on the mean velocity U_b and pipe diameter D. The corresponding friction Reynolds numbers, based on the friction velocity u_τ and pipe radius R, are $Re_\tau = 320$ and $Re_\tau = 1142$. In both cases, a no-slip boundary condition is applied at the wall and a periodicity condition along the axial direction; this latter option, when applied to overly constrained domain (i.e., pipe) lengths deserves further attention. Indeed, in recent years, several DNS studies have put considerable attention to the description of large and very large-scale motions in turbulent pipe flows [15, 16, 17] in order to better clarify their role, as

Journal of Physics: Conference Series **501** (2014) 012020 doi:10.1088/1742-6596/501/1/012020

energy carrying structures, in the determination of higher-order flow statistics. Hence, when periodic boundary conditions are applied in a computational study, the issue of the minimum required pipe length to avoid numerical artifacts due to artificial boundary conditions clearly emerges. Previous experimental studies [18, 19] evidenced how such large scale structures can have wavelengths as long as $16R$ in the outer region of the flow and this has led some DNS studies, e.g., [8], to adopt a pipe length $L_z = 15R$. However, more recent computational works [20] have identified the more stringent requirement $L_z = 8\pi R$ in order for higher order statistics to converge within 1% of the respective values for longer domains (up to $L_z = 20\pi R$); as a matter of fact, the current trend in DNS of turbulent pipe flows [15, 16, 17] is to adopt $L_z = 30R$. Despite this evidence, in the present work we have adopted the axial domain lengths $L_z = 10R$ for the low Reynolds case [7] and $L_z = 15R$ for the high Reynolds one [8], both in accordance with the respective DNS references. This was motivated by two main reasons. The first, more practical one, is that there is abundant empirical evidence [7, 8, 16, 17, 20] of the fact that, for the low order statistics which are analyzed here, no substantial differences arise between the present reference DNS and computations on longer domains or experimental data; hence, the computational cost reduction resulting from a shorter domain length has been readily exploited. The second, more theoretical, reason stems from the fact that the selection of a reference DNS case allows to treat the comparison in purely mathematical terms, without any further reference to the relative target flow, as repeatedly stated in [21](pp. 111 and following). In the present case, this means that the proposed LES approach is tested toward the fully resolved periodic turbulent flow in a pipe with length, respectively, $L_z = 10R$ for the low Reynolds case and $L_z = 15R$ for the higher one; that is, the reference mathematical problem, the one which is actually discretized in the computations, exactly has the same boundary conditions for the present LES and the reference DNS. Also, these are exactly the unique boundary conditions which would allow the present implicitly filtered LES to converge toward the reference DNS with further grid refinements. A possibility, which would still not escape this reasoning, is that the large scale motions which cannot be captured by the present domain sizes (also known as super-grid scales) have such a strong interaction with the modeled sub-grid scales to require a specific SGS modeling approach. While this would invalidate the LES approach as a whole and, to the best of authors' knowledge, has never been investigated before[1], here we simply assume this to be a possible deficiency of the SGS model which, in case, will be the object of future works.

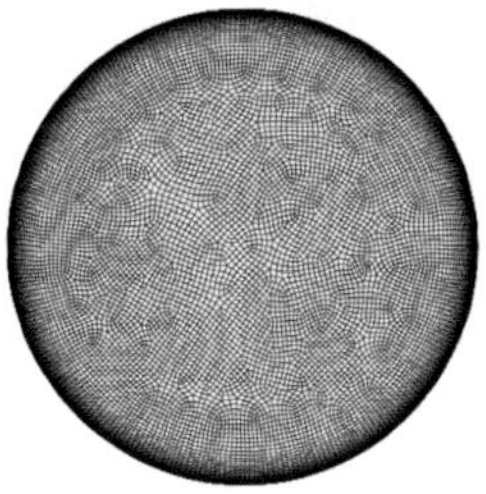

Figure 2. Grid topology.

Table 1. Dimensionless grid parameters.

Case	Δz^+	$R\Delta\theta^+$	Δr^+_{min}	N° DOF
$Re_\tau = 320 - DNS$[7]	6.6	8.4	0.6	$8.2 \cdot 10^6$
$Re_\tau = 320$	50	15	0.1	$0.2 \cdot 10^6$
$Re_\tau = 1142 - DNS$[8]	8.4	7.0	0.4	$6.3 \cdot 10^8$
$Re_\tau = 1142$ Coarse	60	32	0.1	$2.4 \cdot 10^6$
$Re_\tau = 1142$ Medium	57	28	0.1	$3.4 \cdot 10^6$
$Re_\tau = 1142$ Fine	54	25	0.1	$5.2 \cdot 10^6$

For what concerns the computational grids, as the test case is selected to be also representative of more complex applications, unstructured hexaedral grids of different resolutions are used, the topology being sketched in figure 2. As summarized in table 1, a single grid resolution is used for

[1] Such a test would require the comparison of LES computations made on domains of different lengths with the relative DNS data from computations on the same domains.

the low Reynolds case, while three different resolutions are tested for the higher Reynolds case. Computational parameters for the reference DNS computations are also reported to highlight the difference, with respect to the present computations, in terms of total degrees of freedom represented. The computations are performed with the commercial, unstructured, FV solver Fluent [22] and the following numerical parameters are common among all the simulations: a non-iterative time advancement with the fractional step algorithm for the pressure-velocity coupling; an implicit second order algorithm for time advancement with a time step selected to maintain the maximum Courant number around 0.1; non-dissipative, second order, central schemes for the spatial discretization of convective and diffusive terms. According to equation (5), no specific modifications are needed for the pressure-velocity coupling, however the pressure is not included in the scale-similar part of the SGS model.

The dynamic mixed model (DM) is tested with $\theta = 0$ and compared with a dynamic Smagorinsky model (DS, $C_{ss} = 0$ and $\theta = 1$) and the reference DNS data [7, 8]. In addition, computations with a classical dynamic Smagorinsky model [10] have been performed, but no appreciable differences with the present implementation were found and are not presented; nonetheless, this confirms that the new dynamic procedure introduced here can safely replace the classical one while attaining a significant reduction in the computational cost of the model. It is also worth mentioning that no spatial average of the dynamic constants is ever performed as it is not explicitly required by the formulation and in no case stability issues emerged during the computation. However, a clipping of the dynamically computed constants was adopted in order to limit the detrimental effects due to the approximate least square procedure; in particular, according to previous results [6], the constants were allowed to vary within the limits $0 \leq C_{ev} \leq 0.18^{2\theta}0.2^{1-\theta}$ and $0 \leq C_{ss} \leq 2$.

The flow is initialized with a power law velocity profile with superimposed random fluctuations. After the statistically steady state is reached, flow samples are collected at each time step for a total sampling time of $80R/u_\tau$ or, approximately, $1200R/U_b$ and $1540R/U_b$ for the low and high Reynolds cases; enough to allow a particle to travel more than 100 times trough the pipe axial length at velocity U_b. Additional spatial averages are performed in the homogeneous direction θ over 4 perpendicular radial profile samples. All the results are made non-dimensional by the friction velocity u_τ and the pipe radius R.

7. Numerical results

Flow statistics for the $Re_\tau = 320$ case are presented first, in figure 3. As both mean and fluctuation profiles show, no substantial differences are found between the proposed DM model and the simpler DS formulation.

This could be explained by two main reasons: the first one is the low Reynolds number of the case, which implies a lack of inertial range structures in the flow and a relative low significance for the scale-similar part of the model; the second one is the relatively high resolution of the grid for this case (see table 1), which makes the contribution of the scale-similar term also unimportant with respect to the resolved part of the flow. However, both models provide a fairly good agreement with the reference DNS data.

Differences among the DM and DS modeling strategies are better highlighted in the $Re_\tau = 1142$ case, whose main statistics are presented in figure 4. This is especially true for the mean axial velocity and the axial velocity fluctuations. Indeed, while none of the models can correctly reproduce the mean DNS velocity profile, the proposed DM model provides a better agreement for both the medium and fine grids (the relative curves being completely overlapped), the error on the resulting Re_D being 8% (it rises to 11% for both the DS Medium and DM coarse cases); the matching with the DNS data becomes almost perfect for the axial velocity fluctuations, as the DM model can substantially reduce the over-prediction of the inner-layer peak on all the grids. Despite some residual effects due to the convergence of the statistics, no particular differences are

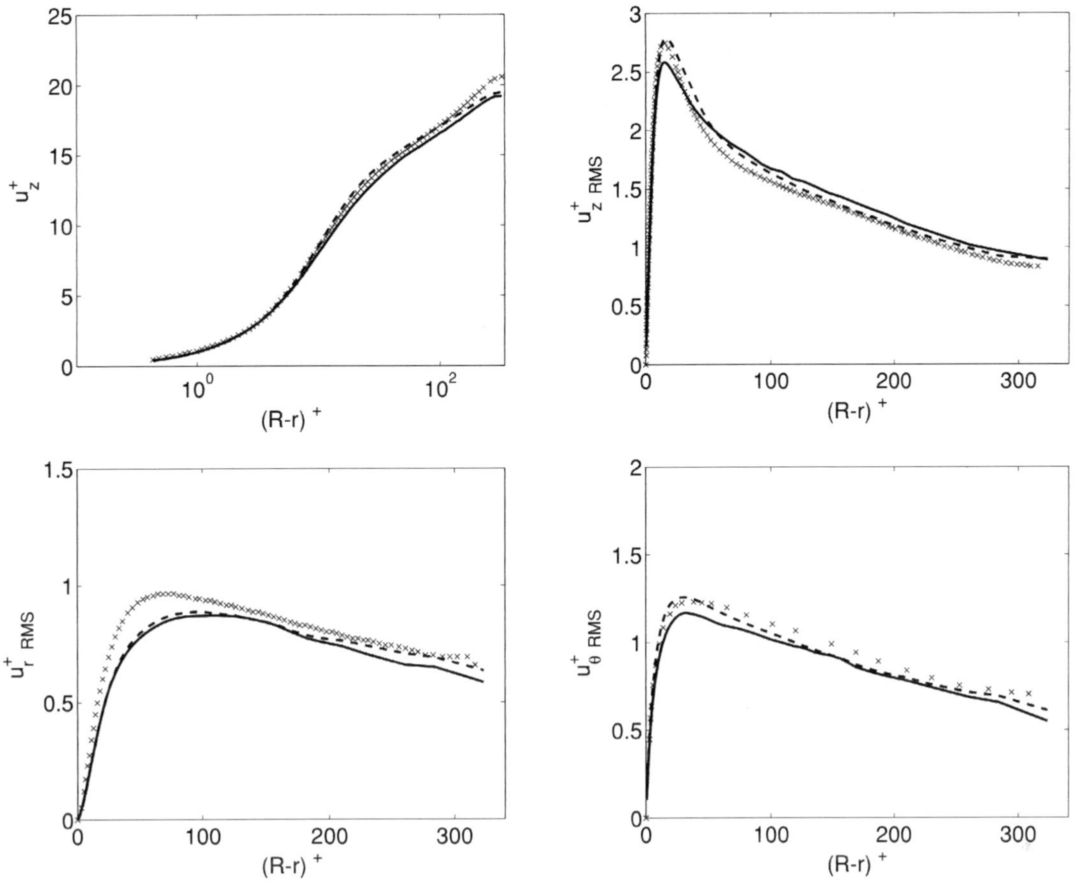

Figure 3. Velocity statistics for the $Re_\tau = 320$ case. Top-left: mean axial velocity, top-right: R.M.S. axial velocity fluctuations, bottom-left: R.M.S. wall-normal velocity fluctuations, bottom-right: R.M.S. tangential velocity fluctuations. Legend: × DNS [7], ——DM, - - - -DS.

found on the remaining velocity fluctuations profiles except, maybe, for the tangential velocity fluctuations peak, which is better predicted by the DS model. What is worth noting is that the DM model can provide the same level of accuracy of the DS model at a fraction of its cost (the coarse grid employed here allowed a 30% reduction in computational costs with respect to the medium grid employed for the DS model).

However, these results also highlight a feature of the proposed model DM which, especially for the mean velocity profile, might appear clumsy, at least, for the present implicitly filtered code: a sort of grid convergence toward a non DNS result. A possible explanation is connected to the non optimality of the adopted grid refinement with respect to an expected grid convergence toward the DNS result; indeed, not only the three grids all have tangential and stream-wise spacings well above the Taylor microscale and within the inertial range (and are thus non-optimal [23] and expected to provide a sort of inertial range convergence [24]), but the near wall radial grid spacing has been kept constant among the three grids, possibly limiting the effectiveness of the adopted grid refinement. A second, less evident, possible explanation is connected to the clipping limits used for the constant C_{ss} and the resulting effectiveness of the scale-similar part of the model. Indeed, the present clipping limits have been determined according to previous results for the turbulent channel flow [6] where, as shown in the left part of figure 5, allowed the

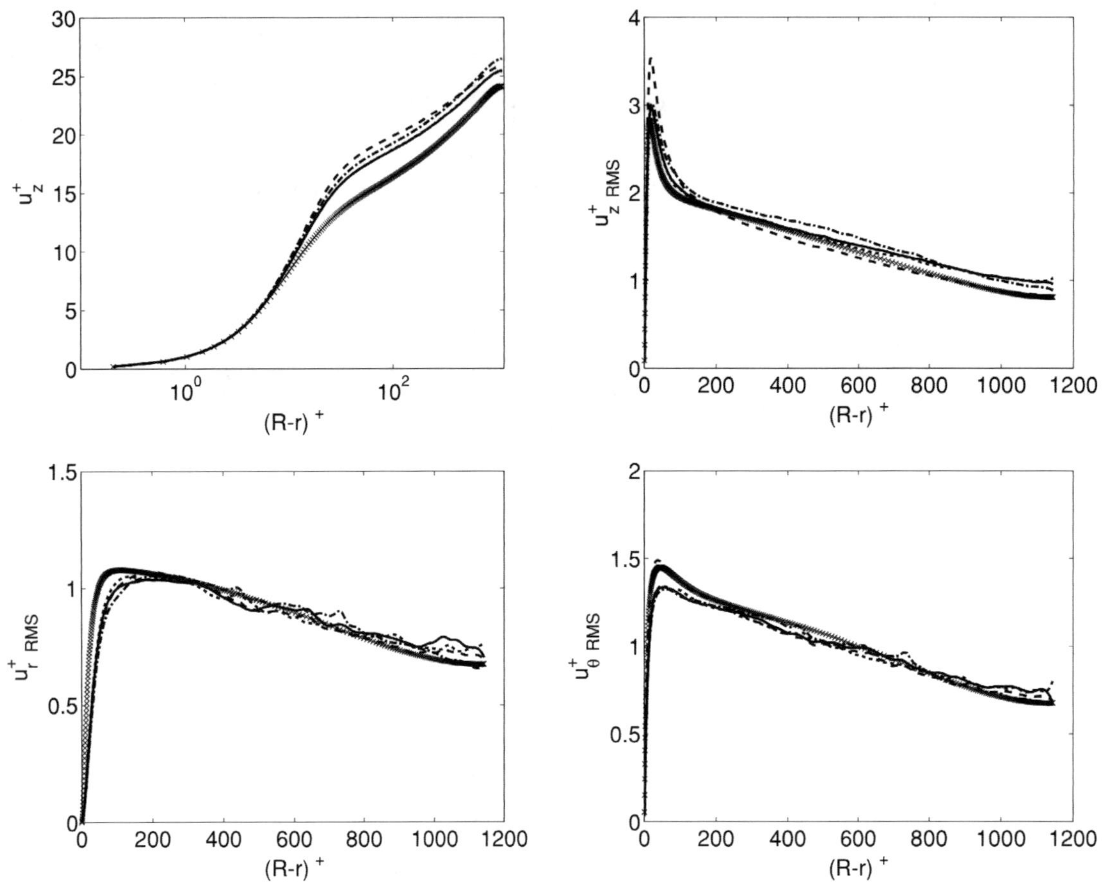

Figure 4. Velocity statistics for the $Re_\tau = 1142$ case. Top-left: mean axial velocity, top-right: R.M.S. axial velocity fluctuations, bottom-left: R.M.S. wall-normal velocity fluctuations, bottom-right: R.M.S. tangential velocity fluctuations. Legend: × DNS [8], ······DM fine, ——DM medium, — · —DM coarse, - - - -DS Medium.

full dynamic range for the constant C_{ss} in the first layer of cells near the walls, where the scale-similar model is more effective; in contrast, for the present turbulent pipe flow at $Re_\tau = 1142$, as shown in the right part of the same figure for the coarse grid (the same results hold also for the two additional grids not reported), the effect of the clipping has strongly limited the dynamic range for the constant near the wall. In particular, it is worth noting that the dynamic range of the unclipped constant for the turbulent pipe is much larger than the interval represented in figure 5 but here it has been restricted for a better graphical representation; for the same reason, the clipped constant for the same case has two strong peaks at 2 and 0 which have been removed from the picture. Hence, this behavior suggests that, most of the time, the scale-similar model is not active at the wall or not as much as would have been required. As the cut in the positive tail of the distribution has been observed also for the lower Reynolds case (not shown), without the same negative effects, it appears probable that the major problem of the present approach for the pipe resides in the clipping of negative constants, which are instead present only for the higher Reynolds case. However, none of these explanation has been completely confirmed yet and further investigations are required.

The nature of the two modeling strategies, DM and DS, can be better appreciated by

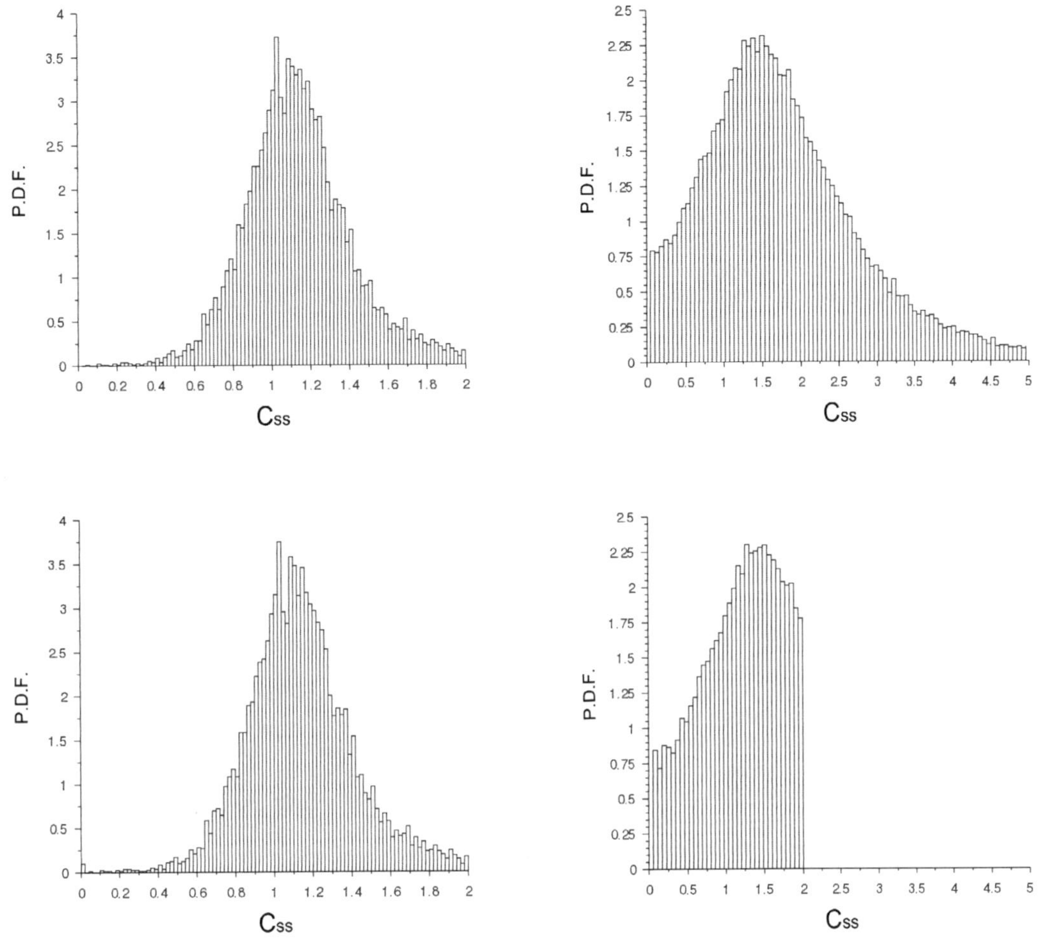

Figure 5. PDF for the dynamic constant C_{ss} at the wall. Left: turbulent channel flow at $Re_{\tau} = 590$ [6], right: turbulent pipe flow at $Re_{\tau} = 1142$, coarse grid; top: constant as computed by the dynamic procedure, bottom: clipped constant.

analyzing the stream-wise spectra of the underlying velocity fields, presented in figure (6) for the common medium grid. For both the velocity components, but especially for the radial one, the DM model provides a considerable recovery of energy at the smallest represented scales with respect to the DS model; however, the model does not seem capable of totally removing the energy pile-up which is evident for the radial velocity component of both the tested models. This fact can be explained by considering the nature of the test filter and its previously cited limitations. Indeed, the extended filter stencil adopted here is actually required in order to obtain a larger filter width and make the dynamic procedure sufficiently independent from the discretization error which, for the computational grids in use, tends to pollute a much wider range of scales with respect to more regular grids[2]. However, for the top-hat like filter used here, the use of a larger filter width also implies an incomplete removal of the energy content at

[2] This was determined by preliminary tests performed with a reduced stencil for the test filter and by additional simulations performed on turbulent channel flows with more regular grids.

the smallest resolved scales, as indeed appears from the velocity spectra. Nonetheless, the same figure also shows that the test filter applied on the resolved velocity field for the DM model can reproduce with a certain accuracy the velocity spectra of the DS model; this heuristically suggests that, in relation to the DM model used here, the present test filter can sufficiently represent the implicit filtering effect due to the FV formulation.

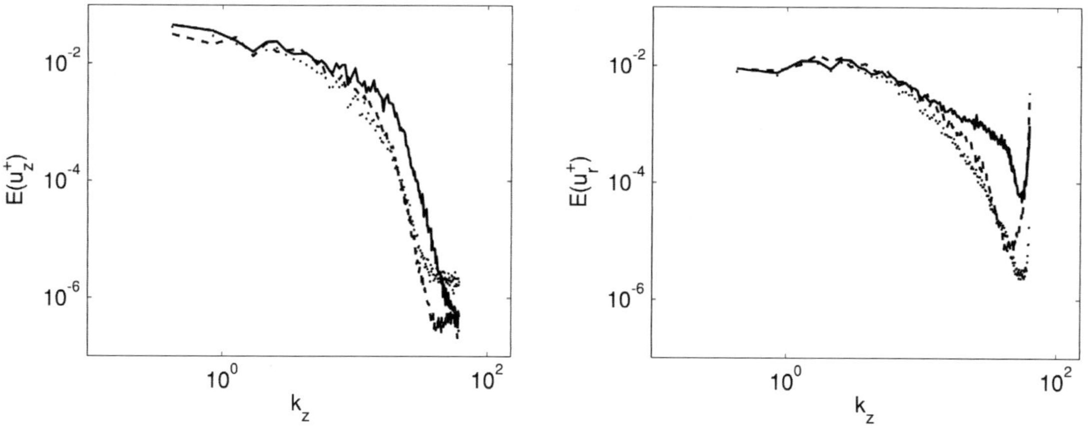

Figure 6. Stream-wise spectra at $r = 0$ for the $Re_\tau = 1142$ Medium case. Left: axial velocity, right: radial velocity. Legend: ——DM Basic (n), ⋯⋯DM Test ($n+1$), - - - -DS Basic (n).

8. Conclusions

We have presented an LES framework which has several advantages with respect to classical ones as it: is fully consistent (i.e., without commutation errors); exactly fits the classical FV discretization as implemented in most solvers; has an associated GI/dynamic procedure which is much cheaper than classical ones without compromising the resulting SGS model effectiveness and stability. The proposed formalization has also allowed an easy derivation of the proper form of scale-similar term to be adopted in this LES framework. The resulting dynamic, two parameters, mixed model proposed here (DM) is still substantially cheaper than a classical dynamic Smagorinsky model (60% according to the number of required test filter operations) while generally improving most of the statistical results. The proposed test case, the turbulent flow in a straight pipe, has highlighted that the advantages of the DM model are more evident for the higher Reynolds number case, while no particular improvements over the simpler DS model are found for the less relevant, low Reynolds number case. The main limitation emerged from the present implementation of the DM model is related to a general lack of effectiveness for the scale similar part of the model when compared to similar approaches (e.g., [11]) and previous analyses [6]. Current results indicate as a possible cause of this the combination of the clipping procedure, which has been shown to be overly constrictive in the present case, and the spatial resolution, which for all the high Reynolds grids adopted here is far above the typical Taylor microscale of a wall bounded turbulent flow. Finally, the spectral results have also pointed out a criticality in the adopted test filter as, according to its top-hat like formulation and extended stencil, it is found unable to completely remove energy on the smallest resolved scales. Nonetheless, it is confirmed a posteriori that the similarity with the basic implicit FV filter, as required by the proposed formulation, is retained.

References

[1] Bouffanais R 2010 *Comput. Fluids* **39** 735–8

[2] Leonard A 1974 *Adv. in Geophys.* A **18** 237–48
[3] Denaro F M 2011 *J. Comp. Phys.* **230** 3849–83
[4] Vreman A W and Geurts B J 2002 *A new treatment of commutation errors in large-eddy simulation* Proc. IX European Turbulence Conf. (Barcelona)
[5] Denaro F M and De Stefano G 2011 *Theor. Comput. Fluid Dyn.* **25** 315–55
[6] Lampitella P, Inzoli F and Colombo E 2013 *Numerical experiments with a new dynamic mixed subgrid-scale model* Proc. of the 9th ERCOFTAC Work. on Direct and Large Eddy Simulation (Dresden), Accepted
[7] Wagner C, Huttl T J and Friedrich R 2001 *Comput. Fluids* **30** 581–90
[8] Wu X and Moin P 2008 *J. Fluid Mech.* **608** 81–112
[9] Vasilyev O V 2001 *Computational Constrains on Large Eddy Simulation of Inhomogeneous Turbulent Complex Geometry Flows* Proc. of the 3rd AFOSR Int. Conf. on DNS/LES (Arlington)
[10] Germano M, Piomelli U, Moin P and Cabot W H 1991 *Phys. Fluids* A **3** 1760–5
[11] Sarghini F, Piomelli U and Balaras E 1999 *Phys. Fluids* **11** 1596–607
[12] Liu S, Meneveau C and Katz J 1994 *J. Fluid Mech.* **275** 83–119
[13] Vreman B, Geurts B and Kuerten H 1996 *Theor. Comp. Fluid Dyn.* **8** 309–24
[14] Sagaut P 1996 *Rech. Aéro.* **1** 51–63
[15] Wu X, Baltzer J R and Adrian R J 2012 *J. Fluid Mech.* **698** 235–81
[16] Lee J H and Sung H J 2013 *Phys. Fluids* **25** 045103
[17] Ahn J, Lee J H, Jang S J and Sung H J 2013 *Int. J. Heat Fluid Flow* Preprint
[18] Kim K C and Adrian R J 1999 *Phys. Fluids* **11** 417–22
[19] Guala M, Hommema S E and Adrian R J 2006 *J. Fluid Mech.* **554** 521–42
[20] Chin C, Ooi A S H, Marusic I and Blackburn H M 2010 *Phys. Fluids* **22** 115107
[21] Jimenéz et al. 1998 *A Selection of Test Cases for the Validation of Large-Eddy Simulations of Turbulent Flows* AGARD Advisory Report **345**
[22] Lampitella P, Colombo E and Inzoli F 2011 *Sensitivity Analysis on Numerical Parameters for Large Eddy Simulation with an Unstructured Finite Volume Commercial Code* Proc. XX AIMETA Conf. (Bologna)
[23] Addad Y, Gaitonde U, Laurence D and Rolfo S *Optimal Unstructured Meshing for Large Eddy Simulations* Proc. of the 1st ERCOFTAC Work. on the Quality and Reliability of Large Eddy Simulations (Leuven)
[24] Pope S 2004 Ten questions concerning the large-eddy simulation of turbulent flows *New J. Phys.* **6** 35

Journal of Physics: Conference Series **501** (2014) 012021 doi:10.1088/1742-6596/501/1/012021

Microtomography-based CFD Analysis of Transport in Open-Cell Aluminum Metal Foams

Paola Ranut[1], **Enrico Nobile**[1], **Lucia Mancini**[2]

[1] Department of Engineering and Architecture, University of Trieste, Piazzale Europa 1, 34127 Trieste (Italy)

[2] Elettra-Sincrotrone Trieste S.C.p.A., S.S. 14 - km 163.5 in Area Science Park, 34149 Basovizza (Trieste) (Italy)

E-mail: pranut@units.it

Abstract. Nowadays, the need for developing more effective heat exchange technologies and innovative materials, capable of increasing performances while keeping power consumption, size and cost at reasonable levels, is well recognized. Under this perspective, metal foams have a great potential for enhancing the thermal efficiency of heat transfer devices, while allowing for the use of smaller and lighter equipments. However, for practical applications, it is necessary to compromise between the augmented heat transfer rate and the increased pressure drop induced by the tortuous flow passages. For design purposes, the estimation of the flow permeability and the thermal conductivity of the foam is fundamental, but far from simple. From this perspective, besides classical transport models and correlations, computational fluid dynamics (CFD) at the pore scale, although challenging, is becoming a promising approach, especially if coupled with a realistic description of the foam structure. For precisely recovering the microstructure of the foams, a 3D X-ray computed microtomography (μ-CT) can be adopted. In this work, the results of μ-CT-based CFD simulations performed on different open-cell aluminum foams samples, for laminar flow regime, will be discussed. The results demonstrate that open-cell aluminum foams are effective means for enhancing heat transfer.

1. Introduction

Heat transfer augmentation is of conspicuous importance in many fields like, to name a few, renewable energy production and conversion, fossil fuels and thermal control of electronic devices. The possibility of enhancing the heat transfer rate will lead to both energy savings and the employment of equipments of reduced size, weight and cost. Metallic foams are among one of the most promising candidates for this purpose, since they might guarantee an increase of the heat transfer rate with an acceptable increment of pressure losses, and the added benefit of lighter and more compact devices.

Metal foams are a class of porous materials with low densities and tortuous and irregular flow passages. They have been available since the 1980s but at the beginning, in view of the high manufacturing costs, were mainly used in the aerospace industry. Nowadays, metal foams are employed also in multifunctional heat exchangers, solar collectors, heat pipes, and so on, and can also serve as fluid distributors in fuel cells and electrolysis applications, or as catalyst supports in reactors. Moreover, in view of their structural strength and mechanical energy absorption capability, they can also be utilized in lightweight structural sandwich panels and car bumpers and, thanks to their acoustic properties, they can also be exploited as sound absorbers and

silencers.
Most commercially available metal foams are made of aluminum, nickel, copper, and metal alloys. Two kinds of foams exist, namely the open-cell and the closed-cell metal foams. As the name implies, in open cell metal foams the flow paths are interconnected, which is beneficial for heat transfer applications. Typically, the pore density of uncompressed open-cell foams varies between 5 to 100 PPI (pores per inch), while the porosity ε ranges from 80% to 97%. The cross section of the fibers is function of the porosity, and varies from a circle at $\varepsilon = 0.85$ to an inner concave triangle at $\varepsilon = 0.97$.
Metal foams have a great potential for enhancing thermal performances of heat exchangers for several reasons, i.e., their high surface area to volume ratio, the high conductivity of the solid ligaments and the enhanced flow mixing due to their tortuous flow paths, which is sometimes referred to as thermal dispersion. Literature provides severals examples of metal foams solutions for heat transfer applications, dealing with both experimental, numerical and analytical studies: metal foam channels [1–7], metal foam covered tubes [8, 9], metal foam filled tubes [4, 10] and double tubes [11], metal foam tube bundle [12, 13], metal foam heat sinks [14, 15], and so on. Mahjoob and Vafai [4] registered an increase in the heat transfer rate of 8-13 times for a metal foam filled tube, and of 15 times for a metal foam channel heat exchanger, at the expense of a limited additional pressure drop. Much more promising are the results of Lu et al. [10], who analytically predicted a heat transfer enhancement up to forty times for the metal foam filled tube. Kim et al. [14] experimentally tested an aluminum foam heat sink and observed a reduction of the thermal resistance by up to 28% compared to a plate fin heat sink, while Seyf and Layeghi [15] proved numerically that a metal foam pin fin heat sink is about 2-3 times more effective in heat removal than a pin fin array, with only a moderate increase in pressure drop. They also noticed a contraction of the recirculation behind the pins with the foam inserts.
Usually, it is common to assimilate the foam to a homogeneous medium having an effective thermal conductivity which depends, to varying degrees, on conduction in both the solid and fluid phase, convection and radiation. At ambient temperature, with this assumption, the heat transfer is mostly governed by thermal conduction; the radiative contribution is important only at higher temperatures. As declared in [10], metal foams are scarcely effective when the thermal conductivity of solid is close to that of the fluid.
The proper characterization and identification of transport and thermal properties of metal foams - hydraulic permeability and effective thermal conductivity- is fundamental for design purposes but far from simple, due to the geometric complexity and the random orientation of the solid phase. Moreover, this information is, at present, quite scarce and mostly based on simplifying assumptions and simpler models, as described in [16]. This lack of information constitutes a fundamental limit in the employment of metal foams in practical and industrial applications.
More recently, Computational Fluid Dynamics (CFD) has demonstrated to be a promising means for predicting the flow through open cell metal foams and recovering their characteristic properties. For modeling flows in porous media, two different strategies can be followed, i.e., a macroscopic or microscopic approach. In the former, also known as *volume averaging* approach, the governing equations are volume averaged, thus they ignore the small-scale details and treat the porous solid-fluid system as homogeneous. In the latter, also known as *representative elementary volume* (REV) approach, the complex (real or idealized) geometry of the medium is taken into account, and the small-scale flow details are captured. This approach, which is potentially more accurate than the macroscopic one but even more costly, has shown encouraging results for metal foams.
Numerical simulations on idealized geometries have been presented by several authors [17–20]. Furthermore, at the moment a significant effort is devoted to using more realistic representations of the foam structure. From this perspective, X-ray computed microtomography (μ-CT) seems a promising and appealing approach for characterizing the 3D foam structure at the pore level

[21–25].
In this paper, the results of microtomography-based CFD simulations performed on three different open-cell aluminum foams samples, having different pore density (10-20-30 PPI), will be illustrated. This work is the natural extension of [16], which illustrated the preliminary results of the μ-CT characterization of the three samples. The high resolution X-ray μ-CT imaging was performed at the *TomoLab station* [26], a cone-beam μ-CT system developed at the Elettra Synchrotron Radiation Facility [27] in Trieste.

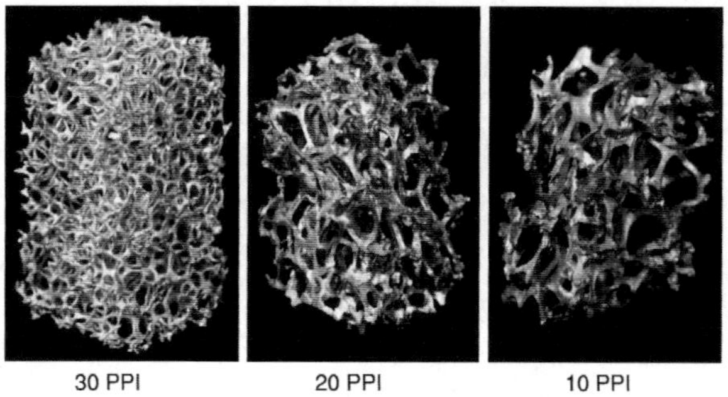

(a)

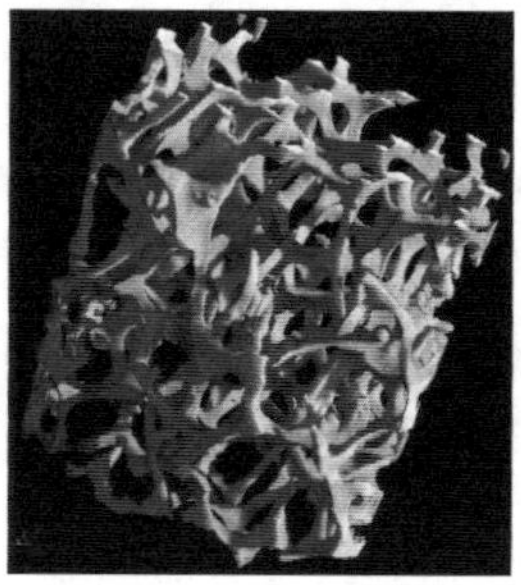

(b)

Figure 1: Aluminum foam samples employed in the μ-CT analyses: (a) photographies of the specimens and (b) reconstructed volume for the 20 PPI sample.

2. Methodology

X-ray μ-CT is a non invasive X-ray based method allowing the 3D reconstruction of an opaque sample by illuminating it in different directions [28]. Several radiographs of the object are recorded at different angles, corresponding to the projection of the linear attenuation coefficient. In absorption μ-CT, the modality employed in this activity, the difference between the linear attenuation coefficients of the two phases determines the contrast of the sample.
Afterwards, with the help of a suitable algorithm, a 3D image of the object is reconstructed and then, in the case of two-phase materials (as metallic foams), a segmentation process can be applied thus to distribute the voxels, i.e., the elemental digital units forming the reconstructed volume, in only two kinds of elements, that is solid and void, by selecting a proper threshold value in the histogram of the intensities.
For this activity, three aluminum foams samples, shown in Fig. 1(a), having different pore densities (10-20-30 PPI) and manufactured by m.pore GmbH [29], were analyzed. From each sample, a specimen of nearly $12 \times 12 \times 20$ mm^3 was extracted and characterized by means of high resolution 3D μ-CT. The metal foam samples were imaged in the following conditions: current = 133 μA, voltage = 60 kV, 0.5 mm thick Al filter, exposure time/projection = 2 s, source-to-sample distance = 80 mm, source-to-detector distance = 220 mm. For each sample 2400 projections were recorded over a 360° angular rotation. A 2×2 binning was applied to the detector pixels.
The sample volumes were reconstructed with an isotropic voxel size of 9.09 μm by using the commercial software COBRA 7.4 (EXXIM) [30]. It is worth noting that the 9.09 μm-resolution of the tomographic scan is significantly higher than most of the works published in the literature. In Fig. 1(b), the reconstructed volume for the 20 PPI specimen is represented. The choice of performing a high resolution μ-CT allowed us to well capture the surface peculiarities of each sample. The tomographic scan revealed a smooth and regular surface for the 30 PPI foam, and an increased roughness as the pore density decreases. Moreover, in the case of the 10 PPI foam, some spherical lumps of metals can be distinctly recognized.

From each reconstructed volume, a smaller representative cubic volume element was extracted, whose dimensions depended on the PPI of the sample: 548 voxels for the 30 PPI sample, 848 for the 20 PPI sample and 1000 for the 10 PPI one (1voxel side = 9.09μm). In the case of the 30 PPI sample, the size of the domain was chosen by comparing the results on domains of increasing size. For the other two samples, the dimensions were selected in order to have a number of foam cells comparable to the 30 PPI, and compatibly with the available computational resources.
In order to generate a volume file usable in the CFD simulations, the foam surface was triangulated with the commercial software VGStudio Max 2.0 [31], and a STL (STereoLitography) file was exported.
The computational domain, sketched in Fig. 2(a), consists of a square duct having an inlet region, an outlet region and an intermediate foam region. The numerical simulations were performed along the three spatial directions, in order to investigate the isotropy of the samples.

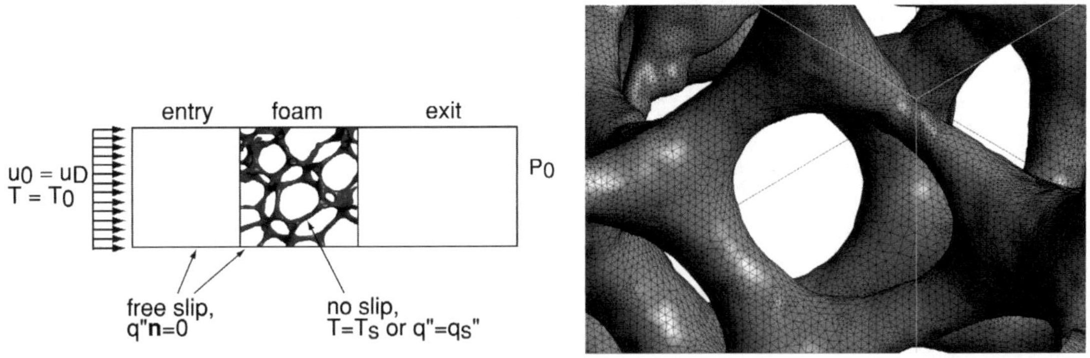

(a) (b)

Figure 2: (a) Sketch of the computational domain and (b) detail of the surface mesh for the 30 PPI sample.

Geometrically exact, feature-preserving grids were generated with the commercial software ANSYS ICEM CFD: the central foam region was meshed with an unstructured grid, while a structured mesh was employed for the inlet and outlet regions. The meshing of the foam volumes was not straightforward at all, since its quality was related to that of the surface triangulation. The final meshes were chosen after thorough grid independence analyses. The central foam region of the 30 PPI sample was discretized with about 1.4×10^6 nodes, while those of the 20 and 10 PPI foams with about 2.4×10^6 and 4.1×10^6 nodes, respectively. A detail of the surface mesh for the 30 PPI sample is illustrated in Fig. 2(b).
For the CFD simulations the commercial software ANSYS CFX 13.0 was used. ANSYS CFX is a node-centered, Finile Element-based Finite Volume code. By default, the pressure-velocity coupling is implemented by using the Rhie-Chow algorithm. The linear solver uses an Algebraic Multigrid method based on an additive correction multigrid strategy, and the fully coupled momentum and mass conservation equations are solved simultaneously. The Navier-Stokes equations were solved by using the High Resolution scheme for the advection term.
The boundary conditions applied are summarized in Fig. 2(a). The working fluid was air with constant thermophysical properties, evaluated at 25°C. Only the laminar flow regime was simulated, and the Reynolds number, based on the nominal pore diameter of the foam and the superficial velocity, was made to vary between 1 and 100. This interval corresponds to Reynolds numbers based on the square root of permeability, Re_K, varying between 0.1 and 13, similar to those considered in [25].

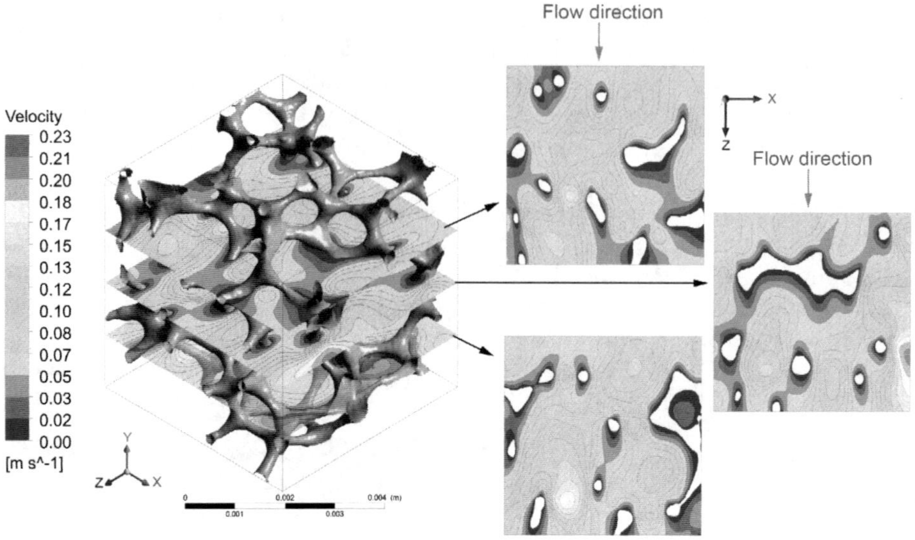

Figure 3: Visualization of the velocity contours inside the 30 PPI foam at three different sections.

3. Resuls and discussion

3.1. The flow field

For illustration purposes, the velocity contours at three different sections, for the case of the 30 PPI foam, are given in Fig. 3.

Figure 4(a) plots the pressure distribution along the z direction for the 30 PPI sample at varying Reynolds numbers. As expected, the pressure drop, ΔP, across the foam increases with the Reynolds number. To better compare the pressure trends for the whole set of Reynolds numbers simulated, in Fig. 4(b) the normalized pressure profiles are depicted. These profiles have been computed as $\tilde{P} = 1 - (P - P_{in})/(P_{in} - P_{out})$, being P_{in} and P_{out} the pressure at the entrance and exit of the foam region, respectively. As it can be seen, as the velocity increases, a linear trend is still established, but some local fluctuations appear.

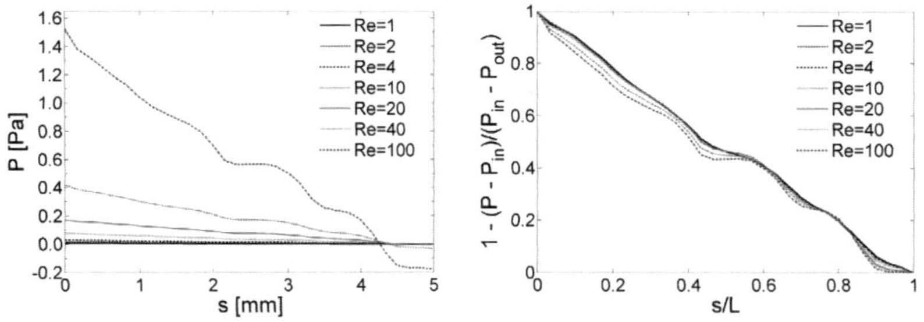

(a) (b)

Figure 4: (a) Pressure and (b) normalized pressure distributions, along the z direction for the 30 PPI sample, at varying Reynolds numbers.

Figure 5 compares the pressure profiles for the three foam samples, evaluated along the three spatial directions for Re=10. As it can be seen, the pressure drop reduces as the pore density decreases, since a larger cross sectional area is available for the flow. For the 30 PPI sample

Journal of Physics: Conference Series **501** (2014) 012021 doi:10.1088/1742-6596/501/1/012021

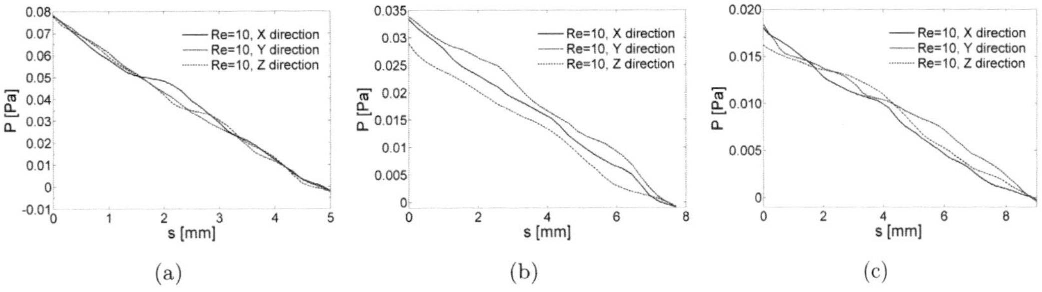

Figure 5: Pressure distribution along the three spatial directions at Re=10 for the (a) 30 PPI, (b) 20 PPI and (c) 10 PPI foam foam.

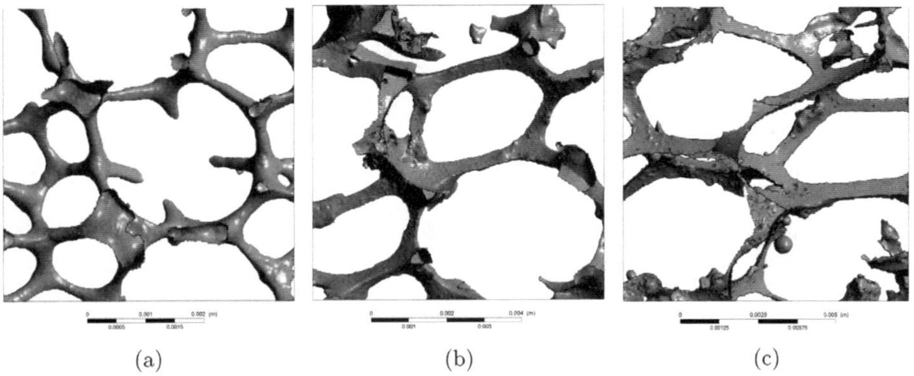

Figure 6: Detail of the foam structure along the z direction. (a) 30 PPI, (b) 20 PPI and (c) 10 PPI sample.

the value of ΔP and the pressure profiles along the three spatial directions are very similar and, therefore, the foam can be considered isotropic. On the contrary, for the other two samples, even though an almost linear trend can be observed, the deviations between the three spatial directions are amplified. This fact might be ascribed to the structure of the foams, illustrated in Fig. 6: while in the 30 PPI sample the cells are almost identical along the three spatial coordinates, in the other two samples they appear elongated along the z direction, thus producing the lower pressure drop. It should be investigated whether this structure is an intrinsic characteristic of the foams, or the result of the manipulation and cutting process.
In addition, in order to verify whether the lower pressure drop along the z direction was related to an insufficient length of the domain, for the 20 PPI sample some other tests were performed, by using a larger volume extended in the z direction. The values of the pressure drop along z computed on the new domain differ only marginally from those evaluated on the original domain, and still remain lower than those computed along the other two directions. Therefore, the anisotropy of the foam is not related to the chosen domain but, rather, it is an intrinsic property of the medium.

3.2. Evaluation of the permeability

For calculating the permeability of the foam along the three directions, the Darcy-Forchheimer equation can be written both in its dimensional form:

$$\nabla P = -\frac{\mu}{K} u_D - C\rho u_D^2 \tag{1}$$

and non-dimensional form:

$$\frac{\nabla P d_{nom}^2}{\mu u_D} = -\frac{d_{nom}^2}{K} - C d_{nom} Re \tag{2}$$

being K the permeability of the medium $[\mathrm{m}^2]$, C the form drag coefficient $[\mathrm{m}^{-1}]$, d_{nom} the nominal diameter of the foam and u_D the superficial velocity, calculated by dividing the volumetric flow rate by the total cross sectional area. While the linear term of Eq. (1) results from viscous effects, predominant at low Reynolds numbers, the quadratic one accounts for inertial effects which can not be neglected at higher velocities. Both C and K strongly depend on the structure of the porous medium.

The results demonstrate that, while at lower Reynolds numbers the pressure drop is a linear function of velocity, at higher Reynolds numbers the inertial effects become dominant and the pressure drop can be correctly described by the Dupuit-Forchheimer law. In Fig. 7, the dimensionless pressure gradient along the three space directions is plotted. The trends further confirm the isotropy of the 30 PPI sample, while predict a slight anisotropy for the 20 and 10 PPI foams.

The computed values of permeability and of the Dupuit-Forchheimer coefficient are summarized in Tab. 1, while Tab. 2 reports some of the values published in the literature: as it can be appreciated, the present values of K are comparable with the literature data. It is worth pointing out that, for a fixed pore density, the experimental values of K published in the literature show a wide dispersion.

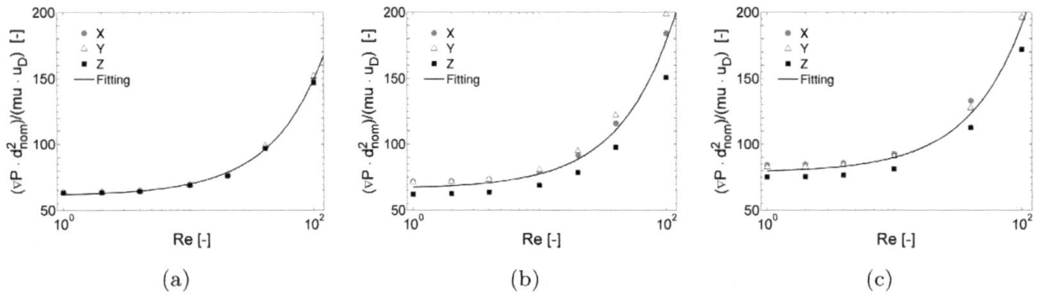

Figure 7: Spatial variation and least square fitting of the non-dimensional pressure gradient, for the (a) 30 PPI, (b) 20 PPI and (c) 10 PPI sample.

3.3. The thermal field

For the heat transfer simulations, two different limiting boundary conditions (b.c.) were specified on the foam wall, i.e., a constant temperature, $T_s = 323\mathrm{K}$, and a constant heat flux, $q_s'' = 100$ $\mathrm{W/m}^2$. As remarked in [19], a constant solid temperature may be compatible with the fully developed flow conditions. In Fig. 8, the variation of the heat transfer coefficient $h = q_s''/(T_s - T_{bulk})$ is illustrated: the heat transfer coefficient increases with the pore density and is higher in the case of the constant temperature b.c. Furthermore, the trend of h is not monotonic and manifests a minimum for a certain value of the Reynolds number: this behaviour is well prominent for the

Journal of Physics: Conference Series **501** (2014) 012021 doi:10.1088/1742-6596/501/1/012021

Table 1: Values of permeability K [m^2] and of the Dupuit-Forchheimer coefficient C [m^{-1}], for the three foam samples (Av.=average). ε is the porosity of the foam.

	30 PPI $\varepsilon = 0.929$		20 PPI $\varepsilon = 0.927$		10 PPI $\varepsilon = 0.944$	
	K×10^{-7}	C	K×10^{-7}	C	K×10^{-7}	C
x	0.872	389.0	1.965	309.4	2.906	218.6
y	0.859	392.9	1.946	346.3	3.058	255.2
z	0.858	370.6	2.222	240.5	3.234	187.9
Av.	0.863	384.2	2.045	298.7	3.066	220.6

Table 2: Summary of some permeability values published in the literature.

PPI	Porosity	$K \times 10^7$	Reference
10	0.9131	1.731	[25], num.
10	0.92	1.04	[1], exp.
10	0.926	4.29	[6], exp.
10	0.949	1.49	[32], exp.
20	0.9126	1.606	[25], num.
20	0.92	0.76	[1], exp.
20	0.9245	1.1	[32], exp.
20	0.935	2.42	[33], exp.
30	0.9	1.644	[34], exp.
30	0.93	0.63	[5], exp.

constant temperature b.c. while, for the constant heat flux b.c, in some directions a monotonic trend can be seen. Therefore, since a quadratic curve scarcely fits the numerical data, a new fitting relation was conceived, $h = a_1 [\log(Re)]^{a_4} + a_2 Re^{-a_5} + a_3$, providing very satisfactory results for all the samples.

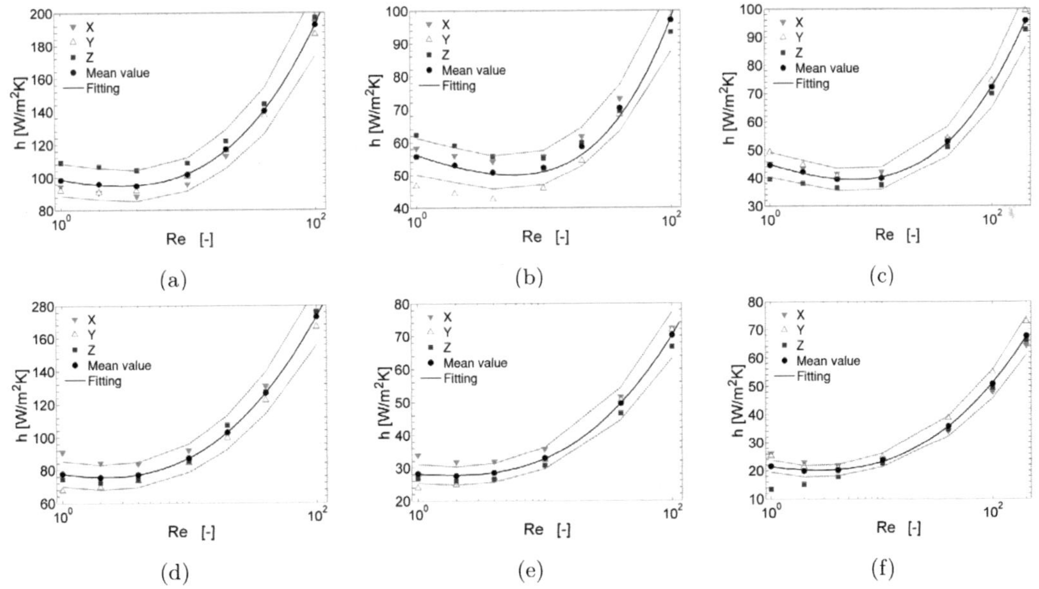

Figure 8: Heat transfer coefficient, for the case of constant temperature (top) and constant heat flux (bottom) boundary condition, for the (a,d) 30 PPI, (b,e) 20 PPI and (c,f) 10 PPI foam. The dot lines represent a 10% variation around the mean values.

3.4. Estimation of the effective thermal conductivity

For computing the effective thermal conductivity of the foam, k_{eff}, only the central foam region was simulated, and an arbitrary temperature gradient was fixed along one direction. The other lateral faces were kept adiabatic, and the conduction equation was solved in both the fluid and the solid region. The momentum equation was not solved, since the fluid (air or water) was considered as static.

The effective thermal conductivity was computed as:

$$k_{\text{eff}} = \frac{-\int \mathbf{J} \cdot d\mathbf{A}}{\dfrac{\partial T}{\partial x_i} A} = \frac{-\left(\int_s \mathbf{J} \cdot d\mathbf{A}_s + \int_s \mathbf{J} \cdot d\mathbf{A}_f\right)}{\dfrac{\partial T}{\partial x_i}(A_s + A_f)} \tag{3}$$

in which $\mathbf{J}$ and $d\mathbf{A}$ are the heat flux and the outward pointing area vector, respectively, and the subscripts s and f refer to the solid and the fluid, respectively.
Also in this case, the degree of isotropy of the foam seems to be related to the pore density of the medium. In fact, while for the 30 PPI sample the conductivities differ only marginally along the three spatial directions, for the other two samples their values diverge to a larger extent. As already stated in [25], this outcome might be connected to the size of the domains employed. However, the choice of performing a high resolution μ-CT limited the size of the scanned samples and preclude the employment of significantly larger domains for the CFD simulations. Besides this, for all the foams, the highest value is encountered along the z coordinate which, in the case of the 20 and 30 PPI foams, corresponds to the direction of elongation of the cells.
In view of the above considerations, in Tab. 3 only the mean values of the effective thermal conductivity, averaged among the three spatial coordinates, are reported, as in [25]. As it can be seen, the values of k_{eff} evaluated with water are slightly larger than those with air, thus proving that the thermal conductivity of the fluid influences, but not in a significant way, the effective thermal conductivity of the foam. In addition, it is worth noting that the values of k_{eff} evaluated in this study agree well with the values of k_{eff} published in the literature, some of which are summarized in Tab. 3.

Table 3: Prediction of effective conductivity values with air or water as the interstitial fluid and summary of some the values published in the literature.

PPI	Porosity	$k_{\text{eff,air}}$	$k_{\text{eff,water}}$	Reference
10	0.944	3.49	4.13	Present
20	0.927	4.97	5.76	Present
30	0.929	5.31	5.99	Present
10	0.9131	5.34	7.05	[25], num.
10	0.949	3.9	4.8	[32], exp.
20	0.9126	6.59	7.36	[25], num.
20	0.949	4	4.95	[32], exp.
40	0.9198	5.35	6.08	[25], num.
40	0.937	4.5	5.35	[32], exp.

4. Concluding remarks

In this paper, the results of accurate pore scale CFD simulations of flow and heat transfer in three foam aluminum samples, for laminar flow regime, have been presented. The 3D structure of the foams has been recovered through high-resolution X-ray computed microtomography, which allowed to capture the peculiar features of the foams' surfaces.
For all the samples, the flow is well described by the Darcy-Forchheimer law. In fact, while at low Reynolds numbers the pressure drop varies linearly with the velocity, at higher Reynolds numbers inertial effects are found to contribute.
For all the samples, the computed values of the flow permeability and the effective thermal conductivity agree well with the data, experimental and numerical, available in the literature. Moreover, the simulations demonstrate that the effective thermal conductivity of the foam is mainly given by the conductivity of the solid ligaments, since the contribution of the fluid is only marginal.

The flow and thermal simulations prove that the 30 PPI foam exhibits an isotropic behaviour, while a certain degree of anisotropy is detected for the 20 and 10 PPI foams. This fact may likely be ascribed to the structural characteristic of these two samples, characterized by an elongation of the foam cells along a specific spatial direction.
In view of the good results obtained, the present strategy could be applied to other porous materials, appealing for engineering applications.

Acknowledgment
This work was financially supported by Italian government, MIUR grant PRIN-2009KSSKL3.

References

[1] Kim S, Kan B and Kim J H 2001 *Int. J. Heat Mass Tran.* **44** 1451–1454
[2] Boomsma K, Poulikakos D and Zwick F 2003 *Mech. Mater.* **35** 1161–1176
[3] Tadrist L nad Miscevic M, Rahli O and Topin F 2004 *Int. J. Thermophys.* **21** 453–464
[4] Mahjoob S and Vafai K 2008 *Int. J. Heat Mass Tran.* **64** 3701–3711
[5] Kurtbaş I, Celik N and Dinçer I 2010 *Energy* **35** 451–460
[6] Mancin S, Zilio C, Rossetto L and Cavallini A 2011 *ASME J. Heat Transf.* **133** 0609041–0609049
[7] Mancin S, Zilio C, Diani A, Rossetto L and Cavallini A 2012 *Exp. Therm. Fluid Sc.* **36** 224–232
[8] Yang C, Nakayama A and Liu W 2012 *Int. J. Therm. Sci.* **54** 98–108
[9] Odabaee M and Hooman K 2011 *Int. Commun. Heat Mass* **38** 838–843
[10] Lu W, Zhao C and Tassou S 2006 *Int. J. Heat Mass Tran.* **49** 2751–2761
[11] Zhao C, Lu W and Tassou S 2006 *Int. J. Heat Mass Tran.* **49** 2762–2770
[12] Raju K and Narasimhan A 2007 *ASME J. Heat Transf.* **129** 273–281
[13] Odabaee M and Hooman K 2012 *Appl. Therm. Eng.* **36** 456–463
[14] Kim S, Paek J and Kang B 2003 *IEEE T. Compon. Pack. T.* **26** 262–267
[15] Seyf H and Layeghi M 2010 *ASME J. Heat Transf.* **132** 0714011–0714019
[16] Ranut P, Nobile E and Mancini L 2012 *Proc. of the 30th UIT Heat Transfer Conference* ed Lazzari S and Rossi di Schio E (Società Editrice Esculapio) pp 61–66
[17] Boomsma K, Poulikakos D and Ventikos Y 2003 *Int. J. Heat Fluid Flow* **24** 825–834
[18] Krishnan S, Murthy J and Garimella S 2006 *ASME J. Heat Tr.* **128** 793–799
[19] Kopanidis A, Theodorakakos A, Gavaises E and Bouris D 2010 *Int. J. Heat Mass Tran.* **53** 2539–2550
[20] Wu Z, Caliot C, Flamant G and Wang Z 2011 *Int. J. Heat Mass Tran.* **54** 1527–1537
[21] Petrasch J, Meier F, Friess H and Steinfeld A 2008 *Int. J. Heat Fluid Flow* **29** 315–326
[22] Magnico P 2009 *Chem. Eng. Sc.* **64** 3564–3575
[23] Coquard R and Baillis D 2009 *Acta Mater.* **57** 5466–5479
[24] Gerbaux O, Buyens F, Mourzenko V, Memponteil A, Vabre A, Thovert J F and Adler P 2010 *J. Colloid Interface Sci.* **35** 155–165
[25] Bodla K, Murthy J and Garimella S 2010 *Numer. Heat. Tr. A-APPL* **58** 527–544
[26] `http://www.elettra.trieste.it/lightsources/labs-and-services/tomolab/tomolab.html`
[27] `http://www.elettra.trieste.it`

Journal of Physics: Conference Series **501** (2014) 012021 doi:10.1088/1742-6596/501/1/012021

[28] Kak A and Slaney M 1988 *Principles of Computerized Tomographic Imaging* (IEEE Press)
[29] http://www.m-pore.de
[30] http://www.exxim-cc.com
[31] http://www.volumegraphics.com/en/products/vgstudio-max.html
[32] Bhattacharya A, Calmidi V and Mahajan R 2002 *Int. J. Heat Mass Tran.* **45** 1017–1031
[33] Liu J, Wu W, Chiu W and Hsieh W 2006 *Exp. Therm. Fluid Sci.* **30** 329–336
[34] Kamath P, Balaji C and Venkateshan S 2011 *Int. J. Heat Mass Tran.* **54** 5231–5241

Appearance of quasiperiodicity within a period doubling route to chaos of a swaying thermal plume

D Angeli[1], M A Corticelli[1], A Fichera[2] and A Pagano [2]

[1] Dipartimento di Ingegneria Enzo Ferrari, Università degli Studi di Modena e Reggio Emilia, Via Vignolese 905, 41125 Modena (Italy)

[2] Dipartimento di Ingegneria Industriale, Università degli Studi di Catania, Viale Andrea Doria 6, 95125 Catania (Italy)

E-mail: [1]`diego.angeli@unimore.it`

E-mail: [2]`apagano@diim.unict.it`

Abstract. The birth, evolution and disappearance of quasiperiodic dynamics in buoyancy-driven flow arising from an enclosed horizontal cylinder are analysed here, by numerical means, in the limit of the 2D approximation. The governing equations are solved on orthogonal Cartesian grids, giving special treatment to the internal, non-aligned boundaries. Thanks to the adoption of a high level of refinement of the Rayleigh number range, quasiperiodicity was observed to emerge from a periodic limit cycle (P1), and to turn into its omologous orbit with doubled period (P2), eventually evolving into a classical period-doubling route to chaos, for further increases of the Rayleigh number. The present study gives a deeper insight to what appears to be an imperfect period doubling bifurcation through a quasiperiodic T2-torus. The approach used is based on the classical tools for time series analysis. The distribution of the power spectral densities is used to search for and characterise the existence of relations between the frequencies of the P1, T2 and P2 dynamics. The topology of the orbits, as well as their evolution within the quasiperiodic window, are analysed with the aid of phase space representation and Poincaré maps.

1. Introduction

The strong dependence on the system geometrical configuration and on the thermal boundary conditions is known to determine a great variety of complex buoyancy-driven flows in confined enclosures. In fact, great interest has been given to the non-linear dynamics of thermal convection in basic enclosure configurations, with the heat source either corresponding with the bottom or side walls of the cavity [1, 2]. Several theoretical and experimental studies have also been devoted to the horizontal annulus, as summarized in [3]. For more complex configurations, though most of the effort has been mainly focused on heat transfer performances [4–7], the use of optical techniques such as Particle Image Velocimetry (PIV), has recently allowed the experimental study of the flow field and of its influence on the overall heat transfer[8, 9]. Several authors also dealt with natural convection from enclosed heat sources with the use of state-of-the-art numerical techniques [6, 7, 10–12].

The onset of chaotic dynamics has been addressed in several works. Indeed, the description of the transitional phenomena that drive the dynamics of the flow inside confined cavities, typically from stationary solutions to chaos as the Rayleigh number increases, represents one of the main research topics in the study of buoyant flows. Different kinds of instabilities and bifurcation scenarios have been reported for different cavity shapes and, for a given shape, for different values of the characteristic aspect ratio. Details of transition vary greatly from one case to another; however, a number of established routes to chaos are found to be common to many types of dissipative dynamical systems [13]. As several types of possible flow regimes are encountered, the simple set of acronyms and abbreviations proposed by Gollub and Benson [14] and extended in [3] is adopted in the following in order to facilitate their description.

With respect to the study of the instabilities and bifurcations of a buoyant plume arising from an enclosed heat source, most of the attention has been devoted to the case of the horizontal annulus. The numerical work of Yoo et al. [15] is of particular interest, although limited to the case of 2D narrow gap annuli (with radius ratio $R = 1.2$) filled with a very low Prandtl number fluid ($Pr = 0.02$). Reported results show that the transition to chaotic flow occurs through the breaking of a quasi-periodic orbit. Later, Yoo and Han [16] investigated the case of a large-gap annulus ($R = 2$), for a slightly larger value of Pr, $Pr = 0.1$, and following an entire route to chaos. The occurrence of a period-tripling bifurcation is observed after the transition to oscillatory flow. The three-period phase (P_3) is interrupted by a window of quasi-periodicity (QP_2); subsequently, a period-doubling cascade (P_6, P_{12}) leads to a first nonperiodic phase (N_I). Before the final transition to chaos (N_{II}), a relaminarisation phase is observed, consisting in a small window of periodicity (P_4) whose frequency is related with the fundamental frequency of the first periodic regime by a factor 4. A very thorough experimental analysis of the transition to chaos in an air-filled annulus with $R = 2.36$ was provided by Labonia and Guj [17]. Quantitative and qualitative measurements were performed by interferometry, smoke visualization and hot wire anemometry. The route to chaos appears to be characterised by a sequence of Hopf transitions, separated by windows of quasiperiodic behaviour.

With respect to other type of cavities with internal heat sources, considerable attention has been devoted in the last two decades to the investigation of the flow and heat transfer characteristics. The issues of flow stability and bifurcations have been addressed only for long square or rectangular-sectioned cavity with two types of internal source: wires of small diameter and line heat sources [18–21], and horizontal circular or rectangular cylinders [5, 10, 22–25]. In [18] Desrayaud and Lauriat thoroughly investigated the dynamical behaviour of the buoyant plume arising from a horizontal line source, modeled as a point source in a 2D rectangular cavity with adiabatic side walls and isothermal top and bottom walls. The analyses were performed taking air as the heat carrier fluid ($Pr = 0.71$), for several values of the enclosure aspect ratio, and for various depths of immersion of the source along the vertical axis of the cavity. Depending on these two parameters, various dynamics were detected. In particular, for square cavities and moderate depths of immersion, the onset of unsteadiness occurred by the subcritical transition from steady-state to a periodic swaying motion of the buoyant plume. The correspondent route to chaos passed through phase-locking and intermittency states. The progressive increase of the dimensionless dominant frequency appeared to be fairly resolved, thus demonstrating the overall consistency of the results. For higher values of both parameters the scenario changed dramatically. Above the thermal plume, an unstably stratified region was maintained at rest thanks to the hydrostatic gradient, until a pitchfork bifurcation occurred, leading to a non-symmetric bending of the plume, which entrained the above fluid layer. The subsequent transition to oscillatory flow was characterised by hysteresis phenomena, revealing the occurrence of subcritical Hopf bifurcations.

Recently, Angeli et al. [22] investigated by means of 2D finite-volume simulations the nonlinear dynamics of the buoyancy-induced flow of air (Pr=0.7) heated from a horizontal cylindrical source centred in a square enclosure. The study was developed under various geometrical configurations, as defined by the aspect ratio $A = H/L$, i.e. the ratio between the maximum gap height and the length of the side walls. For the cavities with aspect ratio $A = 5$

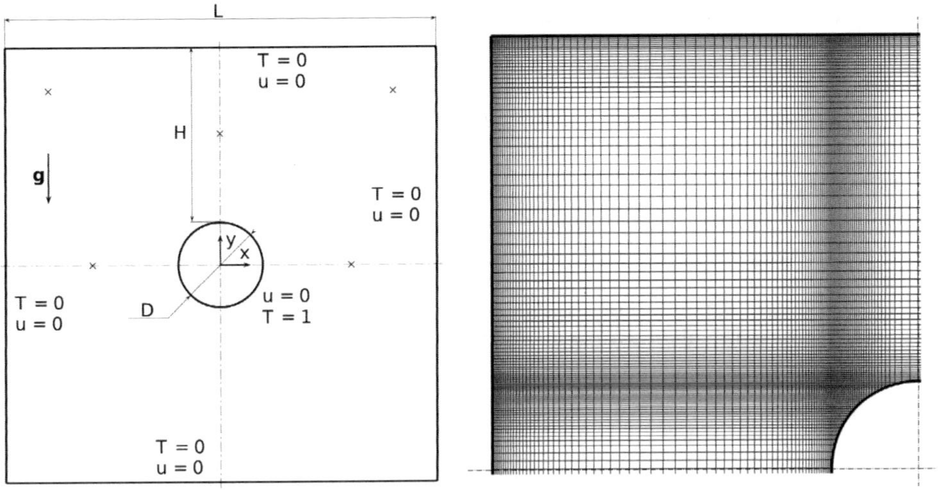

Figure 1. Left: schematic of the system under consideration; (×) symbols indicate locations of the sampling points. Right: quadrant of the computational grid.

and $A = 10$, the flow asset and early bifurcations were found to be primarily determined by the formation of Rayleigh-Benard-type rolls above the heat source. These steady-state, multicellular patterns were seen to undergo symmetry-breaking pitchfork bifurcations, eventually leading to periodic states.For $A = 3.3$ and $A = 2.5$ the dynamics were found to be governed by the oscillations of a swaying thermal plume originating around the cylindrical heat source.

Aiming at an accurate assessment of the entire scenario leading to deterministic chaos, in [26] a single value of the aspect ratio, $A = 2.5$, was analysed in detail adopting a thorough refinement of the range of Ra in order to observe the a sequence of bifurcations leading to chaos. In this study it was shown that the bifurcation scenario can be summarized by a period-doubling route to chaos, originated by a supercritical Hopf bifurcation and followed up to the P_{128} limit cycle. Though this scenario can be considered relatively common for thermofluid dynamic systems, one main peculiarity of the reported bifurcation path is the existence of a window of quasiperiodic behaviour in the range $1.735 \times 10^5 < Ra < 1.790 \times 10^5$, which appears as an imperfect occurrence of the first period doubling, i.e. from a P_1 to a P_2-limit cycle, at the beginning of the mentioned period-doubling cascade. In particular, what appears unusual in the reported scenario is that the disappearance of the quasiperiodic torus corresponds to the birth of a stable P_2 limit cycle and, above all, to the subsequent period doubling cascade. Such a behaviour, atypical for autonomous systems and not yet reported for thermofluid dynamic systems. The aim of the present study is to give further details on the origin, evolution and disappearance of the quasiperiodic T_2-torus mainly through the analyses of the power spectral density distributions and of the topology of the limit cycles and of the torus involved in this restricted sequence of bifurcations.

2. Problem statement

The problem is stated in terms of the incompressible Navier-Stokes formulation. The Oberbeck-Boussinesq approximation is enforced, all the fluid properties being consistently assumed as constant, apart from density in the buoyancy term.

The governing equations are tackled in their non-dimensional form. The gap between the top of the cylinder and the upper cavity wall is indicated in [22] as the most suitable scale length, $H_{ref} = H$. In fact, it reduces the dependence of the solution ranges on the aspect ratio, in particular, for what concerns the heat transfer rate and the first transitions between different regimes. Moreover, the region above the cylinder is subject to the maximum inverse thermal gradient, *i.e.* to the highest buoyancy force acting on the fluid system, and this is again related to the reference length H.

Temperature is non-dimensionalized according to a reference temperature T_{ref}, *i.e.* the temperature difference between the cylinder and the cavity walls ($T_{ref} = T_S - T_W$), and the following velocity scale is chosen:

$$U_{ref} = \sqrt{g\beta T_{ref} H_{ref}} \tag{1}$$

where g denotes the gravitational acceleration and β is the thermal expansion coefficient of the fluid. The continuity, momentum, and energy equations are given the following form:

$$\nabla \cdot \mathbf{u} = \mathbf{0} \tag{2}$$

$$\frac{\partial \mathbf{u}}{\partial t} + \mathbf{u} \cdot \nabla \mathbf{u} = -\nabla p + \frac{Pr^{1/2}}{Ra^{1/2}} \nabla^2 \mathbf{u} + T\hat{\mathbf{g}} \tag{3}$$

$$\frac{\partial T}{\partial t} + \mathbf{u} \cdot \nabla T = \frac{1}{(RaPr)^{1/2}} \nabla^2 T \tag{4}$$

where t, $\mathbf{u}$, p and T represent the dimensionless time, velocity vector, pressure and temperature, respectively, and $\hat{\mathbf{g}}$ is the gravity unit vector. The Rayleigh and Prandtl numbers are defined as:

$$Ra = \frac{g\beta T_{ref} H_{ref}^3}{\nu\alpha} \tag{5}$$

$$Pr = \nu/\alpha \tag{6}$$

where ν and α represent the momentum and thermal diffusivity, respectively. A constant value $Pr = 0.7$ is assumed for air. With reference to figure 1, the following non-dimensional boundary conditions are imposed:

$$T = 0, \quad \mathbf{u} = \mathbf{0} \tag{7}$$

at the enclosure walls, and:

$$T = 1, \quad \mathbf{u} = \mathbf{0} \tag{8}$$

on the cylinder surface.

3. Numerical methods

The numerical technique adopted is based on a Finite Volume implementation of a second order Projection Method, following [27]. Time-discretizations of the conservation equations are performed according to a three-level scheme, which is fully implicit for the diffusive terms, and explicit Adams-Bashforth for the advective terms. Such a practice is second order accurate in time.

Spatial derivatives are approximated with second order central differences on staggered, non-uniform Cartesian grids. A direct resolution of the discrete momentum and energy equations at each time-step is made possible by means of Approximate Factorization, while the Poisson problem associated with the pressure-velocity coupling [27] is solved through a fast Poisson solver, based on Matrix Decomposition.

The 2D modelling of arbitrarily irregular boundaries on Cartesian grids is achieved thanks to the original scheme developed in [28]. The technique involves a local modification of the 5-point computational stencil where boundary segments intersect the stencil arms. The variables on the modified stencil are mapped on the global grid, by means of a linear operator determined by geometrical features and boundary conditions. The overall accuracy of the method is virtually preserved, as well as the computational efficiency of the Cartesian approach.

The Cartesian grid employed, made up by 190 × 190 elements, is shown in figure 1. Along both grid directions, a variable spacing was used, and special care was put on the grid sizing of both near-wall areas and internal domain regions, in view of the work objectives. The criteria adopted for the choice of grid spacing and time step size are outlined in detail in [26], while a grid independency test for a periodic flow is reported in [22].

In order to analyze the system dynamics in the vicinity of bifurcation points, Ra was increased monotonically with suitable steps, each simulation starting from the final frame of the preceeding one. All the simulations were protracted until a fixed dimensionless time span was covered, large enough for an asymptotic flow to be attained [26].

Time histories of dimensionless temperature T, and velocity components along x and y directions, respectively u_x and u_y, were tracked for 5 locations, reported in figure 1, and analysed by means of classical tools for nonlinear time series analysis, such as phase-space representations, power spectra and Poincaré maps.

4. Results and discussion

In the present section, the dynamical behaviour of the system under investigation is described for the cavity of aspect ratio A = 2.5, characterized by a thermal plume whose swaying motion is dependent on the intensity of the buoyancy force, i.e. on the forcing term Ra [29]. Reported results refer to the system variables simulated at point (0; 0.5) and are representative of the dynamics observed in the other points. In particular, the dynamics in the other control points are equivalent, in terms of both spectral analysis and phase-space topology, to those described in the following.

For increasing values of the Rayleigh number, the adoption of an accurate scanning of the bifurcation parameter allowed the observation of the entire sequence of bifurcations driving the plume from stationary to chaotic motion as summarized in table 1 and discussed in detail by Angeli and Pagano [26]. From the analysis of table 1 it clearly emerges that the bifurcation scenario is, in general, well described by a Feigenbaum cascade. In other terms, for increasing Ra-values the periodicity of the plume oscillation undergoes a series of doublings, which occur closer and closer to one another until chaos appears. Nonetheless, it is worth observing that this paradigm appears distorted by the occurrence, at the beginning of the period doubling, of a window of Ra-values characterized by the quasiperiodic oscillation of the thermal plume. In

Table 1. Flow regimes and corresponding ranges of Ra throughout the transition to chaos [26].

flow regime		$Ra \times 10^{-5}$
steady-state	S	≤ 0.66200
periodic	P_1	$0.66200 \div 1.73500$
quasiperiodic	QP	$1.74000 \div 1.79270$
periodic	P_2	$1.79280 \div 1.89750$
periodic	P_4	$1.89800 \div 1.93670$
periodic	P_8	$1.93675 \div 1.94730$
periodic	P_{16}	$1.94735 \div 1.94950$
periodic	P_{32}	$1.94955 \div 1.94990$
periodic	P_{64}	$1.94995 \div 1.95002$
periodic	P_{128}	$1.95002 \div 1.95010$
chaos	N	≥ 1.95025

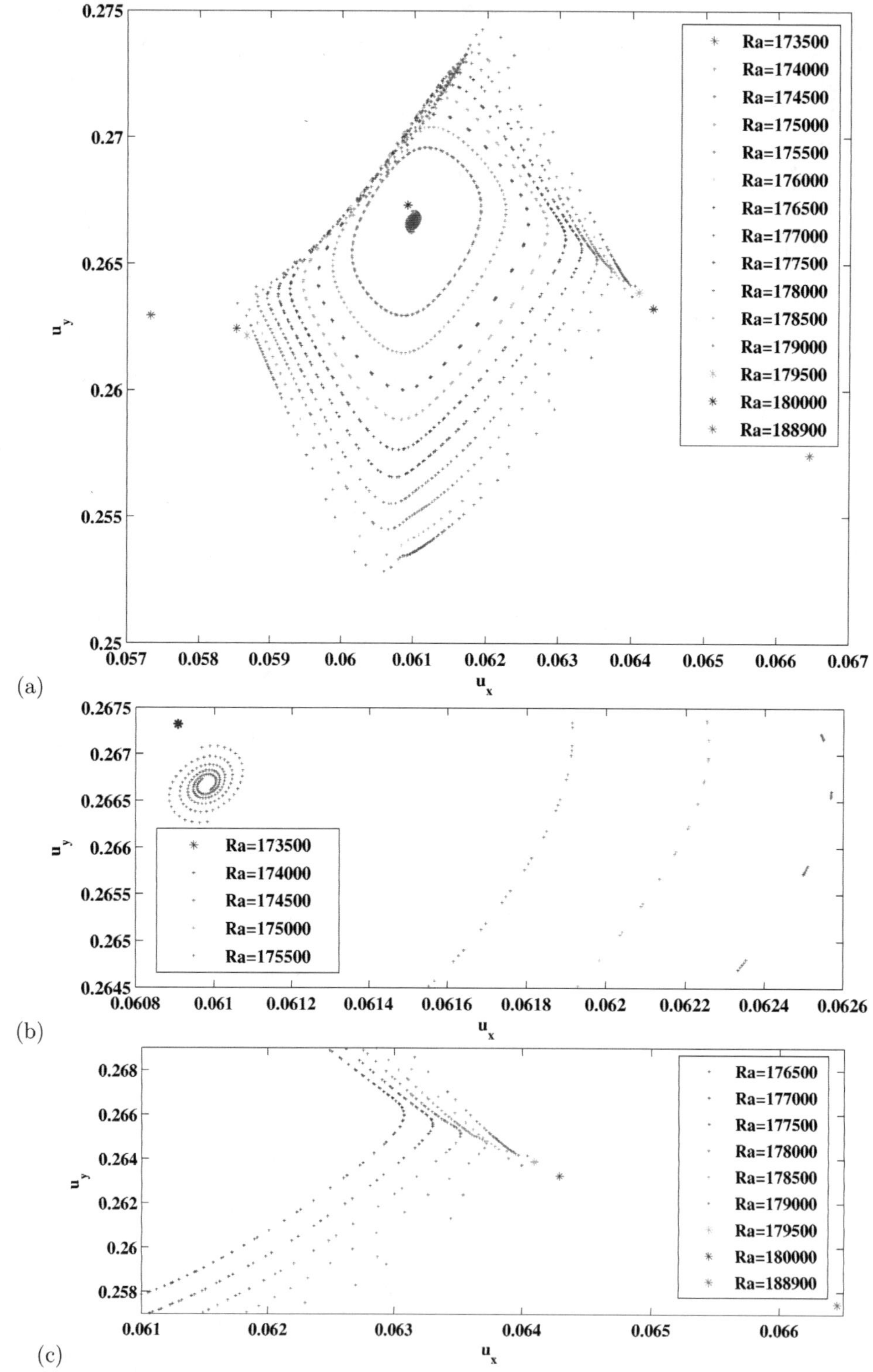

Figure 2. Sequence of Poincaré maps at point (0; 0.5) (a) spanning the whole range of Ra considered; (b) enlarged view of the $P_1 \rightarrow QP_2$ transition range; (c) enlarged view of the $QP_2 \rightarrow P_2$ transition range.

this contest, the present study aims at offering a deeper insight on the plume dynamics within the window of quasiperiodic dynamics and in its immediate neighborhoods.

As it is shown in table 1, for $Ra \geq 1.735 \times 10^5$ the formerly stable P_1 limit cycle becomes unstable. In fact, a new attractor of the system dynamics appears in phase space in the shape of a dense T_2 torus, typical of quasiperiodic dynamics characterised by two incommensurate frequencies. The bifurcation of the P_1 limit cycle to the T_2 torus implies the occurrence of a Neimark-Sacker bifurcation [30]. Considering that the classical Feigenbaum doubling scenario is determined by a series of flip bifurcations of limit cycles [30], the reported observation implies that the first of these bifurcations is replaced by a Neimark-Sacker bifurcation followed by the collapse of the T_2 torus onto a P_2 limit cycle. The Neimark-Sacker bifurcation leads to the permanence of the unstable P_1 limit cycle in the wake of a ghost entrapped inside the torus and acting as an axis for it.

A clear documentation of this behavior is represented by the evolution of the topology of the Poincaré maps of the T_2 torus reported in figure 2a and in its local detail reported in figure 2b. The Poincaré sections were obtained by cutting the attractors with the secant plane normal to the T-coordinate, at its mean value. The exact location of the intersections was determined by linear interpolation of the position of two points on the opposite sides of the secant plane. The interpolation error is negligible due to the high resolution of the simulated time series. From the analysis of these figures, in fact, it emerges that all of the pseudo-elliptic maps, i.e. the characteristic traces of T_2 tori onto the Poincaré surface of section, are centered on the stationary point, marginally stable, of the map at $Ra = 173500$. As a consequence, one main consideration that can be drawn is that, referring to the torus axis, only marginal variations occur with the increase of the Rayleigh number, i.e. the torus amplitude and its general topology in phase space are not sensitively altered. In fact, the most important influence of the growing energy associated to the increase of Ra within the range of quasiperiodic behavior is indeed represented by the progressive enlargement of the torus trace onto the Poincaré plane. In other terms, the higher is the value of the forcing term, Ra, the larger is the torus trace, which, in turn, is associated to the relative importance of the second incommensurate frequency with respect to the primary one, inherited from the unstable limit cycle and associated to the dominant feature of the motion around the torus axis.

At the same time, another important effect associated to the increase of Ra is the progressive deformation of the transversal section of the torus. In particular, the Poincaré section, while remaining dense, progressively changes its shape. From the analysis of the general behavior in figure 2 and of the specific behaviors at $Ra = 174500$ in figure 3 and at $Ra = 178500$ in figure 4, it is possible to observe that the pseudo-elliptical shape, which is a typical feature of a Poincaré traces at the onset of the quasiperiodic dynamics on a torus, is progressively deformed, eventually turning into a pseudo-rectangle, with a marked local deformation of its edges. Moreover, the edges of the pseudo-rectangular Poincaré trace assume the behavior of saddle points on the Poincar map (see figure 2c and 4c-e). From the plots, the progressive and relatively slow accumulation of the traces of the trajectory approaching them on one direction and the abrupt repelling action on the direction pseudo-orthogonal to the first appear evidently. It is then possible to explain the disappearance of the torus and the birth of the P_2 limit cycle as a consequence of the destruction of a couple of opposite edges of the pseudorectangular Poincaré trace, and of the stabilization of the remaining couple of saddles, i.e. of the two branches of the stable P_2 limit cycle on the 3D state space.

Some further details, substantially confirming the description reported so far on the topology of the T_2 torus in the 3D state space and of its Poincaré section in 2D, can be drawn from the analysis in the frequency domain. As it was shown in [26], within the range of quasiperiodic behavior, two incommensurate frequencies f_1 and f_2 exist, which are only marginally influenced by the value of Ra, and so it happens to their ratio, f_2/f_1. The main effect of the increase

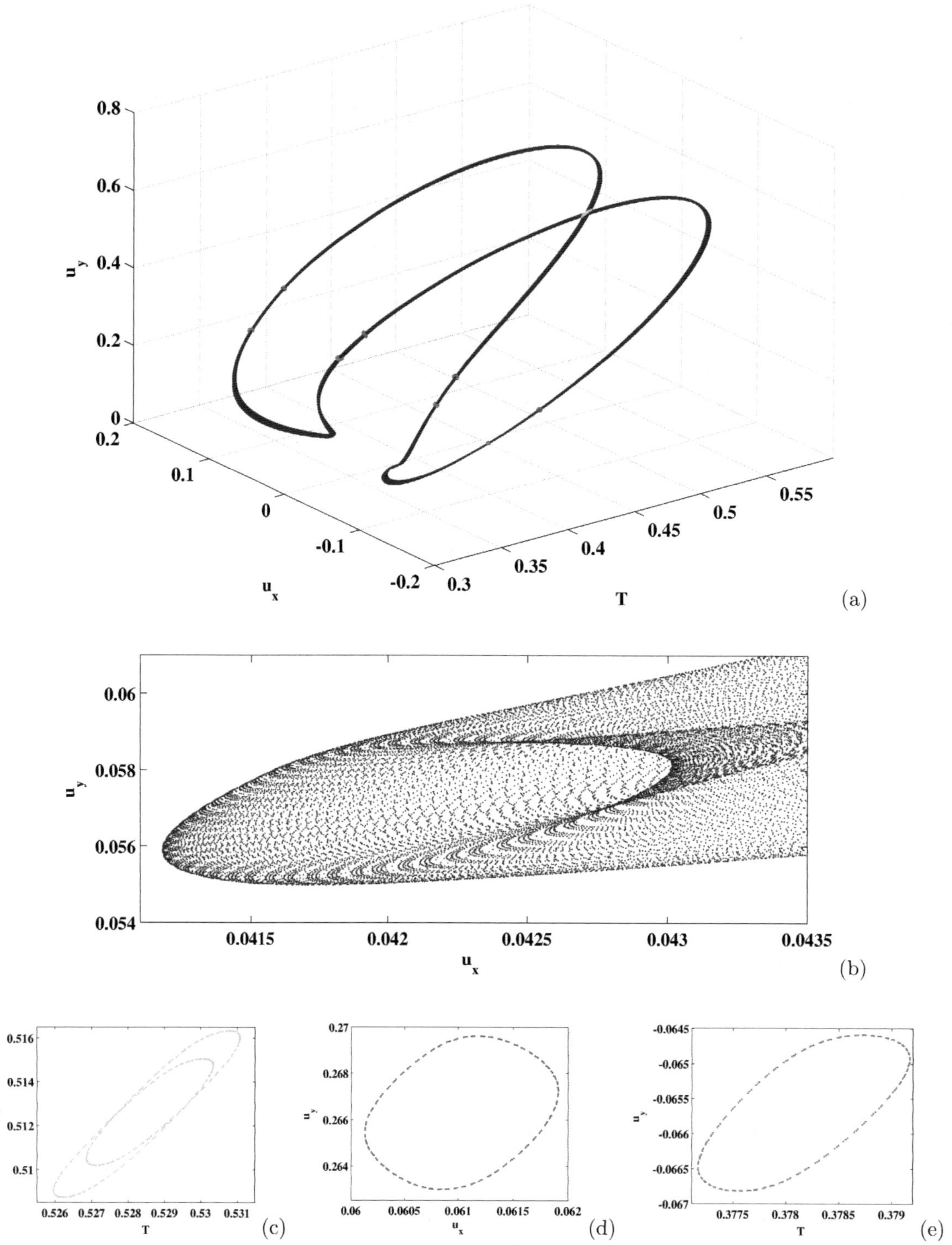

Figure 3. Phase plots of the quasiperiodic dynamical behaviour at point (0; 0.5) for $Ra = 174500$: (a) attractor in the state space T-u_x-u_y; (b) particular evidencing the structure of the T_2 torus; (c), (d), (e) Poincaré maps.

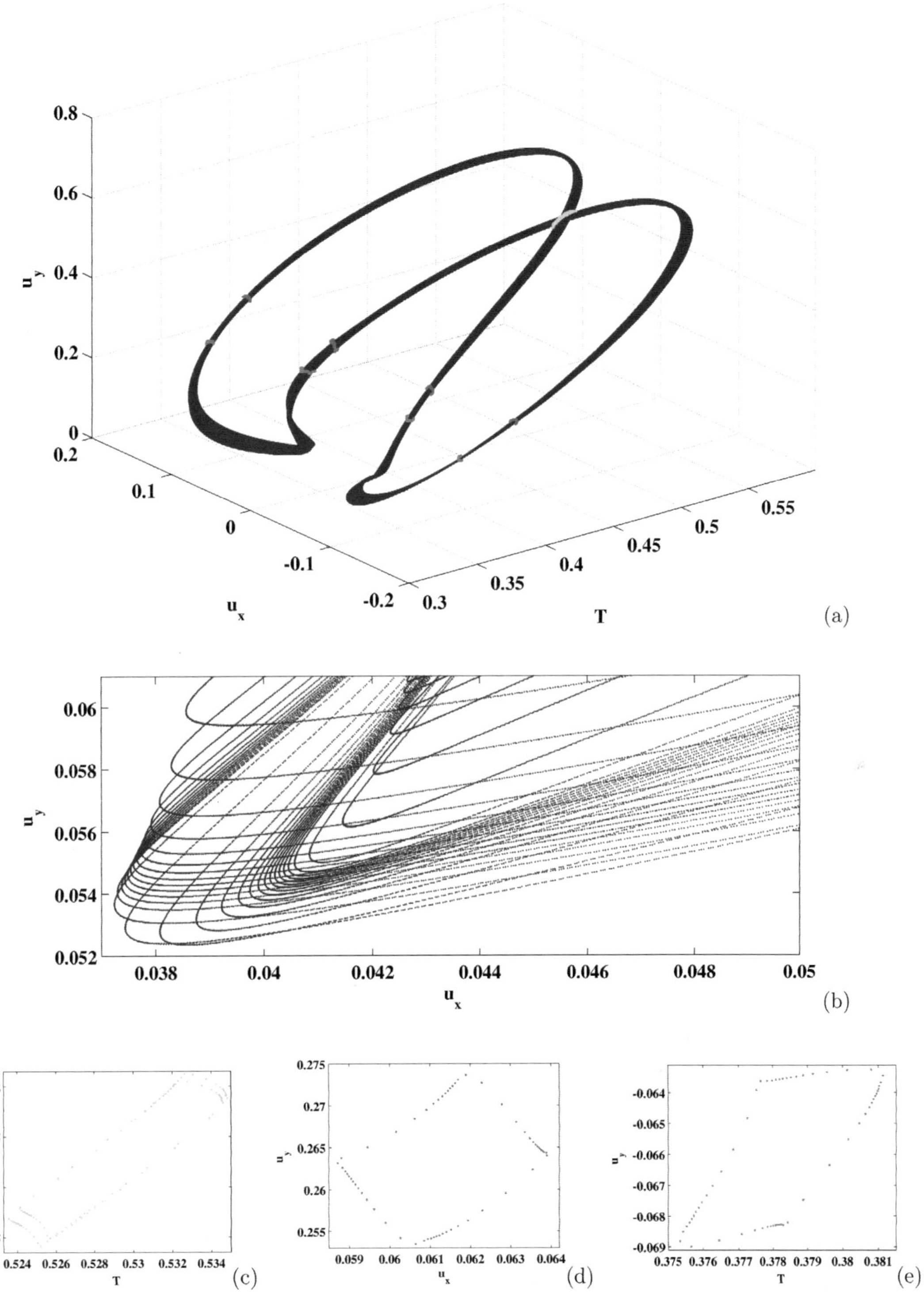

Figure 4. Phase plots of the quasiperiodic dynamical behaviour at point (0; 0.5) for $Ra = 178500$: (a) attractor in the state space T-u_x-u_y; (b) particular evidencing the structure of the T_2 torus; (c), (d), (e) Poincaré maps.

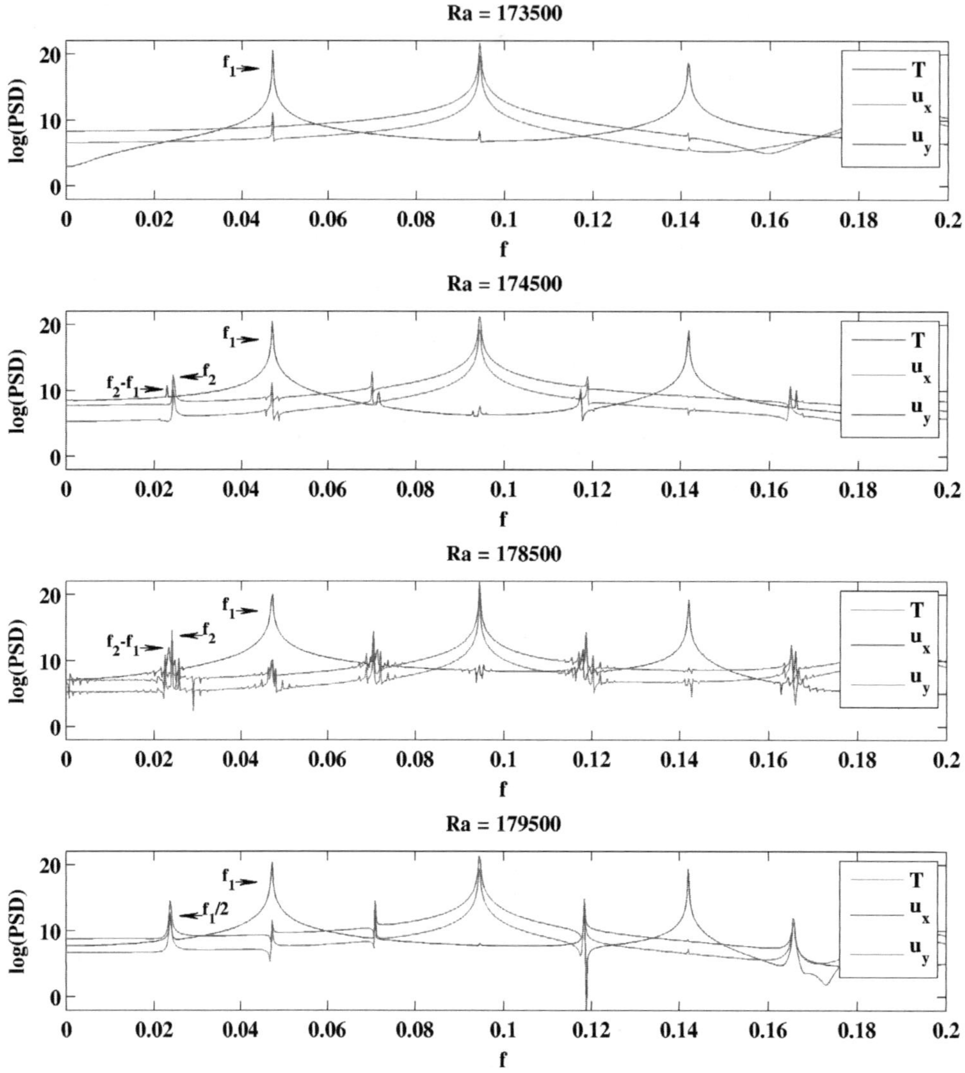

Figure 5. PSD distribution of T, u_x and u_y at point (0;0.5) for selected Ra-values

of Ra on the power spectral density (PSD) can be drawn from the analysis of figure 5. In accordance with previous observations on the progressive enlargement of the Poincaré traces, it can be observed that the power associated to the secondary frequency, f_2, and to its harmonic combinations with f_1, clearly increases with Ra. The analysis of the PSD also helps to stress the nature of imperfect period doubling, i.e. of the missed flip bifurcation of the P_1 limit cycle, that gives rise to the reported Neimark-Sacker bifurcation. In fact, within the entire range of quasiperiodic behavior:

- the secondary frequency f_2 is always very close to $f_1/2$;
- the main harmonic combination of f_1 and f_2 emerging in the PSD for all of the simulated time series is always the frequency $f = f_1 - f_2$;
- the amplitude of $f = f_1 - f_2$ is comparable with that of f_2, and, for increasing Ra, further harmonic combinations of the two main frequencies arise in the PSD
- accordingly, the trajectories distribute themselves onto the T_2-torus through an almost

parallel and alternated winding on opposite sides of the torus itself.

As a final observation, it is worth noting that, for increasing Ra, the difference between f_2 and the harmonic $f_1 - f_2$ progressively decreases, until they collide for $Ra = 1.7928 \times 10^5$, leading to the disappearance of the quasiperiodic torus and to the birth of a stable P_2 limit cycle, as reported in figure 2a and 2c.

5. Concluding remarks

The occurrence of quasiperiodicity preluding a sequence of period-doubling bifurcations leading to deterministic chaos was analysed in detail by numerical means, for a natural convection flow originating from a horizontal cylindrical source, centred in a square enclosure of aspect ratio $A = 2.5$.

The analysis of long-term, highly resolved time series, extracted from numerical predictions of the asymptotic system behaviour, allowed for the disclosure of significant features of such a peculiar bifurcation pattern. The quasiperiodic orbit originates from the destabilization of a P_1 periodic orbit, through a Neimark-Sacker bifurcation, and then disappears giving way to a second periodic orbit P_2, whose fundamental frequency is equal to half of the frequency of the P_1 orbit. The topological evolution of the T_2 torus in the 3D state space, and of the related traces in the Poincaré maps, revealed the progressive creation of singularities in the shape of the attractor, in the form of four accumulation curves, corresponding, in the Poincaré section, to the four corners of a pseudorectangle. Eventually, two of such singularities disappear, while the remaining two survive in the P_2 orbit. Furthermore, the analysis of the time series in the frequency domain indicate that the halving of the fundamental frequency f_1 of the P_1 orbit results from the collision between the secondary frequency f_2 of the quasiperiodic flow and the harmonic $f_1 - f_2$.

The observed behaviour suggests that, in the present case, the Neimark-Sacker bifurcation replaces the first flip bifurcation of a classical Feigenbaum sequence, and may then be viewed as an imperfect period doubling, a feature yet unprecedented in confined convective flows.

References

[1] Yang K T 1988 *J. Heat. Transf.* **110** 1191–1204
[2] Le Quéré P 1994 *Procs. 10th Int. Heat Transf. Conf.* vol 1 ed Hewitt G F (Brighton, UK: Hemisphere Publishing Corporation) pp 281–296
[3] Angeli D, Barozzi G S, Collins M W and Kamiyo O M 2010 *Int. J. Therm. Sci.* **49** 2231–2492
[4] Warrington J R O and Powe R E 1985 *Int. J. Heat Mass Transf.* **28** 319–330
[5] Cesini G, Paroncini M, Cortella G and Manzan M 1999 *Int. J. Heat Mass Transf.* **42** 1801–1811
[6] Kim B S, Lee D S, Ha M Y and Yoon H S 2008 *Int. J. Heat Mass Transf.* **51** 1888–1906
[7] Lee J M, Ha M Y and Yoon H S 2010 *Int. J. Heat Mass Transf.* **53** 5905 – 5919
[8] Paroncini M and Corvaro F 2009 *Int. J. Therm. Sci.* **48** 1683–1695
[9] Butler C, Newport D and Geron M 2012 *Exper. Therm. Fluid Sci.* Article in press
[10] Peng Y, Chew Y T and Shu C 2004 *Phys. Rev. E* **67** 026701/1–026701/6
[11] Bararnia H, Soleimani S and Ganji D D 2011 *Int. Comm. Heat Mass Transf.* **38** 1436 – 1442
[12] Nabavizadeh S A, Talebi S, Sefid M and Nourmohammadzadeh M 2012 *Int. J. Therm. Sci.* **51** 112 – 120
[13] Bergé P, Pomeau Y and Vidal C 1988 *L'Ordre dans le Chaos* 2nd ed (Hermann, Editeurs des Sciences et des Arts)
[14] Gollub J P and Benson S V 1980 *J. Fluid Mech.* **100** 449–470
[15] Yoo J S, Choi J Y and Kim M U 1994 *Num. Heat Transf. A* **25** 103–115
[16] Yoo J S and Han S M 2000 *Fluid Dyn. Res.* **27** 231–245
[17] Labonia G and Guj G 1998 *J. Fluid Mech.* **375** 179–202
[18] Desrayaud G and Lauriat G 1998 *J. Fluid Mech.* **252** 617–646
[19] Deschamps V and Desrayaud G 1994 *J. Thermophys. Heat Transf.* **8** 84–91
[20] Lauriat G and Desrayaud G 1994 *Sādhanā* **19** 671–703
[21] Barozzi G S and Corticelli M A 1999 *Procs. 6th UK Nat. Conf. Heat Transf.* (IMechE *Conference Transactions* vol 1) ed Hahane E W P, Heidemann W and Spindler K (Edimburgh, UK: Professional Engineering Publishings) pp 69–74

[22] Angeli D, Levoni P and Barozzi G S 2008 *Int. J. Heat Mass Transf.* **51** 553–565
[23] Ghaddar N K 1992 *Int. J. Heat Mass Transf.* **35** 2327–2334
[24] Moukalled F and Acharya S 1996 *Journal of Thermophys.* **10** 524–531
[25] Shu C and Zhu Y D 2002 *Int. J. Num. Meth. Fluids* **38** 429–445
[26] Angeli D and Pagano A 2013 *Int. J. Therm. Sci.* **68** 20–31
[27] Gresho P M 1990 *Int. J. Num. Meth. Fluids* **11** 587–620
[28] Barozzi G S, Bussi C and Corticelli M A 2004 *Numerical Heat Transfer B* **46** 56–77
[29] Bouafia M and Daube O 2007 *Int. J. Heat Mass Transf.* **50** 3599–3615
[30] Kuznetsov Y 1998 *Elements of Applied Bifurcation Theory* 2nd ed (*Applied Mathematical Sciences* vol 112) (New York: Springer-Verlag)

Characterization of a mini-channel heat exchanger for a heat pump system

A Arteconi[1,3], G Giuliani[2], M Tartuferi[2] and F Polonara[2]
[1]Università degli Studi eCampus, via Isimbardi 10, Novedrate (CO), 22060, Italy
[2]Università Politecnica delle Marche, Dipartimento di Ingegneria Industriale e Scienze Matematiche, via Brecce Bianche 1, Ancona, 60131, Italy

E-mail: a.arteconi@univpm.it

Abstract. In this paper a mini-channel aluminum heat exchanger used in a reversible heat pump is presented. Mini-channel finned heat exchangers are getting more and more interest for refrigeration systems, especially when compactness and low refrigerant charge are desired. Purpose of this paper was to characterize the mini-channel heat exchanger used as evaporator in terms of heat transfer performance and to study the refrigerant distribution in the manifold. The heat exchanger characterization was performed experimentally by means of a test rig built up for this purpose. It is composed of an air-to-air heat pump, air channels for the external and internal air circulation arranged in a closed loop, measurement sensors and an acquisition system. The overall heat transfer capacity was assessed. Moreover, in order to characterize the flow field of the refrigerant in the manifold of the heat exchanger, a numerical investigation of the fluid flow by means of CFD was performed. It was meant to evaluate the goodness of the present design and to identify possible solutions for the future improvement of the manifold design.

1. Introduction

Mini-channel finned heat exchangers are getting more and more interest for refrigeration systems, especially when compactness and low refrigerant charge are desired [1]. Several studies investigated performance and operative characteristics of such systems. Many authors presented heat transfer and pressure drops correlations [2 - 7]. In particular maldistribution of the refrigerant is a challenge in this type of heat exchangers, especially for evaporators where the entering fluid is usually in two-phase condition. The design of the distribution manifold plays an important role in how the gas and liquid phases distribute, together with pressure drops in the manifold and air distribution on the external side [8]. Among the main findings in the literature, Kulkarni et al. [9] found that mass flow maldistribution cannot be controlled through changing either the port/header diameter or the refrigerant state, but it can only be controlled by minimizing pressure gradients along header. Regarding the inlet configuration, Kim et al. [10] stated that the best gas distribution is obtained for normal inlet configuration, followed by vertical and parallel inlet configuration, and the effect of mass flow or quality on gas distribution is not significant. Moreover the protrusion of multiports into the header makes incoming liquid bypass the frontal channels yielding better flow distribution [11]. Qi et al. [12] studied the effect of fin shape, revealing that the louver fin which is flat at fin top is more suitable for mini-channel evaporator than corrugated louver fin in reduction of air side pressure drop and, for the

[3] To whom any correspondence should be addressed.

Journal of Physics: Conference Series **501** (2014) 012023 doi:10.1088/1742-6596/501/1/012023

same purpose, a simpler refrigerant side flow path arrangement is preferable. However the main effect on heat exchange is due to the refrigerant maldistribution more than air maldistribution (30% performance reduction against 9% respectively) [13], even if both aspects have to be taken into account. Different approaches were used in the abovementioned works, based on experimental tests and/or numerical simulations, among them Selleri et al. [14] developed a multi-objective optimization to find the best design configuration for a mini-channel heat exchanger.

Purpose of this paper was to characterize the mini-channel heat exchanger used as evaporator in terms of heat transfer performance and to study the refrigerant distribution in the manifold. The heat exchanger characterization was performed experimentally by means of a test rig built up for this purpose. Moreover, in order to characterize the flow field of the refrigerant in the manifold of the heat exchanger, a numerical investigation of the fluid flow by means of CFD was performed. It was meant to evaluate the goodness of the present design with the main objective of identifying useful aspects for the future improvement of the manifold design.

2. Experimental investigation

The heat exchanger characterization was performed experimentally by means of a test rig built up for this purpose. It is composed of an air-to-air heat pump, air channels for the external and internal air circulation arranged in a closed loop, measurement sensors and an acquisition system. In the following sections a detailed description of the test rig and of the obtained results is provided.

2.1. The test rig set up

The test rig is composed of an-to-air reversible heat pump, using R410A as refrigerant fluid, with air channels for the external and internal air circulation arranged in a closed loop. In figure 1 a schematic (a) and a 3D layout (b) of the test rig are shown. The channels are made of expanded polyurethane covered with aluminum sheets, that provides thermal insulation and avoids air leakages. The air is blown inside the channels by means of fans. The fan on the indoor side is provided with an inverter, thus it is possible to vary the air flow rate for the experimental tests purposes. The aim of the closed air channel circuits is to simulate the indoor and outdoor environment. Each channel collects the air at the outlet of the heat pump and, depending on the considered side (outdoor or indoor) and operative mode (heating or cooling), heats or cools it up to the necessary inlet conditions and then sends it back into the heat pump. In this way it is possible to test the system at different outdoor and indoor conditions. The air heating inside the channels is made by means of a 12kW electric resistance, while the air cooling by means of a water cooling circuit. Both the systems are regulated with a PID controller that keeps the air input temperature at the desired set-point values by varying the power in the former and the water flow rate in the latter system.

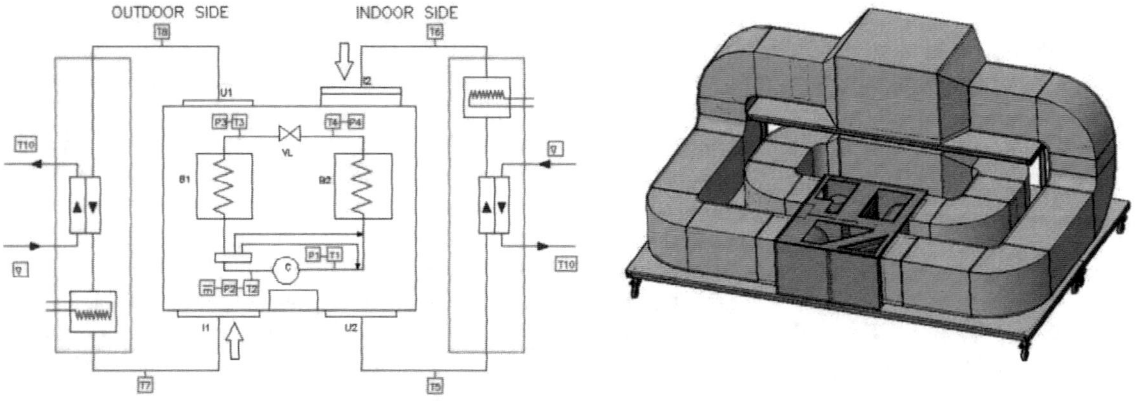

(a) (b)

Figure 1. A schematic (a) and a 3D layout (b) of the test rig.

The mini-channel heat exchanger is set inside the heat pump at the indoor side, working as evaporator or condenser on the basis of the operative mode. The test rig is equipped with different sensors (see figure 1a) in order to collect all the information about the working parameters and to assess the performance of the analyzed heat exchanger: pressure transducers on the heat pump refrigerant circuit, copper-constantan T thermocouples on the air, refrigerant and water circuits and flow meters for the refrigerant (Coriolis flow meter) and the cooling water circuits (turbine flow meter). In table 1 the sensors accuracy is reported. The data acquisition system is used to record the measurements and evaluate and control the experimental system.

Table 1. Sensors accuracy

Sensor	Accuracy
Thermocouples	± 0.5°C
Pressure transducers	1%
Turbine flow meter	1%
Coriolis flow meter	0.75%

2.2. *The mini-channel heat exchanger*

The heat exchanger is composed of 60 multiport extruded aluminum tubes (420 mm in height, 25 mm in width and 2 mm in thickness). Each tube consists of 15 rectangular shaped mini-channel (hydraulic diameter of a channel: 1.346 mm). Between two tubes there are corrugated fins with a pace of 1.4 mm and a thickness of 0.1 mm. The frontal area, perpendicular to the air flow, is 0.256 m^2, while the total external exchange area is 8.612 m^2 and the internal exchange area is 2.069 m^2. The heat exchanger has only one passage and at the extreme sides there are two cylindrical manifolds (diameter 33 mm) where the tubes are inserted and brazed. The heat exchanger is vertically mounted and the manifolds are horizontally placed, because for evaporator usage this configuration is preferred to facilitate the air-side condensate drainage [8]. The bottom manifold has 7 equally distributed inlets, while the top manifold has 2 outlets. The higher number of inlets is requested for a better distribution of the two phase refrigerant entering the heat exchanger working as evaporator. In figure 2 a schematic of the heat exchanger and a detail of an inlet nozzle of the bottom manifold are represented. In figure 3 a picture of the heat exchanger installed in the test rig is provided.

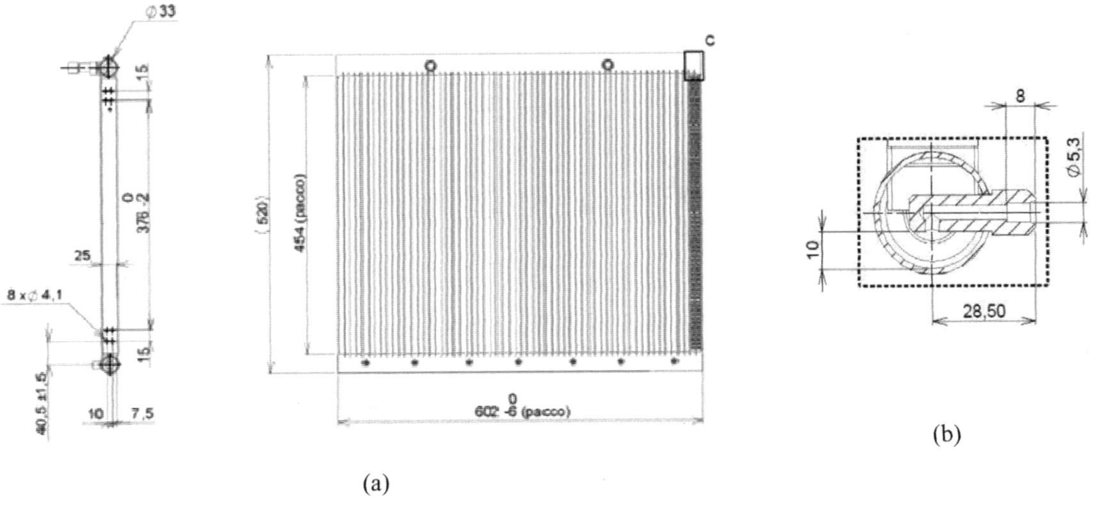

Figure 2. A schematic of the mini-channel heat exchanger (a) and a detail of the inlet nozzle (b).

Figure 3. A picture of the mini-channel heat exchanger.

2.3. The experimental results

The experimental tests were aimed at characterizing the mini-channel heat exchanger into the heat pump system. The purpose was to draw curves that plot the exchanged heating rate with varying the external conditions (inlet air temperature and inlet refrigerant temperature). In this paper only the results from cooling operative mode are reported, when the mini-channel heat exchanger, set on the indoor side, plays as evaporator, that is also its most critical role.

The heat pump is an air-to-air reversible heat pump using R410A as refrigerant. In figure 4 it is possible to see the schematic of the system with the position of the installed sensors and the thermodynamic cycle on the Ph diagram. The compressor is an hermetic scroll compressor and an inverter was used during the tests to vary its speed and thus the refrigerant flow rate. The condenser is a shell and tube finned heat exchanger and a thermostatic expansion valve is employed to keep the superheating after the evaporator at the desired value (point 2 on the Ph diagram in figure 4; a further superheating up to point A is due to the four way valve before the compressor inlet).

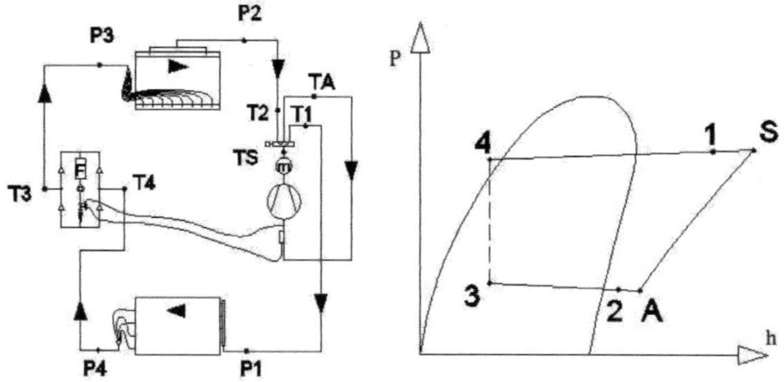

(a) (b)

Figure 4. Heat pump schematic (a) and representation of the thermo-dynamic cycle on the Ph diagram for the cooling mode.

During the tests the superheating was kept constant to the value of 7.7°C as well as the air flow rate at the indoor side (2100 m^3/h at inverter frequency 25Hz), while the air inlet temperature at the indoor side was varied between 21°C to 27°C. Recording the refrigerant flow rate ($\dot{m}_{ref}$) and temperatures (T) and pressures (p), it was possible to calculate the heating rate exchanged in the evaporator ($\dot{P}_{ev}$) as function of the enthalpy difference (values taken from REFPROP 9.0 database [15]):

$$\dot{P}_{ev} = \dot{m}_{ref} \cdot (h_{ref,out} - h_{ref,in}) \quad (1)$$

In Table 2 some of the recorded measures are reported. Considering the sensors accuracy (Table 1) and the uncertainty of the equation of state for the refrigerant used in REFPROP [16], a relative error of 1.2% for $\dot{P}_{ev}$ was estimated using the law of error propagation.

Table 2. Experimental measurements

Test ID	T_{ev} °C	p_{ev} bar	$\dot{m}_{ref}$ kg/h	superheating °C	$\dot{w}_{air}$ m3/h	$T_{air,in}$ °C	$T_{air,out}$ °C
1	5.2	9.4	169.5	7.7	2131.5	20.9	10.6
2	9.5	10.7	154.9	7.8	2131.2	23.8	13.9
3	13.6	12.1	146.4	7.8	2148.9	27.1	17.5
4	7.5	10.1	173.2	7.6	2031.2	24.0	12.9
5	7.6	10.1	173.0	7.8	2024.6	23.9	13.0
6	3.5	8.9	180.5	7.8	1981.8	21.1	9.6
7	12.8	11.8	159.7	7.7	2191.0	26.9	17.2
8	7.1	10.0	153.2	7.9	2075.0	22.1	12.0
9	7.3	10.0	164.5	7.6	2360.0	22.0	12.9
10	11.3	11.3	149.1	7.4	2390.4	24.0	15.7

Plotting the heating rate exchanged in the evaporator versus the air inlet temperature and the difference between the air inlet temperature and the refrigerant evaporating temperature, it was assessed an equation for the interpolating surface:

$$\dot{P}_{ev} = a + b \cdot T_{in} + c \cdot \Delta \mathrm{T_{ev}} + d \cdot \Delta T_{ev}{}^{2} \tag{2}$$

where T_{in} is the air inlet temperature and ΔT_{ev} is the temperature difference between the air inlet temperature and the refrigerant evaporating temperature; the coefficients are: a=-1.696, b=0.009; c=0.898; d=-0.021. In figure 5 two projections of the equation (2) are shown. They represent the trends of the evaporator heating rate with varying the air inlet temperature and the temperature difference between the air inlet temperature and the refrigerant evaporating temperature. It is possible to notice that the effect of the air inlet temperature variation is limited, while, as expected, the temperature difference between the two carrier fluids is more influent.

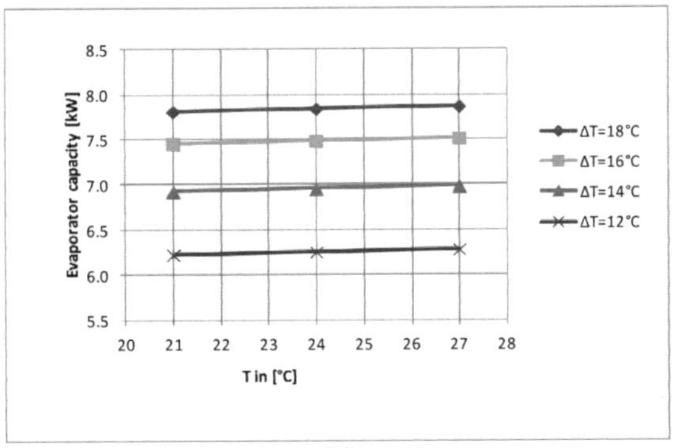

(a)

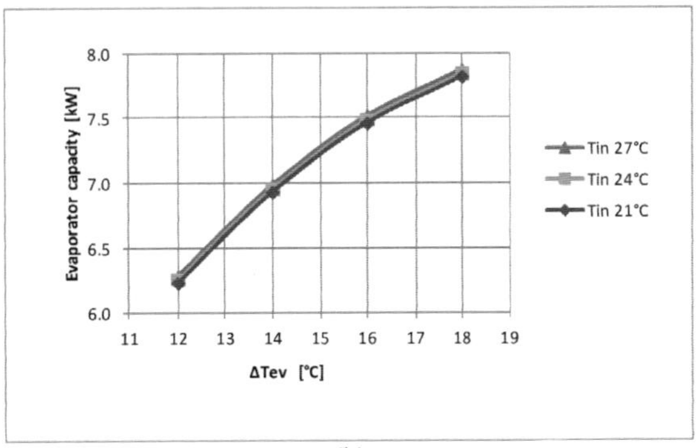

(b)

Figure 5. Evaporator capacity with varying the air inlet temperature, T_{in}, (a) and the temperature difference, ΔT_{ev}, between air and refrigerant evaporating temperature (b).

Moreover, knowing the evaporator capacity, an average overall heat transfer coefficient (UA) for the mini-channel heat exchanger was estimated by means of equations (3) and (4):

$$\dot{P}_{ev} = UA \cdot \Delta T_{ML} \cdot f \tag{3}$$

$$\Delta T_{ML} = \frac{(\Delta T_{\mathrm{ev}} - \Delta T_2)}{\ln\left(\frac{\Delta Tev}{\Delta T2}\right)} \tag{4}$$

Journal of Physics: Conference Series **501** (2014) 012023 doi:10.1088/1742-6596/501/1/012023

where ΔT_{ML} is the logarithmic mean temperature calculated as function of the temperature differences between the two carrier fluids, ΔT_2 and ΔT_{ev}, being ΔT_2 the difference between the air outlet temperature and the refrigerant inlet temperature and ΔT_{ev} the difference between the air inlet temperature and the refrigerant evaporating temperature (figure 6). Such a definition was chosen to approximate the real heat transfer process as much better as possible. The correction factor, f (<1), instead, depends on the fluids inlet and outlet temperatures and on the type of heat exchanger and it takes into account that the heat exchanger is not a counter-current but a cross flow heat exchanger [17]. The obtained UA is equal to 1.158 kW/K with a global uncertainty of 2%.

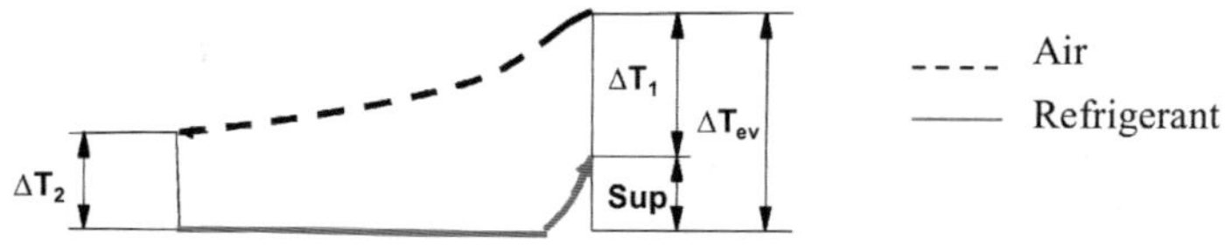

Figure 6. Air and refrigerant temperature trend in the evaporator.

Other tests were conducted with varying the air flow rate (inverter frequency: 20÷27.5 Hz) and a relationship between the evaporator heating rate variation and the air flow rate ($\dot{w}$) variation was found:

$$\frac{\Delta \dot{P}_{ev}}{\dot{P}_{ev,0}} = 0.1 \frac{\Delta \dot{w}_{air}}{\dot{w}_{air,0}} \tag{5}$$

the subscript "0" refers to the standard condition at 25Hz. The relation shows that the exchanged power variation is 10% of the air flow rate variation.

3. CFD analysis of the flow field

Main purpose of the numerical simulation of the fluid flow was to analyze the distribution of the two phase refrigerant in the inlet manifold of the evaporator. The CFD analysis was oriented to obtain a first guess solution of the refrigerant flow field, useful to highlight which elements of the actual geometry of the manifold have a greater responsibility in the non optimal distribution of the refrigerant among the multiports.

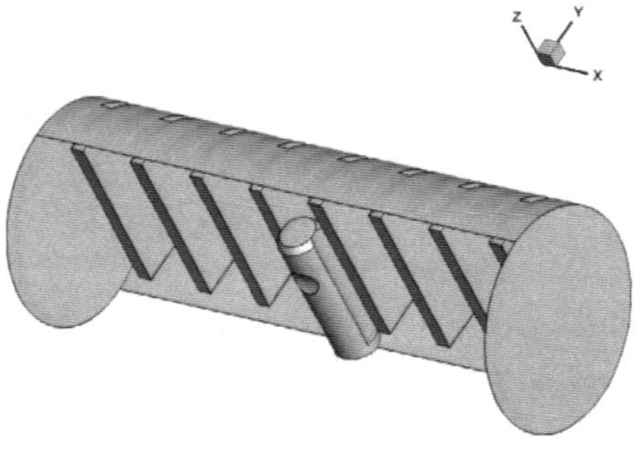

Figure 7. Geometric representation of a portion of the manifold considered in the simulation.

Considering the repetitive pattern around the inlets of the manifold, the computational domain comprises a periodic portion of the manifold, containing only one inlet nozzle and an equal number of multiports for each side (see figure 7). The computational domain was discretized using an unstructured grid of 650.000 nodes: tetrahedral elements were adopted in the core of the fluid volume and a prismatic layer refinement was realized in the near wall region in order to better describe the behavior of the refrigerant entering the multiports channels (Figure 8). The unsteady Reynolds Averaged Navier – Stokes (RANS) equations were solved for the computation of the turbulent flow, using the SST k-ω turbulence model. Time integration of the fluid flow equations was performed with a time step of 10^{-3} s, in order to produce a second order accurate solution of the flow field. Calculations were assessed by means of the CFD commercial code STAR – CCM+ [18].

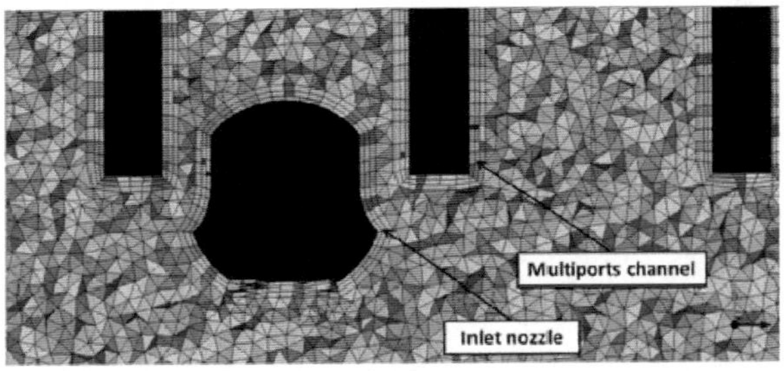

Figure 8. Mesh details in the near wall region

A set of operative conditions, measured during the experimental investigations, were adopted for the numerical model of the refrigerant flow field. The refrigerant, R410A, enters the manifold with a temperature of 276.95 K, a pressure of 0.9 MPa, a quality of 0.359 and a mass flow rate of 165 kg/h, corresponding to an inlet velocity of 6 m/s. Given the pretty high inlet velocity it is possible to consider that the two phase refrigerant flow behaves like a mist [19]. Due to the lack of certainty about the two-phase regime of the refrigerant in this type of application in scientific and technical literature, it was used a simplified approach for modeling the two-phase state of the fluid: this was intended to obtain a first guess reconstruction of the main fluid dynamic structures of the refrigerant flow field in the inlet manifold. Indeed, in the present work, the CFD analysis is aimed at identifying major elements and guidelines to improve the manifold design and achieve a better distribution of the refrigerant among multiports. The adopted model does not reconstruct the interface between liquid and gas phases, assuming equilibrium phase change. The R410A is classified as a quasi-azeotropic mixture, with negligible temperature glide during phase change: thus, it can be modeled as a pure substance. The equation of state of Redlich-Kwong [20] was considered, while the refrigerant thermophysical properties were taken from REFPROP 9.0 database [15].

The numerical results pointed out the symmetry of the flow respect to the inlet nozzle position and, mainly due to the high refrigerant inlet velocity, showed the presence of high vorticity structures among the multiports (figure 9). In figure 9 dark gray areas are associated with fluid moving downward along the vertical direction (negative velocity), thus not contributing to the development of mass flow rate through the multiport inlet sections. This behavior is also evident in terms of the refrigerant mean velocity at the multiports inlet, as shown in figure 10. Indeed a different distribution of the refrigerant between the multiports closer and farther to the inlet nozzle was highlighted: a bigger refrigerant flow passes through the multiports farther from the inlet nozzle (figure 10). Instead, considering a single multiport the flow is higher in the external channels rather than in the inner ones (figure 11).

In order to improve the refrigerant distribution, a new configuration was considered: the inlet nozzle length was reduced while its inlet section was enlarged. This simple modification reduced the inlet velocity of the refrigerant with the advantage of a more uniform distribution of the fluid among the multiports, as shown by the inlet velocity trend in figure 12.

On the basis of the obtained simulation results some conclusions can be drawn to improve the manifold design. In fact the numerical reconstruction of the flow field showed a non homogeneous distribution of the refrigerant into the minichannels and in the different multiports. Possible modifications suggested by such results to improve the distribution could be: reduce the refrigerant inlet velocity, increase the number of inlet nozzles and reduce the protrusion inside the manifold of the multiports closer to the nozzle, in order to reduce the presence of obstacles and stagnation regions in the manifold. The realization of the latter point needs to be verified from a technical, but mainly, from an economic point of view by the heat exchanger producer.

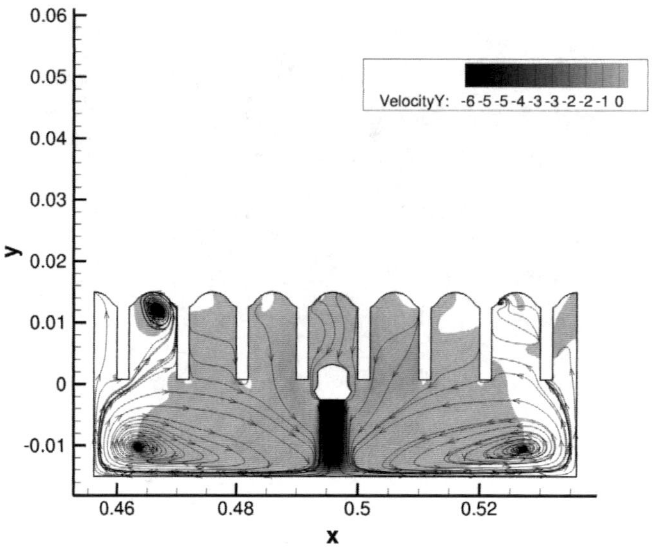

Figure 9. Velocity (m/s) cut-off contour and velocity streamline of the refrigerant in the manifold (view on the vertical middle surface of the manifold).

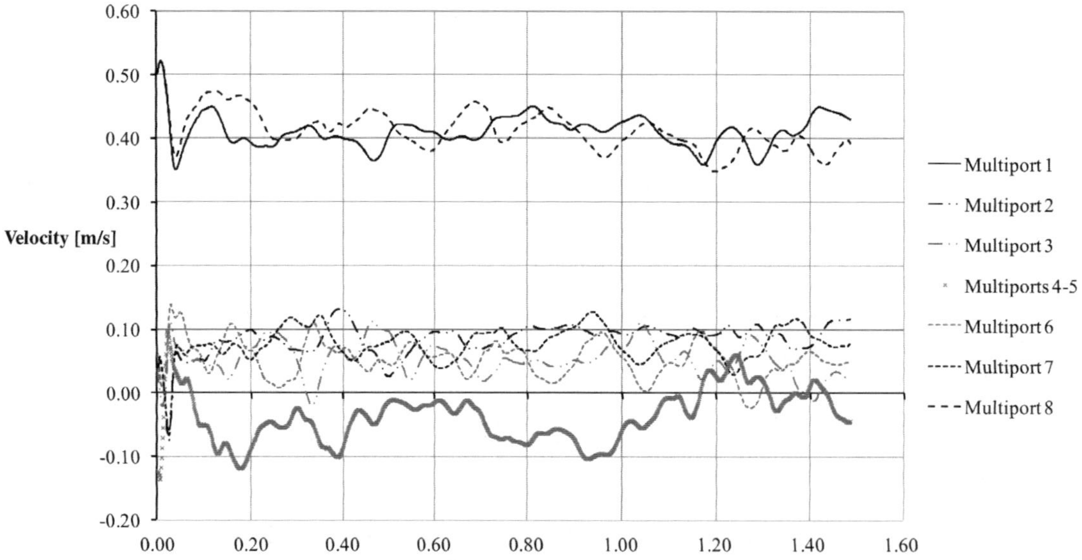

Figure 10. Refrigerant mean velocity (m/s) along the vertical axis (y) at the inlet of the multiports .

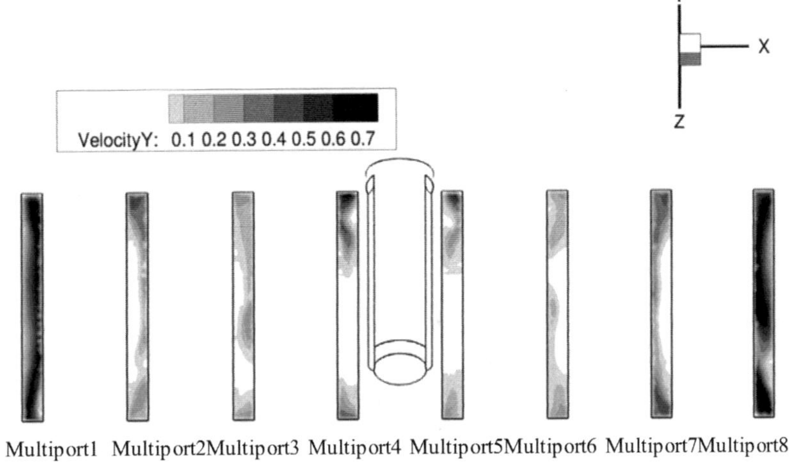

Figure 11. Refrigerant velocity (m/s) cut-off contour on the inlet sections of the multiports.

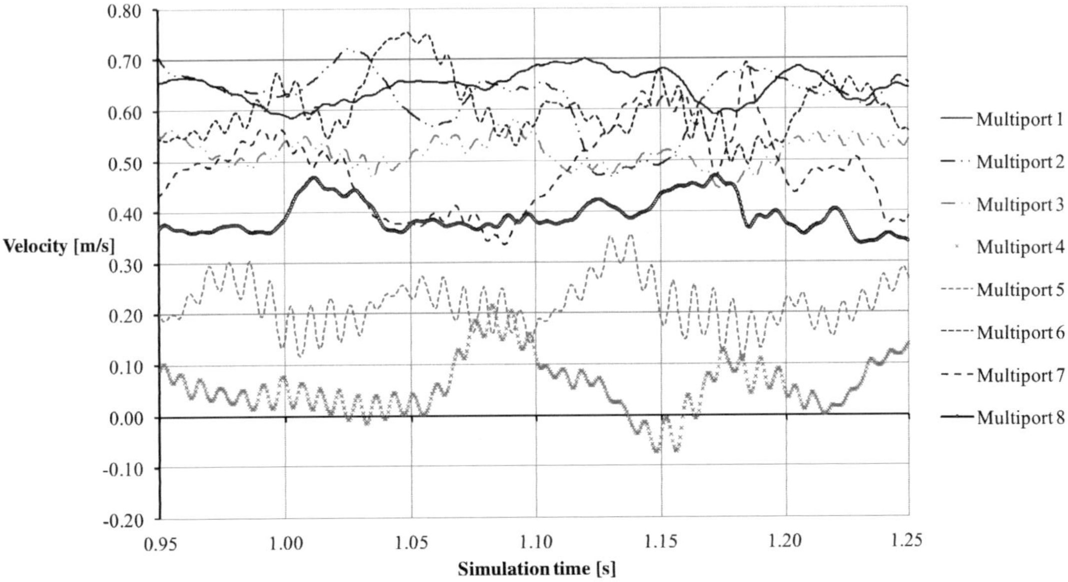

Figure 12. Refrigerant mean velocity (m/s) along the vertical axis (y) at the inlet of the multiports for the manifold with the modified inlet nozzle

4. Conclusions

In this paper a mini-channel heat exchanger used as evaporator in an air-to-air heat pump was studied. The heating rate that can be exchanged was evaluated with varying the inlet fluids conditions. In particular a relationship between such heating rate and the air and refrigerant inlet temperatures was found. Moreover the overall heat transfer capacity was assessed. Eventually, to analyze the role of the inlet manifold for the refrigerant distribution, a CFD study was performed. The obtained results showed a non homogeneous distribution and some hints for the geometry improvement could be drawn. Future developments of this study consist in testing the heat exchanger working as condenser in heating mode and, from the numerical point view, in improving the adopted approach for modeling the two-phase state of the fluid, by means of methods able to the reconstruct the interface between liquid and gas phases.

Acknowledgement
The authors wish to thank VALMEX for the possibility to study their heat exchanger and present the obtained results.

References

[1] Kandlikar S G 2007 A roadmap for implementing minichannels in refrigeration and air-conditioning systems - Current status and future directions *Heat Transf. Eng.* **28** 973-985
[2] Cavallini A, Del Col D and Rossetto L 2012 Heat transfer and pressure drop of natural refrigerants in minichannels (low charge equipment) *Int. J. Refrig.* **36** 287-300
[3] Alagesan V 2012 Flow boiling heat tranfer in Mini and Micro Channels – A state of the art review *Int. J. Chem. Tech. Res.* **4** 1247-59
[4] Kim S-M and Mudawar I 2013 Universal approach to predicting saturated flow boiling heat transfer in mini/micro-channels – Part II. Two-phase heat transfer coefficient *Int. J. Heat Mass Transf.* **64** 1239–1256
[5] Zhang W, Hibiki T and Mishima K 2004 Correlation for flow boiling heat transfer in mini-channels *Int. J. Heat Mass Transf.* **47** 5749–5763
[6] Sun L and Mishima K 2009 Evaluation analysis of prediction methods for two-phase flow pressure drop in mini-channels *Int. J. Multiph. Flow* **35** 47–54
[7] Cheng L and Mewes D 2006 Review of two-phase flow and flow boiling of mixtures in small and mini channels *Int. J. Multiph. Flow* **32** 183–207
[8] Brix W, Kærn RM and Elmegaard B 2010 Modelling distribution of evaporating CO_2 in parallel minichannels *Int. J. Refrig.* **33** 1086-94
[9] Kulkarni T, Bullard CW and Cho K 2004 Header design tradeoffs in microchannel evaporators *Appl. Therm. Eng.* **24** 759-76
[10] Kim N-H, Lee E-J and Byun H-W 2012 Two-phase refrigerant distribution in a parallel flow minichannel heat exchanger having horizontal headers *Int. J. Heat Transf.* **55** 7747–59
[11] Kim N-H, Kim D-Y and Byun H-W 2011 Effect of inlet configuration on the refrigerant distribution in a parallel flow minichannel heat exchanger *Int. J. Refrig.* **34** 1209-21
[12] Qi Z, Chen J and Radermacher R 2009 Investigating performance of new mini-channel evaporators *App. Therm. Eng.* **29** 3561–67
[13] Choi JM, Payne WV and Domanski PA 2003 Effects of nonuniform refrigerant and airflow distribution of finned-tube evaporator performance *Proc. Int. Cong. Refrig.* Washington DC, USA, ICR0040
[14] Selleri T, Najafi B, Rinaldi F and Colombo G 2013 Mathematical modeling and multi-objective optimization of a mini-channel heat exchanger via genetic algorithm *J. Therm. Sci. Eng. Appl.* **5** (3) 10p
[15] REFPROP 9.0 2010 NIST Reference Fluid Thermodynamic and Transport Properties Database
[16] Lemmon EW and Jacobsen RT 2004 Equations of State for Mixtures of R-32, R-125, R-134a, R-143a, and R-152a *J. Phys. Chem. Ref. Data* **33** 593-620
[17] Guglielmini G and Pisoni C 1990 *Elementi di trasmissione del calore* (Milan: Editore Veschi).
[18] STAR - CCM+ 2012 User Guide release 7.04. CD – Adapco
[19] Webb RL and Chung K 2004 Two-phase flow distribution in tubes of parallel flow heat exchangers *Heat Transf. Eng.* **26** 3-18
[20] Redlich O and Kwong JNS 1949 On The Thermodynamics of Solutions *Chem. Rev.* **44** 233–44

Journal of Physics: Conference Series **501** (2014) 012024 doi:10.1088/1742-6596/501/1/012024

Study of laminar separation bubble on low Reynolds number operating airfoils: RANS modelling by means of an high-accuracy solver and experimental verification

A Crivellini[1], V D'Alessandro[1], D Di Benedetto[1], S Montelpare[2] and R Ricci[1]

[1] Dipartimento di Ingegneria Industriale e Scienze Matematiche
Università Politecnica delle Marche
Via Brecce Bianche, 60100 Ancona (AN), Italy
[2] Dipartimento di Ingegneria e Geologia
Università degli Studi "G. D'Annuzio" di Chieti-Pescara
Viale Pindaro 42, 65127, Pescara (PE), Italy

E-mail: a.crivellini@univpm.it, v.dalessandro@univpm.it, d.dibenedetto@univpm.it, s.montelpare@unich.it, ricci@univpm.it

Abstract.
This work is devoted to the Computational Fluid-Dynamics (CFD) simulation of laminar separation bubble (LSB) on low Reynolds number operating airfoils. This phenomenon is of large interest in several fields, such as wind energy, and it is characterised by slow recirculating flow at an almost constant pressure.
Presently Reynolds Averaged Navier-Stokes (RANS) methods, due to their limited computational requests, are the more efficient and feasible CFD simulation tool for complex engineering applications involving LSBs. However adopting RANS methods for LSB prediction is very challenging since widely used models assume a fully turbulent regime. For this reason several transitional models for RANS equations based on further Partial Differential Equations (PDE) have been recently introduced in literature. Nevertheless in some cases they show questionable results.
In this work RANS equations and the standard Spalart-Allmaras (SA) turbulence model are used to deal with LSB problems obtaining promising results. This innovative result is related to: (i) a particular behaviour of the SA equation; (ii) a particular implementation of SA equation; (iii) the use of a high-order discontinuous Galerkin (DG) solver. The effectiveness of the proposed approach is tested on different airfoils at several angles of attack and Reynolds numbers.
Numerical results were verified with both experimental measurements performed at the open circuit subsonic wind tunnel of Università Politecnica delle Marche (UNIVPM) and literature data.

1. Introduction

In this paper an innovative approach for the study of low Reynolds numbers operating airfoils involved in the blade sections of wind turbines is presented. In particular, great efforts are addressed to Laminar Separation Bubble (LSB) prediction, which is particularly difficult to be accurately computed without using large amount of computational resources, those required by the Direct Numerical Simulation (DNS) or by the Large Eddy Simulations (LES) approaches, as well known in literature.

The LSB, illustrated in figure 1, is a local boundary layer separation phenomenon that may occur on aerodynamic bodies operating at Re $\leq 10^6$. It can occur in presence of some

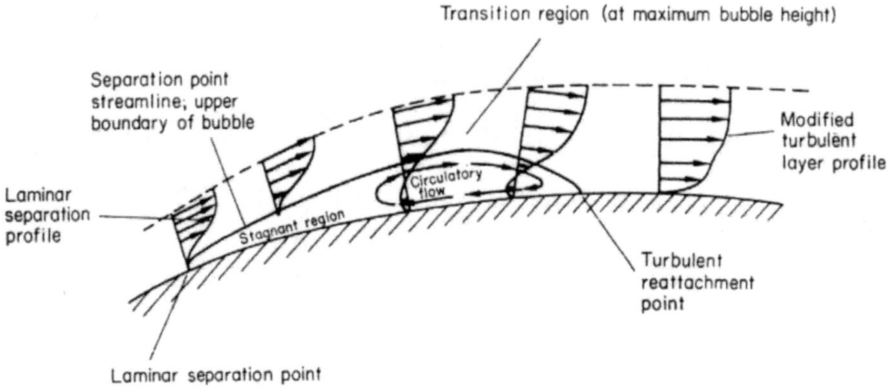

Figure 1. Laminar Separation Bubble features, [1].

conditions that are briefly described: a laminar separation of the laminar boundary layer due to an adverse pressure gradient; a turbulent transition within the separated shear layer; a turbulent reattachment. Under these conditions a region characterised by a slow recirculating flow and by an almost constant pressure is formed. The presence of a LSB may introduce two classes of problems: (i) a decrease in the efficiency of the airfoil, mainly due to the airfoil drag rise; (ii) the appearance of large pressure fluctuations in the case of bubble bursting.

This complex phenomenon is a challenging task of aerodynamics and it has already been widely studied by several authors with both experimental [2, 3, 4, 5, 6, 7, 8, 9, 10, 11] and numerical techniques [12, 13, 14, 15, 16, 17]. However, a computational efficient prediction of the LSBs still remains a difficult task. From the engineering point of view, thanks to their limited computational requests, the main simulation technique feasible to deal with complex configuration, such as wind turbines blades, appears to be the Reynolds Averaged Navier-Stokes (RANS) class of methods. A critical point in applying the RANS approach to low-Reynolds number flows is the presence of laminar regions. In fact in these cases the flow is laminar at least up to the separation point and the transition is difficult to be accurately detected. Widely used turbulence models are designed only for fully turbulent flows, thus they cannot be easily adopted in this context.

However in this work we have used an incompressible high-order discontinuous Galerkin (DG) RANS solver, described in [18, 19] and based on the standard Spalart-Allmaras (SA) model, to simulate the occurrence of LSBs on airfoils.

As recognised by Rumsey [20] the SA model exhibits an apparent transition behaviour if the boundary free-stream condition for $\tilde{\nu}$ is sufficiently small. In our numerical results, reported in [18], the artificial transition appears even if the trip term, considered by Rumsey the main responsible for this undesirable behaviour, is not implemented. Furthermore we have also verified that accurate numerical solutions, either on refined grids or with higher polynomial orders, produce larger laminar regions.

Our approach for the LSBs prediction uses the previous observations, in other words we adopt the standard SA model with the ideal free-stream condition $\tilde{\nu}_\infty = 0$, being $\tilde{\nu}_\infty$ the free-stream turbulent viscosity, without any modification or specific equations to model the turbulent transition. If a fine enough resolution is guaranteed the artificial transition is delayed. Hence the computed solution is induced to remain laminar up to the separation point. Downstream

the flow separation the production of vorticity activates the SA production term and in this way the turbulent transition is modelled. This behaviour allows, as it will be clear below, to capture the LSB phenomena almost correctly.
It is particularly important to remark that the use of the ideal free stream condition, $\tilde{\nu}_\infty = 0$, is not trivial, due to the numerical stability problems induced by the condition itself. Our computations can benefit from this condition thanks to our robust SA model implementation characterised by a non-standard r closure function definition, see §2.
The numerical results obtained with our DG solver were verified with the experimental data available in literature and with the measurements performed at the open circuit subsonic wind tunnel of Università Politecnica delle Marche (UNIVPM). More in depth, two airfoils were investigated: (i) the Eppler 387 (E387) airfoil, to validate both the numerical code and the experimental facilities with literature data; (ii) the WT2 airfoil. The later airfoil was designed at UNIVPM as a medium blade airfoil for a small wind turbine, having a nominal power of 3.5 kW, for possible installations in green docks. The airfoil maximum thickness is $0.139c$, placed at $0.318c$, while the maximum camber is equal to $0.03c$ and is located at $0.448c$, being c the chord length.

2. Governing equations

The complete set of the incompressible RANS-SA Partial Differential Equations (PDEs) system can be written as

$$\begin{aligned} \nabla \cdot \mathbf{u} &= 0, \\ \frac{\partial \mathbf{u}}{\partial t} + \nabla \cdot (\mathbf{u} \otimes \mathbf{u}) + \nabla p - \nabla \cdot \left((\nu + \nu_t) \left(\nabla \mathbf{u} + \nabla \mathbf{u}^T \right) \right) &= \mathbf{0}, \\ \frac{\partial \tilde{\nu}}{\partial t} + \nabla \cdot (\mathbf{u} \tilde{\nu}) - \frac{1}{\sigma} \nabla \cdot (\xi \nabla \tilde{\nu}) &= s, \end{aligned} \tag{1}$$

where $\mathbf{u}$ is the velocity vector, $p = P/\rho$ is pressure divided by the density, ν is the kinematic viscosity and ν_t is the turbulent viscosity, which is computed according to the $\tilde{\nu}$ variable as

$$\nu_t = f_{v1} \max(0, \tilde{\nu}). \tag{2}$$

The SA equation diffusion coefficient ξ and the source term s will be defined below. Even if the SA model was developed only for positive $\tilde{\nu}$ values, its numerical solution can admit un-physical negative values which often produce numerical instabilities and the blow-up of the computation. For this reason the equation for $\tilde{\nu}$ is manipulated with the same approach used for the turbulent kinetic energy in the $k - \omega$ turbulence model implementation proposed by Bassi et al. [21]. Whenever negative $\tilde{\nu}$ values are reached the source terms are set to zero, while the diffusion coefficient is fixed to the kinematic viscosity value, this means

$$\xi = \begin{cases} \nu(1+\chi) & \chi \geq 0 \\ \nu \quad \chi < 0 \end{cases}, \tag{3}$$

$$s = \begin{cases} \left(\frac{\nu\chi}{d}\right)^2 \left(\frac{c_{b1}}{k^2 r} - c_{w1} f_w\right) + \frac{c_{b2}}{\sigma} \nabla\tilde{\nu} \cdot \nabla\tilde{\nu} & \chi \geq 0 \\ 0 \quad \chi < 0 \end{cases}, \tag{4}$$

where $\chi = \tilde{\nu}/\nu$ is the non dimensional turbulent variable.

Note that with positive turbulent viscosity the standard SA equation is recovered. To completely define the PDEs system, equation (1), the following closure functions are introduced.

$$f_{v1} = \frac{\chi^3}{(\chi^3 + c_{v1}^3)}, \qquad f_{v2} = 1 - \frac{\chi}{(1 + \chi f_{v1})},$$

$$g = r + c_{w2}\left(r^6 - r\right), \qquad f_w = g\left[\frac{1 + c_{w3}^6}{g^6 + c_{w3}^6}\right]^{\frac{1}{6}}, \tag{5}$$

$$r^* = \frac{\tilde{\nu}}{\tilde{S}k^2d^2}, \qquad r = \begin{cases} r_{max} & r^* < 0 \\ \min\left(r^*, r_{max}\right) & r^* \geq 0 \end{cases}, \tag{6}$$

where d is the minimum distance from wall, r_{max} is constant positive value and $\tilde{S}$ is a function of both the vorticity magnitude, S, and of the $\tilde{\nu}$ variable

$$\tilde{S} = S + \frac{\tilde{\nu}}{k^2d^2} f_{v2}, \quad \mathbf{\Omega} = \frac{\nabla \mathbf{u} - \nabla \mathbf{u}^T}{2}, \quad S = \sqrt{2\mathbf{\Omega} : \mathbf{\Omega}}. \tag{7}$$

Finally, in order to completely define the SA model, the standard closure constants are introduced

$$\begin{aligned} c_{b1} &= 0.1355, & c_{b2} &= 0.622, & c_{v1} &= 7.1, \\ \sigma &= 2/3, & c_{w1} &= \frac{c_{b1}}{k^2} + \frac{(1 + c_{b2})}{\sigma}, & & \\ c_{w2} &= 0.3, & c_{w3} &= 2, & k &= 0.41. \end{aligned} \tag{8}$$

The resulting turbulence model is the standard SA model without the trip term, that in the standard nomenclature is referred to as $f_{t1}\Delta u^2$. Moreover also the ad-hoc numerical fix term is missing, $f_{t2}c_{b1}\left(\tilde{S}\tilde{\nu} - (\tilde{\nu}/kd)^2\right)$. This term was originally introduced in order to avoid the turbulence transition upstream from where the numerical trip was set. However, it has the drawback of promoting an undesired apparent transition behavior if it is run in a ft simulation, [20].

This model is usually classified as fully turbulent but, as it will be clear below, it can effectively perform even in laminar regions. Note that in our high-order SA model implementation the r closure function, equation (5), is modified by introducing a kind of realizable condition proposed in [18]. The new definition consists in limiting r to a positive value, here $r_{max} = 10$, and handles negative and large values in the same way because the original r, for a positive χ, is negative only if it goes through a singular point with $r \to \pm\infty$.

3. Numerical Discretization

RANS equations and Spalart–Allmaras turbulence model were solved, as introduced in §1, by using the DG method, which is a finite-element method of Galerkin characterized by discontinuous solutions across the cell boundaries. During the last years, DG methods have received increasing interest from the Computational Fluid-Dynamics (CFD) community because they combine flexibility, stability, robustness and accuracy.

As in continuous Galerkin methods, DG methods can increase their order of accuracy by rising the degree k of the piecewise polynomial approximation $\mathbb{P}_k$ describing the solution. Moreover, the resulting high-order accuracy is retained even using unstructured and anisotropic hybrid meshes usually adopted in real–life applications. The discontinuous approximation between neighbouring elements is treated, as in finite volume methods, using numerical fluxes.

In these methods the removal of the continuity constraint enhances the compactness of the scheme, as elements share information only with their closest neighbours through surface flux terms, establishing the implicit time integration to block matrices of lower density. This compactness of the computational stencil also implies a straightforward parallelization on clusters of CPUs.

3.1. Implementation issues

The DG code here employed is able to solve the incompressible RANS equation up to the seventh order in space and up to the fourth order in time. A peculiarity of our implementation is related to the shape finite-element functions, which are defined in the physical space rather than in the reference one. Another distinctive trait of our solver is connected to the convective flux evaluation, which is defined using the artificial compressibility flux introduced in [22]. In this scheme the incompressibility constraint is relaxed by adding an artificial compressibility, but only at the interface flux level, in order to recover the hyperbolicity of the equations. This algorithm allows, as in the compressible case, the definition of a Riemann problem for the inviscid numerical flux evaluation. One advantage of this method, differently by the standard artificial compressibilty algorithm, is that the resulting scheme is time consistent, see [23] for two-dimensional and [24] for three–dimensional unsteady applications. Lastly the viscous fluxes are computed using the well established BR2 scheme, [25].

The space discretized non-linear RANS equations are here solved with a pseudo time-stepping strategy based on a fully–coupled backward Euler implicit scheme. The linear system derived from the DG space discretization was solved using the PETSc library [26] adopting its restarted GMRES preconditioned algorithm in both the matrix–based and matrix-free versions [27].
All the numerical solutions here reported were obtained in parallel on a small Linux Cluster, with 8 AMD Opteron based nodes for a total of 64 CPU cores operating at 2.3 GHz.

4. Experimental Apparatus

Aerodynamic tests were carried out in the open circuit subsonic wind tunnel available at the DIISM Department of UNIVPM and having the following features:

- test section dimensions [cm]: 62(W) 38(H) 150(L);
- inlet test section area contraction ratio: 4.65;
- max velocity: 38 m/s;
- mean turbulence in the inlet section: 0.3%;
- fan power 5.5 kW, inverter controlled.

Flow velocity may be varied from 0 to 35 m/s by controlling the fan motor RPMs with an AC inverter. Several measurements, periodically performed by means of a calibrated hot wire anemometer, showed a mean turbulence level lesser than 0.3% for a section area greater than 90%.
Both wing sections (the one equipped with the E387 airfoil and the one with the WT2 airfoil) were realised by joining the two moulds CNC built: in this way it was possible to reduce the uncertainties related to the manufacturing process, especially for the airfoil nose. Before assembling the two moulds, a complete set of pressure taps were realised in the middle section of the wing section, both on the airfoil extrados and intrados. Each tap was equipped with a plenum chamber, flush mounted below the wing surface, so to reduce the dynamic pressure component and obtain a more reliable static pressure reading. Velocity and turbulence measurements were carried out for each experimental test by a Dantec CTA anemometric system equipped with a pneumatic calibration unit; this is used before every measurement session in order to reduce the velocity overall error, estimated as 0.08 m/s. Pressure measurements were carried out with a Druck LPM 9481 differential pressure transducer having an operating range between 10 Pa and 1000 Pa and a typical error less than 0.1% FS BSL. In order to obtain the pressure coefficient

distribution on the airfoil, this device was sequentially connected to all the pressure taps drilled in the middle section of the wing.

5. Results

5.1. Eppler 387 airfoil

The Eppler 387 airfoil was extensively investigated at $\mathrm{Re} = 10^5$ and $\mathrm{Re} = 3 \cdot 10^5$ at several angles of attack in order to verify the accuracy of both the high-order CFD solver and the experimental facilities available at UNIVPM. Reference values for this test case are: pressure and force coefficient measurements, performed at NASA Langley Research Center, [28], and force coefficients measurements obtained at University of Illinos, [29].

The numerical solutions were computed, up to $\mathbb{P}_5$ polynomial approximation, on two grids. The first one has a C-type topology and it is named G1 in the following. It consists of 2880 computational elements with a piecewise cubic representation of the edge elements. More in depth this grid was obtained by agglomerating 8 elements of multi-blocks structured finer grid. The second grid, named G2 and characterised by 9239 elements and by a circular domain shape, is unstructured, triangular elements, with a boundary layer refinement consisting of quadrilaterals. It was generated using a code developed by Ghidoni et al. [30]. In this case the edges of the elements have a piecewise quadratic representation. Both G1 and G2 grids are able to guarantee $O(y_c^+) = 1$, where y_c^+ is the viscous sub–layer scaled non-dimensional first centroid distance of the elements next to the wall. Both the grids here used are represented in figure 2.

At $\mathrm{Re} = 10^5$ our numerical simulations, performed using both the G1 and the G2 grids, and the experimental measurements are in good agreement with the reference data, figure 3(a). It is worth mentioning that the numerical computations do not exhibit completely satisfactory results in the stall region due to RANS modelling weakness for large separated flows.

At $\alpha = 0°$, figure 3(b), the laminar separation point, predicted by numerical computations, is located upstream with respect to the experimental data. At $\alpha = 4°$, figure 3(c), the UNIVPM numerical and experimental data agree pretty well, however in the NASA reference data the LSB is shifted downstream.

In figure 3(d) the results at $\alpha = 8°$ are reported. Numerical and reference data are in very good agreement, while our experimental results do not show evidence of LSB. Nevertheless, as already showed in [31], our wind tunnel can be considered reliable only up to the angles prior to the stall.

The flow field at $\mathrm{Re} = 3 \cdot 10^5$ has been up to now computed using only the G2 grid. Figure 4 shows the comparison between the numerical results and the reference one and they clearly agree very well. These data highlight what we have observed many times during our computational experience about this specific topic: the LSB prediction capabilities of our approach increase with the Reynolds number and seem to be “optimal” at $\mathrm{Re} = 2 \cdot 10^5 \div 3 \cdot 10^5$. However, it is important to mention that at $\alpha = 8°$ there is no experimental evidence of LSB and that the related simulation was performed adopting a small free-stream $\tilde{\nu}_\infty$ value, $10^{-3}\nu$, otherwise the solver did not converge to a steady state solution. This behaviour seems to establish in some way a further link between the laminar separation and the transitional capabilities, unintentionally inserted in the PDE of the SA model. If the laminar separation does not take place in the flow field and if the apparent transition of the SA model does not move up the natural transition location, in this case experimentally detected at $0.4c$, serious convergence problems of the solver are observed. This circumstance apparently prevents to compute un-physical solutions signalling the need of a larger free-stream $\tilde{\nu}_\infty$ boundary condition to encourage the numerical transition in the attached boundary layer. However, this subject still requires a more accurate investigation.

Journal of Physics: Conference Series **501** (2014) 012024 doi:10.1088/1742-6596/501/1/012024

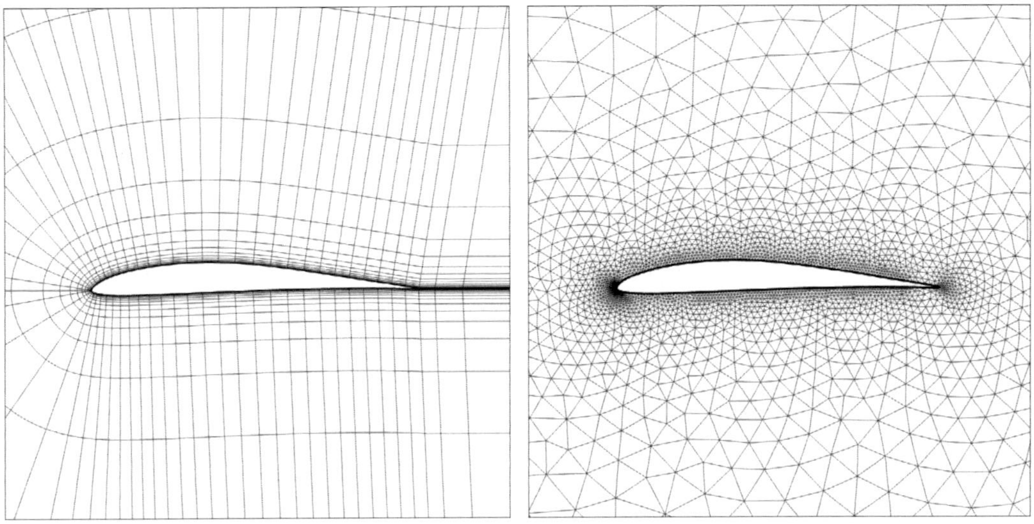

(a) G1, $ne = 2880$. (b) G2, $ne = 9239$.

Figure 2. Computational grids

5.2. WT2 airfoil

A second airfoil was here considered in order to fully prove the almost surprising, since the SA model was not designed for this purpose, effectiveness in handling LSB of the proposed approach. This profile, called WT2, was designed for wind turbine applications and tested at UNIVPM. Numerical computations were performed, at $\mathrm{Re} = 2 \cdot 10^5$, up to $\mathbb{P}_5$ polynomial approximation, see figure 5 for a representation of the flow–field, while in figure 6 the solution behaviour is depicted close and inside the LSB. A C-type structured grid with 2688 elements and a piecewise cubic representation of the edges was used for these computations. Obviously the computational elements are clustered near the airfoil surface in order to accurately solve the boundary layer and, even in this case, the first cell height ensures a y^+ value of about 1. Another time the high-order grid was obtained by agglomerating 32 elements of multi-blocks structured linear grid.
Comparisons between numerical and experimental pressure distribution at $\alpha = -2°, 1°, 3°$ and $5°$ are here reported in figure 7. Also in this case the results appear very convincing and seem to confirm the reliability of the proposed approach for the simulation of the LSB phenomena at moderate Reynolds numbers. Note that at $\alpha = -2°$ laminar separation bubbles are correctly detected at both the pressure and the suction sides.

6. Conclusions

This work assesses the reliability the results of a high-order CFD solver for LSBs simulations. The code is based on a DG method and implements, as described in [18], the SA equation for turbulence modelling. The RANS-SA system is employed in the fully-turbulent mode using the ideal free-stream condition, $\tilde{\nu}_\infty = 0$. In this way the flow behaves as a laminar at least up to the separation point. However, the separated flow exhibits high level of vorticity activating the source terms of the SA equation and producing significative values of turbulent viscosity. This behaviour allows to promote the turbulent transition, without introducing a specific model for this purpose, inducing in the here investigated cases the turbulent boundary layer reattachment. The numerical approach was tested on two airfoils: the E387 and the WT2. In the first case the CFD solver and the wind tunnel facility, available at UNIVPM, were tested with respect to literature results. Lastly the airfoil for small wind turbines WT2, developed within UNIVPM,

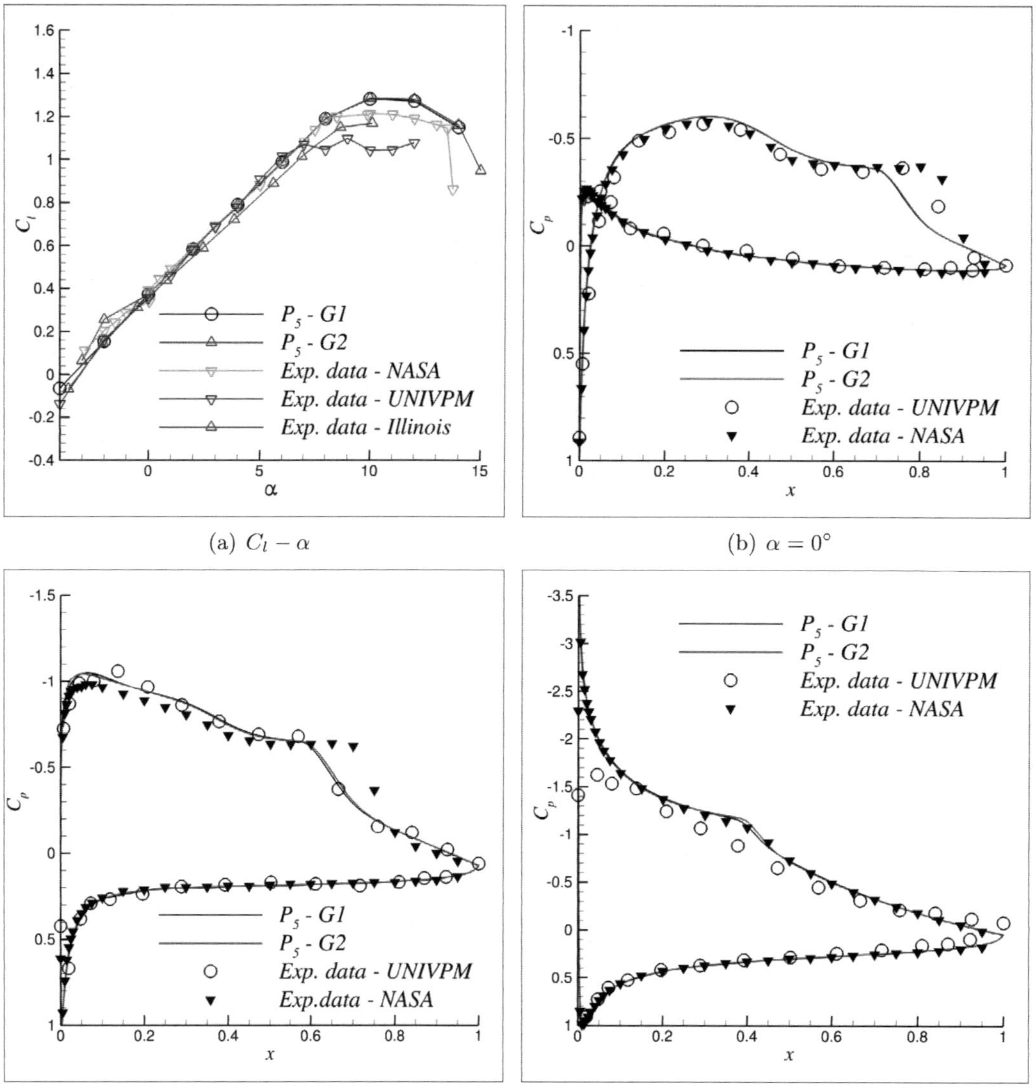

(a) $C_l - \alpha$ (b) $\alpha = 0°$

(c) $\alpha = 4°$ (d) $\alpha = 8°$

Figure 3. E387, $\mathrm{Re} = 10^5$.

was tested both numerically and experimentally.
Very promising results were obtained for both the profiles, especially for the $\mathrm{Re} = 2 \cdot 10^5$ and $\mathrm{Re} = 3 \cdot 10^5$ flow regimes.
Moreover note that the main object of this work was to develop a computational efficient simulation tool suitable for the study and the design of airfoils and to validate its effectiveness. Indeed the aim was not to obtain a "perfect" numerical solution for LSB problems, in this occurrence the use of more computationally demanding approaches like DNS or LES are almost unavoidable. Our goal was, in our opinion, completely reached since each of the computations here reported required no more than 8 CPU cores, nowadays available in almost all workstations, and only few hours of wall clock time to achieve a converged, up to machine precision, $\mathbb{P}_5$ solution which is clearly even uselessly accurate. In fact, the most of the cases could be computed, with no evident changes in the obtained results and with a clear advantage in the CPU time, only

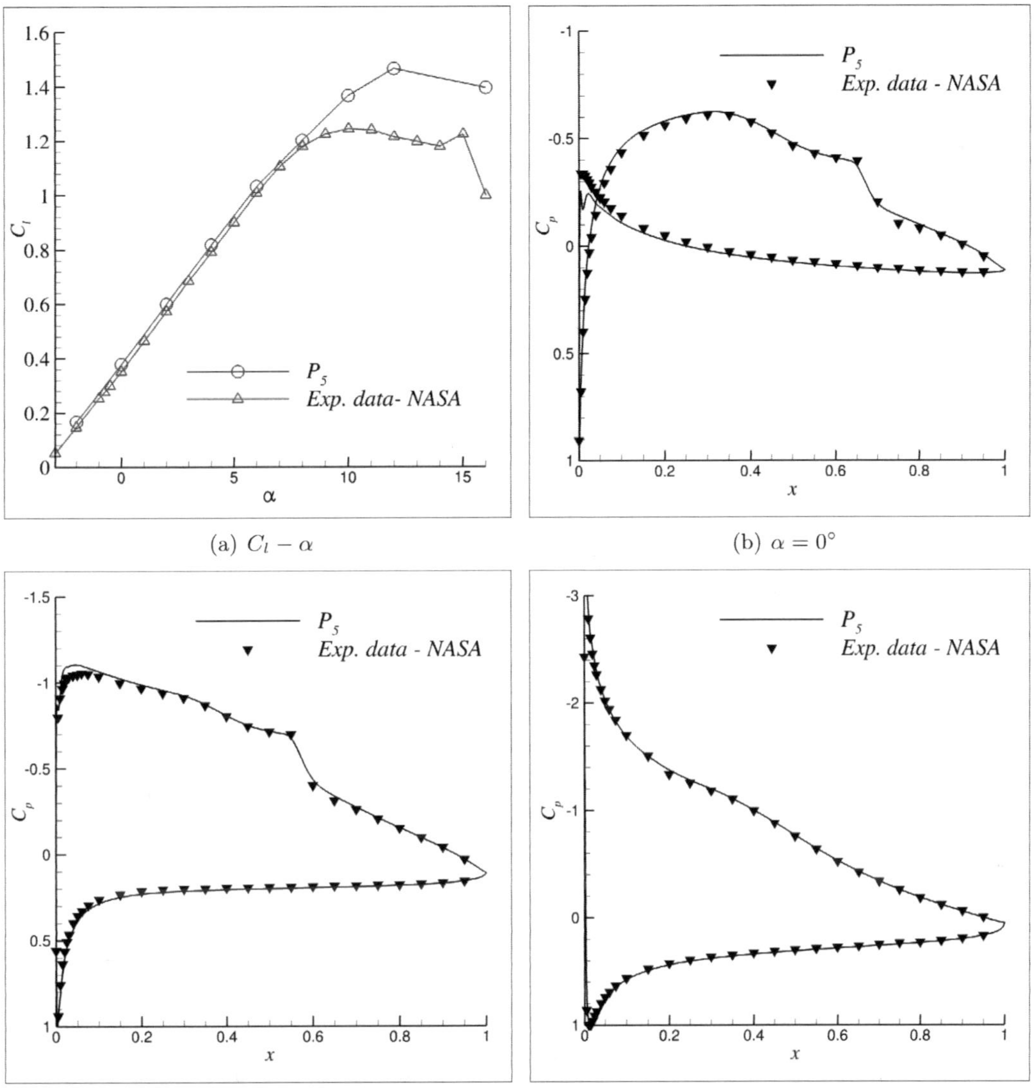

(a) $C_l - \alpha$ (b) $\alpha = 0°$

(c) $\alpha = 4°$ (d) $\alpha = 8°$

Figure 4. E387, $\mathrm{Re} = 3 \cdot 10^5$, G2.

up to the $\mathbb{P}_2$ or $\mathbb{P}_3$ orders of approximation. A paper with a more in-depth numerical and theoretical analysis of the SA behaviour related to the laminar separation inducing the turbulent transition is under preparation, while we are actually considering the application of our model to three-dimensional problems.

Acknowledgments
The Authors sincerely thank Antonio Ghidoni (University of Brescia), who provided the unstructured grid used for the computations of E387 airfoil.

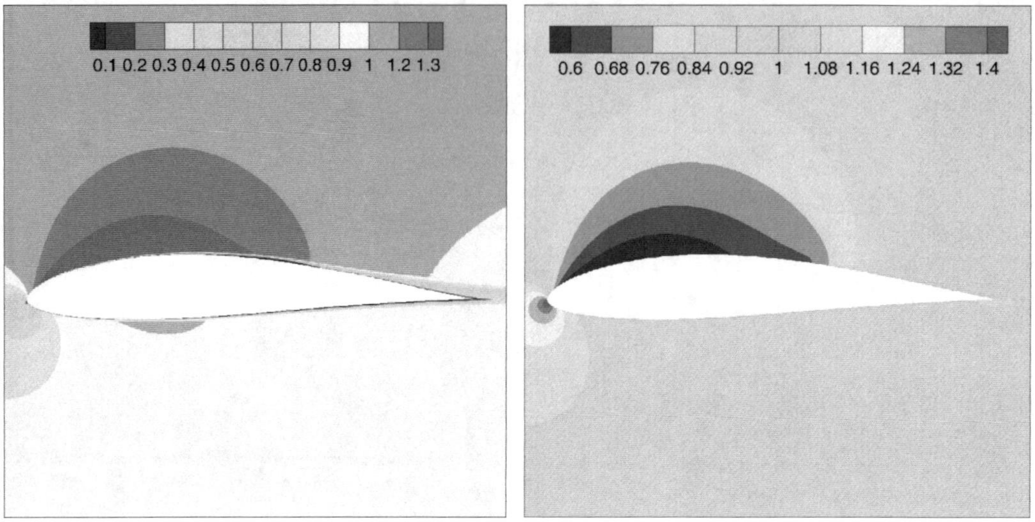

(a) Non–dimensional velocity magnitude contours (b) Non–dimensional pressure contours

Figure 5. WT2, Re $= 2 \cdot 10^5$, $\alpha = 3°$. $\mathbb{P}_5$ solution.

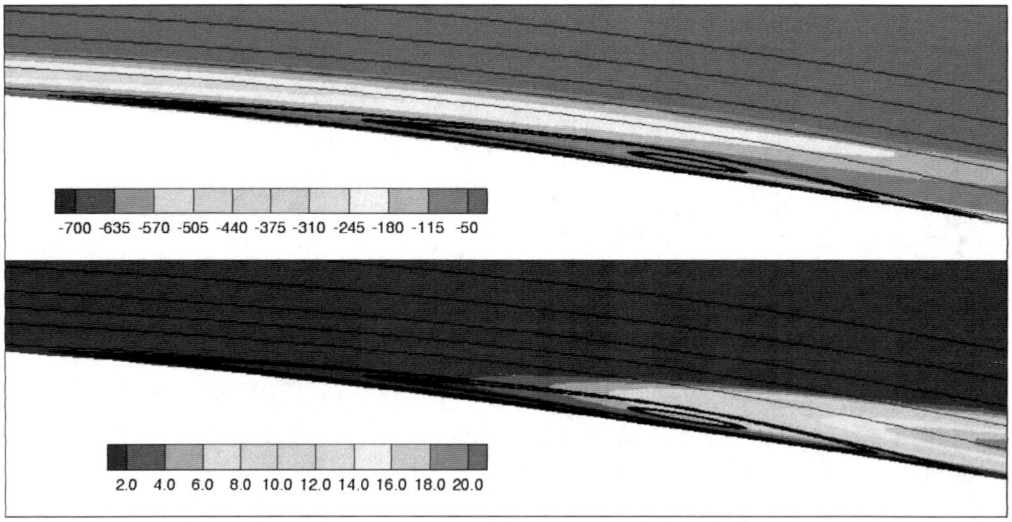

Figure 6. WT2, Re $= 2 \cdot 10^5$, $\alpha = 3°$. $\mathbb{P}_5$ solution. Top: non–dimensional vorticity contours; Bottom: $\tilde{\nu}/\nu$ contours

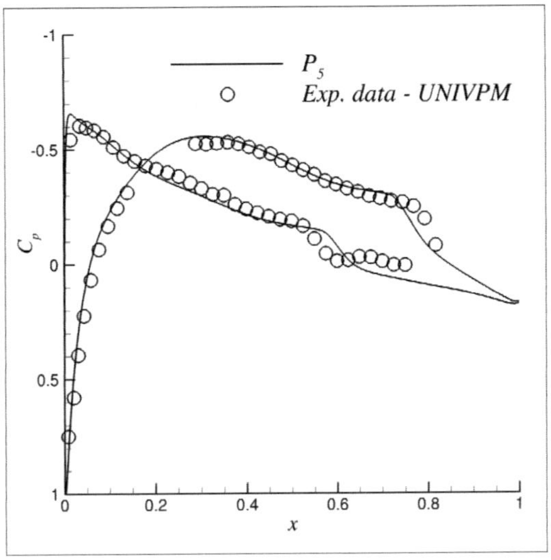

(a) $\alpha = -2°$

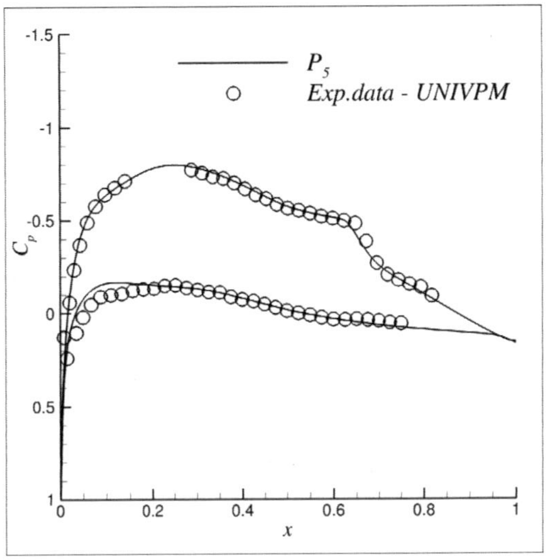

(b) $\alpha = 1°$

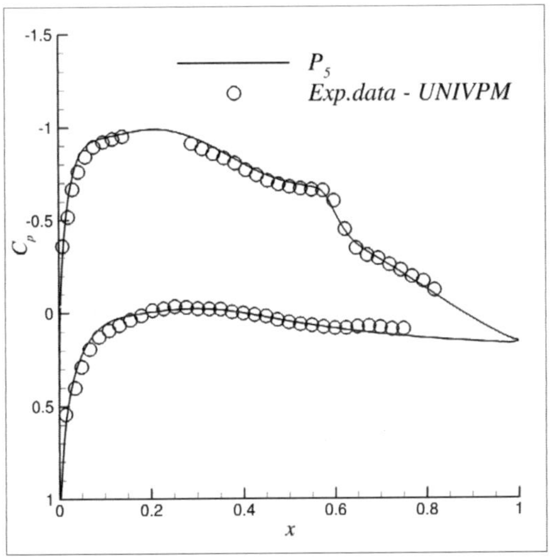

(c) $\alpha = 3°$

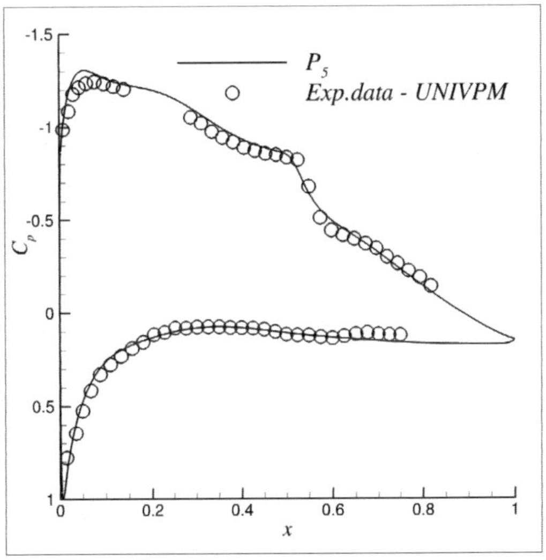

(d) $\alpha = 5°$

Figure 7. WT2, $\mathrm{Re} = 2 \cdot 10^5$, $ne = 2688$.

Journal of Physics: Conference Series **501** (2014) 012024 doi:10.1088/1742-6596/501/1/012024

References

[1] Houghton E and Carpenter P 2003 *Aerodynamics for Engineering Students, 5th edition* (Boston: Butterworth-Heinemann)
[2] Ricci R and Montelpare S 2005 *Int J Therm Sci* **44** 709 – 719
[3] Ricci R, Montelpare S and Silvi E 2007 *Exp Therm Fluid Sci* **31** 349 – 359
[4] Ricci R, Montelpare S and Renzi E 2011 *Int J of Therm Sci* **50** 2091–2103
[5] Zhang W, Hain R and Kahler C 2008 *Exp Fluids* **45** 725 – 743
[6] Tani I 1964 *Prog Aerosp Sci* **5** 70 – 103
[7] Rinioie K and Takemura N 2004 *Aeronaut J* **108** 153 – 163
[8] Hain R, Kahler C and Radespiel J 2009 *J Fluid Mech* **630** 129 – 153
[9] Burgmann S, Brücker C and Shroder W 2006 *Exp Fluids* **41** 319 – 326
[10] Haggmark C, Hildings C and Henningson D 2001 *Aerosp Sci Technol* **5** 317 – 328
[11] Genc M S, Karasu I and Acikel H H 2012 *Exp Therm Fluid Sci* **39** 252 – 264
[12] Howard R, Alam M and Sandham N 2000 *Flow Turbul Combust* **63** 175–191
[13] Kalitzin G, Gould A and Benton J 1996 *34th AIAA Aerospace Sciences Meeting Including the New Horizons Forum and Aerospace Exposition* (Reno, NV: AIAA)
[14] Windte J, Scholz U and Radespiel R 2006 *Aerosp Sci Technol* **10** 484 – 494
[15] Catalano P and Tognaccini R 2011 *Aerospace Science and Technology* **15** 615 – 626 ISSN 1270-9638
[16] Sorensen N 2009 *Wind Energ* **12** 715–733
[17] Lian Y and Shyy W 2007 *AIAA J* **45** 1501–1511
[18] Crivellini A, D'Alessandro V and Bassi F 2013 *J Comput Phys* **241** 388 – 415
[19] Crivellini A, D'Alessandro V and Bassi F 2013 *Comput Fluids* **81** 122 – 133
[20] Rumsey C L 2007 *Int J Heat and Fluid Fl* **28** 1460–1471
[21] Bassi F, Crivellini A, Rebay S and Savini M 2005 *Comput Fluids* 507–540
[22] Bassi F, Crivellini A, Di Pietro D A and Rebay S 2006 *Journal of Computational Physics* **218** 794–815
[23] Bassi F, Crivellini A, Di Pietro D A and Rebay S 2007 *Comput Fluids* **36** 1529–1546
[24] Crivellini A, D'Alessandro V and Bassi F 2013 *Comput Fluids* **86** 442–458
[25] Bassi F, Rebay S, Mariotti G, Pedinotti S and Savini M 1997 *Proceedings of the 2nd European Conference on Turbomachinery Fluid Dynamics and Thermodynamics* ed Decuypere R and Dibelius G (Antwerpen, Belgium: Technologisch Instituut) pp 99–108
[26] Balay S, Buschelman K, Gropp W D, Kaushik D, Knepley M G, McInnes L C, Smith B F and Zhang H 2001 PETSc Web page http://www.mcs.anl.gov/petsc
[27] Crivellini A and Bassi F 2011 *Comput Fluids* **50** 81 – 93
[28] McGhee R, Walker B and Millard B 1988 Experimental Results for Eppler 387 Airfoil at low Re numbers in Langley Low Turbulence pressure tunnel Tech. Rep. TM 4062 NASA
[29] Selig M, Donovan J and Fraser D 1989 *Airfoils at Low Speeds, Soartech 8* (H.A. Stokely Publisher)
[30] Ghidoni A, Pelizzari E, Rebay S and Selmin V 2006 *Int J Numer Meth Fluids* **51** 1097–1115
[31] Montelpare S and Ricci R 2004 *Int J Therm Sci* **43** 315–329

Journal of Physics: Conference Series **501** (2014) 012025 doi:10.1088/1742-6596/501/1/012025

On the influence of curvature and torsion on turbulence in helically coiled pipes

M Ciofalo [+], **M Di Liberto and G Marotta**
Dipartimento DEIM, Università degli Studi di Palermo, Italy
[+] Corresponding author; email michele.ciofalo@unipa.it, tel. +39 320 43 95 854

Abstract. Turbulent flow and heat transfer in helically coiled pipes at Re_τ=400 was investigated by DNS using finite volume grids with up to 2.36×10^7 nodes. Two curvatures (0.1 and 0.3) and two torsions (0 and 0.3) were considered. The flow was fully developed hydrodynamically and thermally. The central discretization scheme was adopted for diffusion and advection terms, and the second order backward Euler scheme for time advancement. The grid spacing in wall units was ~3 radially, 7.5 circumferentially and 20 axially. The time step was equal to one viscous wall unit and simulations were typically protracted for 8000 time steps, the last 4000 of which were used to compute statistics. The results showed that curvature affects the flow significantly. As it increases from 0.1 to 0.3 the friction coefficient and the Nusselt number increase and the secondary flow becomes stronger; axial velocity fluctuations decrease, but the main Reynolds shear stress increases. Torsion, at least at the moderate level tested (0.3), has only a minor effect on mean and turbulence quantities, yielding only a slight reduction of peak turbulence levels while leaving pressure drop and heat transfer almost unaffected.

1. Introduction and literature review on flow and heat transfer in curved pipes and coils

1.1. Problem definition and notation

Curved pipes are commonly encountered in engineering equipment involving fluid flow and heat transfer. For example, helical coils are widely used for heat exchangers and steam generators in power plant because they easily accommodate thermal expansion and may exhibit higher heat transfer rates than straight pipes [1-3].

Figure 1 shows a schematic representation of a single coil of a helical pipe (computational domain) with its main geometrical parameters: coil radius c, pipe radius a, and coil pitch $2\pi b$. Inner and outer sides are indicated by "*I*" and "*O*", respectively; top and bottom sides by "*T*" and "*D*", respectively.

The dimensionless curvature and torsion can be defined as:

$$\delta = \frac{a}{c} \quad (1)$$

$$\lambda = \frac{b}{c} \quad (2)$$

Instead of the torsion λ, the torsion parameter τ (dimensionally length^{-1}) is sometimes used:

$$\tau = \frac{b}{\left(c^2 + b^2\right)} \quad (3)$$

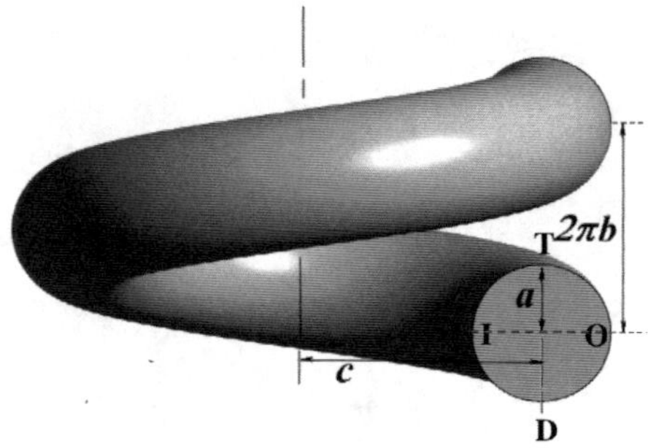

Figure 1. Schematic representation of a helical pipe with its main geometrical parameters: *a*, pipe radius; *c*, coil radius; $2\pi b$, coil pitch. The inner (*I*), outer (*O*), top (*T*) and bottom (*D*) sides of the curved duct are indicated.

Helical coils reduce to curved (toroidal) pipes when λ=0. Several studies, e.g. [4, 5], show that coil torsion λ has only a higher order effect on the flow and on global quantities, such as the friction coefficient, with respect to the first order effect of curvature δ. The effect of curvature is the occurrence of two counter-rotating vortices in the cross section, while the effect of torsion is a loss of symmetry with respect to the equatorial midline *I-O*.

Without loss of generality, the duct can be considered as a right-handed helix. The centreline of this helix, i.e. the main axis of the duct, can be parameterized using the angle of rotation φ as:

$$\mathbf{P}(\varphi) = c\cos\varphi\cdot\mathbf{i} + c\sin\varphi\cdot\mathbf{j} + b\varphi\cdot\mathbf{k} \tag{4}$$

where **i**, **j** and **k** are the unit vectors of a Cartesian reference frame (x_1, x_2, x_3), see figure 2.

Germano [4] introduced an orthogonal reference system for helical pipes, based on the helical coordinates *s* (axial), *r* (radial) and θ (azimuthal), with the prescription that θ be measured from a reference azimuth which rotates anticlockwise (for a right-handed helix) as $-\tau s$ (where τ is the torsion parameter defined above). The position of any point inside the helical pipe is given by the vector

$$\mathbf{X} = \mathbf{P}(s) - r\sin(\theta - \tau s)\cdot\mathbf{N}(s) + r\cos(\theta - \tau s)\cdot\mathbf{B}(s) \tag{5}$$

where **T**, **N** and **B** are the unit vectors along the tangential, normal, and binormal directions of the pipe axis **P**(*s*). By expanding **T**, **N** and **B** in terms of the unit vectors **i**, **j** and **k**, and taking account of Eq. (4), Eq. (5) becomes a rule for transforming Germano coordinates into Cartesian ones. Complete direct and inverse transformations [6] are cumbersome and will not reported here.

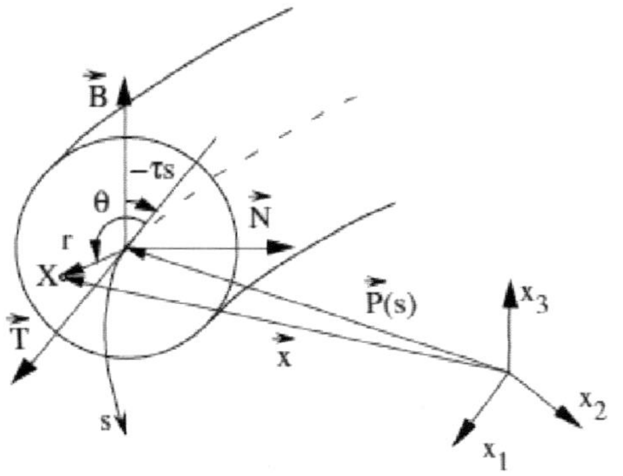

Figure 2. Sketch of the orthogonal helical coordinate system (*s*, *r*, θ), as introduced by Germano [4].

In the present study, as also discussed in section 2.3, instantaneous velocity components and their first- and second-order moments, computed by the code in a Cartesian reference frame, were converted in the Germano coordinate system where they acquire their full physical significance.

In the following, the cross-sectional average of a generic quantity ϕ will be indicated by Φ and its time average by ϕ_{avg} or $\overline{\phi}$, while the fluctuation $\phi-\phi_{avg}$ will be indicated by ϕ' and its root mean square value by ϕ_{rms}. The notation $\langle\phi\rangle$ will be used for the circumferential average of a wall quantity (e.g. the wall shear stress τ_w). The bulk Reynolds number Re will be defined on the basis of the pipe diameter

$2a$ and the time- and cross-section-averaged (bulk) velocity U_{avg} as Re=$2U_{avg}a/\nu$, ν being the kinematic viscosity of the fluid. The friction-velocity Reynolds number will be defined as $\mathrm{Re}_\tau=u_\tau a/\nu$, where $u_\tau=\left(\langle\overline{\tau}_w\rangle/\rho\right)^{1/2}$ is the friction velocity based on the time- and circumferentially-averaged wall shear stress. Wall scales can be built on the basis of u_τ, i.e. u_τ itself for velocity, ν/u_τ for length, u_τ^2 for Reynolds stresses and turbulence energy, etc.; the corresponding normalized quantities will be denoted by a superscript +. The temperature wall scale will be $T_\tau=\langle\overline{q}_w''\rangle/\left(\rho c_p u_\tau\right)$, $\langle\overline{q}_w''\rangle$ being the time- and wall-averaged heat flux; in this way, the wall scale for heat fluxes, including the turbulent (Reynolds) heat flux $\rho c_p\overline{u_i'T'}$, will be $\langle\overline{q}_w''\rangle=\rho c_p u_\tau T_\tau$. The dimensionless temperature T^+ will be computed as $(T_w-T)/T_\tau$; note that T^+ vanishes at the wall and is always positive in the bulk flow, provided the wall heat flux q''_w is regarded as positive if entering the fluid and negative otherwise.

The friction velocity u_τ can also be used to define the Large Eddy TurnOver Time (LETOT) a/u_τ, a scale currently used in direct and large-eddy simulations of turbulence to identify the minimum simulation time required for statistical significance. Since the wall time unit is ν/u_τ^2, one LETOT corresponds to a number of wall time units equal to the friction velocity Reynolds number Re_τ [7].

Finally, the inner (tube) side Nusselt number will be defined as:

$$\mathrm{Nu}=\frac{\langle\overline{q}_w''\rangle 2a}{\chi\left(T_w-T_b\right)} \tag{6}$$

where χ is the fluid thermal conductivity and T_b is the bulk fluid temperature. On the basis of the above-mentioned definitions, one also has Nu=$(2a)/(\chi T_b^+)$.

1.2. Flow field and pressure drop

Flow in curved pipes with zero torsion is characterized by the existence of a secondary circulation in the cross section, caused by the local imbalance between pressure and inertial (centrifugal) forces. The fluid moves towards the outer bend side near the equatorial midplane, returns towards the inner side along two near-wall boundary layers, and then forms two symmetric secondary cells (Dean vortices) having a characteristic velocity scale $U_{avg}\ \delta^{1/2}$. This picture changes only slightly for finite but moderate values of the torsion.

Experimental pressure drop results for a wide range of curvatures and Reynolds numbers were presented by Ito [8], who derived the following correlations for the Darcy-Weisbach friction factor f (four times the Fanning coefficient) in curved pipes ($5\times10^{-4}\le\delta\le0.2$):

$$f=\frac{64}{\mathrm{Re}}\cdot\frac{21.5\cdot\mathrm{De}}{(1.56+\log_{10}\mathrm{De})^{5.73}} \qquad \text{(laminar flow)} \tag{7}$$

$$f=0.304\cdot\mathrm{Re}^{-0.25}+0.029\sqrt{\delta} \qquad \text{(turbulent flow)} \tag{8}$$

in which Re is the bulk Reynolds number and De=Re $\delta^{1/2}$ is the *Dean number*, which accounts for inertial, centrifugal and viscous effects. Note that Eqs. (7)-(8) do not explicitly contain torsion.

Ito's correlations, although dated, have been confirmed to a notable degree by a large bulk of experimental studies; for example, they agree well with the experiments conducted by Cioncolini and Santini [9] in a broad range of curvatures ($0.027\le\delta\le0.143$) and Reynolds numbers (Re$\approx10^3$-7×10^4).

Several authors attempted to characterize the transition to turbulence in curved pipes. Among computational studies, Di Piazza and Ciofalo [10] presented detailed results on the breakdown of steady laminar flow and the transition to turbulence in closed toroidal pipes having curvatures δ of 0.1 and 0.3. They observed a complex transition scenario, characterized by travelling waves, involving periodic, quasi-periodic and chaotic solutions depending on curvature and Reynolds number. Chaotic flow was obtained for Re>~7000 in the case δ=0.1 and for Re>~8000 in the case δ=0.3.

Three-dimensional numerical simulations of unsteady flow in helical and curved pipes were also presented by Friedrich and co-workers [11, 12]. They compared toroidal and helical pipe results for

Re≈5600 (Re_τ≈230) and δ=0.1. Although the authors performed a statistical processing of the computational results (e.g. by computing Reynolds stress), the case they studied was a time-dependent laminar flow rather than a truly turbulent flow [10].

1.3. Heat transfer
As regards heat transfer, many experimental studies were performed in the 1960s and the 1970s on the average heat transfer rate in curved and helical pipes, e.g. by Seban and McLaughlin [13] and Mori and Nakayama [14]; only some of these investigations explored the influence of the Prandtl number on heat transfer, and very few investigated the local heat transfer rate distribution.
Rogers and Mayhew [15] proposed for the Nusselt number in helical pipes the following correlation:

$$\mathrm{Nu} = 0.023\,\mathrm{Re}^{0.85}\,\mathrm{Pr}^{0.4}\,\delta^{0.1} \tag{9}$$

This formula is basically a curvature correction to the Dittus-Bölter correlation; since it does not contain the pipe torsion, it does not distinguish between planar and helical curved pipes and it does not exhibit the correct asymptotic behaviour for small δ, predicting Nu=0 for straight pipes.
In their experimental study on heat transfer in helical pipes, Xin and Ebadian [16] explored two values of curvature, i.e. δ=0.027 and 0.08, Re from $5\cdot10^3$ to $1.1\cdot10^5$, and three different fluids covering a broad range of Prandtl numbers, i.e. air (Pr=0.7), water (Pr=5), and ethylene glycol (Pr=175). Results for air and water (0.7<Pr<5) were approximated by the following correlation:

$$\mathrm{Nu} = 0.00619\,\mathrm{Re}^{0.92}\,\mathrm{Pr}^{0.4}\left(1+3.455\delta\right) \tag{10}$$

with an RMS deviation of 18% for 0.7<Pr<5, $5\times10^3<\mathrm{Re}<10^5$, $0.0267<\delta<0.0884$. Even Eq. (10) does not distinguish between planar and helical pipes, but, unlike Eq. (9), it exhibits a reasonable asymptotic behaviour for straight pipes.

2. Models and methods

2.1. Computational mesh
As mentioned above, the computational domain was a single coil of helical pipe, shown in figure 1 for the case δ=0.3. In particular, two curvatures, δ=0.3 and δ=0.1, and two torsions, λ=0.3 and λ=0 (toroidal pipe), were simulated [18]. The finite volume computational grid was hexahedral and multi-block structured; it was characterized by the parameters N_{RAD} and N_θ as shown in figure 3.

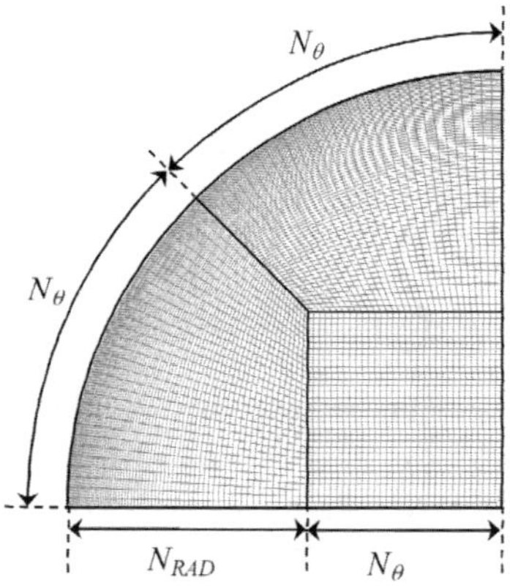

Figure 3. One fourth of the cross section with the computational multi-block hexahedral mesh. The number of cells in the whole cross section is $4N_\theta(N_\theta+2N_{RAD})$.

In the fine grid used for the simulations reported here, the values N_{RAD}=64, N_θ=32 were adopted. With these choices, the cross section was resolved by ~21 000 volumes. In the streamwise direction the domain was discretized by N_S=384 cells for δ=0.3 and by N_S=1152 cells for δ=0.1, so that the total number of control volumes was $N\approx7.86\cdot10^6$ for δ=0.3 and $N\approx2.36\cdot10^7$ for δ=0.1. Geometric refinement was introduced at the wall, with a consecutive cell size ratio of ~1.025 in the radial direction. For Re_τ=400, the first near-wall grid point (volume centre) was at $y^+\approx0.5$ and the viscous sublayer ($y^+\le11$)

was resolved by 10 grid cells. The Kolmogorov length scale $\Lambda_K=(\nu^3/\varepsilon)^{1/4}$, ε being the dissipation of turbulence energy per unit mass, can be expressed for the present configuration (in the average) as $\Lambda_K=a\mathrm{Re}_\tau^{-1/2}\,\mathrm{Re}^{-1/4}$. For Re_τ=400 and Re ranging from 9450 (δ=0.3) to 11800 (δ=0.1), it can be shown that, in wall units, $\Lambda_K^+ \simeq 2$ so that the fine mesh provides a resolution of ~0.5Λ_K to 2.5Λ_K radially, ~10Λ_K streamwise and ~5Λ_K or less spanwise (of course, the largest values are attained near the wall).

2.2. Numerical methods

Simulations were conducted by the ANSYS CFX 13 code [17]. It uses a finite volume approach, a co-located (non-staggered) grid layout and a coupled technique, which simultaneously solves all the transport equations in the whole domain. The linearized system of equations is preconditioned in order to reduce all the eigenvalues to the same order of magnitude. An algebraic multi-grid solver reduces the low frequency error, converting it to a high frequency error at the finest grid level; this results in a great acceleration of convergence.

To simulate fully developed flow and heat transfer, periodic boundary conditions were imposed at inlet-outlet, and no slip conditions for the velocity at the wall. A uniform source term was added to the RHS of the axial momentum equation as the driving force per unit volume balancing pressure drop.

At walls, a constant wall temperature T_w was imposed. A local energy source term was applied to compensate the wall heat flux. Taking account of the definition of the Nusselt number based on the bulk temperature T_b, this local source term must be proportional to the local axial velocity. With this treatment, the bulk temperature and the Nusselt number tend to stable values once a statistically steady state is reached. The Prandtl number was set to 0.86 (representative of saturated water at 58 bar).

Simulations were protracted for ~20 LETOT's a/u_τ, the last 10 of which were used to build averages and statistics, including RMS temperature fluctuations and turbulent heat fluxes. The time step Δt was chosen equal to one wall time unit ν/u_τ^2, so that one LETOT was resolved by Re_τ=400 time steps and a typical simulation included 8000 time steps. The resulting Courant number was less than 1 almost everywhere in all cases; this time discretization is sufficient to capture most of the turbulent variations.

2.3. Post-processing

Simulations were conducted in a Cartesian reference frame (x, y, z). Instantaneous velocities, pressure and temperature on 8 equally spaced cross sections of the pipe were stored at each time step.

The subsequent post-processing (by purpose written Fortran code) included the following phases:

a. interpolating all quantities on a local 2-D polar grid (r, θ) (100 radial nodes, selectively refined towards the wall, and 180 circumferential nodes were used);
b. projecting all velocity components onto the Germano reference frame (s, r, θ);
c. computing time averages and time statistics to obtain velocity fluctuations and Reynolds stresses;
d. further regularizing all the statistics by averaging them over the 8 cross sections stored.

3. Results

3.1. Power spectra

Figure 4 reports spectra of the axial velocity u_s for δ=0.3, λ=0.3. They were computed in the frequency domain for two points located on the equatorial midline of a cross section in the outer and inner bend regions, respectively, at a dimensionless distance from the wall y^+=20. The abscissa is the frequency f^+, normalized by the viscous frequency scale u_τ^2/ν which is the reciprocal of the time step $\Delta t=\nu/u_\tau^2$ (see section 1.1). Therefore, spectra necessarily exhibit a computational cutoff (indicated in the figures) at $f^+_{\max}=(1/2)(\Delta t^+)^{-1}=0.5$. The Kolmogorov length scale $\Lambda_K=a\mathrm{Re}_\tau^{-1/2}\mathrm{Re}^{-1/4}$ (see section 2.1) corresponds to a dimensionless frequency $f_K^+=(1/2)(U_{avg}/\Lambda_K)/(u_\tau^2/\nu)\approx2.88$, also indicated in the figures. The spectrum for the outer bend side, figure 4(a), exhibits a fairly developed inertial subrange with slope -5/3 extending over about one decade. This is followed by a sharp fall at $f^+\approx0.1$, well below the computational cutoff. This suggests that energy-containing scales are sufficiently resolved by the grid, and are actually far larger than the theoretical Kolmogorov scale deduced above. The spectrum for the inner side, figure 4(b), is different; coherently with the fact that the flow there is almost laminarized, it does not exhibit an inertial subrange and falls at a much lower dimensionless frequency of ~3×10^{-2}.

Similar results are obtained for different values of torsion and curvature. These findings support the view that the grid used is sufficient to resolve all significant structures of the turbulent motion.

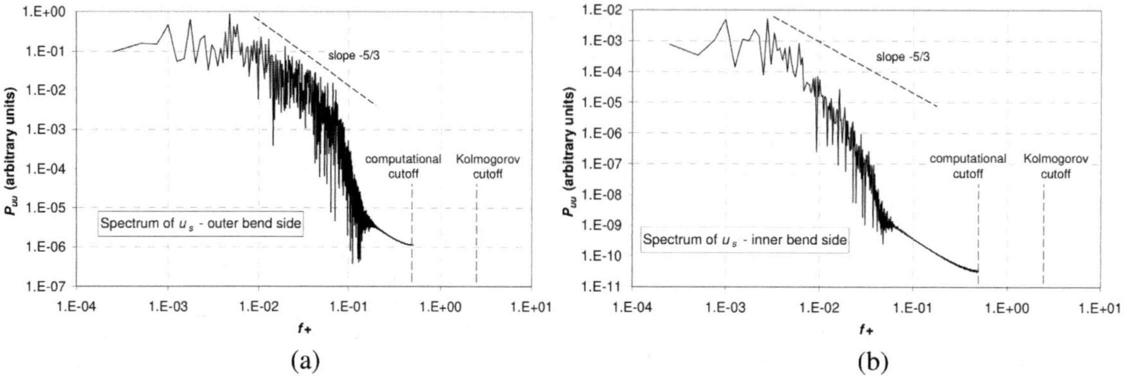

Figure 4. Power spectra of the axial velocity in the frequency domain for the case δ=0.3, λ=0.3. (a) outer bend region; (b) inner bend region. The dimensionless distance from the nearest wall is y^+=20.

3.2. Influence of torsion

Figure 5 compares, for curvature δ=0.3 and torsion λ=0.3 (helical pipe – solid lines) or λ=0 (toroidal pipe – broken lines), radial profiles of time-averaged streamwise velocity u_{savg}^+, time mean temperature T_{avg}^+, total turbulence energy k^+ and axial-radial Reynolds shear stress, all expressed in wall units, along the two orthogonal diametric lines (*I-O* and *T-D*) indicated in figure 1. Note that throughout this paper the cross section is assumed to be viewed from upstream, so that (see figure 1) the inner bend side (*I*) is on the left, while the outer bend side (*O*) is on the right.

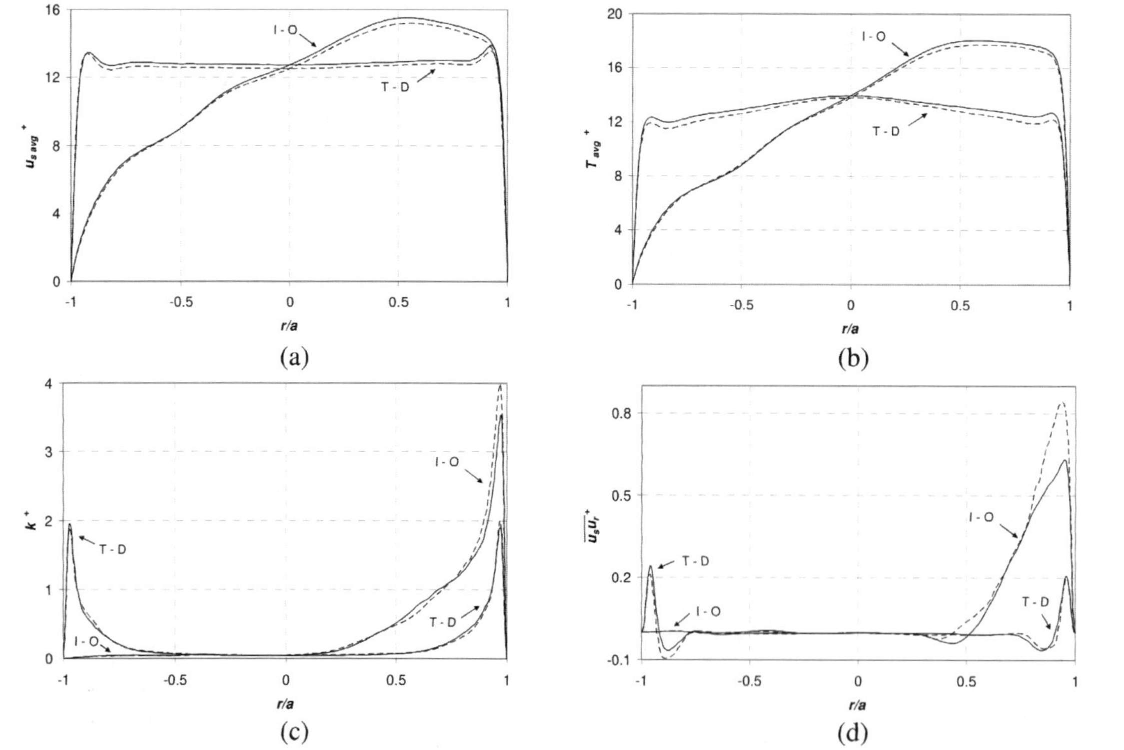

Figure 5. Radial profiles of (a) u_{savg}^+, (b) T_{avg}^+, (c) k^+ and (d) $\overline{u_s u_r}^+$ (in wall units), along the diametric lines *I-O* and *T-D*, for δ=0.3 and λ=0.3 (solid lines) or λ=0 (broken lines).

Average parameters ($u_{savg}{}^{+}$ and $T_{avg}{}^{+}$) are almost identical in the two cases, while turbulent parameters (k^{+} and $u_s u_r{}^{+}$) decrease significantly in the presence of torsion, especially on the outer side ("*O*", $r/a>0$) where the turbulence intensity is highest. The reduction is particularly significant (20-25%) for the tangential Reynolds-stress.
Figure 6 shows a similar comparison for δ=0.1. Conclusions here are similar to those given above for δ=0.3, with the exception that, for this smaller curvature, the reduction in the peak value of the Reynolds shear stress on the outer bend side induced by torsion is more significant (~30-35%).

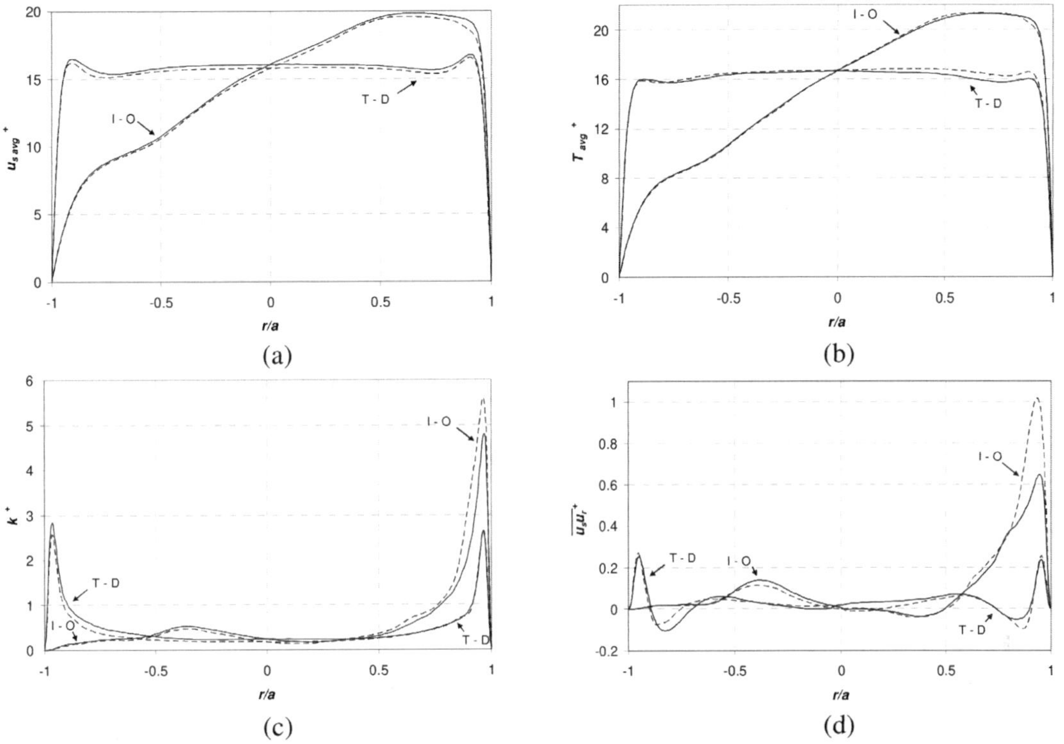

Figure 6. Radial profiles of (a) $u_{savg}{}^{+}$, (b) $T_{avg}{}^{+}$, (c) k^{+} and (d) $\overline{u_s u_r}^{+}$ (in wall units), along the diametric lines *I-O* and *T-D*, for δ=0.1 and λ=0.3 (solid lines) or λ=0 (broken lines).

Figure 7 reports vector plots of the time mean secondary flow for δ=0.3 and λ=0.3 (a) or λ=0 (b).

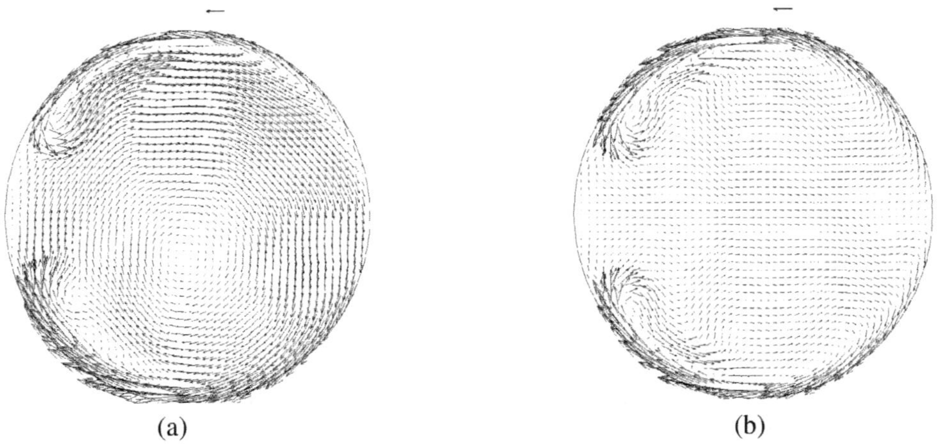

Figure 7. Vector plots of the mean secondary flow for δ=0.3 and λ=0.3 (a) or λ=0 (b). Reference vectors corresponding to u_τ are shown. Inner bend side (*I*) on the left, outer side (*O*) on the right.

The secondary flow for λ=0 (graph b on the right) exhibits the typical Dean circulation pattern, with a broad central region of weak centrifugal flow and two narrow returning (centripetal) secondary flow boundary layers, which end up turning inward and becoming the twin Dean vortices. A similar secondary flow pattern is obtained for λ=0.3 (graph a on the left); here, however, a slight global clockwise rotation of the secondary flow field due to torsion can be noticed, with a loss of the exact top-down symmetry with respect to the equatorial midplane. A weak clockwise recirculation pattern affecting the whole cross section can be observed, with its centre lying slightly below the midplane.
In the following, only results for the cases with finite torsion will be shown. Except for the small differences highlighted above, the results for the cases with zero torsion are very similar.

3.3. Influence of curvature

Figure 8 reports radial profiles (in wall units) of time-mean streamwise velocity u_{savg}^+ (a), mean temperature T_{avg}^+ (b) and rms axial, radial and azimuthal velocity fluctuations u_{srms}^+, u_{rrms}^+, $u_{\theta rms}^+$ (c-d), along the lines *I-O* and *T-D*, for λ=0.3 and δ=0.3 (solid lines) or δ=0.1 (broken lines). As in figures 4 and 5, the inner bend side (*I*) lies on the left and the outer bend side (*O*) on the right.

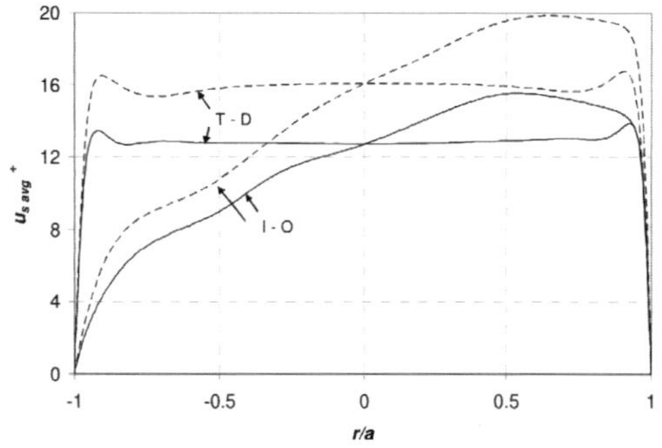

(a)

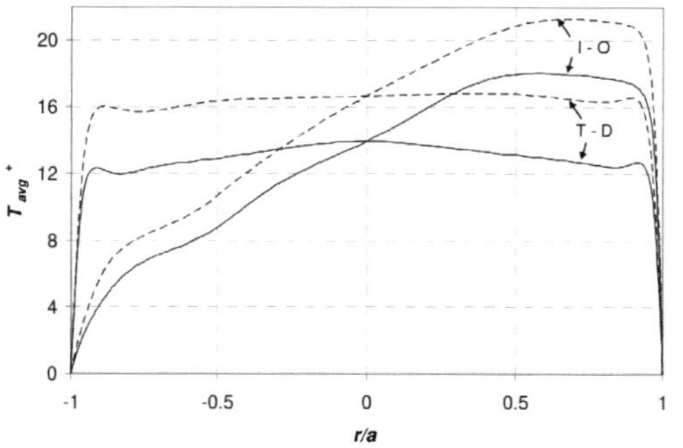

(b)

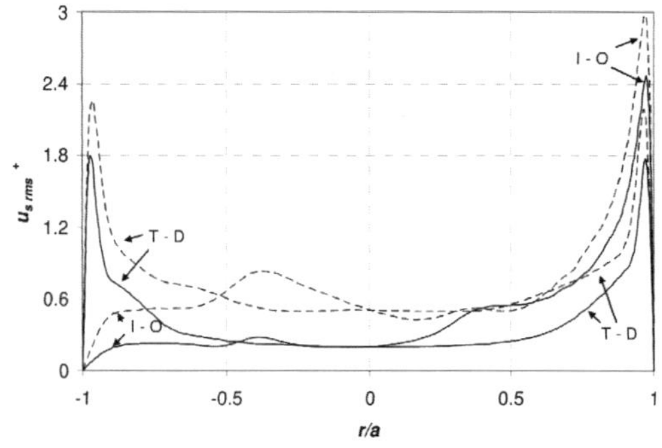

(c)

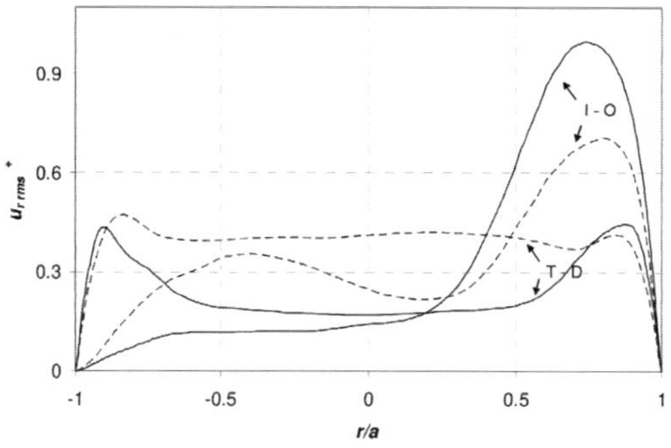

(d)

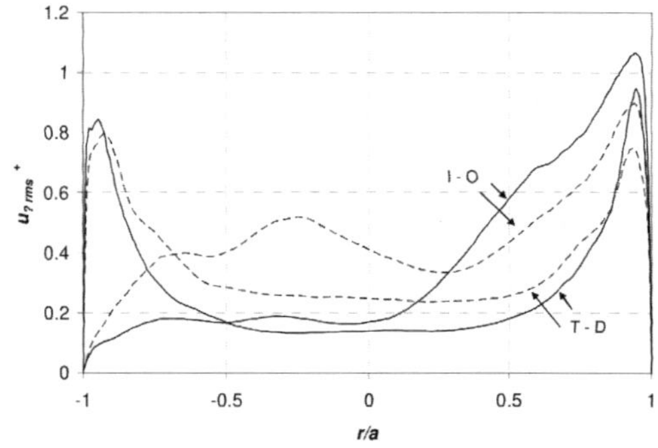

(e)

Figure 8. Radial profiles (in wall units) of time-average streamwise velocity u_{savg}^+ (a), mean temperature T_{avg}^+ (b) and axial, radial and circumferential rms velocity fluctuations u_{srms}^+, u_{rrms}^+, $u_{\theta rms}^+$ (c-e), along the diametric lines *I-O* and *T-D*, for λ=0.3 and δ=0.3 (solid lines) or δ=0.1 (broken lines).

As the curvature δ increases from 0.1 to 0.3, the mean velocity (graph a) and the mean dimensionless temperature (graph b), once measured in wall units, decrease significantly, which corresponds to an increase both in friction and in heat transfer rates for a given flow rate. As the curvature increases, the axial velocity fluctuation decreases significantly in the central region of the channel and, less markedly, in both the inner and outer near-wall layers. The radial and azimuthal velocity fluctuations decrease in the central region of the channel and in the *I* near wall region, change little in the *T* and *D* near-wall regions, but increase considerably in the *O* near-wall region, where the highest fluctuation levels are observed. Since in pipe and channel flow turbulence is mainly produced in the form of axial fluctuations and is then re-distributed among the other components by pressure-strain rate correlations (redistribution terms), the present findings indicate that curvature enhances the importance of redistribution terms and yields a more isotropic turbulence structure.
Vector plots of the in-plane turbulent (Reynolds) heat flux, for λ=0.3 and for δ=0.3 (a) and δ=0.1 (b), are reported in figure 9. A scale vector of length $\langle \overline{q}_w'' \rangle = \rho c_p u_\tau T_\tau$ is also shown.

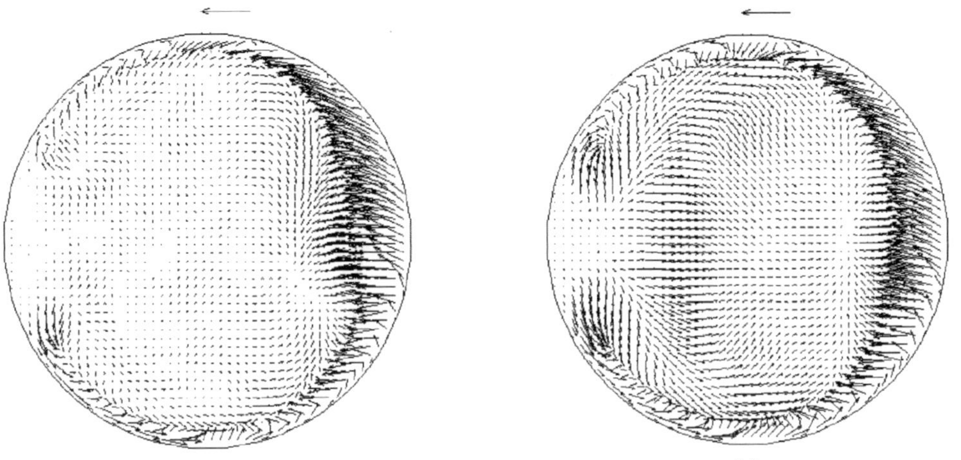

(a) (b)

Figure 9. Vector plots of the turbulent heat flux for λ=0.3 and δ=0.3 (a) or δ=0.1 (b). The scale vector represents the time- and surface-averaged wall heat flux $\langle \overline{q}_w'' \rangle$. Inner bend side (*I*) on the left, outer side (*O*) on the right.

The in-plane turbulent heat flux vector, of components $\rho c_p \overline{u_r' T'}$ (radial) and $\rho c_p \overline{u_\theta' T'}$ (azimuthal), for both cases is large only in the outer region, where its radial component attains peak values well above 1 and contributes significantly to heat transfer from the wall. The turbulent heat flux exhibits negligible values not only near the inner side but also in the Dean vortex regions, which are basically steady structures little affected by turbulent fluctuations.
In the outer bend region the turbulent heat flux is slightly more intense for the higher curvature. On the contrary, in the central and in the Dean vortex regions turbulent fluxes are stronger for the lower curvature, while in the case δ=0.3 these regions are characterized by an almost zero contribution of turbulence to heat transport.

3.4. Wall quantities
Figure 10 represents the distribution of the instantaneous wall shear stress (normalized by its time- and surface-averaged value) over the pipe wall for the four cases.

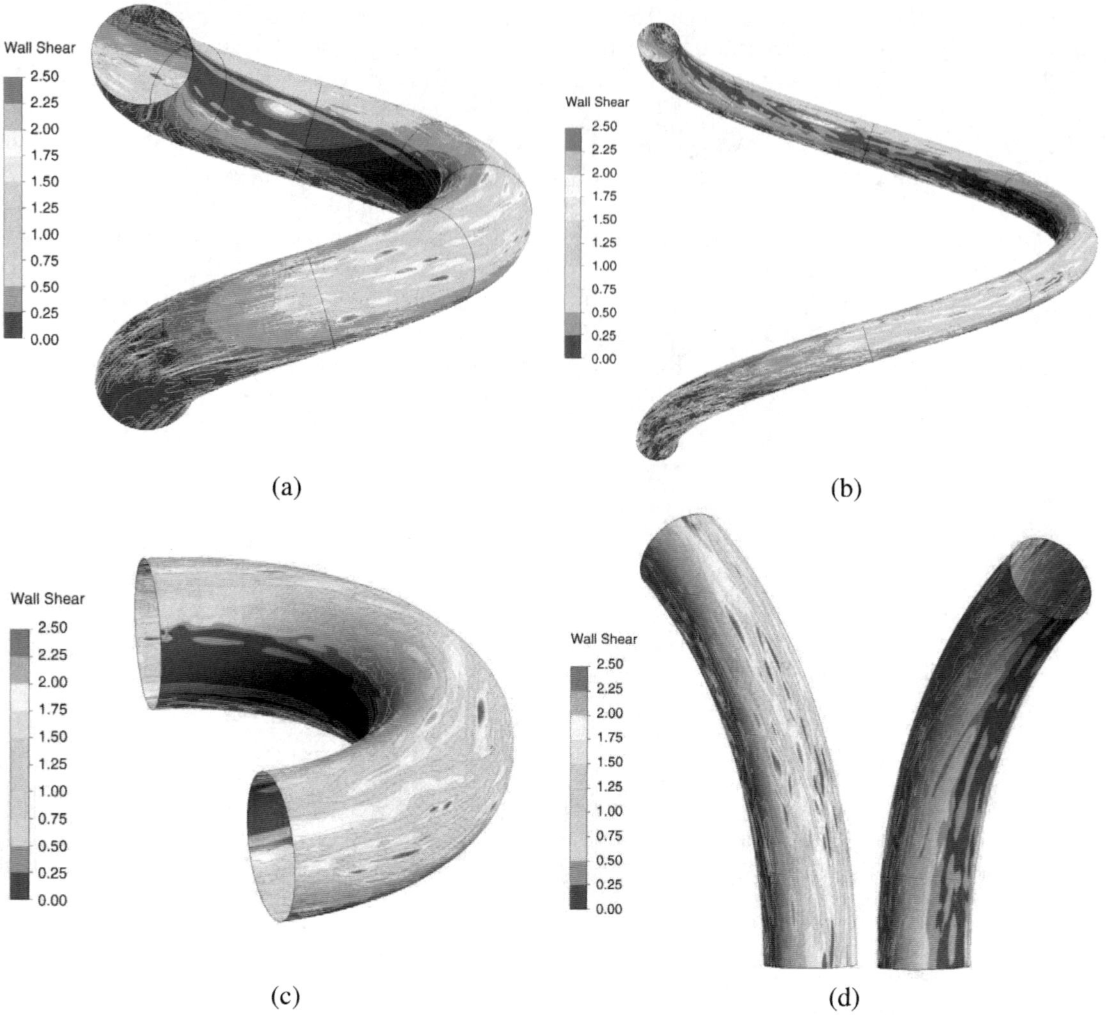

Figure 10. Instantaneous distribution of the wall shear stress τ_w. (a) λ=0.3, δ=0.3; (b) λ=0.3, δ=0.1; (c) λ=0, δ=0.3; (d) λ=0, δ=0.1 (inner and outer side). With reference to the legend on the left, τ_w is normalized by its time- and surface-averaged value $\langle \overline{\tau}_w \rangle$, which is the same for all cases.

All cases exhibit a peculiar pattern. The outer bend side of the pipe, where turbulence levels are high, is covered by a streak pattern not unlike that observed for turbulent flow in straight pipes, and obviously associated with a similar pattern of the near-wall streamwise velocity. The circumferential spacing of the streaks ranges between 100 and 200 wall length units, and their streamwise extent between 500 and 1000. On the other hand, the inner bend side, where turbulence levels are low, exhibits a flat distribution and very low values of the wall shear stress. The difference between the inner and outer sides is more marked for the higher curvature (δ=0.3), and seems to be little affected by torsion.

Similar remarks hold for the following figure 11, which reports the distribution of the instantaneous wall heat flux over the pipe wall for the four cases. Here q'' is normalized by the conductive heat flux, $q''_c=\chi(T_w-T_b)/a$. The outer bend side of the pipe wall is covered by a pattern of thermal streaks, the size and general features of which are similar to those observed for the hydrodynamic streaks in figure 10. The low-turbulence inner bend side exhibits a flat distribution of the local heat flux, which appears to be only slightly higher than the conductive reference heat flux q''_c.

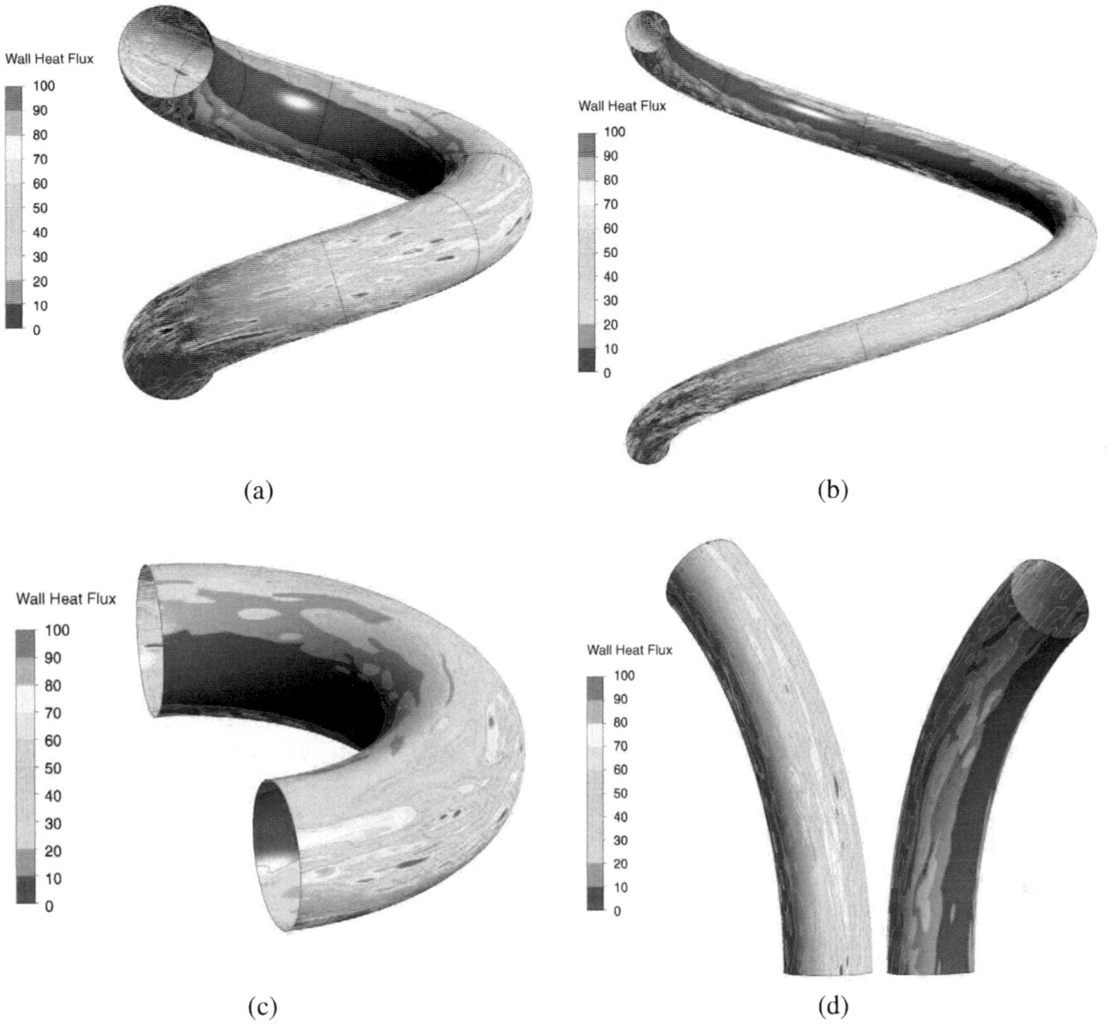

Figure 11. Instantaneous distribution of the wall heat flux q'' (a) λ=0.3, δ=0.3; (b) λ=0.3, δ=0.1; (c) λ=0, δ=0.3; (d) λ=0, δ=0.1 (inner and outer side). Here q'' is normalized by the conductive heat flux, $q''_c=\chi(T_w-T_b)/a$.

4. Conclusions

Direct numerical simulations were performed for fully developed turbulent flow with heat transfer in curved and helical pipes. Four cases, characterized by torsion λ=0.3 or 0 and curvature δ=0.1 or 0.3, were examined. In all cases, the friction velocity Reynolds number was 400 and the Prandtl number was 0.86. The ANSYS CFX 13 computer code was used for all the numerical simulations. Computational grids with up to ~23 million nodes were used; the overall CPU time required for this study was close to 5×10^8 core-seconds.

All cases exhibited a strong asymmetry between the outer bend side, characterized by high turbulence levels, high shear stress and high heat transfer rates, and the inner bend side, characterized by an almost steady flow and by very low levels of wall shear stress and heat transfer. The time-mean flow exhibited a secondary recirculation pattern similar to that observed in high-Reynolds number laminar flow, with the appearance of twin Dean vortices having their centres in the inner region of the cross section.

Increasing the curvature led to a considerable increase of frictional losses and heat transfer rates; it also yielded an increase in turbulence re-distribution, so that peak values of the axial velocity fluctuation decreased whereas peak values of the radial and azimuthal fluctuations increased. Another

effect of increasing the curvature was that fluctuations tended to be confined to the near-wall outer region, while the flow in the core of the pipe became almost laminar.
Torsion, at least at the moderate level tested (0.3), was found to have only a minor effect on mean and turbulence quantities, yielding only a slight reduction of peak turbulence levels while leaving pressure drop and heat transfer almost unaffected. This supports the established practice of using friction and heat transfer correlations which contain only curvature but not torsion.

References

[1] Salimpour M R 2009 Heat transfer coefficients of shell and coiled tube heat exchangers *Exp. Thermal Fluid Science* **33** 203–7.
[2] Yang G and Ebadian M A 1996 Turbulent forced convection in a helicoidal pipe with substantial pitch *Int. J. Heat Mass Transfer* **39** 2015-22.
[3] Di Piazza I and Ciofalo M 2010 Numerical prediction of turbulent flow and heat transfer in helically coiled pipes *Int. J. Thermal Sciences* **49** 653-63.
[4] Germano M 1982 On the effect of torsion in a helical pipe flow *J. Fluid Mech* **125** 1-8.
[5] Chen W H and Jan R 1992 The characteristics of laminar flow in helical circular pipe *J Fluid Mech* **244** 241-56.
[6] Hüttl T J, Chauduri M, Wagner C and Friedrich R 2004 Reynolds-stress balance equations in orthogonal helical coordinates and application *Z. Angew. Math. Mech.* **84** 403-16.
[7] Di Liberto M and Ciofalo M 2013 A study of turbulent heat transfer in curved pipes by numerical simulation *Int. J. Heat Mass Transfer* **59** 112-25.
[8] Ito H 1959 Friction factors for turbulent flow in curved pipes *J. Basic Engineering* **81** 123-34.
[9] Cioncolini A and Santini L 2006 An experimental investigation regarding the laminar to turbulent flow transition in helically coiled pipes *Exp. Thermal Fluid Sci.* **30** 367-80.
[10] Di Piazza I and Ciofalo M 2011 Transition to turbulence in toroidal pipes *J. Fluid Mech.* **687** 72-117.
[11] Hüttl T J and Friedrich R 2001 Direct numerical simulation of turbulent flows in curved and helically coiled pipes *Computers & Fluids* **30** 591-605.
[12] Friedrich R, Hüttl T J, Manhart M and Wagner C 2001 Direct numerical simulation of incompressible turbulent flows *Computers & Fluids* **30** 555-79.
[13] Seban R A and McLaughlin E F 1963 Heat transfer in tube coils with laminar and turbulent flow *Int. J. Heat Mass Transfer* **6** 387-95.
[14] Mori Y and Nakayama W 1967 Study of forced convective heat transfer in curved pipes *Int. J. Heat Mass Transfer* **11** 37-59.
[15] Rogers G F C and Mayhew Y R 1964 Heat transfer and pressure loss in helically coiled tubes with turbulent flow *Int. J. Heat Mass Transfer* **7** 1207-16.
[16] Xin R C and Ebadian M A 1997 The effects of Prandtl numbers on local and average convective heat transfer characteristics in helical pipes *J. Heat Transfer* **119** 467-73.
[17] ANSYS Europe Ltd. 2006 *ANSYS CFX manual.*
[18] Marotta G 2012 *Simulazione numerica diretta della turbolenza in condotti elicoidali* MSc Thesis in Energy and Nuclear Engineering, University of Palermo.

Journal of Physics: Conference Series **501** (2014) 012026 doi:10.1088/1742-6596/501/1/012026

Electro-osmotic flows inside triangular microchannels

P Vocale*[1], M Geri[2], GL Morini[2] and M Spiga[1]

[1]Department of Industrial Engineering, University of Parma, Parco Area delle Scienze 181/A, 43124 Parma, Italy.

[2]DIN, Alma Mater Studiorum Università di Bologna, Viale Risorgimento 2, 40136 Bologna, Italy.

E-mail: pamela.vocale@unipr.it; michela.geri@unibo.it; gianluca.morini3@unibo.it; marco.spiga@unipr.it.

Abstract. This work presents a numerical investigation of both pure electro-osmotic and combined electro-osmotic/pressure-driven flows inside triangular microchannels. A finite element analysis has been adopted to solve the governing equations for the electric potential and the velocity field, accounting for a finite thickness of the electric double layer. The influence of non-dimensional parameters such as the aspect ratio of the cross-section, the electrokinetic diameter and the ratio of the pressure force to the electric force on the flow behavior has been investigated. Numerical results point out that the velocity field is significantly influenced by the aspect ratio of the cross section and the electrokinetic diameter. More specifically, the aspect ratio plays an important role in determining the maximum volumetric flow rate, while the electrokinetic diameter is crucial to establishing the range of pressures that may be sustained by the electro-osmotic flow. Numerical results are also compared with two correlations available in the literature which enable to assess the volumetric flow rate and the pressure head for microchannels featuring a rectangular, a trapezoidal or an elliptical cross-section.

1. Introduction

The progress of lab-on-a-chip technology has allowed an unprecedented development of several fields, both in fundamental and applied research. Microfluidic systems enable systematic, high-throughput manipulation of different classical as well as biological fluids in integrated systems that are designed to perform a number of different operations at ones. A crucial issue for these miniaturized experimental platforms is high-precision fluid transport control. Electro-osmotic pumps (EOPs) can easily achieve such task while working with several fluids, and because there are no moving parts, they do not suffer from manufacturing as well as maintenance issues, like mechanical pumps.

Electro-osmosis is a surface electro-kinetic effect that can arise at the interface between a solid and a liquid. It has been proved to be effective in fluid pumping whenever the system has at least one dimension in the micro- or nano-scale. Most materials obtain a surface electric charge when in contact with a polar medium. The surface charge induces a re-distribution of ions in the electrolyte, generating

*[1] To whom any correspondence should be addressed.

Journal of Physics: Conference Series **501** (2014) 012026 doi:10.1088/1742-6596/501/1/012026

what is known as the Electric Double Layer (EDL). When an electric field is applied, the excess counterions in the EDL move under the applied electric force, pulling the liquid with them. Such movement is then conveyed to the fluid bulk by viscous forces, resulting in an Electro-Osmotic Flow (EOF).

The microchannel cross-sectional geometry plays an important role in defining the flow behaviour, since the local net charge density in the EDL significantly influences the electro-osmotic flows. The dynamic behavior of EOFs through simple geometries, such as circular tubes or parallel plates was investigated by many authors [1-10]. However, due to the available manufacturing techniques microchannels feature more complex geometries which have been the focus of many recent works [11-22]. If a KOH-anisotropic etching technique is employed on <100> silicon wafers, it is possible to obtain microchannels having a triangular cross-section with a fixed value of aspect ratio β defined as the ratio of the altitude h and the side b (see figure 1) equal to 0.707; this value depends on the orientation of the silicon crystal planes. However, using different microfabrication techniques it is possible to obtain triangular cross sections having any aspect ratio.

Few papers have specifically addressed the issue of EOFs trough triangular microchannel. Liao et al. [23] presented a numerical analysis of pure and combined pressure-driven/electro-osmotic flows inside equilateral triangular ducts. They investigated the effect of pressure gradient on the fluid behavior, concluding that the positive pressure gradient increases the mass flux of the electrolyte solution and the negative one leads to a decrease in the fluid mass flux. They also analyzed the influence of the length ratio (i.e. the product of the Debye-Hückel parameter and the characteristic channel dimension) on the fluid velocity, showing that as the length ratio increases the fluid mass flux increases as well. More recently semi-analytical solutions of EOFs in polygonal micro-ducts (i.e. equilateral triangular, square, and hexagonal cross-sections) have been derived by Wang and Chang [24] along with simple approximated correlations which enable to evaluate the net flow rate.

Despite these important contributions, a detailed analysis for electro-osmotic flows in triangular geometries is still missing. The present work is intended to numerically investigate both pure electro-osmotic and combined electro-osmotic/pressure-driven flows inside microchannels featuring a triangular cross-section. The effect of the geometric features, the electrokinetic parameters and the pressure gradient on the flow behavior is investigated. Finally, two correlations which provide the characteristic curve of the pump according to the main non-dimensional parameters, are presented.

2. Governing equations

The microchannel considered in the present study is characterized by a cross-sectional geometry that corresponds to an isosceles triangle.

For an isothermal fluid with constant properties and a laminar, fully developed flow, the equation of motion is given by the modified Navier–Stokes equation:

$$\mu\nabla^2 u(\xi,\psi) = -\rho_e(\xi,\psi)E_{ext} + \frac{dp}{d\omega} \tag{1}$$

where u and μ are the fluid velocity and dynamic viscosity, respectively. The term $\rho_e(\xi,\psi)E_{ext}$ denotes the electric force per unit volume acting on the fluid (ρ_e is the net charge distribution while E_{ext} is the applied electric field per unit length, assumed uniform over the channel cross-section); the last term on the right hand side of equation (1) is the pressure gradient along the channel.

The charge distribution can be calculated by means of the Poisson-Boltzmann equation:

$$\varepsilon\nabla^2\phi(\xi,\psi) = -\rho_e(\xi,\psi) = -e\sum_{i=1}^{N} z_i n_{i\infty}\exp\left(-\frac{z_i e\phi(\xi,\psi)}{K_B T_R}\right) \tag{2}$$

ϕ being the electric potential and ε the permittivity of the solution. The second identity is a consequence of the assumption that the charge distribution is basically unperturbed from equilibrium and it is valid as long as there is no overlap of the EDL in the cross section. In the last term of equation (2) e is the magnitude of the elementary charge, z_i is the valence of the i-th ionic specie, $n_{i\infty}$ is the ionic number concentration at the neutral state of the i-th ionic specie (of N species), K_B is the Boltzmann constant and T_R is the thermodynamic temperature, which is a known scalar quantity, since the flow is considered isothermal.

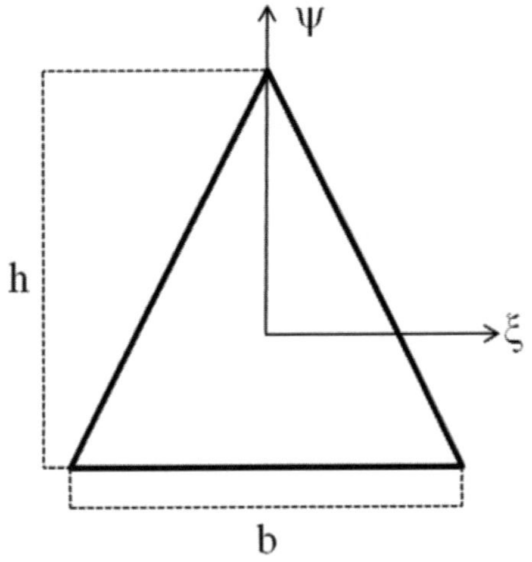

Figure 1. Microchannel cross-sectional geometry.

In a non-dimensional form, the Poisson-Boltzmann equation and the momentum equation can be re-written as follows:

$$\nabla^2\Phi(x,y) = D_e^{\,2}\sinh(\Phi(x,y)) \tag{3}$$

$$\nabla^2 U(x,y) = D_e^{\,2}\frac{(G_{EO} - \nabla^2\Phi(x,y))}{Z} \tag{4}$$

where $\Phi=e\phi/K_BT_R$ is the dimensionless electric potential, $x=\xi/D_h$, $y=\psi/D_h$ and $z=\omega/D_h$, D_h being the hydraulic diameter, are the reduced coordinates, D_e is the electrokinetic diameter, U denotes the non-dimensional fluid velocity defined with respect to the Helmholtz-Smoluchowski velocity [16] and $Z=e\zeta/K_BT_R$ indicates the dimensionless zeta-potential, that is the electric potential at the wall.
The electrokinetic diameter D_e is the product of the Debye-Hückel parameter k and D_h, where k can be calculated by:

$$k = \left(\frac{e^2}{\varepsilon k_B T}\sum_{i=1}^{N} z_i^2 n_{i\infty}\right)^{1/2} \tag{5}$$

The non-dimensional parameter G_{EO} denotes the ratio between pressure surface forces and electrical body forces acting on the fluid:

$$G_{EO} = \frac{\dfrac{dp}{d\omega}}{eE_{ext}\sum_{i=1}^{N} z_i^{\,2} n_{i\infty}} \tag{6}$$

The governing equations are supplemented by the following boundary conditions:

$$\Phi\big|_w = Z \qquad U\big|_w = 0 \tag{7}$$

However, due to the symmetry, the computational domain may be reduced to an half of the triangular cross-section. Therefore equations (3) and (4) are subject to the following boundary conditions on the inner edges:

$$\left.\frac{\partial\Phi}{\partial x}\right|_{x=0} = 0 \qquad \left.\frac{\partial U}{\partial x}\right|_{x=0} = 0 \tag{8}$$

3. Results

The governing equations defined by equations (3) and (4) together with their boundary conditions were solved using the COMSOL Multiphysics™ simulation environment (release 4.2a), where a non uniform mesh with an enhanced refinement in the domain of the EDL close to the walls was employed. Since the analytic solution for triangular microducts is still lacking, to validate the numerical model the results for β=0.866 (equilateral triangular) and G_{EO}=0 (pure electro-osmotic flow) were compared with the semi-analytical solutions for polygonal microchannels available in literature [24].

The accuracy of the solution was checked by carrying out numerical simulations with D_e=200 and G_{EO}=0 on the different grid sizes. In figure 2a the ratio between the value of the dimensionless volumetric flow rate obtained numerically for an equilateral triangle and its value computed by [24] (i.e. V_{ref}) as a function of the mesh elements (N_{el}) is presented. It is evident that for a number of elements N higher than 7314 the value of the volumetric flow rate obtained by using COMSOL becomes independent on the grid size and tends to the same value reported in [24]. Therefore, in all the numerical runs here presented a non uniform mesh characterized by at least 7314 elements was chosen.

In figure 2b, the comparison between the volumetric flow rate obtained numerically for an equilateral triangle (β=0.866) and the values obtained by Wang and Cheng [24] is shown as a function of the electro-kinetic diameter. The perfect agreement between the numerical data and the semi-analytical ones obtained in [24] can be considered a benchmark for the numerical procedure followed in this work.

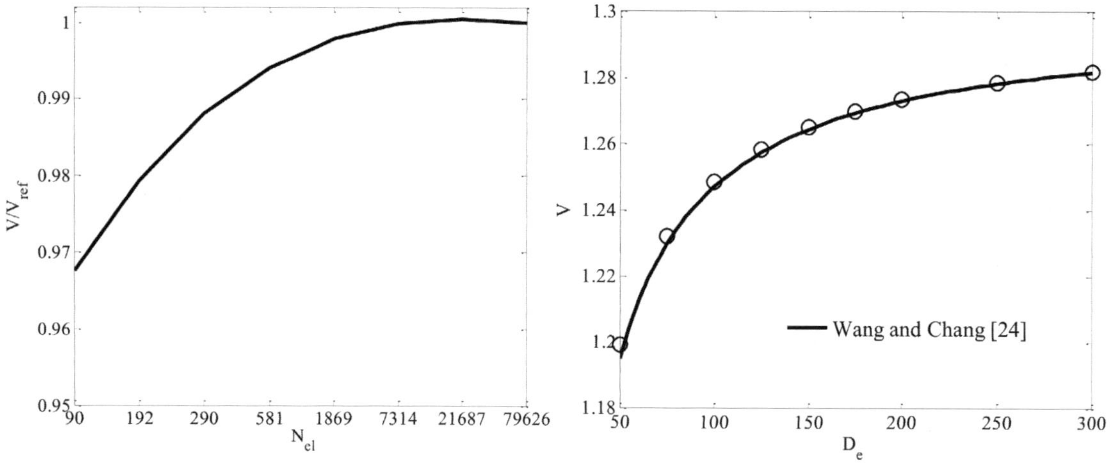

(a) (b)

Figure 2. Equilateral triangle: (a) Mesh independence analysis; (b) Volumetric flow rate as a function of the electrokinetic diameter.

After validating the numerical procedure and checking the accuracy of the numerical solution, several simulations were performed considering different values of the aspect ratio, the electrokinetic diameter and the pressure gradient.

The results here presented were obtained by considering water as electrolyte (concentration set to 10^{-6} M) at 293 K (i.e. thermodynamic temperature T_R). The zeta potential was set to 25.275 mV and hydraulic diameter was varied between 1.9273 μm and 28.910 μm.

The effects of the opposing pressure gradient on the local velocity field are shown in figure 3 for two different values of the aspect ratio, namely β=0.2 and β=0.866 (equilateral triangle), for D_e=100 and Z=1.

When the flow is driven only by the electro-osmotic force (G_{EO}=0), the velocity profile is basically flat, like the ideal plug flow. If an opposing pressure gradient is applied, the velocity profile tends to get a curved shape because the flow driven by the pressure force overcomes the one induced by the electro-osmosis. Trends in figure 3 also point out that as the aspect ratio increases the influence of the opposing pressure gradient becomes less marked.

When the pressure gradient is favorable, the two flow-driving mechanisms (pressure and electric potential) act in the same direction and the velocity get markedly enhanced as the G_{EO} increases. Figure 4 shows the velocity field in such conditions for the same aspect ratios and electrokinetic diameter of fig. 3. It can be seen that the increase in velocity is more pronounced for smaller aspect ratios at any G_{EO}.

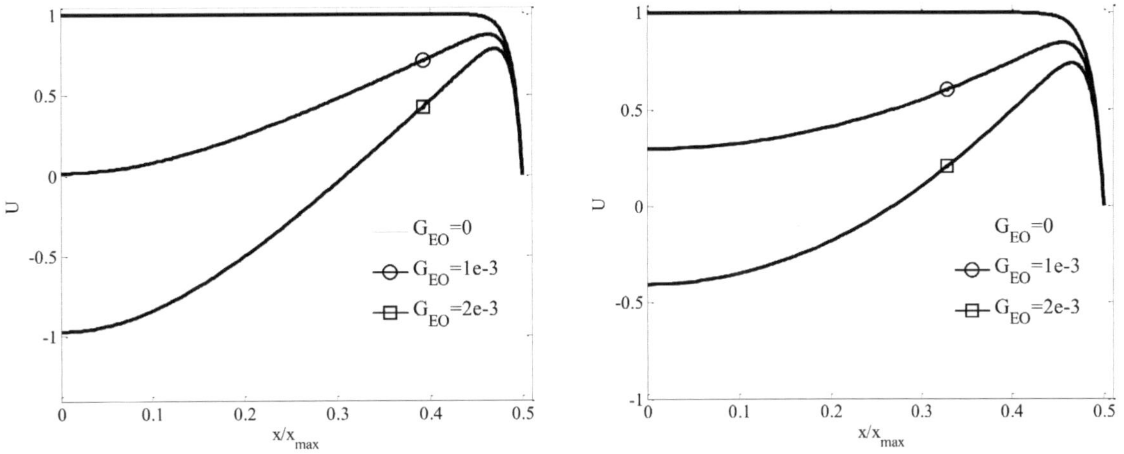

(a) (b)

Figure 3. Velocity profiles at y=y_{max}/2 for positive values of G_{EO}: (a) β=0.2; (b) β=0.866 (equilateral triangle).

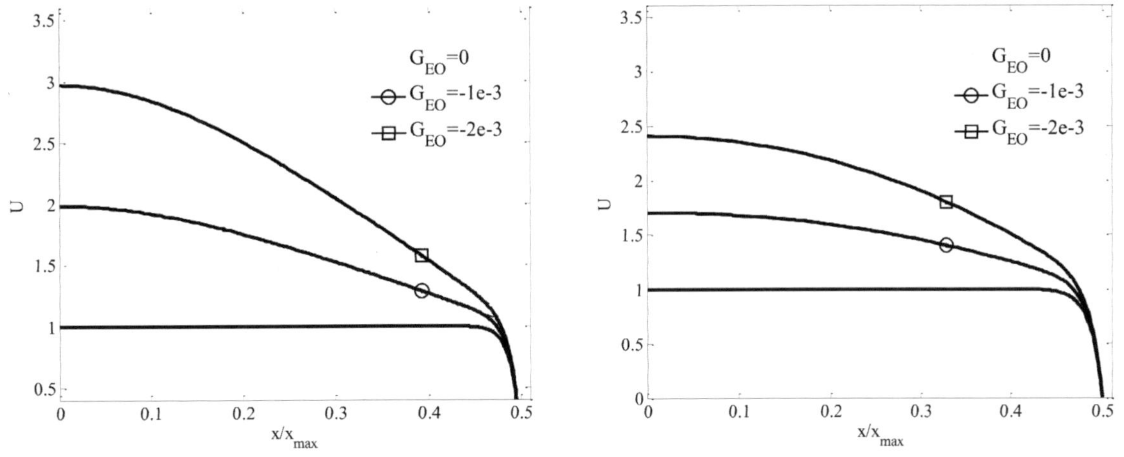

(a) (b)

Figure 4. Velocity profiles at y=y_{max}/2 for negative values of G_{EO}: (a) β=0.2; (b) β=0.866 (equilateral triangle).

The effect of the aspect ratio and the electrokinetic diameter on the maximum volumetric flow rate (i.e. the volumetric flow rate for pure electro-osmotic flow) is shown in figure 5a. It can be seen that V_{max} is significantly influenced by the aspect ratio, as observed by Geri et al. [16] in rectangular and trapezoidal microchannels and by Vocale et al. [19] in elliptic microducts. However, this effect is more evident for small value of the aspect ratio while it becomes less marked for high values of β. On the other hand the electrokinetic diameter plays a less important role in determining the maximum flow rate.

When a pressure field opposes the electro-osmotic flow, the fluid velocity is reduced (figure 3). The flow rate consequently decreases too and it becomes zero when the flow due to the electro-osmosis is balanced by the flow in the opposite direction, driven by the pressure force. The value of the G_{EO} number for which the volumetric flow rate becomes zero is called critical G_{EO} [16]. $G_{EO,c}$ (i.e. critical G_{EO}) as a function of aspect ratio for different values of the electrokinetic diameter is depicted in figure 5b.

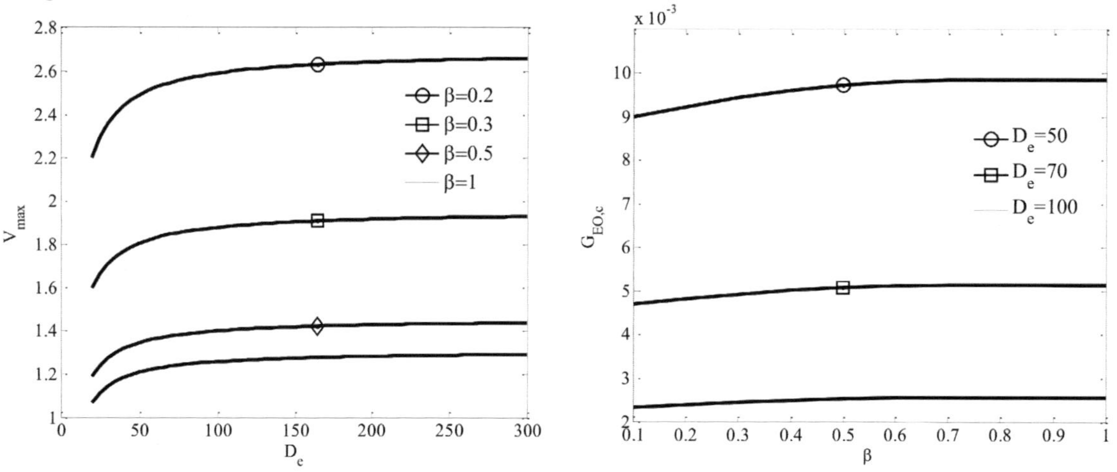

(a) (b)

Figure 5. Influence of the aspect ratio and electrokinetic diameter on flow behavior: (a) Maximum volumetric flow rate; (b) Critical G_{EO}.

It can be seen that the critical G_{EO} is significantly influenced by the electrokinetic diameter, therefore this parameter is very important in evaluating the performance of the EOPs (electro-osmotic pumps) in terms of pressure, as it occurs in microducts characterized by different geometries [16,19].

The maximum volumetric flow rate and $G_{EO,c}$ are important because, as it has been shown [16], for a given geometry, aspect ratio and electrokinetic diameter they allow to determine the linear relationship between the flow rate and G_{EO}, i.e., the characteristic curve of the electro-osmotic pump with chosen features. Knowing the characteristic curve is very important to properly design, optimize and integrate the pump in any microfluidic device. To this end, in the literature there have been suggested a couple of correlations to estimate such values for different geometries of the micro-channel, namely for rectangular, trapezoidal and elliptical cross-sections [16,19]. For the sake of clarity, let us recall here their formulation. The maximum dimensionless volumetric flow rate is given by the following expression:

$$V_{max} = A\left(1-\frac{\tanh\frac{D_e}{2}}{\frac{D_e}{2}}\right)^2 \tag{9}$$

being A the non-dimensional cross-sectional area.

The critical G_{EO}, as a function of the electrokinetic diameter, the aspect ratio and the non dimensional zeta potential, can be evaluated by:

$$G_{EO,c} = 48\frac{Z}{D_e^2}\left(1-\frac{\tanh\frac{D_e}{2}}{\frac{D_e}{2}}\right)^2 g(\beta)F(\beta) \tag{10}$$

This last correlation in particular was modified to use the same expression suggested for rectangular and trapezoidal microchannels on elliptical geometries. While equation (9) works really well also for triangular cross-sections, as shown in figure 6, the correlation for the critical G_{EO} does not. The correction factor is not enough to account for the dependence on the aspect ratio. However, since the functional dependence on the electrokinetic diameter is fully captured, a very good fitting can be achieved just by modifying the function $g(\beta)$, disregarding the area-based correction factor introduced in [19], i.e. setting $F(\beta)=1$. More specifically, it has been found that using a second order polynomial in the form:

$$g(\beta) = a_1\beta^2 + a_2\beta + a_3 \tag{11}$$

allows an estimation of $G_{EO,c}$ with less the 2% error compared to numerical results for triangular geometries. In table 1 the coefficients a_i are shown. Figure (7) shows the comparison between the value of $G_{EO,c}$ evaluated by the correlations (i.e. the correlation proposed in [19] and the modified correlation) and numerical for few different cases. It can be observed that the correlation proposed in the present study fits the data in a better way.

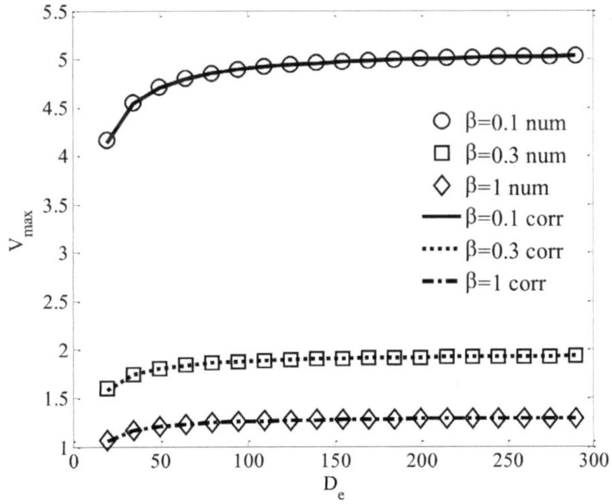

Figure 6. Comparison between the correlation for volumetric flow rate and numerical results.

By means of equations (9) and (10), it is possible to draw the characteristic curve of any electro-osmotic pump with a triangular cross-section for a given aspect ratio, electrokinetic diameter and non-dimensional zeta-potential. The determination of a different dependence on the aspect ratio for different geometries is a simple and effective engineering approach. A more general expression for the critical G_{EO}, based on physical arguments, is currently under investigation.

Table 1. Coefficients of equation (11).

a_1	a_2	a_3
-0.0959	0.1566	0.49318

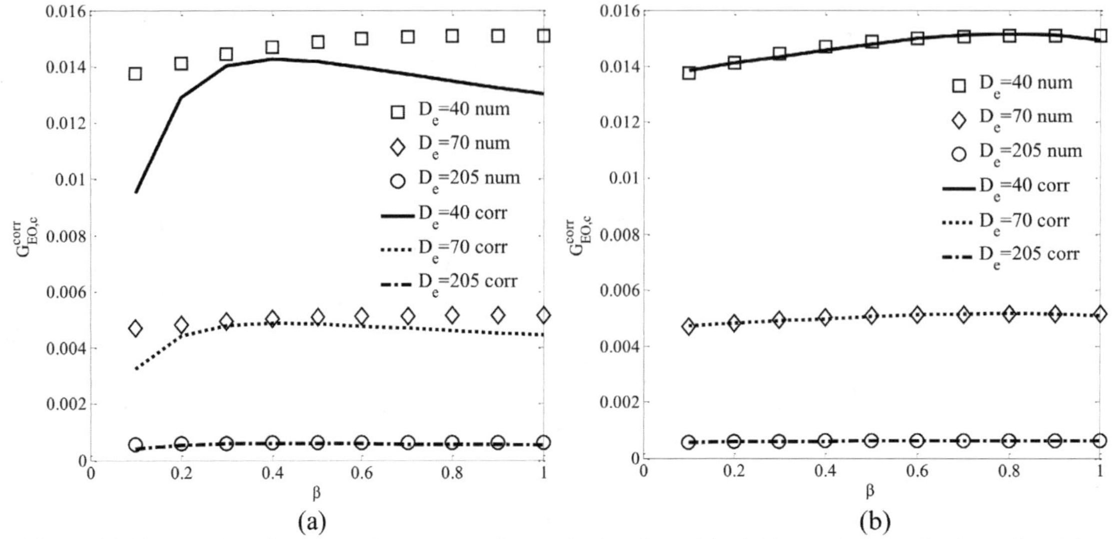

Figure 7. Comparison between the proposed correlation for critical G_{EO} and numerical results: (a) correlation proposed in [19]; (b) correlation proposed in the present study.

4. Conclusions

The dynamic behavior of electro-osmotically driven flow through triangular microchannels has been investigated. Both pure and combined electro-osmotic flows have been analyzed. The fluid has been considered isothermal and with constant properties, undergoing a laminar fully developed flow. The Poisson-Boltzmann equation and the modified Navier-Stokes equation have been solved using COMSOL Multiphysics^{TM}. Numerical results have been obtained for different values of the main parameters that characterize the behavior of electro-osmotic flows.

The results point out that both the aspect ratio β of the cross-section and the electrokinetic diameter D_e significantly influence the velocity field. The maximum volumetric flow rate strongly depends on the value of β, while D_e is crucial in the determination of the maximum pressure that can be overcome by the EOF, here represented by the dimensionless number $G_{EO,c}$.

Finally, two correlations available in the literature for the maximum volumetric flow rate and the critical G_{EO} have been tested against the numerical results. While the functional dependence of the flow rate seems to be already well understood, the dependence of the G_{EO} on the aspect ratio is still theoretically undetermined since the available functions and correction factors do not work for every geometry. In order to get $G_{EO,c}$ in microchannels characterized by triangular geometries, a new expression for the functional dependence on the aspect ratio has been derived using the numerical results. This way, the maximum volumetric flow rate and the critical G_{EO} can be estimated with an error smaller than 2% under a wide range of operative conditions. Characteristic curves of electro-osmotic pumps featuring triangular cross-sectional geometry are therefore available through simple analytical correlations that involve only three fitting coefficients.

Acknowledgments
This work has been carried out thanks to the financial support of the PRIN2009TSYPM7_001 project "Single-phase and two-phase heat transfer for microtechnologies. Heat transfer and fluid flow in microscale".

References
[1] Burgreen D and Nakache F R 1964 *J. Physical Chemistry* **68** 1084–91

[2] Rice C L and Whitehead R 1965 *J. Physical Chemistry* **69** 4017–24
[3] Sørensen T S and Koefoed J 1974 *J. Chem. Soc. Faraday Trans II* **70** 665–75
[4] Qian Y, Yang G and Bowen W R 1997 *J. Colloid Interface Sciences* **190** 55-60
[5] Mala G M, Li D and Dale J D 1997 *Int. J. Heat Mass Transfer* **40** 3079-88
[6] Mala G M, Li D, Werner C, Jacobasch H J and Ning Y B 1998 *Int. J. Heat Fluid Flow* **18** 489-96
[7] Dutta P and Beskok A 2001 *Analytical Chemistry* **73** 5097-5102
[8] Chen C H and Santiago J G 2002 *J. Microelectromechanical Systems* **11** 672-83
[9] Zhao C, Zholkovskij E, Masliyah J and Yang C J. 2008 *Colloid Interface Science* **326** 503-510
[10] Jamaati J, Niazmand H and Renksizbulut M 2010 *Int. J. Thermal Sciences* **49** 1165-1174
[11] Yang C and Li D 1997 *J. Colloid Interface Sciences* **194** 95-107
[12] Yang C, Li D and Masliyah J H 1998 *Int. J. Heat Mass Transfer* **41** 4229-49
[13] Mala G M, Yang C and Li D 1998 *Colloids Surface* A **135** 109-16
[14] Yang C and Li D 1998 *Colloids Surface A* **143** 339-53
[15] Arulanandam S and Li D 2000 *Colloid Surface* A **161** 89-102
[16] Geri M, Lorenzini M and Morini G L 2012 *Int. J. Thermal Sciences* **55** 114-21
[17] Horiuchi K, Dutta P and Richards C D 2007 *Microfluidics and Nanofluidics* **3** 347–358
[18] Hsu J P, Kao C Y, Tseng S and Chen C J 2002 *J. Colloid Interface Sciences* **248** 176-84
[19] Vocale P, Geri M, Cattani L, Morini G L and Spiga M 2013 *La Houille Blanche – Int. J. Water* **3** 42-49
[20] Vocale P, Geri M, Cattani L, Morini G L and Spiga M 2013 *Int. J. Thermal Sciences* **72** 92-101
[21] Xuan X and Li D 2005 *J. Colloid Interface Science* **289** 291-303
[22] Goswami P and Chakraborty S 2011 *Microfluidics and Nanofluidics* **11** 255–267
[23] Liao Q, Wen T Y and Zhu X 2008 *Applied Thermal Engineering* **28** 1463-70
[24] Wang C Y and Chang C C 2011 *Electrophoresis* **32** 1268–72

Journal of Physics: Conference Series **501** (2014) 012027 doi:10.1088/1742-6596/501/1/012027

Use of the μPIV technique for an indirect determination of the microchannel cross-section passage geometry

G Puccetti[1], B Pulvirenti and GL Morini
DIN – Alma Mater Studiorum Università di Bologna,
Laboratorio di Microfluidica, Via del Lazzaretto 15, Bologna 40131, I

E-mail: giacomo.puccetti2@unibo.it

Abstract. In this work the possible use of the μPIV technique for the experimental determination of the microchannel cross-section geometry has been investigated by means of a blind test in which a series of experimental measurements obtained using glass microchannels having a declared rectangular cross-section with a depth of 100 μm and width of 300 μm and a square microchannel with a 300 μm side have been compared with the direct SEM visualisation of the real cross section of the microchannels. For the μPIV measurements water is used as working fluid. The laminar fully developed 2D velocity profile has been reconstructed by moving the focal plane of the microscope objective from the bottom to the top of the microchannel. The results shown in this paper demonstrate that the real cross section geometry of the microchannel can be predicted by minimizing the difference between the theoretical and the experimental 2D velocity profiles. When the right passage geometry is determined, the average difference between the theoretical and the experimental velocity is within 4-6%.

1. Introduction

Starting from the first successful experiment made by Santiago et al. in 1998 [1], Micro Particle Image Velocimetry (μPIV) has assumed an increasing importance for the determination of the velocity profile of liquid flows through micro-passages becoming nowadays the most employed technique [2, 3, 4] for the analysis of liquid velocity field in microdevices. An updated and critical review about the progress of the μPIV technique was made by Lindken et al. [5].

Zheng and Silber-Li [6] have shown how the μPIV technique can be used to reconstruct the 2D velocity profile within a rectangular microchannel having a height of 19.1 μm and a width of 54 μm. The tested microchannel was fabricated using the soft-lithography technique on hydrophobic PDMS and then bonded on a hydrophilic glass substrate. They measured the velocity profile in 14 horizontal planes along the vertical direction of the microchannel at Re = 1.8 and Re = 3.6. They demonstrated that, except for the profiles adjacent to the upper and lower walls, the experimental data obtained are in good agreement with the theoretical profiles. The relative deviations they were able to achieve between experimental data and theoretical values in each horizontal plane were always less than 3.6%.

The high accuracy obtainable with the μPIV technique when laminar flows at low Reynolds numbers are considered has suggested the use of the μPIV as an indirect technique for the determination of the cross section of a micropassage. Silva et al. [7], starting from the 2D

[1] To whom any correspondence should be addressed.

reconstruction of the velocity field in a microchannel, have suggested a indirect method in order to determine the shape and the dimensions of a microchannel by searching iteratively the cross section which minimizes the difference between the volumetric flow rate supplied by means of a syringe pump and the volumetric flow rate yielded by the integration of μPIV velocity profiles.

Starting from the original idea of Silva et al. [7] in this work the use of the μPIV technique for the prediction of the microchannel cross-section geometry has been deeply investigated by means of a series of experimental measurements obtained using two different glass microchannels. At the end of the experimental campaign, the real cross section of the microchannels has been determined by means of destructive tests using a SEM visualisation and a comparison with the predictions obtained by using the μPIV measurements is shown.

2. Experimental Apparatus

The μPIV apparatus used in this work is composed by a pulsed Nd:YAG (λ = 532 nm) monochromatic laser (1) (*NanoPIV, Litron Lasers*): the laser emits two pulses with a duration of each pulse of the order of 5 ns and a time delay between light pulses variable from hundreds of nanoseconds to few seconds.

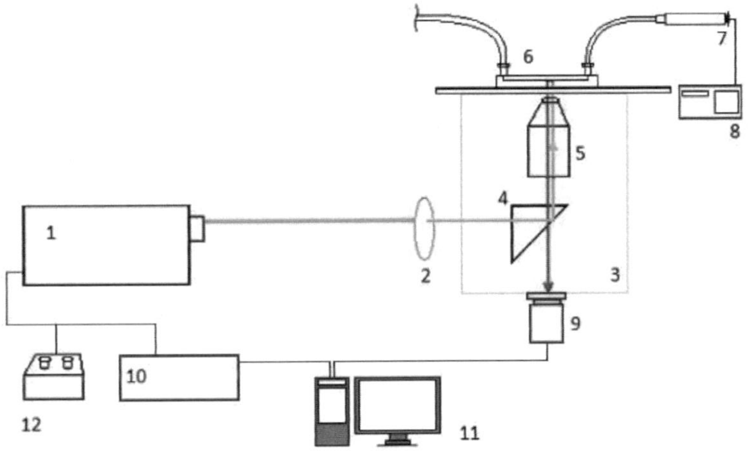

Figure 1. Lay-out of the μPIV experimental apparatus.

The illumination light is delivered to the inverse microscope (*EclipseT2000, Nikon*) (3) through beam-forming optics (2), which consists of a variety of optical elements that sufficiently modify the light so that it fill the back of the objective lens and thereby broadly illuminate the microfluidic device. The illumination light is reflected upward towards the objective lens by an antireflective coated mirror designed to reflect wavelength 532 nm and transmit 560 nm (4). An air immersion lens (5) with a numerical aperture NA = 0.75 and magnification M = 20 is used. A sensitive large-format interline-transfer CCD camera (1376 x 1040 pixels) (*Sensicam, PCO*) (9) is used to record the particle image fields. The interline transfer feature of the CCD camera allows for two image frames to be recorded back-to-back within a 500 ns time delay.

Fluorescent seeding (*Invitrogen, FluoSphere*) is dispersed in the working fluid (water). The particle tracer presents a diameter of 1.0 μm, orange coloration, a carboxylate coating and emitting a radiation wavelength of 560 nm when it is illuminated with a green light (λ = 532 nm). The seeded working fluid is contained in a syringe (*Hamilton Gastight #1010*) (7) of 10.0 ml volume. The pressure required to flow the fluid within the microchannel (6) is delivered by a syringe pump (*Harvard Apparatus PHD 4400*) (8). A *Dantec* synchronization unit (10) connected with the laser, the CCD camera and a computer (11), guarantees the tuning between the light pulse and the acquisition of the images. The right illumination intensity of the laser beam is set from a remote control (12). With this configuration the depth of correlation DOC, calculated according to the equation proposed by Lindeken et al. [5], is estimated to be about 8 μm.

2.1. Mixture preparation.

The concentration C of the tracer within the working fluid is calculated as ratio between the total volume of the seeding ($V_{p,tot}$) and the total volume of solution (V_{sol}):

$$C = \frac{V_{p,tot}}{V_{sol}} \tag{1}$$

In this case V_{sol} has been approximated with the volume of working fluid since the volume of tracer is in the order of few microliter, five order of magnitude lower than the volume of the working fluid. $V_{p,tot}$ can be calculated as:

$$V_{p,tot} = \frac{4}{3}\pi N\left(\frac{d_p}{2}\right)^3 \tag{2}$$

Herein, N is the total number of tracer particles within the working fluid and d_p is the particle diameter. The total number of tracer particle N is given by:

$$N = n_p c \tag{3}$$

where, n_p is the number of microsphere per volume of aqueous suspension (usually expressed in particles per ml) and c is the volume of aqueous suspension. The values of c, N, $V_{p,tot}$, V_{sol} and C used during the experimental tests described in this paper are given in table 1.

Table 1. Seeding characteristics and tracer concentration used in the tests.

d_p: 1.0 μm	n_p: 2.7x10^{10} particles/ml	LI[a]: 540nm	LE[b]:560nm	
c (μl)	N (#)	$V_{p,tot}$ (μl)	V_{sol} (ml)	C (ppm)
25	6.75x10^8	0.35	10	35.33

[a] LI is the tracer wavelength absorption
[b] LE is the tracer wavelength emission

2.2. Microchannels

In this work two different commercial glass microchannels (*Translume*) have been used. The first one presents a declared rectangular cross-section with a depth of 100 μm and width of 300 μm. The second one has a declared square cross-section with a side of 300 μm. The two straight microchannels have the same length (38 mm). Both microchannels are made by laser etching on a substrate of high quality fused silica and then bonded with the same material. The roughness of the inner walls is declared to be <0.1% by the manufacturer.

3. Theoretical determination of the velocity field

For the determination of the theoretical velocity field within a microchannel, let's consider a steady state, forced laminar, hydro dynamically fully developed flow in a duct with constant cross-section in the motion direction. Three assumption have been made: i) the fluid is Newtonian and its physicals properties are constants along the whole duct; ii) natural convection is negligible and the flow can be considered as isothermal; iii) the pressure gradient is constant along the flow direction. Under these conditions and considering an orthogonal Cartesian system (x,y,z), the dimensionless momentum Navier Stokes equation can be written [8]:

$$\frac{\partial^2 V}{\partial \tilde{x}^2} + \frac{\partial^2 V}{\partial \tilde{z}^2} + \tilde{P}_d = 0 \tag{4}$$

where the following dimensionless parameters are used:

$$\tilde{x}=\frac{x}{L_{ch}} \qquad \tilde{z}=\frac{z}{L_{ch}} \qquad V(\tilde{x},\tilde{z})=\frac{v(x,z)}{W_{avg}} \qquad \tilde{P}_d=\frac{D_h^2}{\mu W_{avg}}\frac{\partial p_d}{\partial \tilde{y}} \tag{5}$$

Herein, L_{ch} is the width of the channel (long side) and taking H as the height of the channel (short side) is possible to define the aspect ratio $\beta = H/L_{ch}$, $v(x,z)$ is the dimensional value of the velocity, W_{avg} is the average velocity of the working fluid on a cross section of the channel and $\partial p_d/\partial \tilde{y}$ is the pressure drop over the flow direction. The differential problem is closed by the boundary conditions, which in this case coincide with the no-slip conditions at the walls (the glass has not undergone to hydrophobic treatment).

$$\begin{gathered} V(0,\tilde{z})=V(1,\tilde{z})=0 \\ V(\tilde{x},0)=V(\tilde{x},\beta)=0 \end{gathered} \tag{6}$$

For ducts with a rectangular cross-section, the analytical solution of equation (4) with no-slip conditions at the walls is the following (Morini [9]):

$$V(\tilde{x},\tilde{z})=\sum_{n=1,odd}^{\infty}\sum_{m=1,odd}^{\infty}\upsilon_{n,m}\,sin(n\pi\tilde{x})\,sin\left(\frac{m\pi\tilde{z}}{\beta}\right) \tag{7}$$

where the series coefficients $\upsilon_{n,m}$ are defined as:

$$\upsilon_{n,m}=\frac{\pi^2}{4mn(\beta^2n^2+m^2)}\left(\sum_{j=1,odd}^{\infty}\sum_{k=1,odd}^{\infty}\frac{1}{j^2k^2(\beta^2j^2+k^2)}\right)^{-1} \tag{8}$$

For geometries different from the rectangular, equation (4) must be solved numerically. In this work the solution of equation (4) has been obtained by using FlexPDE.

In order to compare the theoretical profiles with the experimental data the dimensionless velocity must be multiplied by the average velocity which can be obtained as ratio between the volumetric flow rate imposed by means of the syringe pump (Q) and the cross-section area, (S_{cs}):

$$v(x,z)=V(\tilde{x},\tilde{z})\frac{Q}{S_{cs}}=V(\tilde{x},\tilde{z})W_{avg} \tag{9}$$

4. Experimental procedure

The first step of the experimental procedure is the determination of the upper and lower wall position of the microchannel. Due to the adherence of the tracer particles to the wall it is possible to individuate the position of the walls by checking the fluorescent particles attached on the wall varying the focal plane by means of the microscope. Moving upwards and downwards the objective focal plane, it can be found the plane where the particles appears as fixed. When these particles, visible as stable bright points, are detected it means that a microchannel wall is actually on focus. Starting from the lower wall assumed as a reference point for the reconstruction of the real cross section of the channel, the objective focal plan is shifted upward in order to determine the position of upper wall of the channel. The displacement found from the movement of the focal plane between the lower and the upper wall is generally different from the real depth of the microchannel due to the refraction index of the working fluid (water, *i.e.* n_w = 1.333). As stated by Lindken et al. [5], to achieve a proper experimental investigation, the operating parameters, such as the time interval Δt between the image pair, the size of interrogation area and the flow rate, must be consistent with each other. The settings of these parameters has been done in order to satisfy two conditions: i) the displacement of the fluorescent particles at the maximum velocity does not exceed ¼ of the interrogation cell width; ii) at least five particle pairs must be present into each interrogation cell. The setting of the main operative parameters is shown in table 2.

Table 2. Value of the parameters set.

Microchannel	Declared size of the microchannel (μm x μm)	Time interval between the image pair (μs)	Size of interrogation area (overlap) (pixel x pixel)	Flow rate (m^3/s)
#1	100x300	50	64x64 (50%)	5.6 x10^{-10}
#2	300x300	50	64x64 (50%)	2.22 x10^{-9}

The images were acquired in grayscale and then filtered to obtain an high contrast value between the tracer particle and the background, as seen in figure 2. A series of 500 images, for each different plan, has been elaborated in order to avoid the influence of the Brownian motion and other unwanted stochastic micro-effects on the reconstruction of the velocity profiles. For each plane, a vector map containing 40 column of 28 velocity vectors is obtained applying an average cross-correlation among the 500 pairs of images.

Figure 2. Image of the fluorescent particles after the filtering

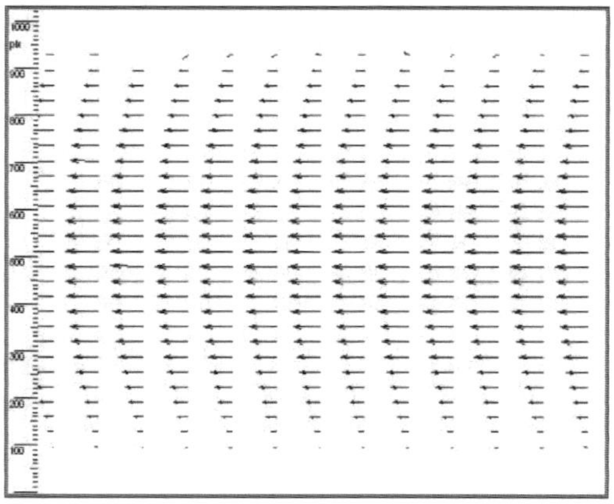

Figure 3. Vector map of the flow field in an horizontal plan

To perform the matching between the experimental data and the theoretical velocity profiles determined by solving equation (4), the velocity, expressed in pixels per unit time, determined by means of μPIV is converted in meters per second in the following way:

$$u(x,z) = \frac{s_{pix} L_{ch}}{\Delta t L_{pix}} \tag{10}$$

where $u(x,z)$ is the dimensional velocity, s_{pix} is the displacement in pixels of the bright point between the two images, Δt is the time interval between the two images, L_{ch} is the width of the channel in meters and L_{pix} is the width of the channel in pixels. The velocity uncertainty has been estimated to be equal to 3% by using the following equation:

$$\frac{\Delta u}{u} = \left(\left(\frac{\Delta s_{pix}}{s_{pix}} \right)^2 + \left(\frac{\Delta L_{ch}}{L_{ch}} \right)^2 + \left(\frac{\Delta L_{pix}}{L_{pix}} \right)^2 \right)^{1/2} \tag{11}$$

where Δs_{pix} has been assumed of 0.1 pixel [7], ΔL_{ch} has been taken of 1 μm and ΔL_{pix} has been estimated to be of the order of 13 pixels in these experiments. The relative error on the time delay Δt has been neglected in equation (11) because the uncertainty on it is several orders lower than the value of the same.

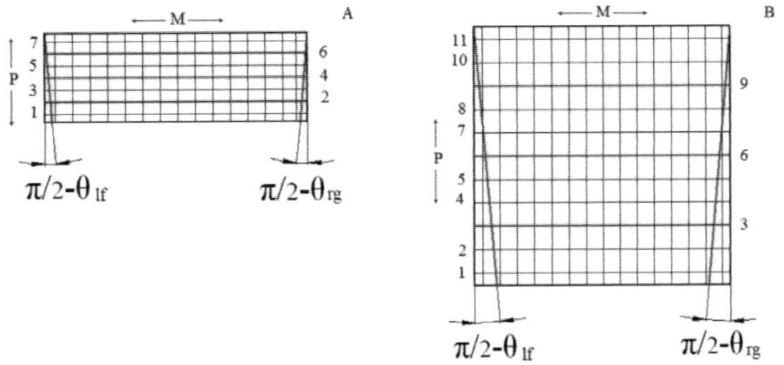

Figure 4. Used grid for the reconstruction of the 2D velocity profile within microchannel #1 (A) and microchannel #2 (B).

5. Experimental results and discussion

As shown in figure 4, for the geometry reconstruction of the 2D velocity profile, P different planes have been used (7 for the rectangular channel and 11 for the square one). These planes have been acquired starting from the central plane (p=4 for the channel #1 and p=6 for the channel #2) with an increasing distance of ± 13.65 μm for the channel #1 and 27 μm for the channel #2. The displacements have to be considered as the real distance between the different planes, *i.e.* the refraction index has already been taken into account. The vertical displacement of the focus plane has an accuracy of 1 μm that corresponds in an accuracy of 1.33 μm on the vertical localization of the plane. For each plane, a vector map containing N_v = 40 columns of M velocity vectors is made (see figure 3). Since the flow is in steady-state conditions, each column of velocity vectors contains the same information. It is possible to select one column of vectors among the N_v columns associated to a fixed plane *p*, choosing the one who results from the optimization of the normalized root mean square error over the plane:

$$e_p = \min\left(\sqrt{\frac{1}{M}\sum_{m=1}^{M}\left(\frac{u^n_{p,m} - v_{p,m}}{v_{p,m}}\right)^2}\right)_n \tag{12}$$

Herein $u^n_{p,m}$ is the experimental velocity vector in the *m-th* position of the *n*-th column on the plane *p*, and $v_{p,m}$ is the value of the theoretical velocity (calculated by using equation (7) for a rectangular cross-section) in the *m*-th position of the plane *p*.

With this method, for each plane M velocity vectors have been determined: in particular, 28 velocity vectors both for the rectangular and square geometry have been associated to the interrogation cells used to divide the field-of-view. In this way a 2D velocity field made by PxM points (7x28 for microchannel #1 and 11x28 for microchannel #2) has been determined.

Varying the aspect ratio of the rectangular microchannel's cross-section from which the theoretical curves have been obtained the relative error between the experimental and numerical velocity changes. Therefore it can be deduced that the relative difference e_p defined by equation (12) can be considered as a function of the geometric shape of the cross-section [7].

In figure 5 the comparison between the experimental values of velocity determined by μPIV measurements and the theoretical profiles determined by using equation (7) and equation (9) for microchannel #1 (a,b,c) and #2 (d,e,f) is shown. The experimental data are shown in figure 5 with the corresponding error bars for three different planes: plane 4 and plane 6 correspond to the central plane of the microchannel #1 and #2 respectively (see figure 4 and table 3 and 4). Plane 6 and 2 are far +27 μm and - 27 μm respectively from the central plane of microchannel #1. Plane 9 and 3 are far +81 μm and - 81 μm respectively from the central plane of the microchannel #2.

The experimental data are in good agreement with theoretical profiles for the upper planes, both for microchannel #1 and #2. On the contrary, when it gets close to the bottom wall, the matching between the theoretical and the experimental values became worse and the relative difference e_p grows up, as it has been reported in the first column of table 3 and table 4 for microchannel #1 and #2 respectively.

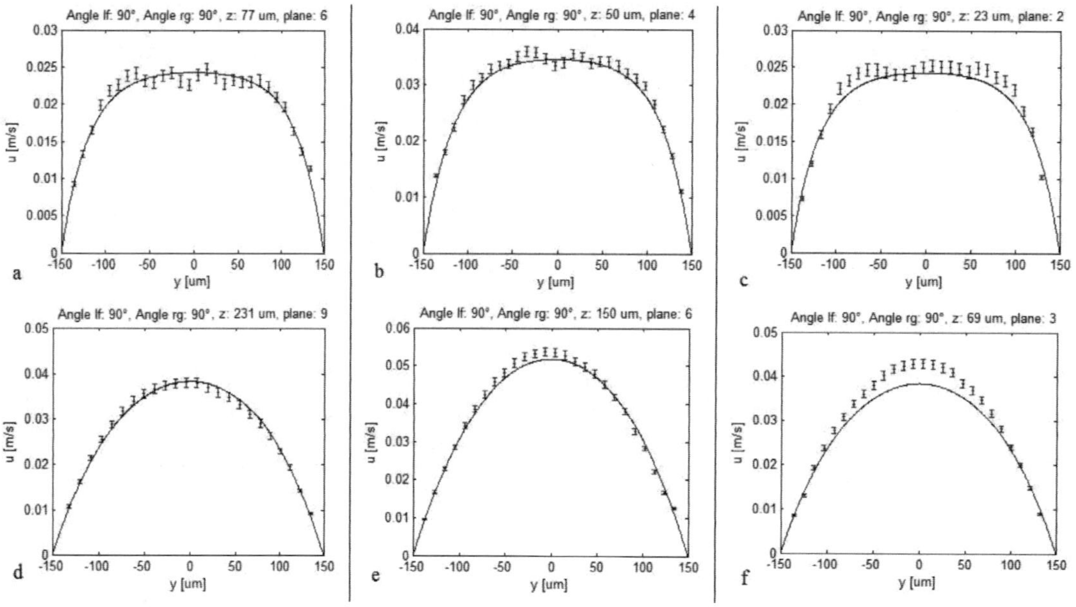

Figure 5. Matching between experimental data and theoretical curves for the microchannel #1 (a,b,c), and for the microchannel #2 (d,e,f,)

It is possible to observe that the relative difference e_p varies for microchannel #1 from a value of 5.0% on the central plane up to 11.2% on the plane close to the upper wall and 15.4% on the plane close to the bottom wall, reaching the minimum value of 4.6% on the planes 6 and 5. For microchannel #2 the relative difference is in between 3 and 6% in the central part of the microchannel and it grows up to 9.6% and 15.0% close to the upper and bottom wall respectively. This result can be explained by considering the following aspects:

1. the μPIV technique gives accurate velocity profiles when the focal plan drops in the central part of a microchannel, as evidenced by Zheng and Silber-Li [6]; on the contrary, when the focal plane is moved close to the bottom wall of the channel the velocity determination becomes more difficult due to the combined effect linked to the influence of the larger velocity gradients on the seeding particles and the reduced quantity of light collected on the focal plane when the water thickness increases.
2. The real geometry of the microchannel is not exactly rectangular as declared by the manufacturer but the width of the channel changes by decreasing from the upper wall to the bottom wall of the microchannel.

In order to exclude the first hypothesis, the measurements have been repeated by increasing the number of particles and the quantity of light collected on the focal plane by increasing the laser power but the experimental data have confirmed the previous results. With the aim to verify the second hypothesis, as suggested by Silva et al. [7], the μPIV experimental data have been used to predict the cross section of a microchannel by minimizing the value assumed by the relative difference e_p. In fact, e_p can be considered as a function of the cross-section geometry of the microchannel; by changing the microchannel geometry one can modify the theoretical velocity profile by solving numerically equation (4) and the relative difference e_p varies as consequence for a fixed set of experimental data. The value of the relative difference e_p has been determined by considering the cross section as trapezoidal with a value of the apex angles between the upper wall and the inclined side equal to θ_{lf} and θ_{rg} on left and right side respectively (see figure 4). The values of e_p obtained for each plane by varying θ_{lf} and θ_{rg} from 90° to 84° are reported in table 3 and table 4 for microchannel #1 and #2 respectively.

Table 3. Relative error as a function of the cross-section of microchannel #1 (300x100 μm)

Plane, p [#]	z [μm]	Relative error e_p [%] for different θ_{lf} and θ_{rg} [°]						
		90/90	88/88	87/87	87/86	86/86	85/85	84/84
7	90.9	11.2	11.4	11.5	11.6	11.8	12.0	12.3
6	77.3	4.6	4.4	4.6	4.6	4.8	5.2	5.7
5	63.6	4.6	4.3	4.2	3.9	4.1	4.7	5.4
4	50	5.0	3.4	2.6	2.3	2.5	4.4	5.1
3	36.4	4.7	4.5	2.8	2.3	2.8	3.1	4.2
2	22.7	6.1	5.0	4.1	3.9	3.7	4.1	4.9
1	9.1	15.4	11.0	11.3	11.3	10.8	10.8	10.0
e_{avg} [%]		7.4	6.3	5.9	5.7	5.8	6.2	6.9

Table 4. Relative error as a function of the cross-section of microchannel #2 (300x300 μm)

Plane, p [#]	z [μm]	Relative error e_p [%] for different θ_{lf} and θ_{rg} [°]					
		90/90	89/89	88/88	87/87	86/86	85/85
10	258	9.6	10.4	10.7	11.5	12.3	13.4
9	231	2.8	3.5	4.3	5.9	7.9	9.1
8	204	2.3	2.0	2.7	4.4	6.5	8.4
7	177	2.8	1.0	1.7	3.5	5.7	6.5
6	150	4.2	1.8	1.2	2.6	5.4	6.3
5	123	4.2	2.9	1.1	1.9	4.2	6.3
4	96	5.9	3.9	2.3	1.2	3.2	5.9
3	69	7.8	5.7	4.0	2.1	1.6	4.0
2	42	15.0	12.2	9.4	6.9	5.3	3.3
e_{avg} [%]		6.1	4.8	4.2	4.4	5.8	7.0

The values quoted in table 3 and 4 demonstrate that for each microchannel there exists a specific combination of θ_{lf} and θ_{rg} which minimize the average value of the relative difference between the numerical and experimental velocity field defined as follows:

$$e_{avg} = \frac{\sum_{p=1}^{P} e_p}{P} \tag{13}$$

where P is the number of planes used for the determination of the velocity profiles through the microchannel (7 for microchannel #1, 9 for microchannel #2, see figure 4).

In table 4, the results obtained for the planes 1 and 11 have not been reported because it was felt they have not physical substance, therefore the average error e_{avg} on the section has been calculated over 9 floors instead of 11.

By observing the data quoted in table 3 and 4 it is evident that the experimental data seem to indicate that the microchannels have a trapezoidal cross section characterized by a decreasing width from the upper wall to the bottom wall. In fact, it is possible to observe that the cross section geometry which minimize the average relative difference between theoretical and experimental values is the trapezoidal geometry with θ_{lf} / θ_{rg} equal to 87°/86° for microchannel #1 and 88°/88° for microchannel #2. By selecting for the microchannels these two cross sections the

relative difference e_p decreases from 15.4% down to 11.3% on the plane close to the bottom wall of the microchannel #1 and, more important, this value becomes similar to the value obtained on the plane close to the upper wall. The same result is obtained for microchannel #2 where e_p decreases from 15.0% to 9.4% on the plane close to the bottom wall.
As an example, in figure 6, the comparison between the experimental data obtained for microchannel #2 and the theoretical velocity profiles obtained by imposing the angles θ_{lf} / θ_{rg} equal to 88°/88° which minimize the average value of the relative difference e_{avg} is shown on the planes 3 (close to the bottom wall), 6 (central) and 9 (close to the upper wall). It is evident that in this case the agreement between theoretical and experimental values is strongly increased and the difference is on each plane within the experimental uncertainty.

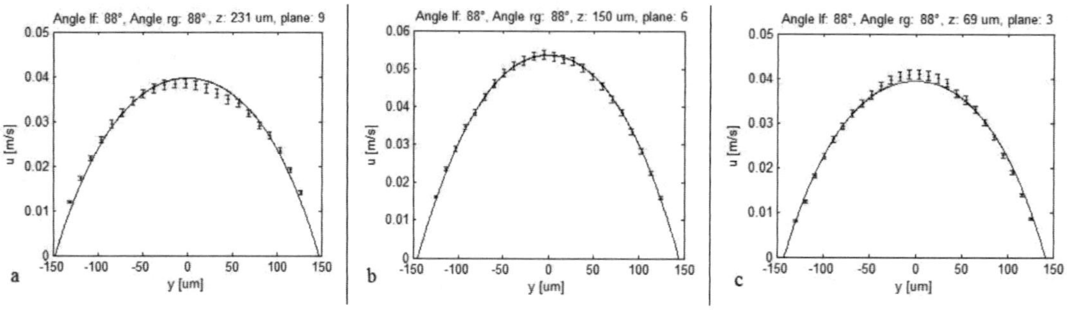

Figure 6. Matching between experimental data and theoretical velocity profiles for microchannel #2 on plane 9, 6 and 3 for θ_{lf} /θ_{rg} equal to 88°/88°.

In order to verify the real cross section of the tested microchannels, the glass microchannel #2 has been submitted to destructive tests by means of which several samples, prepared with an LKB 7800 knife maker and then coated with a golden layer, have been obtained to visualize the real channel cross section by means of a SEM microscope (*Philips SEM EM 515*). In figure 7 a SEM image of the cross sectional area of microchannel #2 with a magnification of 14x is shown. In figure 8 the same image of the cross-section has been taken with a magnification of 212x. As it is possible to see from figure 8, the shape of the cross-section is not perfectly squared, as declared by the manufacturer, but presents a trapezoidal shape in agreement with the indications of the μPIV measurements. In addition figure 8 highlights that the upper side of the microchannel presents a curvature which was not taken into account in this analysis and it can be responsible of the residual deviation of the experimental values of velocity from the theoretical ones. In table 5 are summarized the geometrical features of the microchannel #2 in terms of ratio between the upper and lower wall (L_u and L_b) and the depth (H) of the microchannel and in terms of apex angles.

Table 5. Comparison between measured and predicted geometrical features of microchannel #2.

Microchannel	L_u /H [-]		L_b /H [-]		θ_{rg}/θ_{lf}	
	Experimental	*Predicted*	*Experimental*	*Predicted*	*Experimental*	*Predicted*
#2	1	1	0,91	0,93	87,5°/87,5°	88°/88°

By observing the data reported in table 5 it is possible to conclude that the real data confirms the predictions obtained by using the μPIV velocity profiles and in particular the SEM measurements have confirmed that the real cross section of the microchannel #2 is slightly trapezoidal with an apex angle of 88° in disagreement with the indication of the manufacturer. These results confirm that μPIV measurements can be used as non-destructive test for transparent channels in order to predict the real geometry of a unknown micro-passage.

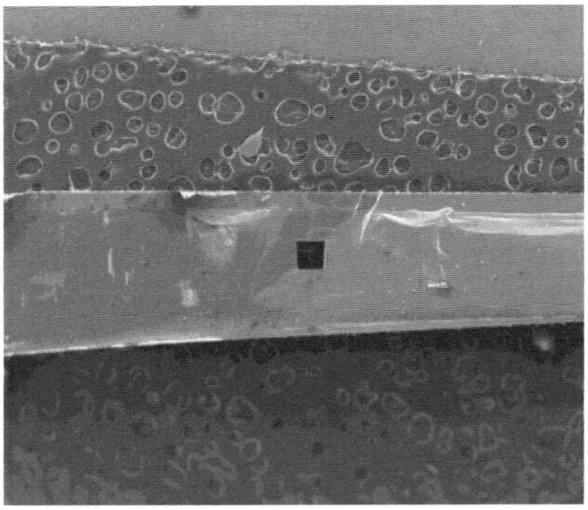

Figure 7. SEM image with magnitude x14 of the channel #2 cross-section

Figure 8. SEM image with magnitude x212 of the channel #2 cross-section

6. Conclusions

In this work the μPIV technique has been used in order to determine the shape of a micropassage starting from the reconstruction of the 2D velocity field within the channel. It has been shown how the average difference between the theoretical and experimental velocity profiles is a function of the cross section geometry and the minimisation of this parameter, obtained by changing the shape of the micropassage, allows the individuation of the geometrical characteristics of the cross section. The results obtained by using this method have been compared with the indications, obtained by means of destructive tests, based on a SEM visualization of the real cross section of the tested microchannels: the SEM visualizations confirmed that the real cross section of the microchannel #2 is slightly trapezoidal with an apex angle of 88° in disagreement with the indication of the manufacturer and in agreement with the predictions obtained by using the μPIV measurements and the procedure described in this paper. These results confirm that μPIV measurements can be employed to predict the real geometry of a unknown micro-passage avoiding the use of destructive test methods.

Acknowledgments

The research leading to these results has received funding from the European Community's Seventh Framework Programme (ITN-FP7/2007-2013) under grant agreement n° 215504 and the financial support of the Italian MIUR in the PRIN 2009 framework. The authors are grateful to Dr. Maria Roberta Randi for helpful support on acquiring SEM images.

References

[1] Santiago J G, Wereley S T, Meinhart C D, Beebe D J and Adrian R J 1998 *Exp. Fluids* **25** 316
[2] Meinhart C D, Wereley S T and Santiago J G 1999 *Exp. Fluids* **27** 414
[3] Wereley S T, Gui L and Meinhart C D 2002 *AIAA J.* **40** 1047
[4] Sinton D 2004 *Microfluidics and Nanofluidics* **1** 2
[5] Lindken R, Rossi M, Große S and Westerweel J 2009 *Lab. on a Chip* **9** 2551
[6] Zheng X and Silber-Li Z 2008 *Exp. Fluids* **44** 951
[7] Silva G, Leal N and Semiao V 2009 *Chem. Eng. Res. Des.* **87** 298
[8] Morini G L and Spiga M 1998 *Microscale Therm. Eng.* **2** 273
[9] Morini G L 2000 *Int. J. Heat and Mass Transf.* **43** 741

Dropwise condensation on superhydrophobic nanostructured surfaces: literature review and experimental analysis

A Bisetto[1,*], D Torresin[2], M K Tiwari[3], D Del Col[1] and D Poulikakos[3]

[1]Dipartimento di Ingegneria Industriale, Università di Padova, Via Venezia 1, 35131 Padova, Italy
[2]ABB Corporate Research, CH-5405 Baden - Dättwil, Switzerland
[3]Laboratory of Thermodynamics in Emerging Technologies, Mechanical and Process Engineering Department, ETH Zurich, 8092 Zurich, Switzerland

E-mail: alberto.bisetto@dii.unipd.it

Abstract. It is well established that the dropwise condensation (DWC) mode can lead up to significant enhancement in heat transfer coefficients as compared to the filmwise mode (FWC). Typically, hydrophobic surfaces are expected to promote DWC, while hydrophilic ones induce FWC. To this end, superhydrophobic surfaces, where a combination of low surface energy and surface texturing is used to enhance the hydrophobicity, have recently been proposed as a promising approach to promote dropwise condensation. An attractive feature of using superhydrophobic surfaces is to facilitate easy roll-off of the droplets as they form during condensation, thus leading to a significant improvement in the heat transfer associated with the condensation process.

High droplet mobility can be obtained acting on the surface chemistry, decreasing the surface energy, and on the surface structure, obtaining a micro- or nano- superficial roughness. The first part of this paper will present a literature review of the most relevant works about DWC on superhydrophobic nanotextured substrates, with particular attention on the fabrication processes. In the second part, experimental data about DWC on superhydrophobic nanotextured samples will be analyzed. Particular attention will be paid to the effect of vapour velocity on the heat transfer. Results clearly highlight the excellent potential of nanostructured surfaces for application in flow condensation applications. However, they highlight the need to perform flow condensation experiments at realistic high temperature and saturation conditions in order to evaluate the efficacy of superhydrophobic surfaces for practically relevant pure vapor condensation applications.

1. Introduction

Vapor condensation is an important phenomenon in nature with direct applications to many fields ranging from the chemical to the power industries. Thus, enhancing thermal transport during condensation is a subject which has received a lot of interest in the literature, being at the center of numerous scientific investigations.

It is well established that the dropwise condensation (DWC) mode can lead up to several hundred percent enhancement in heat transfer coefficients as compared to filmwise condensation (FWC), as firstly underlined by Schmidt et al. [1], who recognized that during DWC heat transfer coefficients are 5 – 7 times higher than those obtained with filmwise mode, and more recently by Rose [2]. The main

* To whom any correspondence should be addressed.

mechanism responsible for heat transfer increase in dropwise condensation over filmwise one is related to droplets mobility. The sweeping and renewal mechanism present in the droplet growth process during DWC directly leads to an augment of both heat and mass transfer coefficients. Moreover, the absence of a liquid layer adjacent to the wall avoids the introduction of an additional thermal resistance which adversely affects the thermal and mass transport.

DWC occurs on surfaces which are not strongly wetted by the liquid, thus avoiding drops spreading out over them. Surfaces with strong hydrophobicity are believed to allow effective DWC, due to the promotion of the drop movement and departure and consequently accelerating surface renewal, allowing small drops to form. Thus, water-repellent (hydrophobic) surfaces should promote dropwise condensation, while water-attracted (hydrophilic) ones should promote FWC [3].

To this end, superhydrophobic surfaces, where a combination of low surface energy and surface texturing is used to enhance the water-repellency, have recently been proposed as a promising approach to promote dropwise condensation [4-7]. Surface wettability is defined by the contact angles of a drop sitting over it. As it can be seen in figure 1, taken from [8], for a static drop the equilibrium contact angle Θ_q is taken as a reference, while for moving drops the advancing and receding contact angles Θ_d and Θ_e are considered. The difference among the last two gives the contact angle hysteresis. Superhydrophobic surfaces present high contact angles, greater than 150°, and low water contact angle hysteresis. These surfaces can be produced by combining two factors: hierarchical roughness scale and hydrophobic coating. These two elements allow water droplets over the surface to easily roll-off, thus improving heat transfer associated with the condensation process [9-13].

When referring to humid air condensing over a superhydrophobic micro-/nano- rough surface, droplets mobility can be promoted by air trapped between the droplets and the texture of the surface, i.e. by promoting the so-called Cassie state of a droplet on a superhydrophobic surface [14], schematically shown in figure 2 [10]. On the contrary, during condensation of a pure vapor it is assumed that the vapor trapped between surface textures condenses in contact with the cold wall. If steam condenses both on the top and between surface textures, droplets sitting on the top of the posts in a Cassie state can coalescence with water between the textures transiting to a Wenzel state, which is characterized by drops penetrating the asperities of the rough surface, as it can be seen in figure 2 [10]. Water droplets which are in a Wenzel wetting regime are strongly pinned on the superhydrophobic substrate, thus they present high adhesion and low mobility. The increase in droplet adhesion is strongly associated with an increase in the droplet contact angle hysteresis. The Wenzel state with its high hysteresis is especially favored by a reduction in the liquid surface tension [15-16] and an increase in the ambient humidity [17-18].

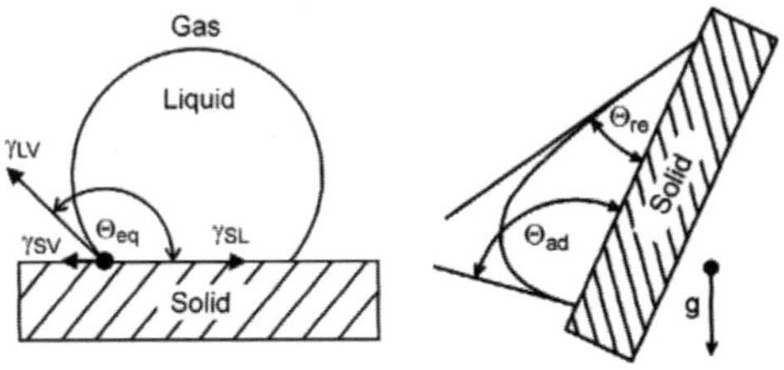

Figure 1. Static and dynamic contact angles over a surface [8].

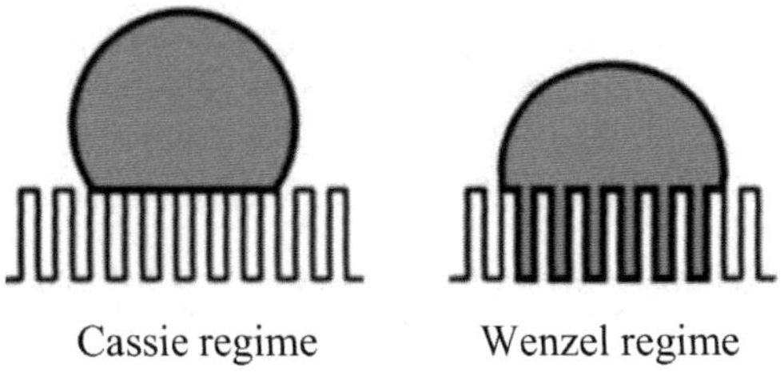

Figure 2. Wetting regimes over a superhydrophobic rough surface [10].

Over the past few decades, considerable attention has been paid toward the development of suitable DWC promoters. The method most employed so far to produce hydrophobic surfaces is to coat the metallic substrate with a thin layer of a material with small surface energy, such as organic substances, polymers, and noble metals. In the recent years, thanks to the developments in the material science, new techniques have been discovered which allow constructing superhydrophobic surfaces by mimicking the surface texture of the lotus leaves, such as micromachining, micro-contact-printing, chemical etching in aqueous solutions and deep radiative ion etching. Nano-engineered superhydrophobic surfaces appear to be a viable and promising solution to promote dropwise condensation.

2. Literature review

In this section a short review of the most relevant works about DWC on superhydrophobic surfaces is presented.

Sarkar et al. [19] demonstrated superhydrophobicity on micropatterned aluminum surfaces obtained by chemical etching with hydrochloric acid (HCl). After the etching treatment the surfaces were coated with an ultrathin rf-sputtered Teflon film using Ar plasma in an inductively coupled plasma reactor. The optimum etching time to obtain high superhydrophobic properties on the surface was 2.5 minutes (contact angle up to 164° and contact angle hysteresis down to 2.5°). More recently, the same authors [20] investigated wetting characteristics of micro-nanorough substrates of aluminum chemically etched using dilute HCl, comparing them by depositing hydrocarbon and fluorinated-hydrocarbon coatings via plasma enhanced chemical vapor deposition. The authors found a contact angle equal to 135° for the hydrocarbon coated surface, and equal to 165° for the fluorinated-hydrocarbon coated one.

A particular treatment to vary the surface wettability is the one proposed in [21], where the creation of superhydrophobic properties on aluminum alloy surfaces treated with a simple one-step process is reported. This process incorporatres low surface energy compounds in the solution used for the creation of surface micro-roughness. Clean Al coupons were superhydrophobized by dipping them in beakers containing a mixture of sodium hydroxide (NaOH) and fluoroalkyl-silane (FAS-17) at different FAS-17 to NaOH ratios. The authors verified that the formation of a rough microstructured surface, in combination with a modified surface chemistry, leads to the creation of superhydrophobic properties. In particular with 1M NaOH and 400 mM FAS-17 the authors obtained a contact angle equal to 162° and a contact angle hysteresis equal to 4°. An alternative approach is the one proposed in [22], where a method for fabricating superhydrophobic aluminum surfaces by chemical etching without hydrophobic treatment is developed. Washed aluminum substrates were treated with 1 mol L^{-1} NaOH aqueous solution for 10 min at room temperature, and immersed in a mixed acid solution of hydrochloric (36 wt%) and acetic acids (CH_3COOH - 99.6 wt%) at the desired ratio of volumes for the desired time. The authors found that water contact angle increases with increasing etching time, until a critical value equal to 5 min, and with increasing acetic acid concentration, until a critical value equal to 1mL.

An innovative method to render aluminum substrates superhydrophobic was proposed in [23]. The authors treated the samples by consecutive immersion in boiling water, to roughen them, and Teflon sputtering, to create the hydrophobic coat. To achieve different roughness levels, the duration of the boiling process was varied. They demonstrated that high superhydrophobic properties (contact angle equal to 164° and contact angle hysteresis equal to 4°) can be obtained after 5 minutes immersion in boiling water.

Dietz et al. [6] reported a study of the dynamic condensation on a superhydrophobic sample characterized by cupric hydroxide ($Cu(OH)_2$) nanostructures coated with 1-hexadecanethiol. The $Cu(OH)_2$ nanostructures were fabricated by sputtering 6 micrometers of copper on a silicon substrate. The substrate was then immersed for 30 minutes in a 50 mL solution of 2.5 M sodium hydroxide (50% weight by weight) and 0.09 M of ammonium persulfate previously stirred for 15 hours. Thus, the $Cu(OH)_2$ structures self-assembled. Next, 30 nm of gold was sputtered on the $Cu(OH)_2$ structures followed by a 1 s immersion in 1-hexadecanethiol to reduce the surface energy. Droplet departure frequency was investigated on the superhydrophobic surface and on a nonstructered hydrophobic surface, finding that the droplets tend to depart the textured surface at reduced diameter. Since droplets with diameter less than 10 μm provide the most significant contribution to the heat transfer during dropwise condensation and the formation of new droplets occurs only once large drops have departed from the surface, the authors concluded that the larger surface renewal frequency of the superhydrophobic surface leads to an increase in the heat transfer coefficient.

In [14] wetting modes of water droplets on a microtextured superhydrophobic surface are analyzed. Microtextures were created by ultraviolet microreplication against a nikel master structure. The authors showed that an irreversible transition from the Cassie to the Wenzel wetting regime can be observed if steam condenses inside the surface texture or if DWC is promoted at high external pressure. The authors underlined that if textures are filled with water the surface dramatically losses his water-repellent properties, thus transition between the two regimes should be properly avoided by accurately designing the surface structure.

Lan at al. [24] investigated DWC on a mirror-polished (SAM-2) and a nanostructured (SAM-1) copper substrate prepared with Self-Assembled Monolayers (SAM) coatings on n-octadecyl mercaptan. SAM-1 surface was first immersed in a 30% hydrogen peroxide (H_2O_2) solution for 3.5 h, during which an oxide layer formed on the copper substrate. Then it was immersed in a 2.5 mmol L^{-1} solution of n-octadecyl mercaptan in ethanol for 1 h. Thus, a SAM film of n-octadecylmercaptan formed on the surface. SAM-2 was similarly prepared without oxidation and etching process in H_2O_2 solution. Figure 3 shows the trends of Lan et al. [24] heat transfer data, comparing them against values predicted with the Nusselt theory for filmwise condensation. The authors found that both SAM-1 and SAM-2 surfaces increase heat transfer with respect to the predicted values for FWC, however, although SAM-1 presents better superhydrophobic performances than SAM-2 (i.e. higher contact angles), the authors reported underperformances of the superhydrophobic oxidized surface as compared to the mirror polished one. The nanostructured substrate does not improve DWC heat transfer performance as expected by the higher contact angle, because during pure steam condensation process condensate fills the cavities of SAM-1 surface, thus implying that the DWC takes place on a composite solid-condensate-liquid surface, compromising SAM-1 performances. Consequently, the DWC heat-transfer coefficients of SAM-1 surface are lower than those of SAM-2.

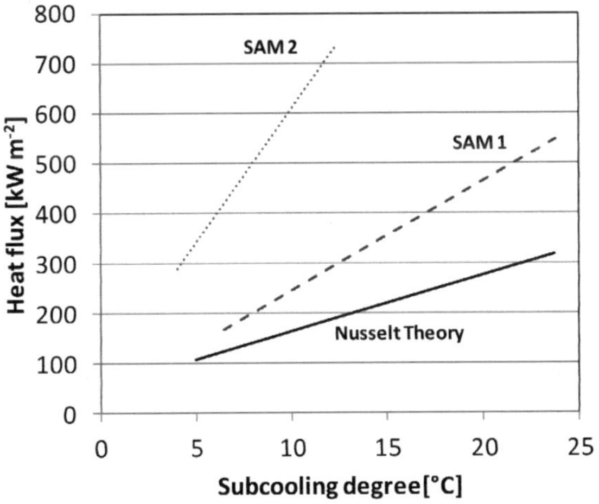

Figure 3. Heat flux trends from [24].

Chen et al. [4] analyzed continuous DWC on a two-tier roughness superhydrophobic surface characterized by carbon nanotubes (CNTs) deposited on micromachined posts, mimicking the structure of the lotus leaves. Squarely positioned pillars were formed by deep reactive ion etching on a Si surface, which was then coated with a thin layer of chromium (100 nm) and then nickel (20 nm) as catalyst, before CNTs were grown by plasma enhanced chemical vapor deposition. The substrate was finally hydrophobicized either by a 10 nm layer of parylene C coating or by a 10 nm layer of gold coated with a monolayer of 1-hexadecanethiol. Comparing experimental results obtained on the two-tier surface with those obtained on a one-tier microtextured one, the authors found increased wetting-repellency performances. In [25] an analysis of a two-tier superhydrophobic surface (similar to the one presented in [4]) is performed. The authors found that during pure vapor condensation process a thick condensate film was formed on the textured surface, dramatically decreasing water mobility.

Autonomous removal of condensate drops on a two-tier roughness superhydrophobic surface (similar to the one proposed in [4]) is reported in [5]. Comparing experimental results obtained during DWC on a horizontally oriented hydrophobic substrate with those obtained in the superhydrophobic one the authors found that, while on the former condensate drops continued to grow and to accumulate, on the second droplets were autonomously removed without the action of any external force. The authors found that self-propelled jumping motion, at a speed as high as 1 m s^{-1}, of condensate drops can be obtained with DWC on superhydrophobic surfaces, leading to autonomous removal of dropwise condensate which has the potential to enhance condensation heat transfer (because of increased droplets mobility).

Varanasi et al. [12] attempted to spatially control heterogeneous nucleation of water by modifying the local intrinsic wettability of a surface. They prepared a textured surface consisting of an array of hydrophobic posts with hydrophilic tops. The hydrophilic regions were made up of the native oxide on a Si wafer, while the hydrophobic ones were modified with fluorosilane. Textures were fabricated via micro contact printing using a prefabricated stamp. The authors demonstrated that nucleation and subsequent growth of droplets occurs preferentially on the hydrophilic post tops of the hybrid surface when compared to the random nucleation of droplets on an identically textured superhydrophobic surface with uniform wettability.

Ko et al. [26] fabricated carbon fibers (CFs) network surfaces with high aspect-ratio nanostructures having superhydrophobic and water-repellent properties. To fabricate superhydrophobic nanostructured surfaces, oxygen plasma etching and coating with hydrophobic material were applied to CFs made of carbonized polyacrylonitrile supported by carbon patches that are mainly composed of carbon powder. Oxygen plasma etching was performed to form the nanostructures on the CFs, and hydrophobic material coating was accomplished with radiofrequency plasma-enhanced chemical vapor deposition system. Aspect ratios were controlled by the oxygen plasma treatment duration. The authors found that, on the superhydrophobic CFs networks with high aspect-ratio nanostructures, a nucleated droplet can maintain its spherical shape and has difficulty to merge with other drops that nucleate in the structure spaces. This phenomena decreases the possibility of transition from the Cassie to Wenzel regime under super-saturated vapor conditions. Thus, during pure steam dropwise condensation on high aspect-ratio surfaces, droplets maintain their high mobility and heat transfer performances are not decreased. This work underlines that micro-nano hierarchical (multiscale) roughness is a key structural factor for sustaining condensed drops in a Cassie state on a superhydrophobic surface during pure steam condensation.

Similarly, in [27] DWC on a multiscale nanograssed micropyramid architecture surface is analyzed. It allows to have global superhydrophobicity as well as locally wettable nucleation sites. Nanograssed micropyramid arrays were fabricated using a combined anisotropic wet-etching and deep reactive ion etching (DRIE) process. After fabricating silicon pyramid microstructures by tetramethylammonium hydroxide etching, the authors performed a modified DRIE process for nanograss coating. Vertically-aligned uniform nanopillar arrays (i.e. nanograss) can be produced by regulating the alternating etching and passivation cycles. To generate relatively smooth patches into the nanograss coating, the authors intentionally over-etched the nanopillar tops in each etching cycle so as to reduce the height of the nanopillars. As a result, the vertical walls underneath the base of the micropyramids were exposed in smooth morphology while the inclined sidewalls and the spaces between the micropyramids were uniformly covered by nanograss. The presence of wettable patches facilitates drop growth and preferential coalescence whereas the global superhydrophobicity promotes spontaneous drop departure without pinning to the substrate. Moreover, synergistic co-operation between the hierarchical roughness structures contributes directly to a continuous process of nucleation, coalescence, departure, and re-nucleation, enabling sustained dropwise condensation over prolonged periods. The authors showed that by harnessing heterogeneous wettability and hierarchical roughness features in multiscale structures, it is possible to enhance both drop nucleation and drop departure on a condensing surface.

A new approach was shown in [28], where the authors demonstrated that pinning of condensate droplets can be drastically reduced by designing hierarchical micro-nanoscale texture on a surface and by impregnating it with an appropriate lubricant, immiscible with the condensed liquid. The condensing surfaces were well-defined cubical microposts of silicon fabricated by standard photolithography, hydrophobized by a low-energy silane coated over them using solution based deposition. Two lubricants, a fluorinated oil and a ionic liquid, were impregnated into the textured surfaces by dipping them into a bath of lubricant and then withdrawing at controlled rates using a dip coater. The authors demonstrated that, carefully selecting an immiscible liquid with desirable surface energy and spreading coefficient with respect to water, condensation of water droplets on lubricant-impregnated nanotextured surfaces occurs with enhanced droplet mobility compared to non-impregnated superhydrophobic surfaces. Increased mobility is related to condensed droplets floating on the lubricant with minimal pinning to the surface.

With reference to new coating materials, recently Azimi et al. [29] investigated rare-earth oxides (REOs) ceramics, demonstrating that these promote dropwise condensation, repel impinging water droplets and sustain hydrophobicity even after exposure to harsh environments. The authors tested these materials through condensation experiments, exposing smooth silicon wafers sputtered with a thin layer of REOs to saturated steam. Their experiments showed that the REOs are sufficiently hydrophobic to promote dropwise condensation. The robustness of REOs hydrophobicity was positively demonstrated subjecting these coats to a high temperature environment (1000°C) as well as abrasive wear, measuring water contact angles before and after exposure.

Ma et al. [30] investigated steam DWC heat transfer characteristics in the presence of NCGs on a superhydrophobic nanostructured surface. A roughness-induced superhydrophobic vertical surface was prepared with n-octadecyl mercaptan SAM. The authors demonstrated that NCGs can be trapped in the cavities of the nanostructured surface by the condensate droplets (the more the higher is NCGs concentration), this way leading to a transition from Wenzel to Cassie wetting modes. Thus, the superhydrophobicity of the nanostructured surface can be sustained and enhanced during steam condensation in the presence of NCGs, especially at high concentration.

Yin et al. [31] comparatively investigated the effect of environmental factors on superhydrophobic behaviors of surfaces with different low-surface-energy coatings under controllable condensation conditions. Aluminum substrates were treated through a chemical etching process by immersing them in a mixed acid solution of HF, HCl and deionized water for different times, to construct textured morphology. Then, the rough surfaces were treated with fluoro-polymer (PFO), palmitic acid (PA) and RTV (room temperature vulcanization) silicon rubber to obtain superhydrophobic samples. Analyzing PFO-treated surfaces etched for different times the authors saw that, although they were all superhydrophobic after increasing the reaction time over 5 s, the highest contact angle (162°) was obtained on the surface etched for 15 s (S15). Focusing on S15 surface treated with the three different coats it has been shown that PFO-coated one was better both in static and dynamic angles than the others. The authors evaluated influences of surface subcooling degree and relative humidity (RH) on contact and sliding angles for the three S15 surfaces during condensation tests, comparing the results also with those obtained with smooth-coated substrates. Under high RH (>60%), contact angles decrease as well as sliding angles increase, indicating an increased wettability of all superhydrophobic surfaces due to the incremental amount of condensed water at low surface temperatures. However, PFO and PA coated surfaces showed good condensation resistance because of rolling drops even at the lower surface temperatures under RH of 90%. Moreover, the authors found that superhydrophobic behaviors of all surfaces were well resumed after evaporation of condensed water.

High temperature saturated vapor flow condensation tests on superhydrophobic nanotextured copper surfaces presented by Torresin et al. [32] show that condensing drops form and penetrate into the surface texture and attain a Wenzel state, but the vapor shear-induced droplets roll-off compensates for the reduced drops mobility and enhances the overall thermal transport. These results underlined the potential of nanostructured surface for application in flow condensation conditions. This work was aimed at showing the effect of vapour flow during pure vapour condensation.

3. Experimental analysis of flow condensation

In this paper the scalable wet-chemical technique used by Torresin et al. [32] to nanotexture copper surfaces, followed by surface functionalization to render them superhydrophobic, is presented in detail. Flow condensation tests performed by the authors to experimentally analyze DWC on the treated samples are investigated.

3.1. Superhydrophobic sample preparation

The two-step method herein shown to modify copper surfaces wettability is very simple, requires less than two hours and can be performed at ambient conditions without special equipment.

Before starting the surface treatments each copper sample is mechanically polished and ultrasonically cleaned in ethanol (99% Solvent grade) and Millipore DI water for 5 min each.

The first step of the treatment consists in dipping the polished copper surface in a mixture solution of 2.5 M L^{-1} NaOH and 0.1 M L^{-1} $(NH_4)S_2O_8$ for 12 min. During the entire immersion process the mixture was continuosly stirred. At the end of the process the sample is fully rinsed with deionized water and dried in a nitrogen stream. The first step leads to the formation of wire-like nanostructures on the surface.

In the second step, the modified copper substrate is immersed, at room temperature, in an ethanol solution of 1 mM L^{-1} 1H,1H,2H,2H-persluorodecanethiol (molecular formula $CF_3(CF_2)_7CH_2CH_2SH$) for 30 min followed by washing in ethanol for one hour. With this step, surface energy change was imparted.

The untreated polished surface was hydrophilic, with a water contact angle equal to 86±2°. After the two-step process the surface presents a contact angle of 159±2° and a sliding angle lower than 2° [32].

3.2. Test apparatus

A thermosyphon two-phase test rig has been developed to study steam condensation heat transfer. A schematic of the test loop is reported in figure 4, taken from [32].

Steam is generated in a stainless steel cylindrical boiler and directed by a stainless steel insulated pipe into the test section, which consists of a Teflon minichannel (D_h = 3.6 mm) fitted with test surfaces of two cylindrical copper specimens (D = 20 mm), over which condensation occurs. The copper specimens are cooled on the backside. After the test section, the remaining vapor and the condensed liquid pass through a secondary condenser, which completes the condensation process, and are sent back to the evaporator through a gravity drain.

Particular care was taken to avoid NCGs from entering into the test rig. Before filling the system with deionized water, air in the test rig was pumped out using a vacuum pump. A hydraulic accumulator was used to adjust the operative pressure and to maintain the flow loop in overpressure as compared to external ambient, thus avoiding any entry of non condensable gases.

3.3. Test section

A detailed description of the test section, shown in figure 5, is given in [32]. It mainly consists of a rectangular shape minichannel grooved into a Teflon block and covered with a glass plate to allow direct visualization of the process. The Teflon block presents two holes where the copper cylindrical samples are accommodated, with the top surface aligned with the bottom of the minichannel. On the back side each specimen is in contact with a water cooling system, described in [32].

Journal of Physics: Conference Series **501** (2014) 012028 doi:10.1088/1742-6596/501/1/012028

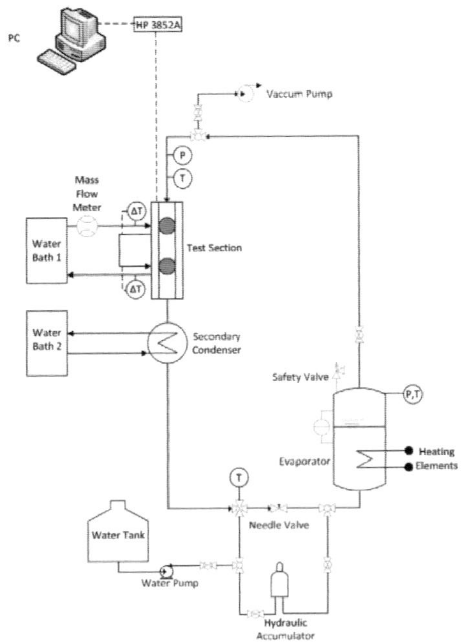

Figure 4. Schematic diagram of the experimental two-phase test loop.

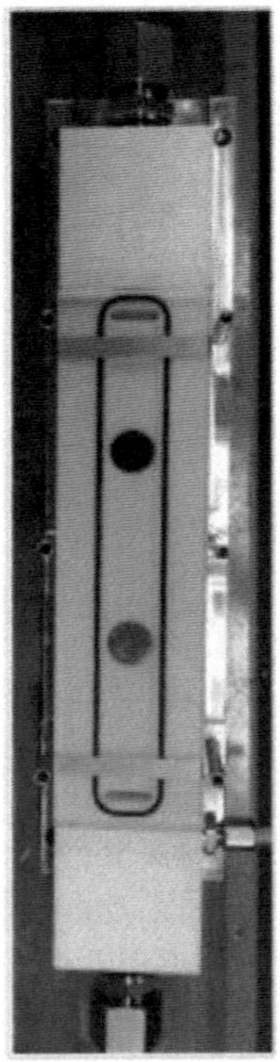

Figure 5. Picture of the test section.

Each cylindrical element is fitted with five T-type thermocouples in holes precisely drilled into the samples. From the acquired temperature profiles in the copper cylinder it was possible to evaluate the surface temperature T_{wall} , the heat flux and the condensation heat transfer coefficient [32]

$$q = -k \, {}^{dT}\!/_{dz} \tag{1}$$

$$h = q \,/\, (T_{sat} - T_{wall}) \tag{2}$$

where k is the thermal conductivity of the sample and z is the axial position. Steam condensation tests have been performed at 1.43 bar, on copper specimens treated with the two-step process as well as on a naturally oxidized polished copper sample (for comparison purposes).

3.4. Experimental results

Figure 6 shows heat flux and heat transfer coefficient data for three different vapor velocities (6, 12 and 18 m s^{-1}). For all the tested velocities steam condenses over the nanostructured superhydrophobic surface in dropwise mode.

The heat flux increases both when increasing the wall subcooling and the flow rate. The vapor steam imposes a drag force on the condensing droplets, which move if it overcomes their capillary adhesion to the substrate. Thus, increasing vapor velocity means reducing droplets departing diameter, thus enhancing thermal performance. Heat transfer coefficient behaves differently with velocity and subcooling: it seems to depend strongly on vapor velocity, while the influence of the subcooling degree is not significant. This highlights the positive influence of vapor flux on the performance of nanostructured superhydrophobic surfaces.

Journal of Physics: Conference Series **501** (2014) 012028 doi:10.1088/1742-6596/501/1/012028

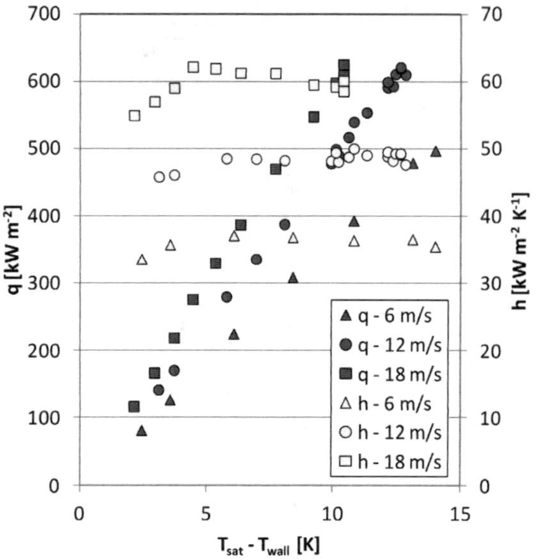

Figure 6. Heat flux (left axis) and heat transfer coefficient (right axis) versus wall subcooling during DWC at different vapor velocities on the superhydrophobic copper sample. Data from [32].

Nanostructured surface was tested for several days to verify its mechanical and chemical robustness. In [32] a comparison between heat fluxes measured over six consecutive working days at T_{sat} = 110°C and v_{vapor} = 12 m s^{-1} is shown. Over the first five days no differences were found, and experimental values of heat flux are well reproduced over the whole range of subcooling degrees. Clear differences are noticeable on the last day, when heat fluxes are 35% lower that the formers. This is because condensation mode changes from dropwise to filmwise between the 5th and the 6th working day, because of deterioration of the hydrophobic layer and mechanical degradation of the surface morphology.

Heat fluxes during filmwise condensation over a non-textured sample are compared in [32] against DWC data acquired with the treated surface, showing that for the first five days, at the same operating conditions, the thermal performance of the superhydrophobic sample is better than that of the oxidized one. After six working days however, the treated surface no longer maintains its superhydrophobic properties, and thermal performances of the oxidized copper sample become better.

4. Conclusions

1. A short literature review about dropwise condensation on superhydrophobic micro-/ nanotextured surfaces shows that:

a) Dropwise condensation mode enhances heat transfer performance as compared to filmwise mode, mainly because of high droplets mobility.

b) Superhydrophobic nanotextured surfaces, developed combining hierarchical scale roughness and hydrophobic coating, are a viable solution to promote DWC. Superhydrophobicity (i.e. high contact angles) can be obtained through different surface treatments.

c) If pure steam condenses over a superhydrophobic nanostructured surface, water droplets could pin on the substrate, reducing their mobility. Thus, during pure steam DWC, superhydrophobic nanotextured surfaces can perform worse than non-textured ones. High aspect-ratio substrates, as well as the presence of vapor stream, help to sustain heat transfer performance. On the contrary, during DWC in the presence of non-condensable gases, these are trapped between the surface textures, thus helping to sustain condensate drops mobility.

2. In this paper a simple two-step method to modify copper surface wettability is reported. The treated samples were used to experimentally analyze heat transfer during dropwise condensation of pure steam at 1.43 bar at varying vapor velocity. The present experimental analysis shows that:

a) DWC heat flux increases when increasing both surface subcooling and vapor velocity, while DWC heat transfer coefficient seems to be dependent only on vapor velocity. In fact, vapor velocity helps to maintain high droplets mobility even during DWC of pure steam.

b) Tests performed over different days showed that the treated surface maintains DWC for 5 working days, over which deterioration of the hydrophobic layer and mechanical degradation of the substrate lead to condensation mode change from dropwise to filmwise, with a reduction of heat transfer performances.

c) While maintaining DWC the superhydrophobic surface performs better than a non-textured oxidized one; when transition to FWC occurs, the superhydrophobic surface performs worse.

References

[1] Schmidt E, Schurig W and Sellschopp W 1930 *Forsch. Ingenieurwesen.* **1** 53-63.
[2] Rose J W 2002 *P. I. Mech. Eng. A-J Pow.* **216** 115-28.
[3] Rainieri S, Bozzoli F and Pagliarini G 2009 *Exp. Heat Transfer* **22** 163-77.
[4] Chen C H, Cai Q, Tsai C, Chen C L, Xiong G, Yu Y and Ren Z 2007 *Appl. Phys. Lett.* **90** 173108.
[5] Boreyko J B and Chen C H 2009 *Phys. Rev. Lett.* **103** 184501.
[6] Dietz C, Rykaczewski K, Fedorov A G and Joshi Y 2010 *Appl. Phys. Lett.* **97** 033104.
[7] Kim S and Kim K J 2011 *J. Heat Trans-T ASME* **133** 081502.
[8] Rausch M H, Leipertz A and Froba A P 2010 *Langmuir* **26** 5971-5.
[9] Graham C and Griffith P 1973 *Int. J. Heat Mass Tran.* **16** 337-46.
[10] Dorrer C and Rühe J 2007 *Langmuir* **23** 3820-4.
[11] Narhe R D and Beysens D A 2004 *Phys.l Rev. Lett.* **93** 076103.
[12] Varanasi K K, Hsu M, Bhate B, Yang W and Deng T 2009 *Appl. Phys. Lett.* **95** 094101.
[13] Feng J, Qin Z and Yao S 2012 *Langmuir* **28** 6067-75.
[14] Lafuma A and Quere D 2003 *Nature Materials* **2** 457-60.
[15] Tiwari M K, Bayer I S, Jursich G M, Schutzius T M and Megaridis C M 2010 *ACS Applied Materials and Interfaces* **2** 1114-9.
[16] Yuyang L, Xianqiong C and Xin J H 2009 *J. Mater. Chem.* **19**.
[17] Yin L, Zhu L, Wang Q, Ding J and Chen Q 2011 *ACS Applied Materials and Interfaces* **3** 1254-60.
[18] Jung S, Tiwari M K, Doan N V and Poulikakos D 2012 *Nature Communications* **3**.
[19] Sarkar D K, Farzaneh M and Paynter R W 2008 *Mater. Lett.* **62** 1226-9.
[20] Sarkar D K, Farzaneh M and Paynter R W 2010 *Appl. Surf. Sci.* **256** 3698-701.
[21] Saleema N, Sarkar D K, Paynter R W and Chen X G 2010 *Applied Materials and Interfaces* **2** 2500-2.
[22] Chen Z, Guo Y and Fang S 2010 *Surf. Interface Anal.* **42** 1-6.
[23] Jafari R and Farzaneh M 2011 *Applied Physics A* **102** 195-9.
[24] Lan Z, Ma X, Wang S, Wang M and Li X 2010 *Chemical Eng. Journal* **156** 546-52.
[25] Cheng J, Vandadi A and Chen C L 2012 *Appl. Phys. Lett.* **101** 131909.
[26] Ko T J, Her E K, Shin B, Kim H Y, Lee K R, Hong B K, Kim S H, Oh K H and Moon M W 2012 *Carbon* **50** 5085-92.
[27] Chen X, Wu J, Ma R, Hua M, Koratkar N, Yao S and Wang Z 2011 *Advanced Functional Materials* **21** 4617-23.
[28] Anand S, Paxson A T, Dhiman R, Smith J D and Varanasi K K 2012 *ACS Nano* **6** 10122-9.
[29] Azimi G, Dhiman R, Kwon H M, Paxson A T and Varanasi K K 2013 *Nature Materials Letters* **10**.
[30] Ma X, Wang S, Lan Z, Peng B, Ma H B and Cheng P 2012 *J. Heat Trans-T ASME* **134** 021501.
[31] Yin L, Wang L, Ding J, Wang Q and Chen Q 2012 *Appl. Surf. Sci.* **258** 4063-8.
[32] Torresin D, Tiwari M K, Del Col D and Poulikakos D 2013 *Langmuir* **29** 840-8.

Journal of Physics: Conference Series **501** (2014) 012029 doi:10.1088/1742-6596/501/1/012029

Dynamic coupling of kinetic and continuum solvers for micro gas flow computations

O Rovenskaya and G Croce
University of Udine, via delle Scienze 208, 33100 Udine, Italy

E-mail: olga.rovenskaya@uniud.it, giulio.croce@uniud.it

Abstract. A hybrid method, dynamic coupling the direct numerical solution of the Bhatnagar-Gross-Krook (BGK) kinetic equation and hydrodynamic Navier-Stokes equations is presented. The decomposition of the physical domain into kinetic and hydrodynamic sub-domains is based on the local Knudsen number and macroparameters gradients. The size of these domains will change during the evolution depending on the current value of the criteria. The solution is advanced in time simultaneously in both kinetic and hydrodynamic domains: the coupling is achieved by matching half fluxes at the interface of the kinetic and Navier-Stokes domains, thus taking care of the conservation of momentum, energy and mass through the interface. Solver efficiency is increased via MPI (Message Passing Interface) parallelization. The accuracy and properties of the proposed method is assessed via successful computation of the flow through a slit, at pressure ratio of 0.5 and for wide range of Knudsen number.

1. Introduction

The coexistence of rarefied continuum flow regime areas and relatively small elements in which rarefaction effects become important is a typical feature of many complex gas flows in micro systems. Thus, kinetic solvers, once limited to niche application in aerospace and vacuum technology, have become of relevant interest for the engineering community. In fact, in micro flows the mean free path of gas molecules is comparable to the characteristic scales of the system. These domains are naturally described by kinetic equation for the velocity distribution function, which involve a considerable effort in terms of CPU time and memory requirements, due to the discretization in both physical and velocity spaces. The continuum domains are well described by the fluid, Euler or Navier–Stokes (NS) equations in terms of average gas flow velocity, gas density and temperature. These equations are more efficient, but less accurate in critical rarefied areas. The development of hybrid solvers combining kinetic and continuum models has, thus, become an important area of research over the last decade. Potential applications of such solvers range from gas flows in micro systems to the aerospace applications, such as high altitude flights. The key parameter defining the choice of the appropriate physical model is related to the local Knudsen number. Major challenges in the development of hybrid code are the identification of kinetic and continuum domains and the choice of the coupling technique.

Different methods presented so far in the open literature can be classified into three categories. The first relies on domain decomposition in physical space: the computational domain is thus decomposed into kinetic and continuum sub-domains using appropriate criteria [1-6]. The second is based on domain decomposition in velocity space, where fast and slow particles are treated separately [7]. The third category includes hybrid models: both kinetic and fluid equations are solved in the entire domain, using velocity distribution function to compute transport coefficients for the fluid equations [8]. Most of the published works fall into the first category.

Typically, particle methods such as DSMC or Molecular Dynamics are used in regions with strong deviations from equilibrium, and a continuum fluid (Euler or NS, depending on problem features) solver is used elsewhere [3-6]. Nonetheless, the Direct Numerical Solution (DNS) of kinetic equations is a viable alternative to DSMC and may be preferable to DSMC, for coupling purposes, since both kinetic and continuum models use similar numerical techniques. Recent effort to combine DNS with a NS solver used a priori decomposition of the domain, e.g. [2], or combined the DNS of the Boltzmann equation with kinetic schemes of continuum fluid dynamics [1].

The present methodology is a development of the work presented in [2], and requires the decomposition of the physical domain into kinetic and continuum sub-domains by computing gradient-length Knudsen number Kn_{GL}, based on the local Knudsen number and macroparameters gradients [9]. The size of these domains is here dynamically updated during the transient depending on the current Kn_{GL}. The solution is advanced in time simultaneously in both kinetic and continuum domains: the coupling is achieved by matching half fluxes at the interface of the kinetic and NS domains, thus taking care of the conservation of momentum, energy and mass through the interface. This allows the combination of existing in-house codes for numerical solution of both BGK based on the discrete velocity method and a finite-difference finite volume scheme for the NS equations. Furthermore, solver efficiency is increased via MPI (Message Passing Interface) parallelization.

The validation and properties of the proposed approach are assessed via the computation of the gas flow through a slit with both the hybrid and pure kinetic method, for pressure ratio of 0.5 and wide range of Knudsen number. The results are discussed in terms of both accuracy and computational efficiency.

2. Coupling algorithm and numerical methods

The coupling strategy between kinetic and continuum solvers is completely general, and can be applied to the full Boltzmann kinetic equation or its different models and NS or Euler equations. For the sake of simplicity, however, the collision integral is here replaced by the BGK model [10], which can be written as:

$$\frac{\partial f}{\partial t}+\xi\frac{\partial f}{\partial \mathbf{x}}=J_{BGK}(f,f)=\frac{p}{\mu}(M-f) \tag{1}$$

$$M(\rho,\mathbf{V},T)=\rho(\frac{m}{2\pi kT})^{3/2}\exp[-\frac{m\mathbf{c}^2}{2kT}] \tag{2}$$

where $f = f(t, \mathbf{x}, \xi)$ is the velocity distribution function, i.e. the probability of finding a molecule with velocity $\xi = (\xi_x, \xi_y, \xi_z)$ in the position $\mathbf{x} = (x, y)$ at the time t. $J_{BGK}(f, f)$ is the collision integral, M is the local Maxwell distribution function, ρ is the gas density, $\mathbf{c} = \xi - \mathbf{V}$ is the relative speed of a single molecule against a background gas with averaged velocity $\mathbf{V} = (u, v)$, p is the local pressure and μ is the viscosity at local temperature T, k is the Boltzmann constant, m is the molecule mass.

The problem is recast in terms of non-dimensional variables using the inlet reservoir equilibrium values as reference ones: density ρ_0, temperature T_0, most probable velocity $v_0 = (2kT_0/m)^{0.5}$, height of the slit H. For the hard-sphere molecular model the non-dimensional viscosity coefficient is $\mu/\mu_0= (T/T_0)^{0.5}$. Thus, non-dimensional variables are ξ/v_0, $\mathbf{x}/H$, $f/(\rho_0 v_0^{-3})$, μ/μ_0, ρ/ρ_0, T/T_0 and p/p_0. For sake of simplicity, in the following the dimensionless quantities keep the same designations as the dimensional ones. In the non-dimensional form the BGK -model equation is thus written as follows:

$$\frac{\partial f}{\partial t}+\xi\frac{\partial f}{\partial \mathbf{x}}=\frac{8}{5Kn_0\sqrt{\pi}}\rho\sqrt{T}\left(M-f\right)=\delta\,\sigma(\rho,T)\left(M-f\right) \tag{3}$$

Journal of Physics: Conference Series **501** (2014) 012029 doi:10.1088/1742-6596/501/1/012029

where $\sigma(\rho, T) = \rho T^{0.5}$, δ is the rarefaction parameter, related to the reference Knudsen number Kn_0 based on the height of the slit H and the mean free path λ_0 at reference conditions:

$$\lambda_0 = \frac{16\mu_0}{5\rho_0}\sqrt{\frac{m}{2\pi k T_0}}, \quad \delta = \frac{8}{5Kn_0\sqrt{\pi}} \tag{4}$$

The macroscopic density, momentum and internal energy per unit mass, respectively, are defined as:

$$\rho = \int f d\xi \qquad \rho(u,v)^T = \int (\xi_x, \xi_y)^T f d\xi \qquad \rho e = \frac{1}{2}\int \mathbf{c}^2 f d\xi \tag{5}$$

The present approach is based on the idea that NS equations are the first order approximation of the Boltzmann equation, according to the Chapman-Enskog theory [11]. The non-dimensional Chapman–Enskog Navier-Stokes velocity distribution function f_{CE} can be written as:

$$f_{CE} = M\left(1 + Kn_0 f_1\right) \tag{6}$$

$$f_1 = -\frac{1}{p}\left(2\left(c_i c_j - \frac{\mathbf{c}^2}{3}\delta_{ij}\right)\frac{\partial u_i}{\partial x_j} + \frac{\partial T}{\partial x_i}\left(\frac{\mathbf{c}^2}{T} - \frac{5}{2}\right)c_i\right) \tag{7}$$

where f_1 is the correction term and M is the Maxwellian. Substituting f_{CE} in equation (1) and multiplying by the collision vector $\varphi(\xi) = (1, \xi, \xi^2/2)$, integration over the whole velocity domain R^3 recovers the usual conservation laws of mass, momentum and energy:

$$\frac{\partial}{\partial t}\left(\int_{R^3} f_{CE}\varphi^T d\xi\right) + \frac{\partial}{\partial \mathbf{x}}\int_{R^3} f_{CE}\xi\varphi^T d\xi = 0 \tag{8}$$

The use of f_{CE} ensures that equation (8) is equivalent to the usual conservative form of NS equations:

$$\frac{\partial \mathbf{U}}{\partial t} + \frac{\partial \mathbf{F(U)}}{\partial \mathbf{x}} = 0 \tag{9}$$

$$\mathbf{U} = \left(\rho, \rho\mathbf{V}, e + \mathbf{V}^2/2\right)^T = \int f_{CE}\varphi^T d\xi \tag{10}$$

$$\mathbf{F}(\mathbf{U}) = \int \xi f_{CE}\varphi^T d\xi \tag{11}$$

where **U** is the vector of macroscopic values, **F(U)** is the flux.

The choice of switching criterion between kinetic Ω_K and continuum Ω_{NS} regions is important because the wrong domain decomposition could even lead to a non-positive velocity distribution function [5, 9]. In the present work we use gradient-length Knudsen numbers [9]:

$$Kn_{GL\,P} = Kn\frac{\nabla P}{P_{ij\max}} \tag{12}$$

where P represents the parameter of interest, namely density, velocity, or temperature. The actual continuum breakdown parameter is then the maximum of these:

$$Kn_{GL}(x,y) = \max(Kn_{GL\rho}, Kn_{GL\,V}, Kn_{GL\,T}) \tag{13}$$

Schwartzentruber and Boyd [9] suggest that if Kn_{GL} is smaller than a threshold value ε, namely $\varepsilon = 0.05$, the NS equations offer a good approximation of the flow. Thus, at each time step the BGK equation (3) is solved only on a kinetic domain Ω_K, which is a subdomain of physical space (x_i, y_j)

satisfying the condition $Kn_{GL}(x_i, y_j) \geq \varepsilon$. Definition equation (13) is slightly different from the criteria used in [2], based on the norm of the gradient-length Knudsen numbers: in particular, equation (13) provides, for any given flow field and threshold parameter ε, a slightly larger kinetic region and, thus, was here considered better suited to take care of any possible rarefaction effect even during the transients.

The solution of the kinetic equations on a cell i requires information on the distribution function for incoming particles from neighbouring cells. On the coupling interface I_c this information has to be provided by the continuum solution; thus, assuming that equation (8) holds true in the continuum subdomain we can impose a Chapman–Enskog distribution, equation (6) :

$$f|_{I_c}(\mathbf{x}) = f_{CE} \qquad \text{if } \xi \cdot \vartheta(\mathbf{x}) < 0 \tag{14}$$

where $\vartheta(\mathbf{x})$ is the outward normal vector to the boundary of Ω_K. Assuming node x_i in the continuum domain and node x_{i+1} in the kinetic one, macroscopic values ρ, V, T appearing in $f_{CE}(x_i)$ are computed at the grid point x_i and the evaluation of parameters gradients involves also values in the neighbour to x_i points: x_{i-1}, x_{i+1}.

At solid walls Maxwell diffuse reflecting boundary condition with the full accommodation is applied:

$$f|_w(\mathbf{x}) = \omega(\mathbf{x}) M(1, \mathbf{V}_w, T_w) \qquad \text{if } \xi \cdot \vartheta(\mathbf{x}) < 0 \tag{15}$$

The parameter $\omega(\mathbf{x})$ is determined so as to avoid a mass flux across the wall. All the particles coming off the surface are emitted with the Maxwell distribution functions corresponding the zero mean flow velocity, the temperature is equal to the wall temperature T_w, and the density is calculated from the condition of equality of the fluxes of particles coming on and off the wall. At the symmetry line the specular boundary condition is imposed.

For continuum domain Ω_{NS} we impose total (equilibrium) pressure p_0 and temperature T_0 at inlet region, pressure p_e at exit region. At the coupling interface I_c the following boundary condition is imposed:

$$\mathbf{F}(\mathbf{U}) \cdot \vartheta(\mathbf{x})|_{I_c} = \mathbf{F}_i(\mathbf{U}) \cdot \vartheta(\mathbf{x}) + \mathbf{F}_o(\mathbf{U}) \cdot \vartheta(\mathbf{x}) \tag{16}$$

where $\mathbf{F}_i(\mathbf{U})\, \vartheta(\mathbf{x})$ and $\mathbf{F}_o(\mathbf{U})\, \vartheta(\mathbf{x})$ are the incoming and outcoming half fluxes. Thus, the coupling between kinetic and NS solvers is achieved by imposing the incoming into Ω_{NS} domain half flux $\mathbf{F}_i(\mathbf{U})$ $\vartheta(\mathbf{x})$, predicted by the kinetic solver:

$$\mathbf{F}_i(\mathbf{U}) \cdot \vartheta(\mathbf{x}) = \int_{\xi \cdot \vartheta(\mathbf{x}) > 0} \xi \cdot \vartheta(\mathbf{x}) f(t, \xi, \mathbf{x}) \varphi^T d\xi \tag{17}$$

where f is the solution of kinetic equation for molecules exiting from Ω_K. The outcoming flux $\mathbf{F}_o(\mathbf{U})\, \vartheta(\mathbf{x})$ from Ω_{NS} to Ω_K domain is defined as:

$$\mathbf{F}_o(\mathbf{U}) \cdot \vartheta(\mathbf{x}) = \int_{\xi \cdot \vartheta(\mathbf{x}) < 0} \xi \cdot \vartheta(\mathbf{x}) f_{CE}(t, \xi, \mathbf{x}) \varphi^T d\xi \tag{18}$$

It can be seen, that the NS model acts on the kinetic one by imposing the incoming Chapman–Enskog function f_{CE} on the coupling interface I_c, while the macroscopic parameters ρ, $\mathbf{V}$, T and their gradients for f_{CE} are computed locally.

2.1. Numerical scheme for the BGK equation

Taking advantage of the two-dimensionality of the flow, the z component of the molecular velocity may be eliminated by the projection procedure. For the sake of simplicity a scheme will be written for general BGK- model equation (3).

To discretize the BGK-model equation we construct in velocity space a two-dimensional grid $\{\xi_\Gamma\} = (\xi_{x\alpha}, \xi_{y\beta})$, $\alpha, \beta = 1, \ldots, N_\xi$ bounded by v_{max}, where $\Gamma = (\alpha, \beta)$ defines velocity grid points indices.

Velocity space nodes are located at Gaussian abscissas, and Gaussian integration is performed via proper Gaussian weight function. In the discrete form the BGK-model equation (3) is the following:

$$\partial_t f_\Gamma + \xi_{x\Gamma}\frac{\partial f_\Gamma}{\partial x} + \xi_{y\Gamma}\frac{\partial f_\Gamma}{\partial y} = \delta\sigma(\rho,T)\left(M_\Gamma - f_\Gamma\right) \tag{19}$$

In the physical space a 2D, non uniform grid is defined by nodes (x_i, y_j), for $i = 1$, i_{max} and $j = 1$, j_{max} and $N_c = i_{\max} \cdot j_{\max}$ cells centred around each node. Denoting $f^n_{\Gamma,i,j}$ as the approximations of $f(t^n, \xi_\Gamma, x_i, y_j)$ the explicit-implicit numerical scheme for the equation (19) can be written as [12]:

$$f^{n+1}_{\Gamma,i,j} = \frac{1}{1+\delta\sigma^{n+1}\Delta t}\left(f^n_{\Gamma,i,j} - \Delta t\xi_{x\Gamma}\frac{\partial f^n_{\Gamma,i,j}}{\partial x} - \Delta t\xi_{y\Gamma}\frac{\partial f^n_{\Gamma,i,j}}{\partial y}\right) + \frac{\delta\sigma^{n+1}\Delta t}{1+\delta\sigma^{n+1}\Delta t}M^{n+1}_{i,j} \tag{20}$$

The transport term in equation (20) is treated explicitly and approximated by a standard finite volume scheme. In particular, the numerical fluxes are determined by the third order MUSCL with the Van Albada limiter. The time step should be limited by the condition:

$$\Delta t_K = CFL / \max_{ij}\left(v_{\max}/\Delta x + v_{\max}/\Delta y\right) \tag{21}$$

The equation (20) is solved on a curvilinear, structured mesh. If we define curvilinear coordinates $\zeta(x, y)$ and $\eta(x, y)$ and a uniform grid $\varsigma_i = i\Delta\varsigma$, $\eta_j = j\Delta\eta$ a scheme very similar (20) may be used.

Since the kinetic part is time consuming, the code is parallelized in order to improve its efficiency. The solution of system (20) is local in $\boldsymbol{\Gamma}$, and therefore completely parallelizable in velocity space. The software code was written in C++ with the use of MPI.

2.2. Numerical scheme for the Navier–Stokes equations

The flux vector in equation (9) **F(U)** may be decomposed into the convective (inviscid) and diffusive (viscous) components:

$$\mathbf{F} = \mathbf{F}^{\mathbf{inv}} - \mathbf{F}^{\mathbf{v}} \tag{22}$$

$$\mathbf{F}^{\mathbf{inv}} = \left(\rho\mathbf{V}, \rho\mathbf{V}\mathbf{V} + p\mathbf{I}, \mathbf{V}(e + \mathbf{V}^2/2 + p)\right)^T \tag{23}$$

$$\mathbf{F}^{\mathbf{v}} = \left(0, \boldsymbol{\tau}, \boldsymbol{\tau}\cdot\mathbf{V} + \mathbf{q}\right)^T \tag{24}$$

where **I** is the identity matrix and **q** is the heat flux vector

Navier-Stokes solver is based on a hybrid finite difference-finite volume method and has formal second order accuracy in space and time [2]. Fluxes are defined via neighbouring value averaging, and an artificial dissipation term is added to prevent checker boarding and numerical instabilities. Artificial dissipation terms are given by a blend of second and fourth order differences, scaled by the maximum eigenvalue of jacobian matrix of vectors $\mathbf{F}^{\mathbf{inv}}$, as suggested in [13]. Second order terms are switched on near discontinuities. Viscous flux vectors are evaluated with second order finite differences at $i+1/2$. The solution is advanced in time via Crank Nicolson integration scheme. The use of the spatially factored ADI scheme originally proposed by Beam and Warming leads, at each time step, to the resolution of two series of block tridiagonal algebraic systems, rather than the original pentadiagonal block system arising from flux discretization.

At the solid wall the Maxwell first order slip boundary condition is used:

$$u_{gas} - u_w = s_p Kn\left(\frac{\partial u_{\eta(\mathbf{x})}}{\partial\varsigma(\mathbf{x})} + \frac{\partial u_{\varsigma(\mathbf{x})}}{\partial\eta(\mathbf{x})}\right) \qquad s_p = \frac{\sqrt{\pi}}{2}\frac{2-\sigma_v}{\sigma_v}\left(1 + 0.1366\sigma_v\right) \tag{25}$$

where $\varsigma(\mathbf{x})$ and $\eta(\mathbf{x})$ are the unit vectors respectively normal and tangential to the solid wall. σ_v is the tangential momentum accommodation coefficient, here chosen equal to one, thus in the computations s_p is close to one. The additional derivative along the tangential direction is essential in capturing even the qualitative behaviour of slip flow along curved walls.

A Dirichlet temperature boundary condition is imposed at the wall. Namely, the wall temperature is fixed at the inlet total temperature value to minimize the effect of viscous dissipation. In the energy equation, the Smoluchowski temperature jump is used:

$$T_{gas} - T_w = s_T Kn \frac{\partial T}{\partial \eta(\mathbf{x})}, \quad s_T = \frac{2-\sigma_T}{\sigma_T} \frac{2\gamma}{\gamma+1} \frac{1}{\mathrm{Pr}} \tag{26}$$

where s_T is the temperature jump coefficient. Since isothermal flow is considered we assume the temperature accommodation coefficient σ_T equals to one.

3. Results and discussion

The two-dimensional flow of a gas through a slit of height H cut in an infinitely thin partition separating two containers is considered. The gas in the containers, far from the slit, is in equilibrium at temperature T_0 and pressures p_0 and $p_e = 10^5$ Pa, with $p_0 > p_e$. The slit is considered as infinite in the z direction. Since the flow is symmetric about $y = 0$, therefore, only a half of the flow, shown in figure 1, will be considered. Two large computational areas of radius R before and after the slit ($-H/2 \le y \le 0$) are included in the computational domain, simulating the upstream and downstream reservoirs. I_c is the interface between inner kinetic sub-domain and outer NS one.

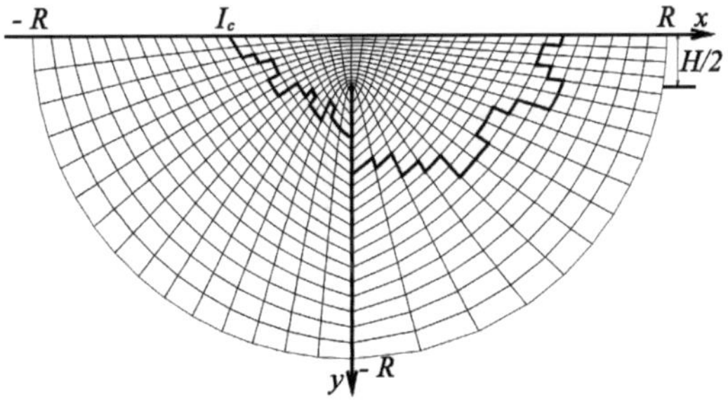

Figure 1. Computational domain sketch.

The gas flows due to a pressure difference p_e/p_0 between upstream and downstream reservoirs. The static inlet pressure p_i is the result of computation, although, due to the low inlet velocities, it almost coincides with p_0. The gas flow through the slit is also determined by the rarefaction parameter δ.

The global characteristic of the flow, dimensionless flow rate W, is introduced as:

$$W = \frac{\dot{m}}{\dot{m}_{fm}} \qquad \dot{m}_{fm} = \frac{p_0}{\sqrt{\pi} v_0} \tag{27}$$

where $\dot{m}_{fm}$ is the analytically deduced mass flow rate in the limit of free molecular regime and the mass flow rate through the slit are computed as:

$$\dot{m} = 2\int_{-1/2}^{0} \rho(0,y) u(0,y) dy \tag{28}$$

The use of circular sectors as plenum, rather than the more common square shape, allows reservoirs of large size with a relatively smaller number of grid points, while preserving a strong refinement near the slit. The accuracy of the numerical results is sensitive to the sizes of the left and right reservoirs: in [2] it was shown that the radius $R = 40H$ is sufficiently large for correct computations.

Journal of Physics: Conference Series **501** (2014) 012029 doi:10.1088/1742-6596/501/1/012029

A non uniform curvilinear grid of 160 nodes in the streamwise direction with minimum grid spacing $0.017H$ and 40 nodes in the transverse direction is used (see figure 1). Grid independence test has been done using the finer grid with 240 × 40 nodes. The mass flow difference between fine and coarse meshes is less than 1%. In the velocity space a 24 × 24 grid is implemented. An increase in the velocity grid points of 64% led to a 1% change in global mass flow.

Since time step is unique for both solvers, it should satisfy the stricter stability (or accuracy) constraint $\Delta t = \min(\Delta t_K, \Delta t_{NS})$. The iteration process is terminated when a relative convergence criterion of 10^{-7} imposed on the non-dimensional flow rate is fulfilled.

The code ran on double processors, quad core systems, using thus up to 8 parallel processes. The CPU time of the hybrid solver is basically determined by the kinetic one and the advantage depends essentially on the ratio between the number of grid nodes on the (expensive) kinetic domain and the number of grid nodes in the (cheap) NS one.

To illustrate effect of switching criteria ε on the computed flow field, we show non dimensional mass flow rate W_h obtained by hybrid simulations for $\delta = 1$ (see table 1). For large ε, the mass flow rate is quite different from the BGK kinetic value, $W_{BGK} = 0.653$. The mass flow rate W_h converges to the correct value when the switching parameter ε decreases below 0.1.

Table 1. Non-dimensional mass flow rate $\delta = 1$.

ε	W_h	W_{BGK}
0.05	0.653	
0.1	0.653	
0.5	0.640	0.653
1	0.621	
2	0.591	

The CPU time per time step of hybrid code is dictated by the time for solving the BGK equation, since the sum of CPU time relative to NS solutions and coupling computations is quite small, relatively to the kinetic solution requirements, at least in the field of interest (i.e., as will be shown, $\delta<20$). For example, if whole domain is considered as kinetic (160 × 40 points) the CPU time is 1.6 s, more than an order of magnitude longer than a full NS time step, while for $\delta = 100$ (approximately 160 kinetic nodes) the CPU time is 0.28 s; for $\delta = 10$ (3000 kinetic nodes) the CPU time is 0.93 s; for $\delta = 5$ (4100 kinetic nodes) the CPU time is 1.2 s.

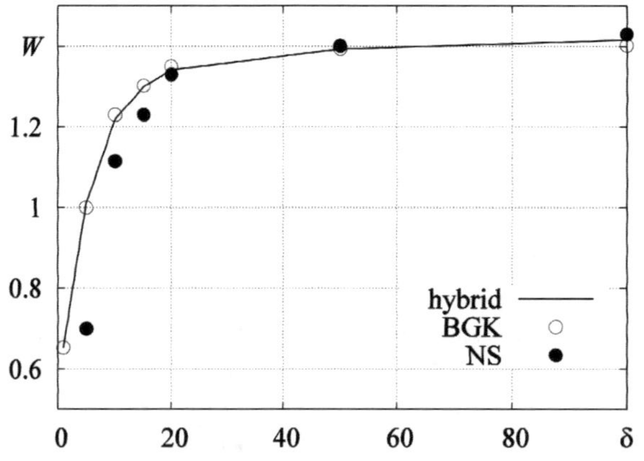

Figure 2. Non-dimensional mass flow rate via δ.

It should be mentioned that parallelization in velocity space allow us to uniformly distribute kinetic points between processors. Each processor works with the same number of kinetic points in physical space. Now the speed up of hybrid code is around 82% was reached for $\delta = 100$ and 25% for $\delta = 5$, but further improvements of code are still possible. An even higher speed up ratio is expected in case the Boltzmann collision integral is used instead BGK model, since the coupling overhead become completely negligible.

In figure 2, non-dimensional flow rates are compared with results computed by the pure BGK and NS solvers for the pressure ratio of 0.5 and rarefaction parameter δ ranging from 1 to 100. For $\delta \geq 20$ mass flow rates obtained by hybrid, BGK and NS solvers are close to each other (maximum difference is less than 2.5%). For smaller δ the difference between NS and BGK flow rates becomes higher and for $\delta = 5$ is around 45%, while hybrid code reproduces mass flow close to kinetic data.

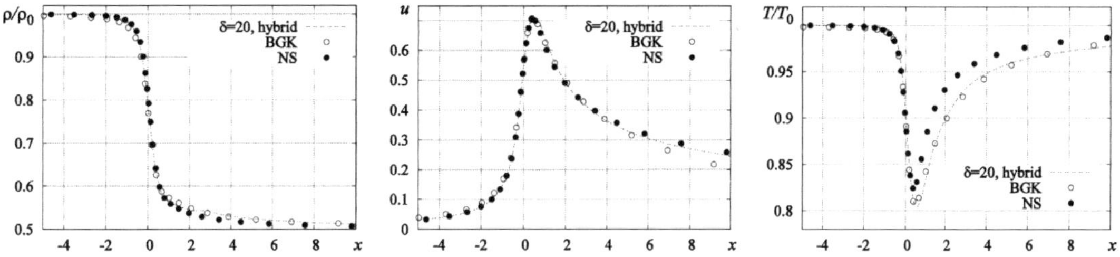

Figure 3. Density ρ/ρ_0, axial velocity u and temperature T/T_0 along the symmetry axis $y = 0$, $\delta = 20$.

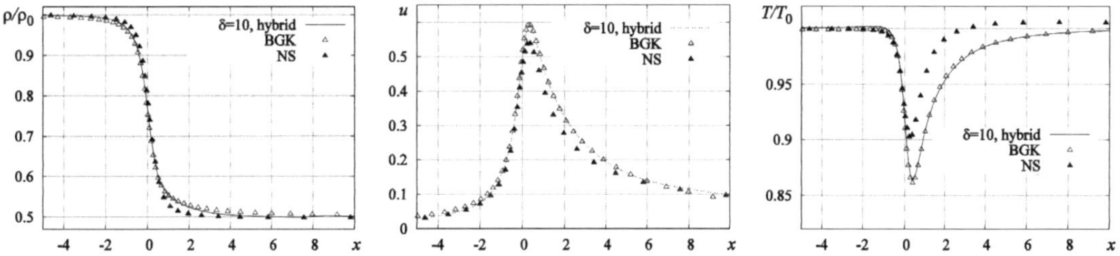

Figure 4. Density ρ/ρ_0, axial velocity u and temperature T/T_0 along the symmetry axis $y = 0$, $\delta = 10$.

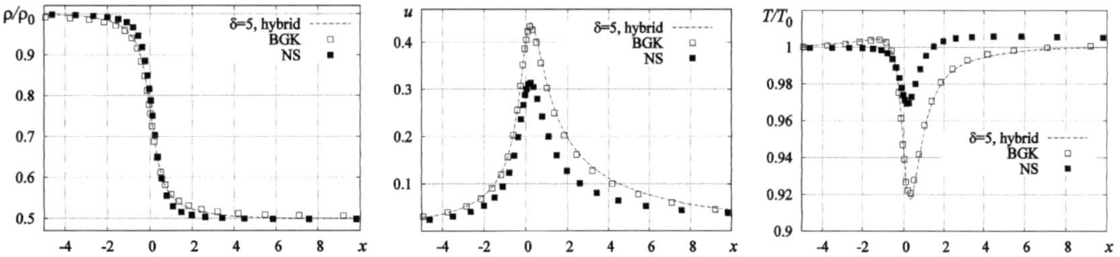

Figure 5. Density ρ/ρ_0, axial velocity u and temperature T/T_0 along the symmetry axis $y = 0$, $\delta = 5$.

The variations of the dimensionless density, temperature, normalized by the inlet values, and velocity along the symmetry axis are shown in figures 3-5 for rarefaction parameters 20, 10 and 5. Coupled solutions are compared with BGK and NS results. Density (or pressure) variations are qualitatively similar in all cases. Before and after the slit density tends to upstream and downstream conditions, while in the slit region it sharply decreases. The axial velocity far upstream is almost zero and it grows in the region around the slit. The velocity increase is smaller at larger rarefaction (smaller δ). The temperature decreases near the slit, while at the rest of the domain it remains very close to the reference temperature. The temperature drop is larger as δ increases. The coupled solutions are close enough to the kinetic ones for all values of rarefaction parameters.

The contour lines of density and Mach number (solid line) near the slit region for different rarefaction are presented in figures 6-8, together with the extension of the kinetic region at convergence (dashed line).

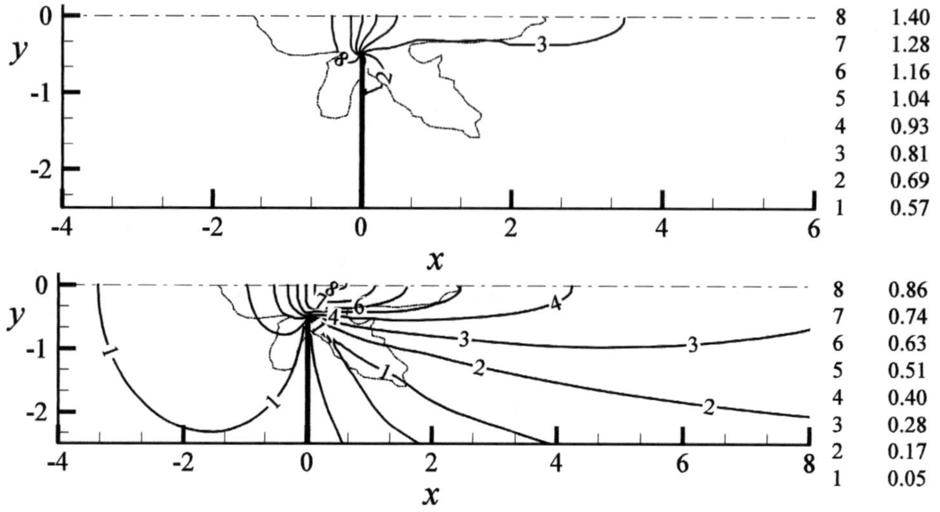

Figure 6. Dimensionless density (top) and Mach number (bottom) near the slit: $\delta = 20$.

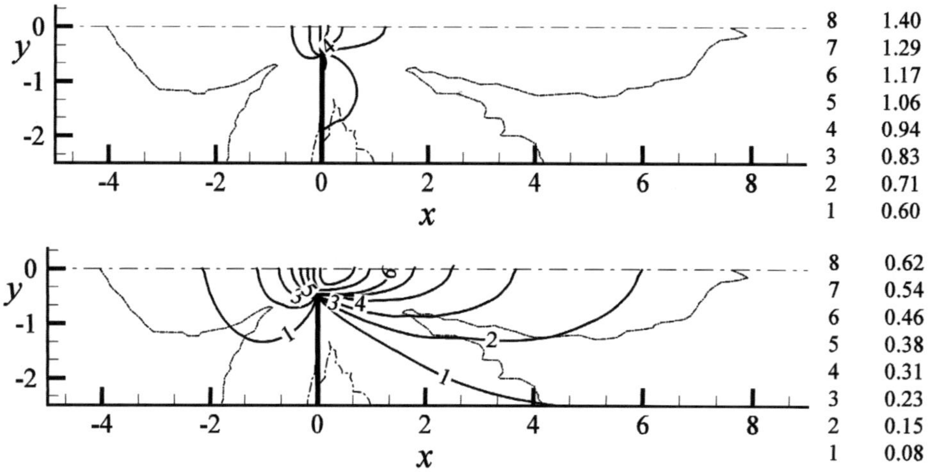

Figure 7. Dimensionless density (top) and Mach number (bottom) near the slit: $\delta = 10$.

For small rarefaction $\delta = 20$ (figure 6) the local Knudsen number near the coupling boundary varies from 0.07 to 0.11. The coupling between kinetic and NS solutions shows a smooth transition along the contour lines crossing the domains interface, especially for the Mach number contour.

Finally, to assess the possibility of hybrid code, the case $\delta = 5$ is shown in figure 8. At the coupling interface the local *Kn* rise up to 0.1; furthermore, its maximum value occurs close to the slit (within the kinetic domain) and is equal to 0.4. NS solution would not, thus, be appropriate for the solution over the whole domain as can be seen in figure 5. The flow solution has now more relevant viscous effects, which smooth the transonic flow structures. Nevertheless, the hybrid solver gives a good prediction of mass flow (see figure 2) and the contour lines crossing domains still appear smooth and regular.

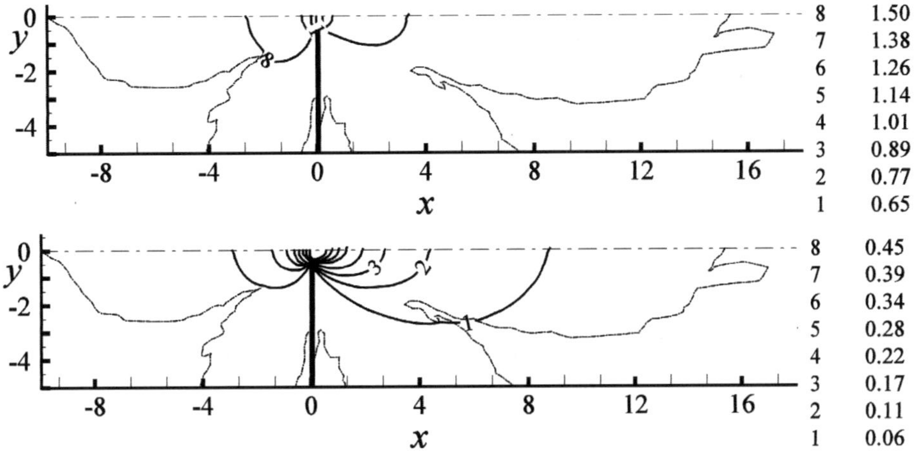

Figure 8. Dimensionless density (top) and Mach number (bottom) near the slit: δ = 5.

4. Conclusion

A hybrid algorithm based on the direct numerical solution of the BGK kinetic equation coupled to a Navier-Stokes model was presented. The solution is advanced in time simultaneously in both kinetic and continuum domains: the coupling is achieved by matching half fluxes at the interface of the kinetic and NS domains, taking care of the conservation of momentum, energy and mass through the interface, and is flexible enough to allow for the use of existing solver on both sides. Mass flow rates and flow parameters are close to full BGK solutions. The capability of the hybrid code to simulate the gas flow through slit for a wide range of Knudsen number, exceeding the range of applicability of pure Navier-Stokes, has been demonstrated. The CPU time savings, with respect to a full BGK solutions, are significant, although strongly depends on the size of the kinetic region.

It should be noticed that the dynamic coupling is more effective then static one, allowing to capture and describe properly the kinetic regions even when they appear, during the convergence transient, in areas not easy to predict a-priori. This allows to get satisfying results even for relatively high Knudsen.

Acknowledgments
The research leading to these results has received partially founding from MIUR grant (PRIN 2009).

References

[1] Kolobov V I, Arslanbekov R R, Aristov V V, Frolova A A, Zabelok S A 2007 *J. of Comput. Phys.* **223** 589–608.
[2] Rovenskaya O, Croce G 2013 *Heat Transfer Eng.* **34** 192-203.
[3] Le Tallec P, Mallinger F J 1997 *Comput. Phys.* **136** 51-67.
[4] Tiwari S 1998 *J. Comput. Phys.* **144** 710-726.
[5] Carlson H A, Roveda R, Boyd I D, Candler G V 2004 AIAA 2004-1180.
[6] Wijesinghe H S, Hornung R, Garsia A L, Hadjiconstantinou H N 2004 *J. Fluids Eng.* **126** 768–777.
[7] Crouseilles N, Degond P, Lemou M 2004 *J. Comput. Phys.* **199** 776-806.
[8] Beylich A E 2000 *Phys. Fluids* **12** 444–465.
[9] Schwartzentruber T E, Boyd I D 2006 *J. of Comput. Phys.* **215** 402–416
[10] Bhatnagar P L, Gross E P, Krook M A 1954 *Phys. Rev.* **94** 511-525.
[11] Chapman S, Cowling T G 1990 *The mathematical theory of nonuniform gases* (Cambridge: Univ. Press).
[12] Filbet F, Jin S 2010 *J. of Comput. Phys.* **229** 7625-7648.
[13] Pulliam T H 1986 *AIAA Journal* **24** 1931-1940

Journal of Physics: Conference Series **501** (2014) 012030 doi:10.1088/1742-6596/501/1/012030

Thermal-hydraulics of internally heated molten salts and application to the Molten Salt Fast Reactor

Carlo Fiorina[1,2,#], Antonio Cammi[1], Lelio Luzzi[1], Konstantin Mikityuk[2], Hisashi Ninokata[1] and Marco E. Ricotti[1]

[1] Politecnico di Milano, Department of Energy, Nuclear Engineering Division – Via Ponzio 34/3, 20133, Milan, Italy

[2] Paul Scherrer Institute, Nuclear Energy and Safety Division, Laboratory for Reactor Physics and Systems Behaviour – PSI, 5232 Villigen, Switzerland

[#] corresponding author: carlo.fiorina@polimi.it / carlo.fiorina@psi.ch

Abstract. The Molten Salt Reactors (MSR) are an innovative kind of nuclear reactors and are presently considered in the framework of the Generation IV International Forum (GIF-IV) for their promising performances in terms of low resource utilization, waste minimization and enhanced safety. A unique feature of MSRs is that molten fluoride salts play the distinctive role of both fuel (heat source) and coolant. The presence of an internal heat generation perturbs the temperature field and consequences are to be expected on the heat transfer characteristics of the molten salts. In this paper, the problem of heat transfer for internally heated fluids in a straight circular channel is first faced on a theoretical ground. The effect of internal heat generation is demonstrated to be described by a corrective factor applied to traditional correlations for the Nusselt number. It is shown that the corrective factor can be fully characterized by making explicit the dependency on Reynolds and Prandtl numbers. On this basis, a preliminary correlation is proposed for the case of molten fluoride salts by interpolating the results provided by an analytic approach previously developed at the Politecnico di Milano. The experimental facility and the related measuring procedure for testing the proposed correlation are then presented. Finally, the developed correlation is used to carry out a parametric investigation on the effect of internal heat generation on the main out-of-core components of the Molten Salt Fast Reactor (MSFR), the reference circulating-fuel MSR design in the GIF-IV. The volumetric power determines higher temperatures at the channel wall, but the effect is significant only in case of large diameters and/or low velocities.

1. Introduction

In recent years, there has been a growing interest in the Molten Salt Reactor (MSR) technology, one of the six nuclear reactor concepts considered in the framework of the Generation IV International Forum as promising candidates for an increasingly sustainable, safe and secure nuclear energy production. A main and characterizing feature of MSRs is the dual role of the molten salts, acting as both fuel and coolant. This requires investigation in view of the little information available in the open literature regarding the thermal-hydraulics of internally heated fluids. Dedicated models and tools have been developed for specific applications like combustion processes (see e.g. [1]), but they are not directly applicable to the case of MSRs. Despite the scarcity of available data and theoretical studies, the presence of an internal heat generation perturbs the temperature field and consequences are to be expected on the thermal-hydraulics of MSRs. Among various aspects requiring investigation, the present paper focuses on the heat transfer phenomena occurring in molten salts flowing in forced convection in a channel.

Some preliminary studies on the subject are available in literature [2-4], but they are in most cases partial treatments and they do not lead to the proposal of correlations to be used for turbulent flows. Recently, a research activity on internally heated molten salts has been undertaken at the Politecnico Di Milano [5-7], focusing in particular on graphite-moderated MSR concepts. In the frame of such research activity, a new correlation has been proposed [5,6], giving a clear indication of the importance of internal heat generation in the heat transfer phenomena occurring in MSRs. As an example, figure 1 shows the Nusselt number in the core channels of the Molten Salt Breeder Reactor (MSBR) [8] as predicted by traditional correlations [9-13], as well as by the correlation proposed in [5]. A non-negligible difference is observed, especially for low Reynolds numbers. A drawback of the correlation proposed in [5,6] is that it has been obtained through interpolation of numerical/analytic results, without relying on a theoretical background and resulting in a complex functional form.

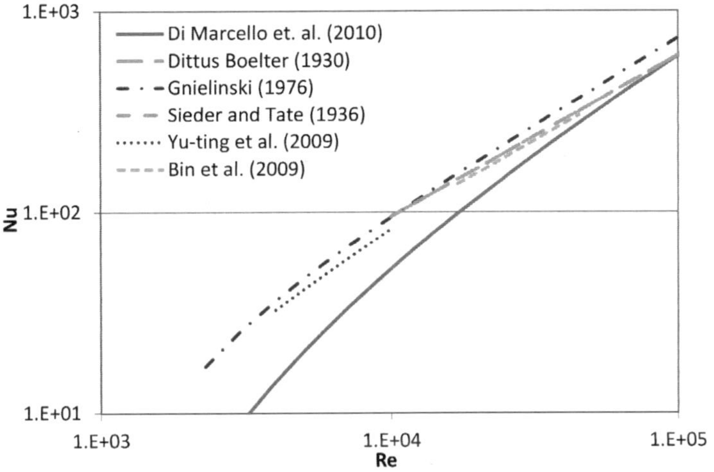

Figure 1. Nusselt number in the average core channel of the MSBR [5,9-13]

The paper is organized as follows. In Section 2, a heat transfer correlation is presented for evaluating the Nusselt number for the turbulent flow of internally heated molten salts in straight circular channels. Section 3 preliminarily discusses a possible experimental set-up that could be used to validate the results presented in Section 2. The proposed correlation is finally used in Section 4 to predict the effect of the internal heat generation on the out-of-core components of the Molten Salt Fast Reactor (MSFR), the reference circulating-fuel MSR design in the GIF-IV. Concluding remarks are provided in Section 5.

2. Heat transfer correlation

An accurate analytic treatment of the heat transfer for internally heated fluids has been developed in the past at the Politecnico di Milano [5-7]. Treatments of this kind, as well as dedicated CFD codes, can be used to investigate the details of heat transfer processes in many engineering applications. Nevertheless, when dealing with complex systems, computational requirements make often impractical the direct application of such techniques. In addition, simpler tools are needed to achieve first general information in preliminary design processes, as in the case of the MSFR out-of-core components. In these contexts, it can be useful to rely on the use of heat transfer correlations. The present section proposes a theoretical derivation of a correlation form for the Nusselt number in channels featured by internally heated fluids. A specific correlation is then obtained for the case of molten salts (Prandtl numbers approximately in the range $7.5<Pr<20$) through interpolation of data provided by the tool described in [5,6].

2.1. Derivation of a general correlation form

In this subsection, the problem of heat transfer in channels featured by internally heated fluids is treated theoretically, and a general form of the Nusselt number correlation is derived by means of the Π-theorem [14] and thanks to some physical considerations. In order to obtain such correlation, the field of investigation is restricted to a smooth channel with circular cross-section, average inlet fluid velocity u_{avg}, hydro-dynamically and thermally developed turbulent flow, uniform internal heat generation Q and constant (inward or outward) wall heat flux j_w. (figure 2).

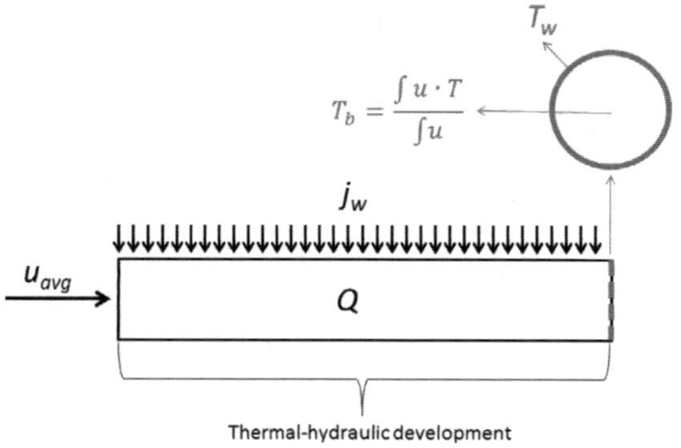

Figure 2. Schematic representation of the investigated heat transfer process

As a first step in the derivation of the heat transfer correlation, it should be recognized that the linearity (with respect to temperature) of the energy equation allows us to treat the physical situation here analysed as the superimposition of two simpler situations:

situation 1. A fluid flow without internal heat generation, but featured by constant wall heat flux j_w, which is the typical case considered by traditional heat transfer correlations;

situation 2. A fluid flowing in an adiabatic channel with volumetric heat generation Q.

Hence, it is possible to compute the temperature difference between wall (T_w) and bulk (T_b) temperatures as:

$$(T_w - T_b)_{Q+j_w} = (T_w - T_b)_Q + (T_w - T_b)_{j_w} \qquad (1)$$

where the subscript $Q+j_w$ refers to the complete situation with both internal heat generation and wall heat flux, while the subscripts Q and j_w indicate that the temperature differences are computed in the simplified situation of internal heat generation alone (situation 2) and wall heat flux alone (situation 1), respectively. The presence of internal heating in situation 2 will result in a higher temperature at the channel wall, where the fluid is slower. In case of an inward wall heat flux the temperature difference $(T_w\text{-}T_b)_{Q+jw}$ will be increased compared to the classical one $(T_w\text{-}T_b)_{jw}$, thus suggesting a worsening of the heat transfer coefficient. In case of an outward heat flux, $(T_w\text{-}T_b)_{Q+jw}$ will instead be decreased (in absolute terms) and the heat transfer coefficient improved.

Introduction of Equation (1) in the definition of heat transfer coefficient leads to (j_w is assumed positive if entering the channel):

$$h_{Q+j_w} = \frac{j_w}{(T_w-T_b)_Q+(T_w-T_b)_{j_w}} = \frac{1}{\frac{(T_w-T_b)_Q}{j_w}+\frac{(T_w-T_b)_{j_w}}{j_w}} = \frac{1}{h_Q^{-1}+h_{j_w}^{-1}} \qquad (2)$$

Attention must be paid to the term h_Q. Although similar to h_{jw} in its definition, it does not represent a heat transfer coefficient, since it includes in the same definition the temperatures of the situation 2 (with internal heat generation and with adiabatic walls) and the heat flux of situation 1 (without internal heat generation). It is now possible to rewrite the previous equation as:

$$Nu_{Q+j_w} = \frac{1}{Nu_Q^{-1}+Nu_{j_w}^{-1}} = Nu_{j_w}\frac{1}{1+\frac{Nu_{j_w}}{Nu_Q}} \qquad (3)$$

Equation (3) implies that the Nusselt number (*Nu*) in case of internally heated fluids and constant wall heat flux Nu_{Q+jw} can be computed by means of traditional correlations for the value of Nu_{jw} with the introduction of the corrective factor:

$$\zeta = \frac{1}{1+\frac{Nu_{jw}}{Nu_Q}} = \frac{1}{1+\frac{h_{jw}}{h_Q}} = \frac{1}{1+\sigma} \tag{4}$$

Assuming constant the properties of the fluid, it is possible to write:

$$\sigma = \sigma(c_p, \mu, \rho, k, D, u_{avg}, Q, j_w) \tag{5}$$

where c_p is the specific heat, μ the dynamic viscosity, ρ the density, k the thermal conductivity, and D the channel diameter. The assumption of constant physical properties is a typical one in the derivation of heat transfer correlations and it is often considered to be of second order importance [9]. However it should be kept in mind that the possibility exists to derive more accurate correlations by including a dependency upon the ratio of the viscosity evaluated at wall and bulk temperatures (see e.g. Ref. [13] for molten salts). Some of the dependencies in Equation (5) can be made explicit, thus reducing the experimental/computational efforts in the derivation of σ. Equations (2) and (4) yield:

$$\sigma = \frac{h_{jw}}{h_Q} = \frac{h_{jw}(T_w - T_b)_Q}{j_w} \tag{6}$$

The terms h_{jw} and $(T_w - T_b)_Q$ are independent of j_w. Moreover, the term $(T_w - T_b)_Q$ is directly proportional to Q - see for example [3]. It follows:

$$\sigma = \frac{Q}{j_w}\sigma'(c_p, \mu, \rho, k, D, u_{avg}) \tag{7}$$

Use of Π-theorem in the previous equation finally leads to:

$$\sigma = \frac{QD}{j_w}\varphi(Pr, Re) \tag{8}$$

Such useful result allows us to characterize the heat transfer in a channel with internal heat generation using only two parameters (Prandtl and Reynolds numbers), similarly to the case of a flow without heat generation. Clearly, this is also based on the assumption of constant physical properties (particularly for viscosity). The overall correlation for the Nusselt number in case of simultaneous wall heat flux and internal heat generation should then assume the form:

$$Nu_{Q+j_w} = \zeta Nu_{j_w} = \frac{1}{1+\sigma} Nu_{j_w} = \frac{1}{1+\frac{QD}{j_w}\varphi(Pr,Re)} Nu_{j_w} \tag{9}$$

Assuming as positive the function $\varphi(Pr,Re)$ (see next subsection and Equation (12)), Equation (9) confirms that the Nusselt number in case of internally heated fluids is increased/decreased in case of outward/inward wall heat flux. The correction is higher when the Q/j_w ratio is higher, and when the channel diameter is increased. In case the wall heat flux is exiting the channel (j_w<0), Nu_{jw} may become negative for high values of the ratio QD/j_w. Physically, this means that the bulk temperature is lower than the wall temperature, although the heat is flowing out of the channel.

2.2. *Data interpolation and application to molten salts*

The results of the previous subsection can be used to derive Nusselt number correlations for internally heated fluids. In particular, if the correlations available in literature for Nu_{jw} are adopted, what is required is the derivation of the function $\varphi(Pr,Re)$. In case of laminar flow, the function $\varphi(Pr,Re)$ is constant and equal to 3/44 [3]. In case of turbulent flows, it can have a complex shape but, by restricting the field of application, it is reasonable to assume a simple dependency like:

$$\varphi(Pr, Re) = a_1 Pr^{a_2} Re^{a_3} \tag{10}$$

where a_1, a_2 and a_3 are constants to be determined. At this point, it is possible to evaluate the function φ and derive the constants a_1, a_2 and a_3. In particular, Equation (9) can be rearranged as:

Journal of Physics: Conference Series **501** (2014) 012030 doi:10.1088/1742-6596/501/1/012030

$$\varphi(Pr, Re) = \frac{j_w}{QD}\left(\frac{Nu_{j_w}}{Nu_{Q+j_w}} - 1\right) \tag{11}$$

and the Nusselt numbers can be computed using the model described in [5,6]. Such model was shown able to reproduce the experimental data presently available with an accuracy comparable to the data uncertainties [5]. Comparison with the numerical results obtained using state-of-the-art CFD codes also showed an excellent agreement (within few percents) [6]. Computing φ for 100 different combinations of Prandtl and Reynolds numbers in the ranges $7.5<Pr<20$ and $10^4<Re<10^5$, and interpolating Equation (10) in the least square sense, the following correlation is obtained:

$$\varphi(Pr, Re) = 1.656 \cdot Pr^{-0.4} Re^{-0.5} \tag{12}$$

Equation (12) interpolates the results given by the model in [5,6] with an average error equal to 4.9% and a maximum error equal to 10.2% (figure 3). Such discrepancies can be considered acceptable for preliminary calculations. Extrapolation to Reynolds numbers as low as 2300 gives discrepancies of about 20%. Adopting Equation (12), the overall correlation for the Nusselt number in case of both internal heat generation and wall heat flux would be:

$$Nu_{Q+j_w} = \frac{1}{1+\frac{QD}{j_w}\varphi(Pr,Re)} Nu_{j_w} = \frac{1}{1+\frac{QD}{j_w}1.656 \cdot Pr^{-0.4} Re^{-0.5}} Nu_{j_w} \tag{13}$$

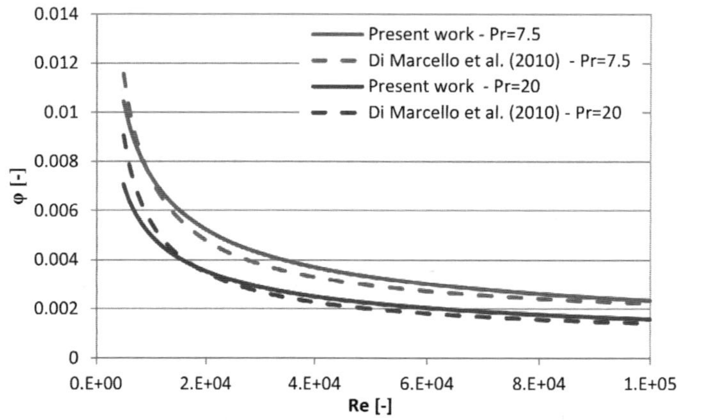

Figure 3. Prediction of the function φ according to di Marcello et al. [5] and using Equation (12)

3. Possible experimental set-up

With the objective of validating the correlation proposed (Equation (12)), the set-up of a proper experimental facility is necessary. Equations (6) and (8) can be rearranged as:

$$\varphi(Pr, Re) = Nu_{j_w} \frac{(T_w - T_b)_Q}{\frac{QD^2}{k}} \tag{14}$$

Assuming Nu_{jw} as known from available correlations (see e.g. [9-13]), what is necessary from an experimental point of view is the evaluation of the term $(T_w\text{-}T_b)_Q$. This requires a facility able to reproduce the condition of an internally heated, thermally and hydro-dynamically developed turbulent flow in a straight, circular and adiabatic channel. The experimental set-up must be suitable for measuring wall and bulk temperatures. Following the assumptions used to derive Equation (12), a uniform (and precisely-known) internal heat generation Q should be guaranteed. In addition, the definition of the dependency of φ upon Prandtl and Reynolds numbers requires the possibility to vary in a known way the fluid properties, as well as the possibility to vary and measure the fluid velocity.

A suitable starting point for the required experimental analyses is the facility adopted by Kinney and Sparrow [2]. This set-up was used to test water (close to room temperature and at atmospheric pressure), which could also be considered for the experimental set-up here under investigation. It would avoid high temperatures (and the related problems of structural materials and instrumentation) and use of toxic molten fluoride salts, while the Prandtl number typical of molten salts could be achieved by using thickening agents. However, attention should be paid to the possible non-Newtonian effects related to the use of thickening agents. In addition, it should be taken into account that simulation of molten salts using a different fluid with the same Prandtl number is theoretically possible only for constant fluid properties. A facility using water will then be useful to assess Equation (12), but it will be impossible to accurately determine additional correction factors allowing e.g. for the variation of viscosity with temperature.

A schematic view of the possible facility is shown in figure 4. A closed loop can be used, with a heat exchanger required for cooling the working fluid after it has been warmed in the test section. Such test section must be long enough to assure condition of full thermal and hydro-dynamic development. Kinney and Sparrow [2] adopted a test section length equal to 85 times its diameter, 40 of which for the sole hydraulic development. As a matter of fact, by analysing their results, it appears that the remaining 45 diameters for the thermal development can be reduced at least to 30 diameters (this reduces by 33% the total power required). Internal heating in the fluid can be obtained through Joule effect by forcing an electrical current to flow into it. This is possible by choosing an electrically insulating material (e.g., polyvinyl dichloride) for the channel wall in the test section, and by placing electrodes at the sides of it. In this way, the current is forced to flow longitudinally in the fluid. Adopting electricity to heat up the water also solves the problem of the knowledge of the volumetric power Q, which can easily be derived by measuring the electric current in the circuit and the voltage difference at the electrodes. Alternate current might help reducing problems of electrolysis, but further evaluations would be needed to exclude major problems in this sense, especially considering the presence of salting agents in the water and the high voltages required (see below). A cylindrical shell is posed outside the duct to reduce thermal losses towards the environment. In this shell, a certain amount of power is dispersed through electrical resistors in order to reach the desired adiabatic boundary condition at the pipe wall. Particular attention is necessary to well insulate the test section in critical zones such as contact points with the structure that supports the channel, near which the temperature field might be distorted by relatively high thermal losses. To overcome such difficulty, the arrangement illustrated in figure 4 can be adopted, where a single beam is used as supporting structure, with the test section connected to it through a silica aero gel support.

The velocity of the fluid can be varied by using an appropriate pump or valves, and it can be measured by means of standard techniques (e.g., Coriolis flow meter). As far as temperature measurements are concerned, for validating Equation (12) it is sufficient to measure wall and bulk temperatures just before the outlet of the heated section (but sufficiently distant from the last electrode, to avoid disturbances due to the distorted electric field). Wall temperature can be achieved using a thermocouple. In view of the electric field in the fluid, and in order to avoid flow disturbances, the thermocouple can be placed outside the channel, drilled partway in a shallow hole through the pipe wall. In fact, the adiabatic condition allows one to consider as approximately constant the temperature throughout the wall. The information about bulk temperatures can be gained using two mixing chambers before and after the test section, possibly compensating the obtained information considering the heat produced after the wall temperature measuring point. This suggests preliminary calculations to assess the electric field configuration and the resulting heating non-uniformity close to the electrodes. Voltage measurement should be taken to assess these calculations after the facility has been set up. A moving and electrically insulated thermocouple could also be placed inside the fluid and adopted to measure the entire radial temperature profile, which could be useful to validate CFD codes, as well as a redundant measurement for the bulk temperature (assuming as approximately known the velocity profile). As regards the irregular electric field close to the electrodes, the distorted heat generation is also expected to perturb the temperature profile, but according to Kinney and Sparrow, the effect of a small perturbation should propagate for a distance equal to 2-3 diameters at most. To confirm this, it is worth considering the use of thermocouples throughout the length of the heated section, at least at wall. This would also allow one to draw information about the thermal development of the fluid.

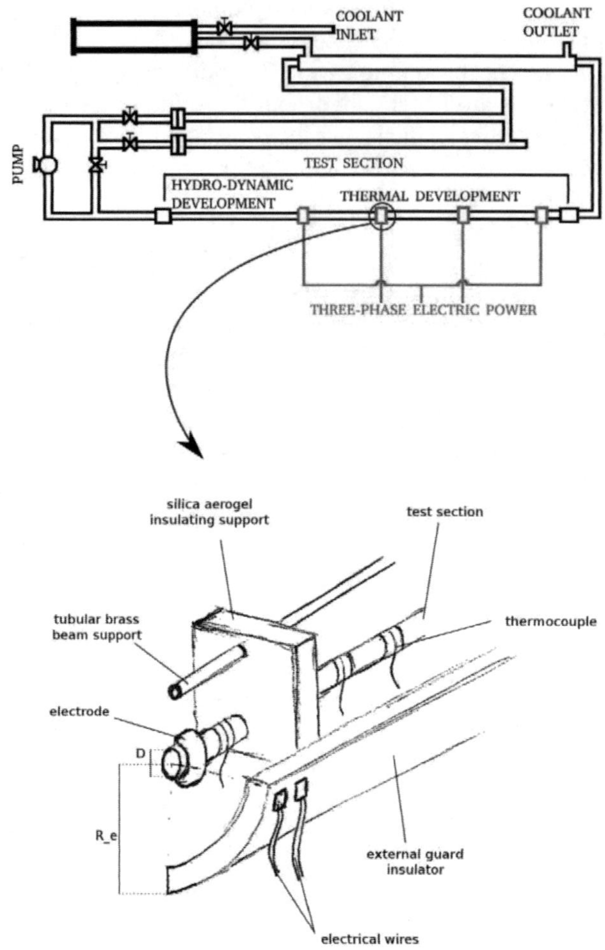

Figure 4. Schematic view of the possible experimental facility

A main difficulty about the experiment is related to the small magnitude of the temperature difference between wall and bulk temperatures. Equation (14) and use e.g. of the Dittus-Boelter correlation [9] for Nu_{jw} yield:

$$(T_w - T_b)_Q = \frac{QD^2\varphi(Pr,Re)}{kNu_{jw}} = \frac{QD^2}{k}\frac{1.656}{0.023}Re^{-1.3}Pr^{-0.8} \tag{15}$$

Taking into account that the proportions of the tube are fixed because of the necessary thermal development of the flow, assuming a 30 diameter long heated section, and assuming the thermal conductivity of the water approximately equal to 0.6 W/(m·K) it is possible to write:

$$(T_w - T_b)_Q = \frac{Q_{tot}}{7.5\pi Dk}\frac{1.656}{0.023}Re^{-1.3}Pr^{-0.8} \cong 1.2[K \cdot m/(SV^2)]\,\sigma_e V^2 Re^{-1.3}Pr^{-0.8} \tag{16}$$

where Q_{tot} is the total power in the heated section, V the effective phase voltage and σ_e the electric conductivity.

The temperature difference is then directly proportional to the Q_{tot}/D ratio, and to the product of the fluid electrical conductivity and the voltage. A minimum diameter of the tube equal to 2.54 cm (1 inch) has been assumed to make the dimensions of the thermocouples reasonably small compared to it. As concerns the electrical conductivity, use of salted water is necessary to have acceptable voltages. Kinney and Sparrow [2] used a 4 molal solution of NaCl in water. With a voltage of 400 V and a total power of 44 kW they were able to attain a $(T_w-T_b)_Q$ on the order of 1 K for Prandtl and Reynolds numbers equal to 4 and $8\cdot 10^4$, respectively. In the present case, Prandtl and Reynolds numbers as high as 20 and 10^5, respectively, are to be investigated. This calls for a highly conductive fluid to keep as low as possible the necessary voltage. Use of NaCl limits the electrical conductivity to about 25 S/m, which has been considered too small for the purposes of the facility. A preliminary investigation allowed us to single out a 4 molal solution of KBr in water as a suitable candidate. KBr is in fact inexpensive and allows reaching a noteworthy 50 S/m electrical conductivity [17]. The effect of KBr addition on the Prandtl number of water is small and goes in the direction of slightly increasing it.

Assuming D=2.54 cm, σ_e=50 S/m, and using the properties of salted water [17], Equations (15) and (16) allow us to obtain power and voltage as a function of $(T_w\text{-}T_b)_Q$. Assuming the worst condition necessary to validate Equation (12) (i.e., Pr=20, Re=10^5), the data reported in figure 5 are obtained. It is clear that high voltages are required to achieve temperature differences above 1 K. For instance, if 1.5 K is considered an acceptable temperature difference to be measured (it caused a 5-10% scatter in the wall-to-bulk temperature difference data in Ref. [2]), 930 V would be required, with a notable total power equal to 260 kW. Velocities in the test section would be high (~10 m/s), while the temperature difference between inlet and outlet is not a concern, being on the order of 10 K.

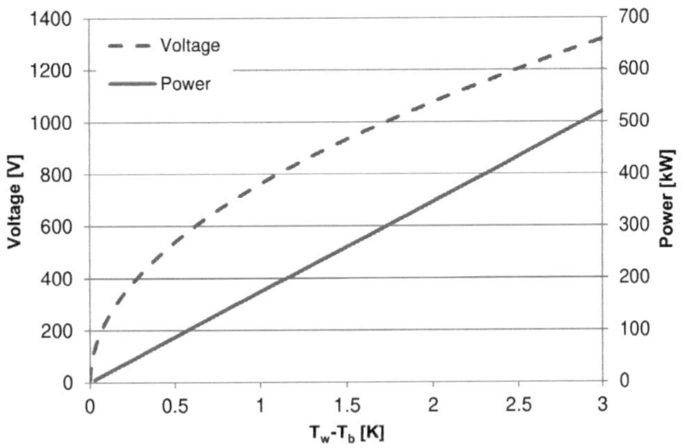

Figure 5. Voltage and power as a function of $(T_w\text{-}T_b)_Q$ – Pr=20, Re=10^5

4. Impact of the internal heat generation on the MSFR out-of-core components

The proposed correlation (Equation (12)) can be used to preliminarily investigate the impact of decay heat on the heat transfer processes in a number of MSFR out-of-core components, including pipes of the primary circuit, tubes of the heat exchanger, and channels of the salt reprocessing system (see e.g. [15,16] for the details about the MSFR design). As observed in subsection 2.1, the effect of the internal heat generation superimposes linearly with the effect of heat transfer at the channel wall. Independent of the wall heat flux, the effect of internal heat generation translates into a wall temperature increased by the term $(T_w\text{-}T_b)_Q$ (see Equation (1)), which can conveniently be used to assess the impact of internal heat generation on a generic component. To this purpose, Equations (6) and (8) can be rearranged to yield:

$$(T_w - T_b)_Q = \frac{QD^2\varphi(Pr,Re)}{kNu_{jw}(Pr,Re)} = \frac{QD^2\varphi(D,u_{avg})}{kNu_{jw}(D,u_{avg})} \quad (17)$$

where the dependency on Re and Pr has been changed to a dependency on diameter and average velocity by using the MSFR salt properties (table 1).

Table 1. MSFR fuel salt properties [15]

Salt density [kg/m^3]	$4094\text{-}8.82\cdot10^{-1}\cdot(T[K]\text{-}1008)$
Salt specific heat [J/kg-K]	$-1111+2.78\cdot T[K]$
Salt thermal conductivity [W/m-K]	$0.928+8.397\cdot10^{-5}T[K]$
Salt kinematic viscosity [m^2/s]	$5.54\cdot10^{-8}\cdot\exp(3689/T[K])$

At this point, it is possible to evaluate the term $(T_w\text{-}T_b)_Q$ by using the correlation (12) for φ and suitable correlations for Nu_{jw}. As mentioned in subsection 2.2, Equation (12) is mainly valid for $10^4<Re<10^5$, but it can be extrapolated to $Re=2300$ with acceptable discrepancies. For Nu_{jw}, the Dittus-Boelter correlation [9] can be used for $Re>10^4$. For $2300<Re<10^4$, the Gnielinski correlation [10] gives instead more accurate results. For a laminar flow ($Re<1000$), Equation (14) becomes:

$$(T_w - T_b)_Q = \frac{QD^2\varphi(Pr,Re)}{kNu_{jw}(Pr,Re)} = \frac{QD^2 3/44}{k48/11} = \frac{QD^2 3}{k192} \tag{18}$$

The internal heat generation Q in the out-of-core components of the MSFR is determined by decay of actinides and fission products, which is on the order ~8 MW/m^3 [16]. Figure 6 shows the results of the numerical evaluation of equations (17) and (18). It only shows results for $Re<10^3$ and $2300<Re<10^4$. The scale of the colorbar has been limited to 120 K in the laminar region to allow a better visualization of results in the turbulent region. In fact, the term $(T_w\text{-}T_b)_Q$ would reach ~300 K at the right corner of the laminar region.

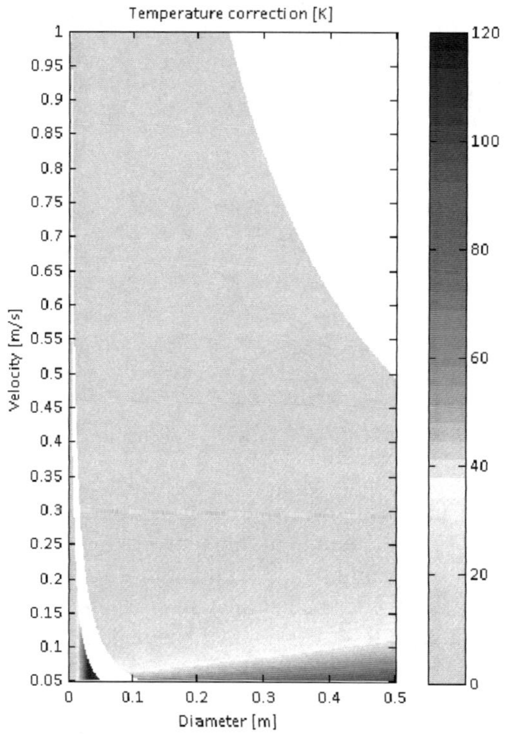

Figure 6. Increment of the wall-to-bulk temperature difference $(T_w\text{-}T_b)_Q$ due to the internal heat generation in the MSFR channels as a function of salt velocity and channel diameters

Figure 6 shows that the internal heat generation has an extremely limited impact on wall temperatures for high velocities. This excludes any significant effects on the pipes of the primary circuit and on the tubes of the heat exchangers, featuring velocities on the order of few meters per second [16]. However, the MSFR design includes a chemical system for fuel reprocessing, with reprocessing rates of the order of few litres or few tens of litres per day [16]. As an example, one can consider a channel used to transfer the fuel salt from the core to the reprocessing system. Adoption of a channel with 5 cm diameter and 5 cm/s of flow velocity would result in a reasonable salt extraction rate of 1 litre every 10 seconds. Considering a channel 2 m long, the internal heat generation would cause an acceptable increment of the bulk temperature of ~50 K, but the wall temperature close to the end of the channel would be approximately 300 K higher than the bulk one (as mentioned above).

5. Conclusions

The problem of heat transfer for internally heated fluids flowing in forced convection in a straight circular channel has been faced on a theoretical ground. After making the assumption of full thermal-hydraulic development of the flow, some physical considerations and application of the Π-theorem have made possible the achievement of a general form for a heat transfer correlation. It has been shown that the effect of internal heat generation can be described by means of a corrective factor to be applied to traditional correlations (e.g., the Dittus-Boelter or Gnielinski correlations) for the Nusselt number. The corrective factor has been demonstrated to be lower than one in case of inward wall heat flux, thus implying an overestimation of the heat transfer coefficient in case traditional correlations were used. In case of outward wall heat flux, the correction factor is instead generally higher than one. The possibility also exists of a negative heat transfer coefficient in case the wall heat flux is small compared to the internal heat generation. It has been shown that the corrective factor can be fully characterized by making explicit the dependency on just Reynolds and Prandtl numbers. On this basis, a preliminary correlation has been proposed for molten salts by interpolating the results provided by an analytic approach previously developed at the Politecnico di Milano.

A possible experimental facility for testing the proposed correlation has been discussed. In spite of some technical difficulty related to the achievement of internal heat generation through Joule effect, the facility appears feasible.

Finally, the developed correlation has been used to carry out a parametric investigation of the effect of decay heat on the out-of-core components of the reference Generation IV Molten Salt Reactor. The volumetric power determines higher temperatures at the channel wall, but the effect is significant only in case of large diameters and/or low velocities. This might be the case of the channels of the salt reprocessing system.

Acknowledgements

The authors acknowledge Mr. Manuele Aufiero (Politecnico di Milano), Mr. Jacopo De Amicis (Politecnico di Milano) and the experts from the SIET Laboratories (Piacenza, Italy) for their valuable suggestions about the presented experimental set-up.

References

[1] FLUENT 6.2 User's Guide, Fluent Inc., 2005
[2] Kinney R B and Sparrow E M 1966 *J. Heat Tranf – T. ASME* **88C** 314-322
[3] Poppendiek H F 1954 *Chem. Eng. Prog. S. Ser.* **50** 93-104
[4] Siegel R and Sparrow E M 1959 *J. Heat Tranf – T. ASME* **81** 280-290
[5] Di Marcello V, Cammi A and Luzzi L 2010 *Chem. Eng. Sci.* **65** 1301-1310
[6] Luzzi L, Cammi A, Di Marcello V and Fiorina C 2010 *Chem. Eng. Sci.* **65** 4873-4883
[7] Luzzi L, Aufiero M, Cammi A and Fiorina C 2012 *Hydrodynamics - Theory and Models*, chap. 6, Jin-hai Zheng Editor, InTech Publisher
[8] Robertson R C 1971 Conceptual design study of a single-fluid molten-salt breeder reactor *Technical report ORNL-4541*
[9] Dittus F W and Boelter L M K 1930 *University of California publications in Engineering* 443-461
[10] Gnielinski V 1976 *Int. Chem. Eng.* **16** 359-367
[11] Sieder E N and Tate G E 1936 *Ind Eng. Chem.* **28** 1429-1435
[12] Yu-ting W, Bin L, Chong-fang M, Meng Y and Hang G 2009 *Exp. Therm. Fluid Sci.* **33** 1128-1132
[13] Bin L, Yu-ting W, Chong-fang M, Meng Y and Hang G 2009 *Int. Commun. Heat Mass* **36** 912-916
[14] Langhaar H L 1962 *Dimensional analysis and theory of models* John Wiley and Sons (New York, US)
[15] Merle-Lucotte E, Heuer D, Allibert M, Brovchenko M, Capellan N and Ghetta V 2011 Launching the Thorium Fuel Cycle with the Molten Salt Fast Reactor *Proc. ICAPP 2011* (Nice, France)
[16] Fiorina C 2013 The Molten Salt Fast Reactor as a Fast-Spectrum Candidate for Thorium Implementation *PhD Thesis* Politecnico di Milano, Italy
[17] Isono T 1984 *J. Chem. Eng. Data* **29** 45-52

Journal of Physics: Conference Series **501** (2014) 012031 doi:10.1088/1742-6596/501/1/012031

Preliminary experimental evaluation of thermal conductivity of ceramic pebble beds

D Aquaro and R Lo Frano
DICI- University of Pisa, Pisa- Italy

E-mail: rosa.lofrano@ing.unipi.it

Abstract.
This paper illustrates the preliminary experimental tests for determining the effective thermal conductivity of ceramic pebble beds versus temperature and compression strains.
Ceramic pebble beds are promising candidates to be used in breeding blankets for nuclear fusion reactor as breeder and neutron multiplier.
The tests were performed with an experimental rig, built at the DICI-University of Pisa, which permits to determine the thermal conductivity of pebble beds in steady state conditions, at several temperatures and compression forces. The values of thermal conductivity of pebble beds are obtained as function of a known conductivity of an alumina disc.
The assessment of the method has been performed determining the effective thermal conductivity of alumina pebbles beds of different diameters. Void fraction and compression strains are the parameters that mainly influence the variability of the thermal conductivity of the beds.

1. Introduction

The breeding blanket of nuclear fusion reactor produces, by means of neutron capture, the tritium necessary for the fusion reaction, slows down the neutrons and permits to extract the thermal power.

Compounds of lithium (lithium orthosilicate (Li_4SiO_4), lithium metatitanate (Li_2TiO_3)) in form of ceramic pebbles are breeder candidates in the design of breeding blankets for nuclear fusion power technology.

These breeders, packed in the form of pebble beds into a box-like structure, are characterized by strong mechanical loading arising from the differential thermal expansion between breeder pebbles and the containing structure.

The knowledge of the effective thermal conductivity of pebble beds is important for a proper thermo-mechanical blanket design [1] and assessment of the heat transfer processes.

The available measurements (in the literature) of the thermal conductivity of ceramic pebbles bed are concentrated mainly on the uncompressed pebble beds [2-3], while very few researches has been performed [1],[4-6] considering the bed subjected to compression forces.

The research program, developed at the Department of Civil and Industrial Engineering (DICI) of the University of Pisa, aims to characterize the thermo-mechanics behavior of ceramic pebble beds. Specifically the thermal conductivity, at several temperatures and values of compression forces (up to 100 kN), was determined experimentally [7].

The experimental tests look like oedometric tests from a mechanical point of view and were carried out on a thin layer of pebbles.

Journal of Physics: Conference Series **501** (2014) 012031 doi:10.1088/1742-6596/501/1/012031

The approach used to evaluate the thermal conductivity is based on the thermal conductivity of an alumina disc (used a reference value) experimentally determined with the hot wire method carried out in a furnace (in the 20 °C-200 °C range) [8].

Numerical simulations were also performed in order to determine the accuracy and consistency of the envisaged method for the evaluation of the pebble bed conductivity.

2. Assessment of thermal conductivity of pebble beds

The pebble bed conductivity depends on the temperature, strain and interstitial gas. Deformations, caused by stresses due to different thermal expansions between beds and other structural materials, may result in variations of the effective thermal conductivity of the beds. The thermal conductivity of pebble beds is determined without any compressive load (zero strain tests) or loading the bed with increasing loads (tests at strain steps). In this latter case, the pebble bed is submitted at progressive strain in correspondence of which the thermal conductivity is evaluated.

Experimental methods for determining the thermal conductivity are the following:

- Pulsed Hot Wire Method (HWM);
- Hot Plate with guard rings Method.

J.Reimann et al. [1] have determined the thermal conductivity of Li4SiO4 and Li2TiO3 pebble beds with different interstitial gas (air and helium) using the pulsed hot wire method. Aquaro et al. [9] applied the 'hot plate method with guard rings' for determining the thermal conductivity of Li4SiO4 and Li2TiO3 pebble beds at room temperature.

In this study a modified 'hot plate method with guardian rings' is used (according to UNI EN 12664 [10]) to evaluate the thermal conductivity of pebble bed in a steady state condition.

A heat flux, q, produced by an electrical resistance, is transmitted to a cylindrical specimen (thick s) of unknown thermal conductivity (λ) reaching a steady state thermal gradient (T_c, T_f upper and lower specimen temperature, respectively). In steady state conditions, the thermal conductivity may be calculated as:

$$\lambda = \frac{qs}{T_c - T_f} \qquad (1)$$

One or more resistances (called 'guard resistances') compensate the heat flux dispersion to the environment during the test execution.

The hot plate method has high accuracy if the heat loss to the environment is minimized; a drawback is, instead, the long duration of the test, necessary to reach steady state conditions.

The experimental device, available at University of Pisa, measures the effective thermal conductivity of ceramic pebbles beds by comparison with a known conductivity of a disc of Alumina. Therefore the experimental measurements are restricted to the temperature gradients through the bed and Alumina.

The main concept of the method is to generate a temperature gradient across the pebble bed and the disc of Alumina. In steady state heat transfer, the heat flux flowing through the alumina disc and pebble bed is the same, for zero heat losses (this is surely true in a region around the centreline).

The thermal conductivity of the pebble bed (λ_{PB}) is calculated as function of that of the alumina (λ_A), by applying the following equation:

$$\lambda_{PB} = \lambda_A \frac{\Delta T_A}{\Delta x_A} \cdot \frac{\Delta x_{PB}}{\Delta T_{PB}} \qquad (2)$$

Where ΔT_A and ΔT_{PB} are the temperature differences across the alumina and the bed. The thickness of the alumina and of the pebble bed are indicated respectively as Δx_A and Δx_{PB}.

Journal of Physics: Conference Series **501** (2014) 012031 doi:10.1088/1742-6596/501/1/012031

3. Description of the experimental device

The experimental equipment, shown in figure 1, consists of the following main parts: measurement cell (containing the pebbles), hydraulic jack, piston, load cell, heating source, heat sink, thermocouples, displacement transducers.

The measurement cell, containing a maximum useful height of 20 mm of the pebble beds, is a cylindrical component, made of stainless steel, having 100 mm inner diameter and 170 mm height.

A disc of alumina, of known conductivity (the value of which was experimentally evaluated by means of hot wire test, as forward illustrated), is directly placed on the upper surface of the pebble bed and it is connected to a copper plate, 20 mm thick, that allows to obtain a uniform radial distribution of the temperature.

The heating source is made of two mica armoured electrical resistances connected to the upper surface of the copper plate, while the heat sink is placed under the bed in order to generate an axial heat flux through the pebble bed.

The pebble bed is resting on 35 mm of thick copper plate, the lower part of which is cooled by air or water. A small duct allows to void the pebble bed or to inject into it pressurized air or different interstitial gases (nitrogen, helium).

Thermocouples are positioned in a similar arrangement on the lower and upper copper plates. In the upper copper plate, five K type thermocouples measure the radial distribution of the temperature: four of them are located on two diameters at 90° each other and at a distance of 36 mm from the fifth one, located in correspondence of the plate centreline.

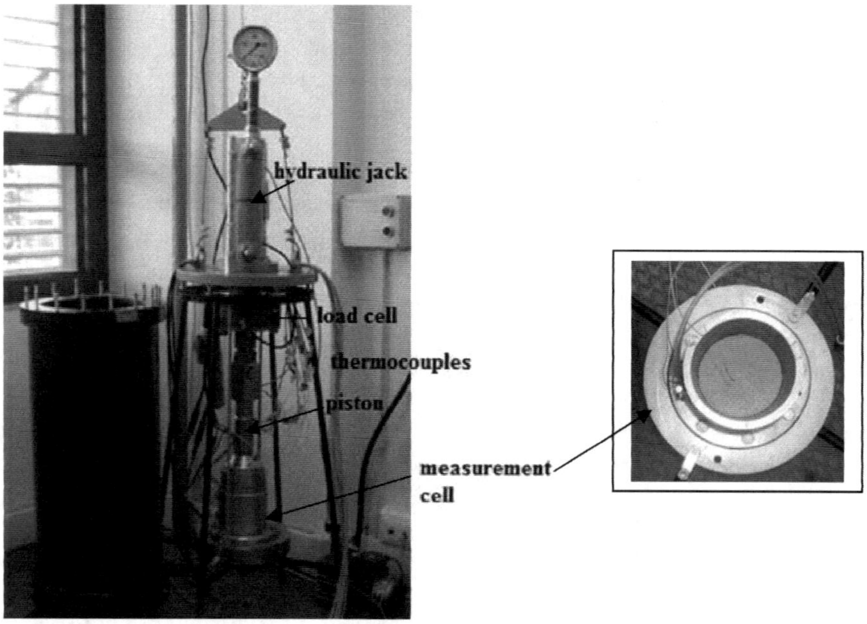

Figure 1. Experimental device: main components and measurement cell with pebbles

Another thermocouple crosses almost all the thickness of the alumina plate and provides a temperature measurement close to the pebble bed upper surface. Furthermore two O-Ring gaskets (capable to withstand temperatures up to 500° C) are mounted on the two mentioned copper plates in order to ensure the cell leak tightness.

Two electrical resistances (both peripherally and centrally positioned, as illustrated in figure 2) are designed to guarantee a minimum thermal gradient of 5 °C between the upper and lower surface of the pebble bed.

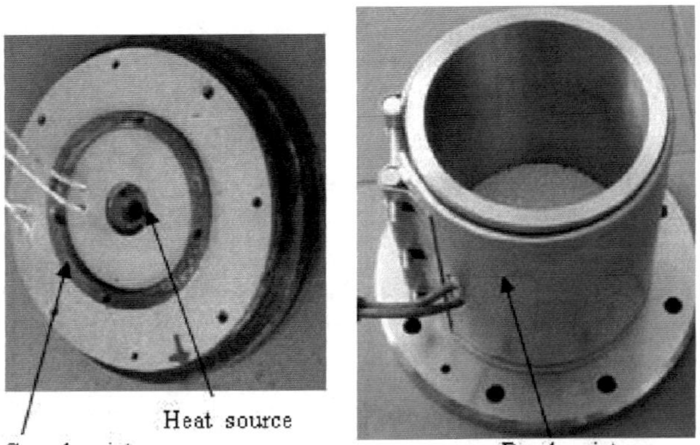

Figure 2. Heating source: electrical resistances.

The peripheral or guard resistance allows to compensate for heat flux radial losses. Around the cylindrical measurement cell a band shaped electric resistance (figure 2), 105 mm height, allows to heat the pebble bed up to 500°C. Moreover a hydraulic jack, connected to a load cell and a stainless steel piston, applies progressively the mechanical load (up to 100 kN) on the bed.

The piston is, in turn, connected through a coupling sleeve (50 mm height and 5 mm thick) to the upper copper plate (shaped in such a way to reduce the heat flux toward the hydraulic jack).

The piston displacement (corresponding to the pebble bed deformation) is measured by means of displacement laser transducer. The measurements of the load, displacement and temperatures were recorded by means of a Data Acquisition System made by National Instrument.

The uncertainty in the measurement of the thermal conductivity of the pebble bed is strictly dependent, in this type of tests, on the accuracy of the alumina conductivity values as well as on the accuracy of temperature measurements. Another source of uncertainty may be also the loss of heat flux in radial direction (reduced because the gradient of temperature is measured on the centerline).

The values of the alumina thermal conductivity, to be used in the equation 2, were determined by an experimental test campaign on an alumina specimen, shown in figures 3, having the same composition of that used in the experimental device.

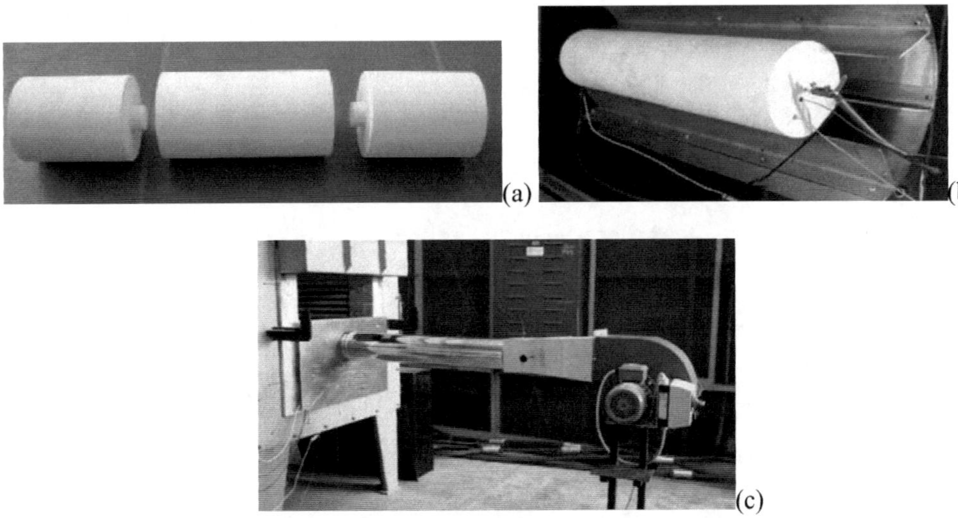

Figure 3. Alumina sample (a) used in 'hot wire test' (b, c) for determining the thermal conductivity.

The alumina thermal conductivity (h in table 1) has been determined at stationary conditions: the temperature gradient arisen across the longitudinal direction of the specimen was measured by two thermocouples by applying the 'hot wire test' method, like represented in figure 3 (b).

The preliminary results obtained (the test campaign is still on-going), summarized in table 1, are considered of meaningful importance to study, analyse and characterize the thermo-mechanical behaviour of the alumina. This phase is necessary and important in the setting up a methodological approach capable, in turn, to determine the thermal conductivity of Li4SiO4 and Li2TiO3 pebbles bed.

Table 1. Experimental values of the alumina thermal conductivity.

h [W/m^2K]	**T** [°C]	λ_A [W/mK]
34.71	36.4	13.82
34.67	39.6	14.71
34.62	44.8	15.53

In previous table 1, h is the heat transfer coefficient between the external surface of Alumina sample and the furnace ambient; T, the average temperature of Alumina; λ_A, the thermal conductivity of Alumina.

The assessment of the experimental method has been done performing tests on Alumina pebble beds. In these tests, the material of disc and pebbles is the same. Therefore the results will depend only on the geometrical characteristics of pebbles.

Preliminarily, two types of pebbles, having average diameter of 1.275 mm and 2.05 mm, respectively, were used. For the first type of pebble, the diameter ranges from 0.75 to 1.8 mm (tests TC1), while for the second ones it ranges from 1.8 and 2.3 mm (tests TC2). Moreover both two TC1 and TC2 were carried out without and with compression load and considering different temperatures.

The test procedure may be summarized as follows:

a) Heating of the pebble bed until an average temperature is reached;

b) Switching on the cooling system in order to generate an axial heat flow through the bed;

c) Start-up of the guard resistance to compensate for heat flux radial losses (depending on the radial distribution of temperature);

d) Application or not (depending on the type of tests) of the compression load;

e) Measurement of the reference temperatures;

f) Calculation of the value of the thermal conductivity of the bed.

4. Numerical simulations

Numerical pre-test simulations were also performed in order to optimize the operating test conditions and to evaluate the accuracy of the thermal conductivity measurements [6].

In this study only a brief description of FEM simulations of the hot plate method (used in the experimental tests) is reported, since it was extensively treated in [6]. The FEM model (implemented by using four nodes axial-symmetric elements) of the experimental device is shown in figure 4.

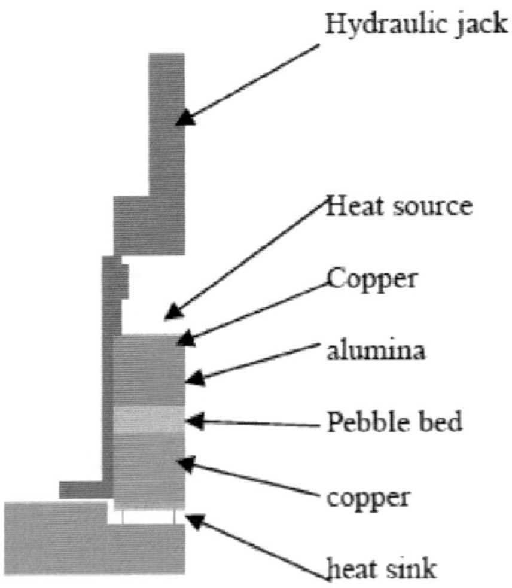

Figure 4. FEM model of the experimental device.

As indicated in [6], the bed was assumed to behave like an isotropic material having the real properties of the pebble bed. The heating sources (simulating the electric resistances shaped as disc, annular disc or band) were represented by means of heat fluxes applied on the correspondent surfaces.

The convection heat transfer in the heat sink was simulated through a suitable heat transfer coefficient (h = 10÷40 W/m^2 °C) and a bulk temperature (T_b), which depends on the cooling conditions. Two boundary conditions were also considered for the remaining surfaces: adiabatic surface (zero heat flux if insulated) or surface cooled by air at 20°C and heat transfer coefficient equal to 10 W/m^2°C (stagnant air).

The results of the numerical simulations, presented in [7], permitted to estimate the accuracy/discrepancy between the values of the thermal conductivity given in input [11-12] in the analyses and those obtained by the method described from the equation (2). As an example, the results, obtained in terms of pebble bed conductivity, highlighted that the discrepancies between the reference conductivity of Li2TiO3 (1 mm diameter)[11] and the values calculated with equations (1) and (2) resulted respectively about 24 % and about 12%. In the case of alumina (1 mm diameter) this discrepancy resulted about 1.26%. Furthermore the heat flux losses in the radial direction resulted higher (about 20%, as shown in figure 5) for the lithium metatitanate than for the alumina (about 4.5% as shown in figure 6).

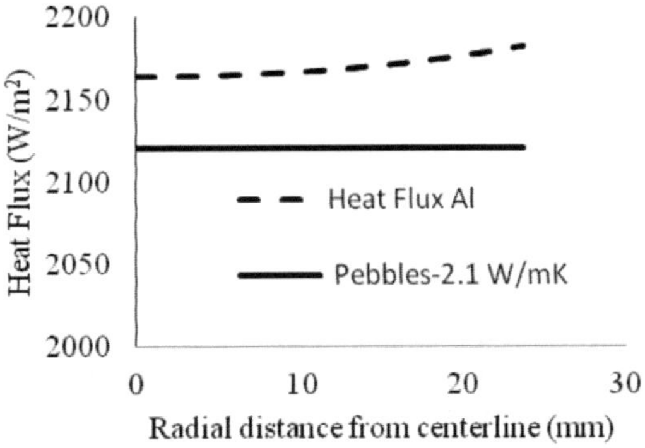

Figure 5. Radial distribution of heat flux for alumina pebble bed.

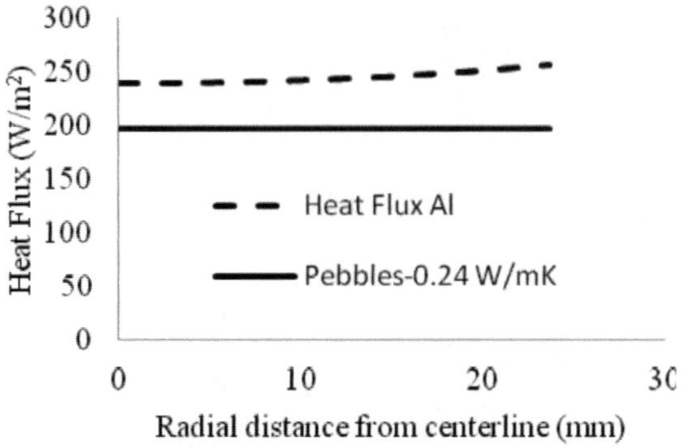

Figure 6. Radial distribution of heat flux in lithium metatinate pebble bed.

5. Experimental results and discussion

The temperature values measured in the alumina and in the bed, were analysed, by applying the equation (2). The results obtained, in terms of thermal conductivity versus temperature, are showed in figures 6 for both two test campaigns TC1 and TC2 (with and without compression load). The tests were performed with air as interstitial gas, while the theoretical packaging factor measured resulted about 92%. The compression force, as already indicated along the text, was 100 kN.

The thermal conductivity, in both two cases, increases along with the temperature of the bed. Moreover in the absence of compression load, the thermal conductivity value seemed to be less influenced by the dimension of pebble, although varying its diameter almost of a factor 2.

In fact Figure 7 (a) shows the ratio of the conductivities for the pebbles TC1 and TC2 versus the temperature. The ratio is almost independent of the temperature in the case of zero axial load, while it increases with the temperature if the bed is compressed.

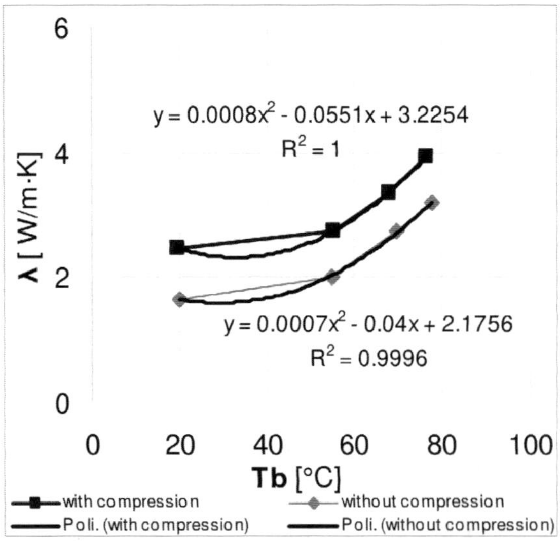

(a)

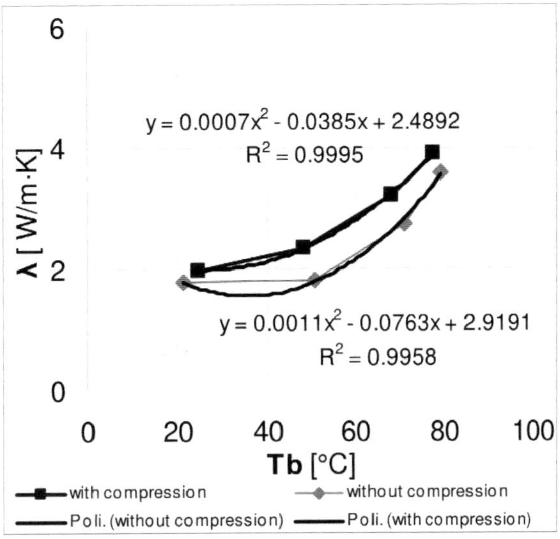

(b)

Figure 6. Thermal conductivity vs. temperature: TC1 (a) and TC2 (b) tests.

The void fraction and the strain (characterizing mainly the experimental data, as also widely addressed in literature) resulted the parameters that mainly influenced the increase of the thermal conductivity of bed. The increase of the temperature is responsible for a pebble deformation that, reducing the void, seemed to induce a compaction of bed.

Figure 7 (b) illustrates the influence of the compression load on the bed conductivity. In this figure the ratio between the Alumina conductivity and that of the beds of different diameters versus temperature is shown.

The compaction of the pebble bed, due to the temperature, determines the decrease of the influence of the diameter as well as of the compression strain on the conductivity of the bed. At 80 °C of temperature, the bed conductivity is 4 times lower than that of the homogeneous material.

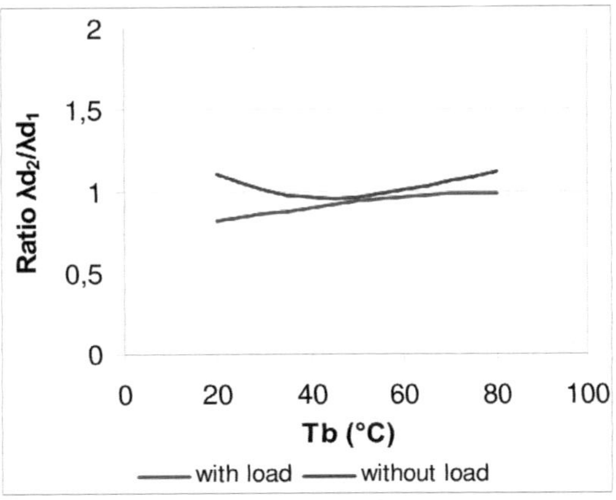

(a)

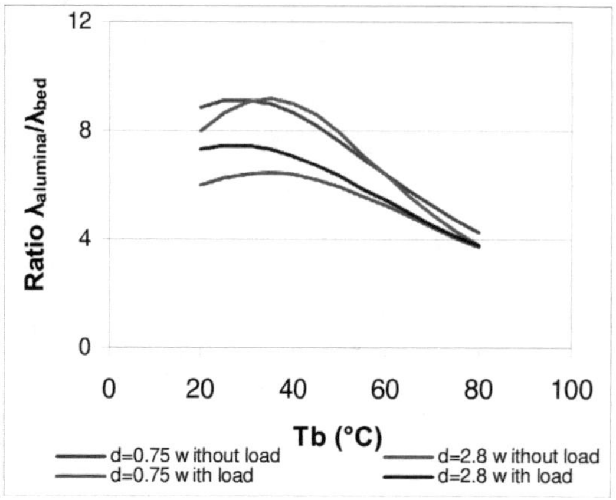

(b)

Figure 7. Thermal conductivity ratio vs. temperature: TC1 (a) and TC2 (b) tests.

6. Conclusion

In this paper the thermal conductivity of ceramic pebble beds, which are one of the candidates in the design of breeding blankets for nuclear fusion reactor has been evaluated. To the aim, the thermo-mechanical characterization of alumina pebble beds was done performing experimental tests at different temperatures and varying the compression force (up to 100 kN). Air was, at this first stage, the interstitial gas. The theoretical packaging factor measured resulted about 92%.

Numerical (pre-test) simulations were also carried out to optimize the operating test conditions: the uncertainty associated to the measurements of the alumina thermal conductivity (of 1 mm diameter) resulted about 1.26%.

The preliminary experimental results, since the test campaign is still ongoing, indicated that:

1) the thermal conductivity increases along with the bed temperature;

2) in the absence of compression load, the thermal conductivity seems to be less influenced by the dimension of pebble even if the diameter of pebbles is greater by, almost, a factor 2;

3) the void fraction and the strain are the parameters that mainly influence the variability of the thermal conductivity of bed;

4) the increase of temperature induces a pebble deformation and, in turn, a compaction of bed, which decreases the influence of the diameter and the compression strain.

The conductivity values increase when the bed is subjected to the compression load highlighting that the compaction phenomenon may induce a further increase of thermal conductivity (caused by the void reduction), even if a saturation effect could be foreseen.

Finally it is worthy to note that the results obtained may contribute to the creation of a database of the thermal conductivity of pebble beds which can be used for the design and analysis of fusion breeding blankets.

References

[1] Reimann J, Hermsmeyer S 2002 Thermal conductivity of compressed ceramic breeder pebble beds *Fus. Eng. Des.* **61-62** 345-351

[2] Enoeda M, Furuya K H, Takatsu S, Kikuchi T, Hatano T 1998 Effective thermal conductivity measurements of the binary pebble beds by hot wire method for the breeding blanket *Fus.*

Tech. **34**

[3] Dalle Donne M et al.1999 Measurements of the effective thermal conductivity of a Li4SiO4 pebble bed, *5th Int. Symp. Fus. Nucl. Tech.*

[4] Tehranian F, Abdou M 1995 Experimental study of the effect of external pressure on particle bed effective thermal properties, *Fus. Tech.* **27** 298-313

[5] Aquaro D, Zaccari N 2006 Experimental and numerical analysis on pebble beds used in an ITER Test Module Blanket *Fus. Eng. Des.* **81** 707-712

[6] Tanigawa H, Hatano T, Enoeda M, Akiba M 2005 Effective thermal conductivity of a compressed Li_2TiO_3 pebble bed *Fus. Eng. Des.* **75-79** 801-805

[7] Aquaro D, Orlandi F 2011 Thermal conductivity of ceramic pebble beds used as breeder in fusion nuclear reactors *Proc. of XXIX UIT Conf.*

[8] Hammerschmidt U 2003 A quasi-steady state technique to measure the thermal conductivity, Int. J. Thermoph. **24** 234-248

[9] Aquaro D, Zaccari N 2007 Experimental and numerical analyses on Li4SO4 and Li2TiO3 Pebble bed used in a ITER test module blanket *J. Nucl. Mat.* **367-370** 1293-1297

[10] Reimann J et al. 2000 Thermal conductivity measurements of deformed beryllium pebble beds by hot wire method *Proc. Int. Conf. on CBBI-9*

[11] Piazza G, Reimann J et al. 2002 Characteristics of ceramic breeder materials for the helium cooled pebble bed blanket *J. Nucl. Mat.* **307-311** 811-816

[12] Hemrick J G et al. 2003 Thermal conductivity of alumina measured with three techniques, *J. Test. Eval.* **31** 438-442

Sensitivity analysis on the effect of key parameters on the performance of parabolic trough solar collectors

Luis S. W. Muhlen[1], Behzad Najafi[2], Fabio Rinaldi[2*], Renzo Marchesi[2]

[1]Scuola di Ingegneria Industriale, Politecnico di Milano, Campus di Piacenza, Via Scalabrini, 76, 29100 Piacenza, Italy
[2] Dipartimento di Energia, Politecnico di Milano, Via Lambruschini, 4, 20156, Milano, Italy

Email: fabio.rinaldi@polimi.it

Abstract. Solar troughs are amongst the most commonly used technologies for collecting solar thermal energy and any attempt to increase the performance of these systems is welcomed. In the present study a parabolic solar trough is simulated using a one dimensional finite element model in which the energy balances for the fluid, the absorber and the envelope in each element are performed. The developed model is then validated using the available experimental data. A sensitivity analysis is performed in the next step in order to study the effect of changing the type of the working fluid and the corresponding Reynolds number on the overall performance of the system. The potential improvement due to the addition of a shield on the upper half of the annulus and enhancing the convection coefficient of the heat transfer fluid is also studied.

1. Introduction

Energy sources that are considered sustainable have become increasingly popular and sought after with the ever growing general concern for the negative effects of human interference in the environment [1-2] and many studies have also been carried out in order to optimize the devices to enhance the efficiency of utilization of such resources [3]. Amongst the technologies being developed in the attempt to control and reverse the current dependency on dangerous and polluting energy sources, the direct use of solar radiation to generate heat has been a successful and widely applicable form of replacing the use of fossil fuels from very small scale installations to large power plants.

The present project intends to study the optimization of the parameters of one of the most used configurations for large scale solar plants: the parabolic trough. At the cost of only being able to reach lower temperatures, restricted at around 550°C for current technologies, this configuration is a promising option in the case of macro scale power generation since several modules can be installed one after the other, allowing the creation of very large plants. The parabolic trough solar system with evacuated receivers is the most economically promising macro scale solar generation available technology today [4].

Studies have been developed for several specific finalities. For chemical plants, it is often desired to reach higher fluid temperatures, while for steam generation the maximum temperature isn't as relevant, allowing the use of cheaper technologies. For electricity generation through the use of steam,

---

*Corresponding author

it becomes very important to achieve high efficiency rates in order to decrease the plant size. A combined solar plant for electricity generation has been presented by Manzolini et al [5], using saturated steam as a low temperature fluid, and then a synthetic oil for super-heating. The use of molten salts is an alternative to try to improve even more the efficiency of the plant, through the use of higher temperatures. A solar plant for industrial purposes was presented by Manzolini et al [6]. Structural parameters were evaluated by Giannuzzi et al [7]. Studies on existing solar plants and their applicability on specific locations have been performed in [8-10] and several analytical correlations have been studied and developed for the parabolic trough solar concentrator. In order to evaluate thermal losses and optimize power plant systems, different 1D to 3D models have been developed [4][11][12-15]. Economic optimizations and analysis have been performed for real case scenarios as well as for hypothetical situations [8][16-18]. Studies on specific improvements on the general layout of the trough collector have been made to provide alternative methods for improving overall efficiency, such as the possibility of using fins inside the collecting tube to promote homogenization of the fluid [19], using capillary tubes to improve the wet surface and reduce the effect of the dual phase condition that may occur at low flow rates [20] or using a thermal isolating material covering the upper half of the tube, where there is no focused light incidence, to reduce radiation losses and reduce natural convection [21].

In the present work, a 1D (axial) analytical approach for simulating the fluid flow and the heat exchange in the collector tube was employed. After validating the developed model using experimental data, the simulations were performed utilizing different working fluids. The effect of geometric parameters was next investigated by analyzing the effect of Reynolds number in the efficiency. Afterwards, the effect of adding a shield on the upper half of the annulus was analyzed where the simulations are performed both for vacuum and atmospheric pressure conditions. Finally, the effect of the strategies leading to an enhancement on the convection coefficient of the heat transfer fluid, as suggested in the literature, was studied and the corresponding simulation were carried out employing both oil and the molten salt as heat transfer fluids.

2. Model Description

The implemented model is largely based on the model proposed by Padilla et al [12]. It is a 1D model that simulates the heat transfer between four elements of the solar collector: the heat transfer fluid (HTF), the absorber tube, the glass envelope and the structural brackets, along with the thermal losses to the environment and the incident solar energy. The total length of the solar collector is divided into elements in which the temperatures and all other properties are considered to be constant.

The absorber and the glass envelope are circular coaxial tubes with constant dimensions across the length of the model. The thermal conductivity of the absorber is assumed to be homogeneous and independent of temperature for the temperature range of the through. It is covered by a thin layer of a coating material with enhanced irradiative heat transfer properties with no effect to the other forms of heat transfer. The HTF properties are functions of temperature, but not pressure. The flow is considered to be turbulent and fully developed for the entire length of the trough. No phase change is simulated, therefore the model doesn't accept saturated vapor as HTF. For the envelope, no radial temperature gradient is considered, only axial. It is considered to have a non-null absorption coefficient, therefore it absorbs part of the solar irradiation. It maintains a controlled atmosphere in the annulus between the absorber and the envelope that can be simulated as a vacuum with any chosen pressure, or even atmospheric pressure to determine the system properties when there is a leakage. The brackets are considered to be fins equally spread along the length of the absorber, with intervals multiple of the adopted mesh size.

Only the following heat transfer mechanisms are assumed to exist: forced convection between the absorber and the HTF, conduction along the width of the absorber, radiation and natural convection between the absorber and the glass envelope, conduction between the absorber and the brackets, natural or forced convection between the envelope and the atmospheric air, radiation between the absorber and the sky and, finally, radiation between the Sun and the absorber and envelope. The model doesn't simulate radiation losses from the absorber to the environment or the sky, as the envelope is assumed to be opaque to the radiation emitted by the coating surface. A representation of the model using the thermal resistances is demonstrated in figure 1.

The model only simulates the steady state condition. Although each section of the mesh can have different temperatures for the envelope and the absorber, the net heat exchange between one section and its neighboring sections is null, condition that is only true if the temperature gradients are constant along the length of the collector. From these assumptions, an energy balance for each section dictates that the sum of all the heat exchanged is null for both the envelope and the absorber.

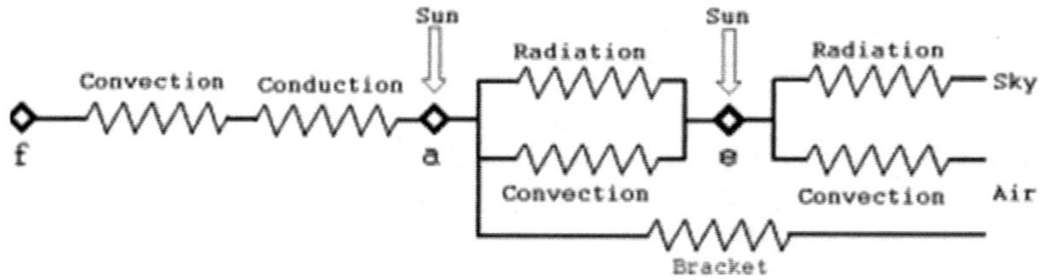

Figure 1. Representation of the model through the use of thermal resistances [9]

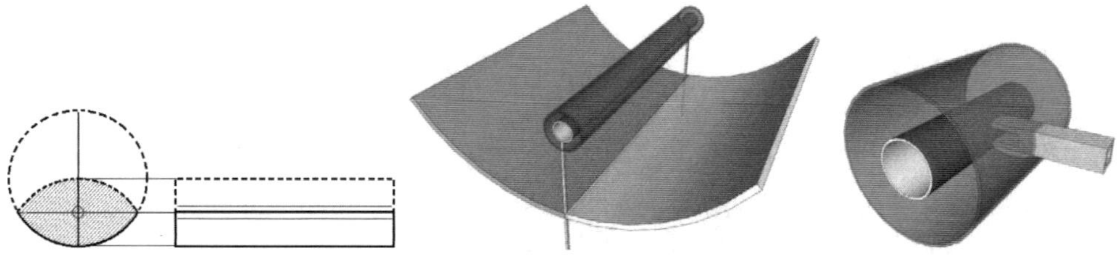

Figure 2. Representation of the modeled mirror

1.1. Convection and conduction between the absorber and the HTF

The convective and conductive heat transfer is determined by:

$$\dot{Q}_{a\to f} = \frac{T_a - T_f}{\frac{1}{\pi \cdot Nu_D \cdot k_f} + \frac{\ln(D_{a,o}/D_{a,i})}{2 \cdot \pi \cdot k_a}} \tag{1}$$

Where the Nusselt number is calculated by the relation given by Gnielinski, with the correction factor recommended by Petukhov and Roizen [22]:

$$Nu_D = \frac{\frac{C_f}{2} \cdot (Re_D - 1000) \cdot Pr}{1 + 12.7 \cdot (\frac{C_f}{2})^{0.5} \cdot (Pr^{2/3} - 1)} \cdot (\frac{Pr}{Pr_w})^{0.11} \tag{2}$$

The friction coefficient is obtained using the Filonenko correlation [23]:

$$C_f = (1.58 \cdot \ln(Re_D) - 3.28)^{-2} \tag{3}$$

1.2. Natural convection between the absorber and the envelope

Two distinct cases were implemented for the natural convection in the annulus, one assuming vacuum and other assuming normal pressures. A vacuum was considered for pressures below 1 Torr, although the recommendation is to use a pressure as low as 0.0001 Torr to reach the free molecule regime that greatly reduces the heat transfer. In the vacuum case, the natural convection heat transfer is given by:

$$\dot{Q}_{a\to e,conv} = h_{a\to e} \cdot \pi \cdot D_{a,o} \cdot (T_a - T_b) \quad (4)$$

Assuming a thermal accommodation coefficient equal to 1, as it is recommended for most solid – gas interactions, the convective heat transfer coefficient is given by [24]:

$$h_{a\to e} = \frac{k_g}{\frac{D_{a,o}}{2} \cdot \ln\left(\frac{D_{e,i}}{D_{a,o}}\right) + \frac{\lambda \cdot (9 \cdot \gamma - 5)}{2 \cdot (\gamma + 1)} \cdot \left(\frac{D_{e,i}}{D_{a,o}} + 1\right)} \quad (5)$$

$$\lambda = 2.331 \cdot 10^{-20} \cdot \frac{(T_a + T_e)}{2} \cdot \frac{1}{P_{an} \cdot D_{mol}^2} \quad (6)$$

For air, the molecular diameter Dmol equals 3.66 x 10-8 cm. While normally the thermal conductivity of the annulus gas is assumed to be only a function of the temperature, in a vacuum scenario it is assumed to be directly proportional to the pressure up to the 1 Torr mark.

In the case where normal pressures are used for the annulus, the following correlation given by Kuehn and Goldstein is used to determine the Nusselt number in the annulus [25]:

$$Nu_D = (Nu_{D,cond}^{15} + Nu_{D,conv}^{15})^{1/15} \quad (7)$$

$$Nu_{D,cond} = \frac{2}{\ln(D_{e,i}/D_{a,o})} \quad (8)$$

$$Nu_{D,conv} = \frac{2}{\ln\left(\frac{1 + 6.659 \cdot Ra^{-0.294}}{1 - 2.441 (Ra \cdot (D_{e,i}/D_{a,o})^3)^{-0.206}}\right)} \quad (9)$$

1.3. Radiation between the absorber and the envelope

All surfaces are assumed to be gray and opaque. Each section of the grid is considered to be closed on the side by adiabatic walls that reflect but do not emit radiation. The view factors [26] are given by:

$$F_{e\to a} = \frac{1}{R} - \frac{1}{\pi \cdot R}\left[\mathrm{acos}\left(\frac{B}{A}\right) - \psi \cdot \mathrm{acos}\left(\frac{B}{R \cdot A}\right) - \frac{B}{2 \cdot L} \cdot \mathrm{asin}\left(\frac{1}{R}\right) + \frac{\pi \cdot A}{4 \cdot L}\right] \quad (10)$$

$$F_{e\to e} = 1 - \frac{1}{R} + \frac{2}{pi \cdot R} \cdot \mathrm{atan}\left(\frac{2 \cdot \sqrt{R^2 - 1}}{L}\right) - \frac{L}{2\pi \cdot R} \cdot \left[\chi \cdot \mathrm{asin}(\varphi) - \mathrm{asin}\left(\frac{R^2 - 2}{R^2}\right) + \frac{\pi}{2}(\chi - 1)\right] \quad (11)$$

Where:

$$\psi = \frac{\sqrt{(A+2)^2 - (2 \cdot R)^2}}{2 \cdot L} \quad (12)$$

$$\chi = \frac{\sqrt{4R^2 + L^2}}{L} \quad (13)$$

$$\varphi = \frac{4(R^2 - 1) + L^2/R^2(R^2 - 2)}{L^2 + 4(R^2 - 1)} \quad (14)$$

$$R = \frac{D_{e,i}}{D_{a,o}} \quad (15)$$

$$L = \frac{h}{D_{a,o}} \quad (16)$$

$$A = L^2 + R^2 - 1 \quad (17)$$

$$B = L^2 - R^2 + 1 \quad (18)$$

The view factor $F_{a\to e}$is found using symmetry and the view factors for the side walls are found using the property that the sum of all view factors for one surface equals one. Assuming that both side walls are a single surface:

$$F_{a\to e} = F_{e\to a} \frac{D_{e,i}^2}{D_{a,o}^2} \quad (19)$$

$$F_{a\to s} = 1 - F_{a\to e} \quad (20)$$

$$F_{e\to s} = 1 - F_{e\to e} - F_{e\to a} \quad (21)$$

The thermal resistances and the heat transfers are given by:

$$R_{a\rightarrow e}=\frac{(1-\varepsilon_a)}{\varepsilon_a\cdot A_a}+\frac{(1-\varepsilon_e)}{\varepsilon_e\cdot A_e}+\frac{1}{A_a\cdot F_{a\rightarrow e}+((A_a\cdot F_{a\rightarrow s})^{-1}+(A_e\cdot F_{e\rightarrow s})^{-1})^{-1}} \quad (22)$$

$$\dot{Q}_{a\rightarrow e,rad}=\sigma\frac{(T_a^4-T_e^4)}{R_{a\rightarrow e}} \quad (23)$$

The Thermal resistance given in equation 22 can be elaborated using Oppenheim circuit shown below

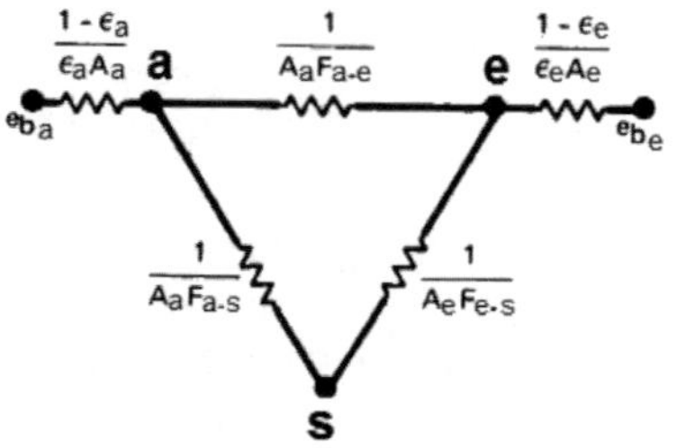

Figure 3. Oppenheim Circuit.

1.4. Conduction through the support brackets

The support brackets are modeled as fins made of carbon steel, composed by two segments with different cross sections. In the first segment, connected to the base of the solar receiver, the cross section is a pair of rectangular plates with dimensions 25.4 x 3.175 mm and 25.4 mm length. The second segment is the support tubes, with cross section formed by a pair of hollow squares with 25.4 mm sides and 3.175 mm thickness. Since the second segment covers most of the length, it is considered to have a length equal to the focal distance of the mirror.

Due to the length of the brackets, an infinite length fin model is used [27]. The temperature at the intersection is given by:

$$\frac{\theta_a}{\theta_b}=\frac{1}{\cosh(m_{b,1}\cdot L_{b,1})+\frac{\zeta_{b,2}}{\zeta_{b,1}}\cdot\sinh(m_{b,1}\cdot L_{b,1})} \quad (24)$$

And the heat loss through the brackets is given by:

$$\dot{Q}_b=\frac{2\cdot\zeta_{b,1}\cdot(T_a-T_{air})\cdot\cosh(m_{b,1}\cdot L_{b,1})-\frac{\theta_a}{\theta_b}}{\sinh(m_{b,1}\cdot L_{b,1})} \quad (25)$$

Where:

$$m_{b,1}=\sqrt{\frac{h_b\cdot Per_1}{k_b\cdot Ac_1}} \quad (26)$$

$$\zeta_{b,j}=\sqrt{h_b\cdot Per_j\cdot k_b\cdot Ac_j}\text{ for } j=1,2. \quad (27)$$

The worst case scenario was adopted to determine the convective coefficient: forced convection with a square orientation instead of a diamond orientation.

$$Nu_b=0.14\cdot Re_b^{0.666} \quad (28)$$

1.5. Convection between the envelope and the atmospheric air

For forced convection, the following correlation was used to determine the Nusselt number:

$$Nu_D=(1+(\frac{Re_D}{282000})^{\frac{5}{8}})^{\frac{4}{5}}\cdot 0.62\frac{Re_D^{0.5}\cdot Pr_{film}^{1/3}}{(1+(\frac{0.4}{Pr_{film}})^{\frac{2}{3}})^{\frac{1}{4}}}+0.3 \quad (29)$$

The heat loss through convection is then calculated as follows:

$$\dot{Q}_{e\rightarrow air,conv} = Nu_D \cdot k_{film} \cdot \pi \cdot (T_e - T_{air}) \quad (30)$$

1.6. Radiation losses from the envelope to the sky

For this analysis several simplifications were made. The sky is considered to be a black body with constant temperature, function of the air temperature. The envelope views only the sky and the mirror, and with the exception of the sides, half of the surface is covered by sky and half by mirror. To calculate the views of the sides, the absorber is placed in the focal distance of the mirror, which is half of the radius of the mirror's circumference, as represented on figure 2.

An average radius is calculated from these assumptions, which is then used to model the sides as circles with the same area. This average radius is given by:

$$\bar{r} = 2\frac{\sqrt{3}}{3\pi} \cdot D_m \cdot \mathrm{atanh}(\sqrt{3} \cdot \tan(\tfrac{\pi}{8})) - \frac{D_{e,o}}{2} \quad (31)$$

Then the view factor of the sides, which is a function of the position x of the segment being analyzed in relation to the edges of the trough, is given by:

$$F_{e\rightarrow s} = \frac{\mathrm{atan}(\bar{r}/x) + \mathrm{atan}(\bar{r}/(L-x))}{\pi} \quad (32)$$

Finally, the view factor for the sky is:

$$F_{e\rightarrow sky} = \frac{(1 - F_{e\rightarrow s})}{2} \quad (33)$$

The heat loss through radiation to the sky is taken from equations given by Padilla et al [12] while the sky temperature is found using the relation given by Swinbank [28]. The Sun light that reaches the mirror is reflected back to the absorber. A series of corrective factors are used to represent the losses due to part of the light reaching the collector not reaching the absorber, as seen in table 1 [12]:

Table 1. Mirror losses coefficients.

Heat collection element shadowing (γ1)	Luz black chrome	0.974
	Luz cermet	0.971
Twisting and tracking error (γ2)		0.994
Geometry accuracy of the collector mirrors (γ3)		0.980
Mirror clearness (γ4)		0.950
Dirt on heat collection element (γ5)		0.980
Miscellaneous factor (γ6)		0.960
Clean mirror reflectivity		0.935

The total heat from the Sun absorbed by the absorber and by the envelope is:

$$\dot{Q}_{sun\rightarrow a} = \gamma \cdot \tau_e \cdot \alpha_a \cdot 1.01 \cdot I_{bn} \cdot \frac{D_m}{2} \cdot \sqrt{3} \quad (34)$$

$$\dot{Q}_{sun\rightarrow e} = \gamma \cdot \alpha_e \cdot 1.01 \cdot I_{bn} \cdot \frac{D_m}{2} \cdot \sqrt{3} \quad (35)$$

Where the term γ is the product of all the terms in table 1. The factor 1.01 is added in the equation to make up for internal reflections in the trough [29]. The solar radiation is considered to reach the trough perpendicularly and geometrical end losses of the collector are not assumed to exist.

The efficiency is calculated as the ratio between the total enthalpy gain (including the gain in kinetic energy) and the total solar energy irradiated on the collector mirror.

3. Numerical solution

The energy balance is performed in the HTF, in the absorber and in the envelope. From the simplifications stated before (steady state and no heat transfer in the axial direction for the absorber and the envelope), the energy balance for the HTF gives an equation that is a function of the temperature of the HTF in the previous section and of the temperatures of the HTF and the absorber in the given section. The balance for the envelope gives an equation that is function only of the temperatures of the absorber and the envelope in the given section, and the balance for the absorber returns an equation that is function of the temperatures of the HTF, the absorber and the envelope in the given section:

Absorber balance
$$\frac{Q_a}{G_{Sun}} - \frac{Q_a}{f_{conv}} - \frac{Q_a}{e_{conv}} - \frac{Q_a}{e_{rad}} - Q_{brack_{cond}} = 0 \tag{36}$$

Envelope balance
$$\frac{Q_e}{rad_{Sun}} + \frac{Q_a}{conv_e} + \frac{Q_a}{rad_e} + \frac{Q_a}{rad_e} - \frac{Q_a}{conv_{air}} - \frac{Q_{ae}}{rad_{sky}} = 0 \tag{37}$$

HTF balance
$$\gamma(T_{m1} - T) + AF_{conv}h - m_f(v_f^{\,2} - v_{f_{m1}}^{\,2}) = 0 \tag{38}$$

For each element, a guess is made for the temperature of the absorber. From this initial guess, the fluid and the envelope temperatures are found using a golden ratio increment line search, using the energy balance for the fluid and the envelope, respectively. Then the guess for the temperature of the absorber is checked using the energy balance of the absorber and all the temperatures found previously. Using again a line search with golden ratio increments, this procedure is repeated until a temperature is found for the absorber that respects its energy balance. A temperature for the HFC at the start of the trough is required as a boundary condition.

4. Model Validation

The model results are validated using the data provided by the Sandia National Laboratory [30], an extensive experimental study performed with a single module from a trough power plant. The available data include Steady State condition final temperatures for the HTF, overall efficiency and heat loss for different HTF flows and initial temperatures, as well as different solar incidences. Tests were also performed with vacuum and with air at ambient pressure in the annulus, and without the glass envelope. The measured data considering the vacuum in the annulus is given in Table 2 and the corresponding validation results are demonstrated in Figure 4.

Table 2. Measured data, vacuum in the annulus.

Test	Direct Normal Irradiation [W/m²]	Wind Speed [m/s]	Air Temperature [ºC]	Inlet Temp. [ºC]	Outlet Temp. [ºC]	Flow Rate [L/min]
1	933.7	2.6	21.2	102.2	124.0	47.7
2	968.2	3.7	22.4	151.0	173.3	47.8
3	982.3	2.5	24.3	197.5	219.5	49.1
4	909.5	3.3	26.2	250.7	269.4	54.7
5	937.9	1.0	28.8	297.8	316.9	55.5
6	880.6	2.9	27.5	299.0	317.2	55.6
7	920.9	2.6	29.5	379.5	398.0	56.8
8	903.2	4.2	31.1	355.9	374.0	56.3

Tests with two different coatings for the absorber are available. Most of the experiments were done using Syltherm 800 as the HTF, and its thermal properties were obtained from its catalogue [31]. The results obtained were overall close to the measured data, with a difference of 2 to 3 [ºC]. Such errors could be a consequence of imprecise values for the mirror efficiencies and are considered to be within acceptable range. It should be mentioned that 2~3°C corresponds to 10~15% in terms of overall energy gain.

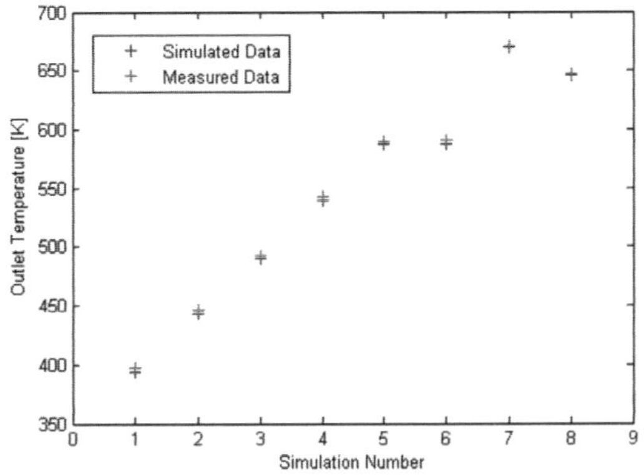

Figure 4 Validation results, vacuum in the annulus.

5. Results and discussions

The results show that, for the utilized model, the temperatures of the HTF and the absorber increase very close to linearly throughout the length of the trough. The temperature of the envelope is almost constant, with a decrease near the ends due to the increase of the ratio of sky and collector viewed by the envelope at the extremities. This happens because the model sees the sides of the collector as sky. In reality, part of this section of the views would be landscape, with a higher temperature than the sky, so the model over estimates those heat losses, even though it is a very subtle overestimation. From these results, the hypothesis that the net heat flux is zero across the length of the absorber and the envelope is acceptable.

The simulations with vacuum at low pressures, around 10-4 Torr, show that the heat losses from the absorber to the envelope are negligible, and most of the solar energy absorbed by the coating surface is transferred to the HTF. The losses to the environment come mostly from the energy absorbed by the envelope. The most important factor that reduces the overall efficiency of the trough is the inefficiencies of the mirror, which this model simplifies as constant factors that multiply the incident solar energy. When the vacuum is broken, the thermal losses from the absorber increase slightly. Conditions that reduce the convective heat transfer coefficient for the HTF can greatly increase the temperature of the absorber, making the thermal losses more relevant. The losses through the support brackets are the highest for most of the simulations made.

The parameters of the considered case study are presented in table 3. As it was expected, higher HTF temperatures increase the losses and decrease the efficiency of the trough. This must be considered when evaluating the benefits of an increased temperature in the overall efficiency of the plant. The temperature distribution of the HTF along the collector length for the case study is shown in Figure 5. The breakdown of the thermal losses in two pressure case is given in table 4.

Table 3 Parameters for example simulations.

Direct Normal Irradiation	900 W/m2
Mass Flow	10 kg/s
Air Velocity	2.6 m/s
Air Temperature	20 ºC
Trough Length	780 m
Mirror Aperture	5 m
Bracket Separation	3.9 m
$D_{a,i}$	0.06 m
$D_{e,i}$	0.1 m

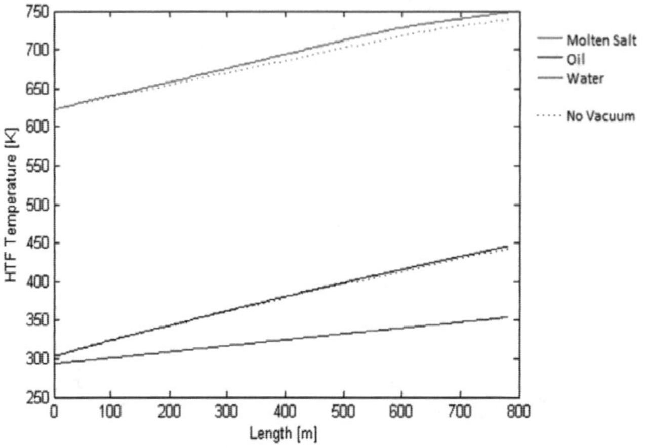

Figure 5. Temperature distribution of HTF along the collector length for different fluids.

Table 4. Break down of thermal losses

Annulus Pressure	Vacuum	atmospheric
$Q_{b_{cond}}$ (w)	7.284. 10^3	7.158. 10^3
$Q_{a-env_{conv}}$ (w)	0.0810	2.973. 10^5
$Q_{a-env_{rad}}$ (w)	2.994.10^4	2.778.10^4
$Q_{air_{conv}}$ (w)	6.4822 .10^4	2.741. 10^5
$Q_{sky_{rad}}$ (w)	2.957.10^4	1.154. 10^5

A sensitivity analysis was performed in order to investigate the effect of the Reynolds number (as a function of both the absorber diameter and the mass flow rate) to the overall efficiency of the trough. Varying the internal absorber diameter from 20 to 100 mm and varying the mass flow from 2 to 34 kg/s, the efficiencies were simulated for both oil and molten salt as the HTF. The results indicate that there is a minimum tolerable mass flow rate that, below which, the efficiency drops to unacceptable levels. This threshold isn't reached in this simulation for the oil, but was present for the molten salt, indicating that when using it as a HTF, especial precautions need to be taken when dimensioning the mass flow rate. Figure 6 shows the efficiency curves as a function of the Reynolds number, for several absorber diameters where oil is used as the HTF.

To simulate the potential improvement due to the addition of a shield on the upper half of the annulus as proposed by Al-Ansary [21], the calculations for the absorption of solar radiation, for the convective losses to the environment and for the radiation losses to the sky were changed. The total mirror aperture was reduced by the value of the envelope diameter, the view factor from the envelope to the sky was reduced to zero and the convection losses were reduced by 25%. These changes simulate the covering of the upper half of the envelope with an isolating, opaque material, which drastically reduces the losses to the environment, but also reduces the amount of incoming solar energy slightly. This change does not simulate the reduction in the natural convection inside the annulus that should also be a benefit of the method proposed. It is expected that the benefit of such a method will vary depending on the losses to the environment, so this analysis was performed for the annulus pressure varying from vacuum to atmospheric pressure, and the adopted working fluid was molten salt, due to the higher operating temperature. The effect of application of shield for different annulus pressures is demonstrated in Figure 7. The HTF temperature profile along the collector in which the shield is used is simulated in Figure 8

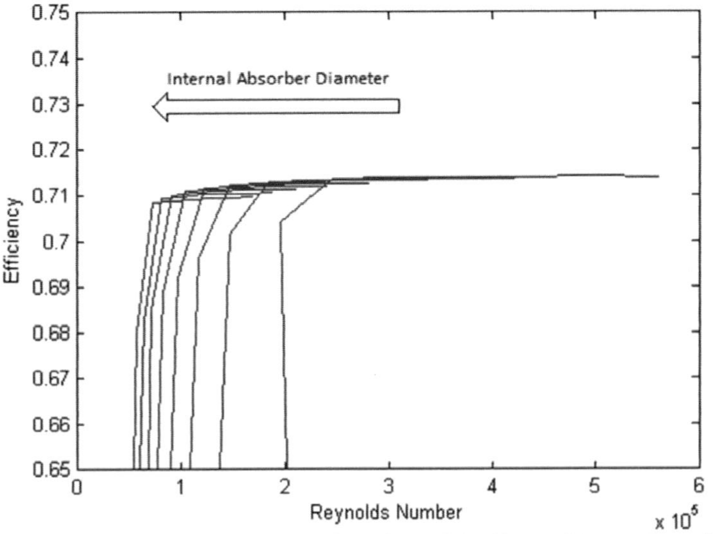

Figure 6. Efficiency curves as a function of the Reynolds number, for several absorber diameters, using molten salt as HTF.

The results indicate that the reduction of the thermal losses doesn't compensate for the reduction of the solar gains. For atmospheric annulus pressure the method gets close to being advantageous, indicating that in a scenario where the losses were much more relevant, this could be a promising technique.

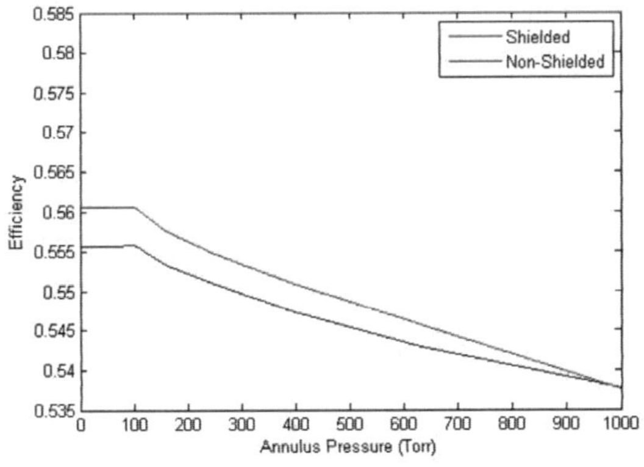

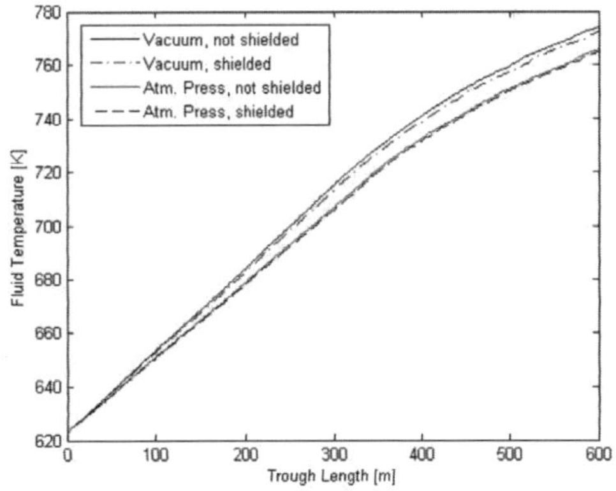

Figure 7. Shield simulation – Efficiency as a function of the annulus pressure.

Figure 8. Shield simulation – HTF temperature along the length of the collector.

Studies suggest methods to boost the efficiency of the trough by increasing the convection coefficient of the HTF [19-20]. To simulate these gains and other possible techniques, a static parameter was added, increasing the convection artificially. Simulations were made with increases of up to 200% for the convection coefficient for standard trough collectors using molten salt and synthetic oil as the HTF. Definition of convection coefficient modifier c_{h_f} which is multiplied by the

Nu number can be represented in the following equation which is expressing the convective and conductive heat transfer between the absorber and the fluid.

$$Q_{a-f_{conv}} = \frac{(T_a - T)}{\frac{1}{\pi Nu_f k_f c_{h_f}}} + \log\left(\frac{Da_{out}}{2Da_{in}} \cdot \frac{k_a}{\pi}\right) \tag{39}$$

This analysis was made simulating a 600 m long trough with a mirror opening of 5 m, mass flow of 6 kg/s and an internal absorber diameter of 0.04 m. Initial temperature for the oil was 30 °C, and for the molten salt it was 350 °C. The effect fluid convection modifier factor on the total efficacy is demonstrated in Figure 9.

The simulations indicate that the efficiencies for the scenario with oil as HTF are almost unaffected by the increase in the convection coefficient. This is most likely due to the smaller thermal losses present in reason of the lower operating temperature, as well as the smaller dependency of the efficiency with the Reynolds number that was described above for this range of diameter and mass flow. On the other hand, the efficiencies were increased considerably for the molten salt scenarios.

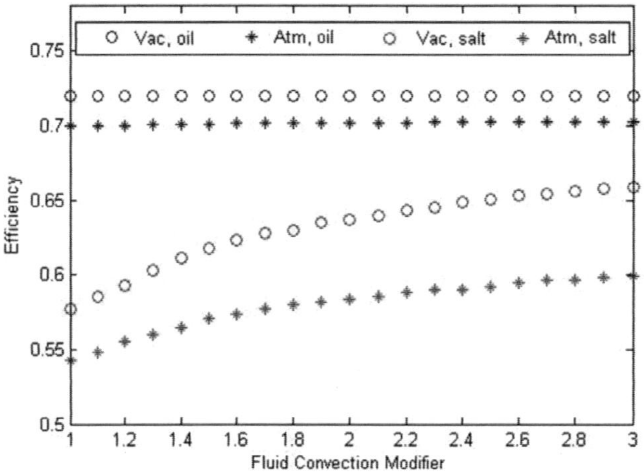

Figure 9. Efficiency as a function of the fluid convection modifier factor.

6. Conclusion

A one-dimensional model for simulating the heat transfers between the heat transfer fluid, absorber and envelope of a solar trough collector and the environment was implemented and the developed model was next validated using the available experimental data. In the next step, simulations were performed for different working fluids and the corresponding temperature profiles were obtained. Sensitivity analysis on the effect of the Reynolds number on the performance of the systems was next carried out and a diagram demonstrating the efficiency curves as a function of the Reynolds number for different internal absorber diameter was provided. Afterwards the effect of adding a shield on the upper half of the annulus, as suggested in the literature, was performed and it was demonstrated that the mentioned strategy can only be advantageous at very high pressures. Finally the effect of enhancing the convection coefficient of the heat transfer fluid was investigated and it was demonstrated that increasing the convection coefficient in case of the oil has a negligible effect while it can considerably increase the efficiency in case of the molten salt. It is noteworthy that since in the this study the analysis on the parameters affecting the losses associated with the mirror and its optical inefficiencies was not performed, an upgrade to the present work is to include a more precise formulation of the mirror parameters and to analyze the corresponding affecting parameters.

Journal of Physics: Conference Series **501** (2014) 012032 doi:10.1088/1742-6596/501/1/012032

Nomenclature

A	Surface area [m²]
Ac	Cross section area [m²]
Cf	Friction coefficient
D	Diameter [m]
F	View factor
h	Convective heat transfer coefficient [W/m².K],
h	Mesh size [m]
I	Solar irradiance [W/m²]
k	Thermal conductivity [W/m.K]
L	Total length [m]
$\dot{m}$	Mass flow rate [kg/s]
Nu	Nusselt number
P	Pressure [Torr]
Per	Perimeter [m]
Pr	Prandtl number
$\dot{Q}$	Heat transfer flux per trough unit length [W/m]
r	Radius [m]
R	Thermal resistance
Ra	Rayleigh number
Re	Reynolds number
T	Temperature [K]
V	Velocity [m/s]
x	Distance in the axial direction [m]

Greek symbols

α	Absorptance
γ	Ratio of specific heats
ε	Emissivity
σ	Stefan Boltzmann constant
θ	Temperature difference with the environment [K]
τ	Transmittance

Subscripts

1	Section 1
2	Section 2
a	Absorber
an	Annulus
b	Bracket
air	Air
bn	Normal beam
cond	Conduction
conv	Convection
e	Envelope
film	Film
f	Fluid
i	Internal, i-th section
m	Mirror
mol	Molecular
o	Outer
s	Perpendicular section, side
w	Wall

References

[1] Shirazi A, Aminyavari M, Najafi B, Rinaldi F, Razaghi M 2012, Thermal–economic–environmental analysis and multi-objective optimization of an internal-reforming solid oxide fuel cell–gas turbine hybrid system. Int J Hydrogen Energy 37 19111-24.

[2] Zago M, Casalegno A, Marchesi R, Rinaldi F 2011, Efficiency Analysis of Independent and centralized heating systems for residential buildings in northern Italy, Energies **4** – Special Issue Energy Efficient Building Design 2115-2131.

[3] Selleri T, Najafi B, Rinaldi F, Colombo G. 2013, Mathematical Modeling and Multi-Objective Optimization of a Mini-Channel Heat Exchanger Via Genetic Algorithm. ASME J. Therm. Sci. Eng. Appl. **5** (3) , art. no. 031013;DOI: http://dx.doi.org/10.1115/1.4023893.

[4] Gong G, Huang X, Wang J and Hao M 2010, An optimized model and test of the China's first high temperature parabolic trough solar receiver, *Solar Energy* **84** 2230–2245

[5] Manzolini G, Giostri A, Saccilotto C Silva P and Macchi E 2011 Development of an innovative code for the design of thermodynamic solar power plants part B: Performance assessment of commercial and innovative technologies, *Renewable Energy* **36** 1993-2003

[6] Manzolini G, Bellarmino M, Macchi E and Silva P 2011, Solar thermodynamic plants for cogenerative industrial applications in southern Europe, Rene. Energy **36** 235-243

[7] Giannuzzi G M, Majorana C E, Miliozzi A, Salomoni V A, Nicolini D 2007 Structural Design Criteria for Steel Components of Parabolic-Trough Solar Concentrators , *Trans. of the ASME 382 / Vol. 129*

[8] Poullikkas A 2009 A Economic analysis of power generation from parabolic trough solar thermal plants for the Mediterranean region—A case study for the island of Cyprus, *Rene Sust. Energy Rev.* **13** 2474–2484
[9] García A F, Zarza E,Valenzuela L andPérez M 2010 Parabolic-trough solar collectors and their applications, *Rene Sust. Energy Rev.* **14** 1695–1721
[10] O.Badran O and Eck M 2006The application of parabolic trough technology under Jordanian climate, *Rene. Energy* **31** 791–802
[11] Manzolini G, Giostri A, Saccilotto C, Silva P and Macchi E 2011 Development of an innovative code for the design of thermodynamic solar power plants part A: Code description and test case, *Rene. Energy* **36** 1993-2003
[12] R.V. Padilla, G.Demirkaya, D.Y.Goswami, E.Stefanakos,M.M. Rahman, Heat transfer analysis of parabolic trough solar receiver, *App. Energy* **88** (2011) 5097–5110
[13].Fadar A E, Mimet A and García M P 2009 Modelling and performance study of a continuous adsorption refrigeration system driven by parabolic trough solar collector, *Solar Energy* **83** pp 850–61
[14] Montes MJ, Rovira A, Muñoz M and Martínez-Val J M 2011 Performance analysis of an Integrated Solar Combined Cycle using Direct Steam Generation in parabolic trough collectors, *App. Energy* **88** 3228–3238
[15] García I L,.Álvarez J L and Blanco D 2011 Performance model for parabolic trough solar thermal power plants with thermal storage, , *Solar Energy* **85** 2443–60
[16] Barberena J G, Garcia, Sanchez M, Blanco M J , Lasheras C, Padrós A and Arraiza J 2011 Analysis of the influence of operational strategies in plant performance using SimulCET, simulation software for parabolic trough power plants, *Solar Energy* **85** 721–735
[17] Rolim M M, Fraidenraich F, and Tiba C 2009Analytic modeling of a solar power plant with parabolic linear collectors, *Solar Energy* **83** 126–133
[18] Morin G,Dersch J, Platzer W , Eck M and Häberle A 2011 Comparison of Linear Fresnel and Parabolic Trough Collector power plants, *Solar Energy* **84** 502-555
[19] Muñoz J and.Abánades A 2011 A technical note on application of internally finned tubes in solar parabolic trough absorber pipes, *Solar Energy* **85** 609–612
[20] Rojas M E, de Andrés M C and González L 2008 Designing capillary systems to enhance heat transfer in LS3 parabolic trough collectors for direct steam generation, *Solar Energy* **82** 53–60
[21] Al-Ansary and H, Zeitoun O 2011 Numerical study of conduction and convection heat losses from a half-insulated air-filled annulus of the receiver of a parabolic trough collector, *Solar Energy* **85** 3036–3045
[22] Petukhov B S and Roizen L I 1964 Generalized relationships for heat transfer in a turbulent flow of gas in tubes of annular section. *High Temp* **2** 65–8.
[23] Kakac S , Shah RK and Aung W. Handbook of single-phase convective heat transfer. New York: John Wiley & Sons; 1987.
[24] Dushman S. Scientific foundations of vacuum technique. 2nd ed. New York: Wiley; 1962.
[25] T.H. Kuehn and R.J. Goldstein, A Parametric Study of Prandtl Number and Diameter Ratio Effects on Natural Convection Heat Transfer in Horizontal Cylindrical Annuli, J. Heat Transfer, Vol. 102, 768-770 (1980).
[26] Modest, M 2003Radiative heat transfer. 2nd ed. Burlington, MA: Academic Press;.
[27] Incropera F P 2006, Fundamentals of heat and mass transfer. 6th ed. Hoboken, NJ: Wiley
[28] Swinbank W. 1963, Long-wave radiation from clear skies. Quar J R Meteorol Soc;89 339–48.
[29] Duffie J A and Beckman WA 1980 Solar engineering of thermal processes. New York: Wiley
[30] Dudley V, Kolb G, Sloan M and Kearney D 1994 SEGS LS2 Solar Collector-Test Results. Report of Sandia National Laboratories.SAN94-1884.
[31]http://www.therminol.com/pages/bulletins/europe/therminol_vp1_anglais.pdf

Journal of Physics: Conference Series **501** (2014) 012033 doi:10.1088/1742-6596/501/1/012033

Borehole Heat Exchangers: heat transfer simulation in the presence of a groundwater flow

A Angelotti[1], L Alberti[2], I La Licata[2] and M Antelmi[2]

[1] Dipartimento di Energia, Politecnico di Milano, v. Lambruschini 4, 20156 Milano, Italy

[2] Dipartimento Ingegneria Civile e Ambientale (D.I.C.A.), Politecnico di Milano, Piazza Leonardo da Vinci 32, 20133 Milano, Italy

E-mail: adriana.angelotti@polimi.it

Abstract. The correct design of the Borehole Heat Exchanger is crucial for the operation and the energy performance of a Ground Source Heat Pump. Most design methods and tools are based on the assumption that the ground is a solid medium where conduction is the only heat transfer mechanism. In turn in regions rich in groundwater the groundwater flow influence has to be assessed, by including the convection effects. In this paper a numerical model of a 100 m U-pipe in a saturated porous medium is presented. The model is created adopting MT3DMS coupled to MODFLOW. A Darcy flow is imposed across the medium. The typical operation of a Borehole Heat Exchanger operating both in winter and in summer is simulated for two years, under different groundwater velocities. The energy injected to and extracted from the ground is derived as a function of the Darcy velocity and compared with the purely conductive case. Temperature fields in the ground at key moments are shown and discussed. From both the energy and the aquifer temperature field points of view, the velocity ranges for respectively negligible and relevant influence of the groundwater flow are identified.

1. Introduction

Shallow geothermal energy is considered an environmental friendly alternative to traditional heating and cooling techniques. Either Ground Source Heat Pumps (GSHPs) or Groundwater Heat Pumps (GWHPs) can be used. In GSHP systems the heat pump is coupled to a closed loop buried into the ground, where a thermal-carrier fluid is circulated, extracting heat from the ground in winter and/or injecting heat into the ground in summer. The most widespread application consists of an arrangement of vertical boreholes (Borehole Heat Exchangers or BHEs), typically 100 m deep, where a single or a double U polyethylene pipe is inserted.

The correct design of the BHEs is crucial for the operation and the energy performance of the GSHP. Several design methods and tools are available, including thumb rules, simplified analytical models and more accurate dynamic simulation tools [1-4]. Moreover recently in Italy a design standard for GSHP and GWHP was published [5]. However most of the available methods are based on the assumption that the ground is a solid medium where conduction is the only heat transfer mechanism. In turn in some regions where GSHP are installed, such as the Padana Plain in Northern

Italy, the ground is extremely rich in groundwater. In such cases the groundwater flow influence on the energy performance of the BHEs has to be assessed, by including the convection effects.

The use of the Moving Line Source solution has recently been proposed to study the interaction between a BHE and a groundwater flow [6, 7]. In this approach the BHE is treated as a line heat source with a given constant heat rate, and then the thermal impact in the ground of the BHE operation is analyzed. However in reality the heat rate exchanged by the BHE is not assigned, since it is influenced by the temperature conditions that develop around it. Therefore an integrated approach, where the BHE operation produces a thermal perturbation into the ground and the temperature field in the ground influences the BHE energy performance, would be more useful.

On the experimental side, studies on BHEs under the influence of a groundwater flow are generally lacking and limited to tests in real conditions, such as in [8]. Yet controlled hydrogeological conditions are necessary in order to have a quantitative accurate assessment of the effects of the groundwater velocity on the energy performance of the BHE.

In the present paper the influence of the groundwater flow on the energy performance of a BHE is studied through a finite-difference numerical model implemented through the code MT3DMS [9] coupled to MODFLOW [10]. An integrated approach is followed, where the exchanged energy and the aquifer temperature field are solved together. An accurate validation of the model by comparing its predictions with the Moving Line Source ones has been carried out and may be found in [11]. The model is used to simulate a BHE operating in both seasons, namely extracting heat from the ground in winter and injecting it into the ground in summer. Two years of operation are considered, in order to understand the impact of the previous year operation on the next one. A broad range of groundwater velocities, well representative for a wide ensemble of hydrogeological systems, is considered. From the simulations the seasonal energy exchanged is derived and compared with the reference case of null groundwater velocity. Temperature conditions in the aquifer are analyzed at key moments. Finally the velocities ranges leading to respectively a negligible and a significant improvement of the BHE performance are clearly identified.

2. The simulation model

2.1. The case study

The case study refers to a typical BHE, consisting of a 100 m polyethylene U-pipe with an inner diameter of 4 cm and a pipe-to-pipe centers distance of 6 cm. The U-pipe is located into a 200 m saturated non-dispersive sandy aquifer, assumed homogeneous. For the sake of simplicity the borehole filling material is assumed to be equal to the surrounding soil. Milano, Northern Italy, conditions are assumed. Therefore the aquifer has an initial uniform temperature of 11.8°C. The BHE is simulated as if connected to a heat pump providing heating and cooling to a building. Five Darcy velocities are considered, namely 0 m/s, 10^{-7} m/s (0.86 cm/day), 10^{-6} m/s (8.6 cm/day), $5 \cdot 10^{-6}$ m/s (46.3 cm/day) and 10^{-5} m/s (86.0 cm/day). Porous medium thermal and hydrogeological properties are listed in table 1.

Table 1. Porous medium thermal and hydrogeological properties

Porosity	θ	0.35
Dispersivity	δ	0 m
Effective thermal conductivity	λ_m	2.3 W m^{-1} K^{-1}
Thermal capacity per unit volume	C_m	2.72 MJ m^{-3} K^{-1}
Hydraulic conductivities (5 cases)	k	2.10^{-4}; 2.10^{-5}; 2.10^{-4}; 10^{-3}; 2.10^{-3} m s^{-1}
Darcy velocities (5 cases)	v	0; 10^{-7}; 10^{-6}; 10^{-5} m s^{-1}

2.2. Model implementation

A MODFLOW/MT3DMS model of the U-pipe in a 3D domain has been implemented. MODFLOW is full 3D finite difference code for groundwater flow simulations developed by U.S. Geological Survey [9], while MT3DMS is a modular three-dimensional multispecies transport code, developed by Alabama University [10], for simulation of advection, dispersion, and chemical reactions of contaminants in groundwater systems. MT3DMS can accommodate very general spatial discretization schemes and is designed for use with any block-centered finite-difference flow model, such as MODFLOW.

Since only a rectangular grid can be used, the real circular geometry of the pipes has been turned into an equivalent square geometry, by imposing the conservation of the total thermal resistance between the thermal-carrier fluid into each pipe and the outer surface of the pipe [12]. Therefore, as shown in figure 1, a square section with 3.36 cm side has been used for the U-pipe. The grid cells have very different lengths, ranging from a minimum of 0.37 cm close to the BHE to a maximum of 10 m close to the domain boundaries. The domain length downstream depends on the velocity in order to allow the proper description of the temperature plumes developed by the BHE. Consequently the domain horizontal dimensions are 60 m x 90 m for the null velocity case, 60 m x 100 m for v = 10^{-7} m/s, 60 m x 200 m for v = 10^{-6} m/s and v = 5.10^{-6} m/s, and finally 60 m x 340 m for v = 10^{-5} m/s.

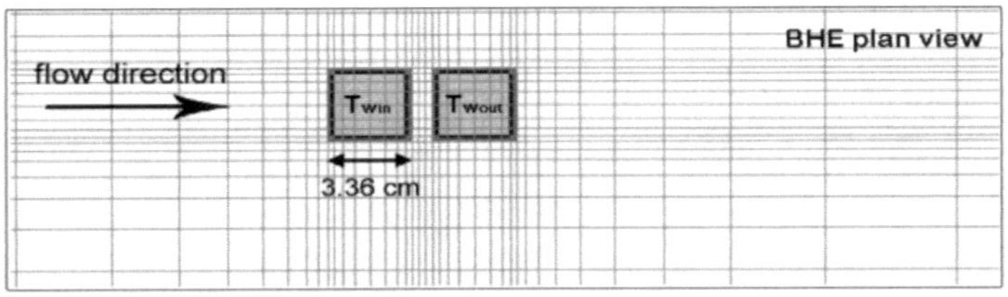

Figure 1. plan view of the numerical model grid: zoom of the zone around the BHE

The boundary conditions given to the model consist of an initial uniform temperature in the medium, a constant unperturbed temperature at the upstream boundary of the medium and a constant hydraulic gradient across the horizontal section. By varying the hydraulic gradient, the Darcy velocity of the groundwater is varied. For the BHE, constant mass flow rate $\dot{m}$ and inlet temperatures T_{wi} are given, as shown in table 2. While in real conditions the inlet fluid temperature may vary according to the GSHP control strategy and building heating/cooling load, a constant inlet fluid temperature approach, although simplified, allows to derive very general results, independent from the specific building and heat pump. As reported in table 2, a 6 months heat injection period during winter, a 3 months heat injection period during summer and two 1.5 months pauses or no operation periods between the two operation seasons can be identified.

Table 2. Operational parameters assigned to the BHE

	Interval	**T_{wi} (°C)**	**$\dot{m}$ (kg/h)**
heat extraction (winter)	15th Oct. – 15th Apr.	1	1000
pause	16th Apr-31st May;	-	-
heat injection (summer)	1st June – 31st Aug.	28	1000
pause	1st Sept – 14th Oct.	-	-

The model predictions have been compared with the Moving Line Source predictions, by running the BHE model in constant heat rate mode, as described more in detail in [11]. The difference between the numerical model and the analytical model temperature distributions in the aquifer was turned into a heat rate difference. It was found that the discrepancy ranges from 2 % for the null velocity case to -

9% for the highest velocity case, namely 10^{-5} m/s. Therefore the numerical model proved to be sufficiently accurate.

3. Simulation results

The energy extracted from and injected into ground by the BHE, respectively in winter and in summer, is shown in figure 2 as a function of the Darcy velocity. It may be noticed that more energy is exchanged during winter (extraction in the graph) than is exchanged during summer (injection in the graph), so that on a yearly basis the aquifer is used as a heat source. In order to observe a significant influence of the groundwater flow, a Darcy velocity equal to 10^{-7} m/s appears to be insufficient. In turn, for velocities equal to 10^{-6} m/s and larger, an important increase in the exchanged energy with respect to the null velocity case can clearly be observed. The increase is generally nonlinear. For null velocity and for 10^{-7} m/s a modest difference between the 1st and the 2nd year exchanged energies may be noticed, while for larger velocities the 2nd year data are almost coincident with the 1st year ones.

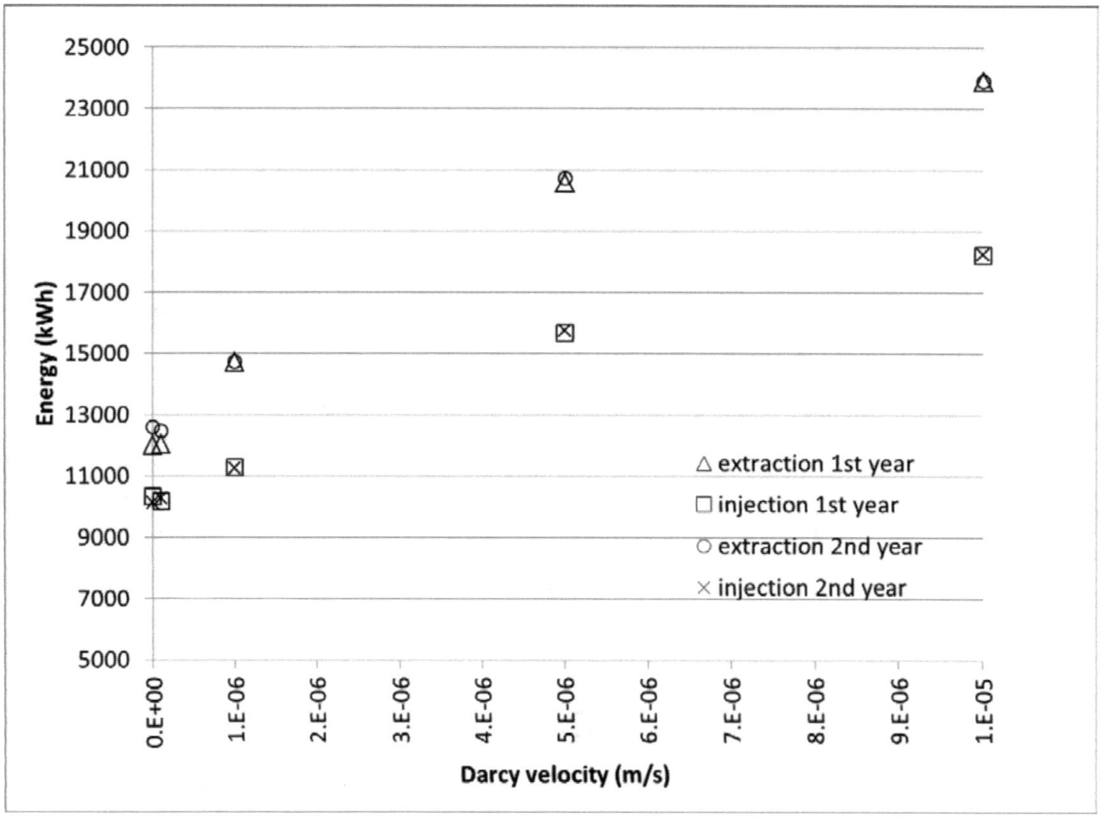

Figure 2. Energy extracted from or injected into the ground at 1st and 2nd year of operation versus Darcy velocity.

The increase in the extracted and injected energy with respect to the null velocity case can be plotted against the Darcy velocity, in a logarithmic scale, as in figure 3. For any given velocity, apart for $v = 10^{-7}$ m/s where negligible differences are found with respect to the null velocity case, considering the 1st year of operation, the extracted heat increases more than the injected heat. It is worth noting that in the 1st year of operation the simulation starts with the heat extraction period and, after a pause, the heat injection period starts. Therefore different temperature conditions are present in the surrounding soil at the beginning of the extraction and injection periods, and this influences the results. In accordance with this remark, figure 3 shows that at the 2nd year of operation for each Darcy velocity the increases in the extracted and injected heat tend to approach.

Focusing then on the results related to the 2nd year of operation, as they appear less influenced by the initial temperature conditions, it may be stated that with respect to the null velocity case for v = 10^{-7} m/s the energy variation is -1% in heat extraction and +2% in heat injection, for v = 10^{-6} m/s it is +17% in heat extraction and +11% in heat injection, for v = 5.10^{-6} m/s it is +65% in heat extraction and +56% in heat injection, and finally for v = 10^{-5} m/s it is +90% in heat extraction and +80% in heat injection. Therefore it may be argued that for velocities larger than 10^{-6} m/s neglecting the convection effects leads to significant errors in the design of a BHE.

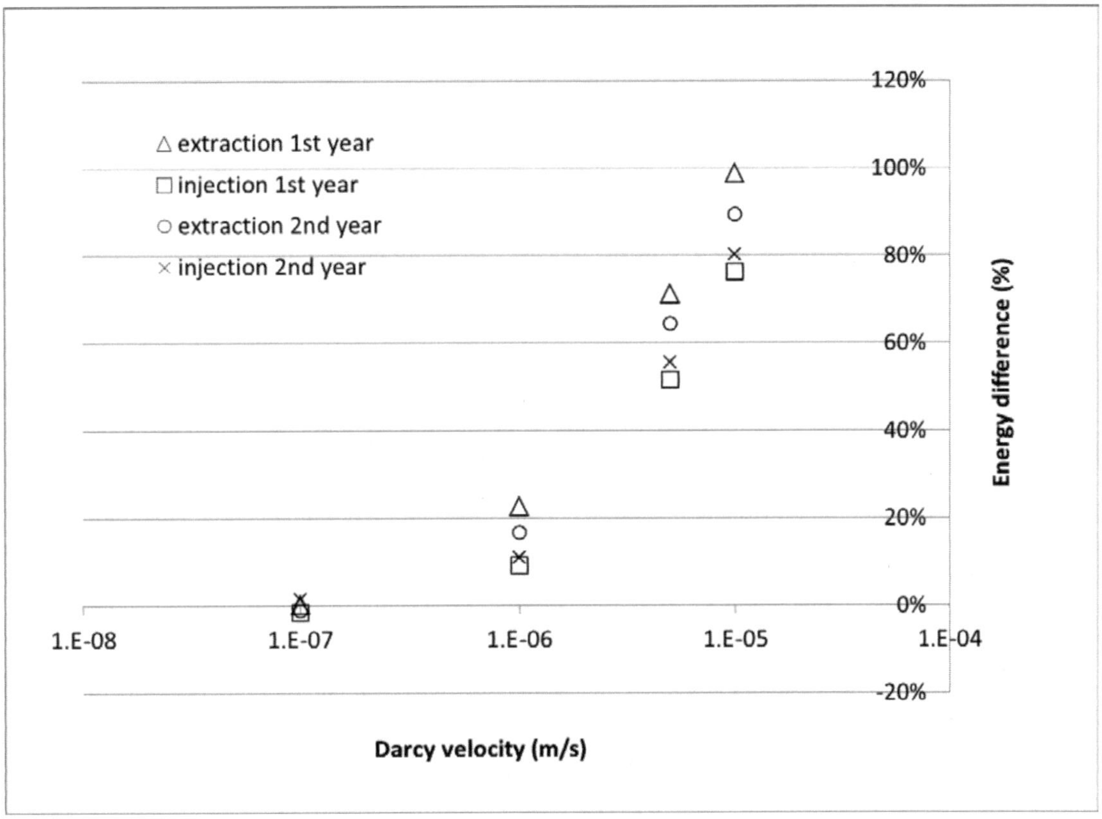

Figure 3. Difference in the exchanged energy with respect to the null velocity case, versus Darcy velocity, in the heat extraction and heat injection periods, at 1st and 2nd year of operation.

In figures 4, 5, 6 and 7 plan views of the temperature field in the aquifer are shown for respectively null velocity, v = 10^{-7} m/s , v = 10^{-6} m/s and v = 10^{-5} m/s. The case v = 5.10^{-6} m/s results in very similar conditions with respect to v = 10^{-5} m/s and therefore for the sake of brevity is not shown. For every case, four moments are considered, namely the beginning of the heat injection period of the 1st year (Ia in figures), the beginning of the heat extraction period of the 2nd year (Ib in figures), the beginning of the heat injection period of the 2nd year (IIa in figures) and finally after two complete years (IIb in figures).

In general the cases of v = 0 (figure 4) and v = 10^{-7} m/s (figure 5) are very similar: the temperature perturbation produced by the BHE operation in the previous season is centred on the BHE at the beginning of the next season, and therefore it has a positive influence on the energy performance in the next period. Moreover, for v = 10^{-7} m/s the temperature contours are only slightly distorted in the groundwater flow direction. With the latter velocity, a limited cold plume due to the previous heat extraction period is visible in Ib and IIb parts of figure 5.

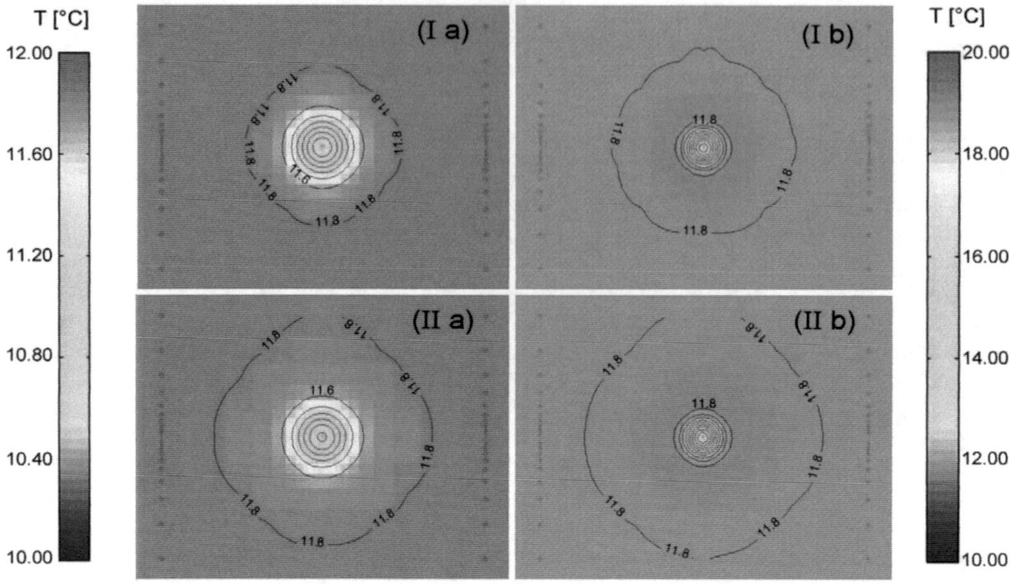

Figure 4: plan views of the temperature field in the aquifer for null velocity: Ia) beginning of heat injection 1st year; Ib) beginning of heat extraction 2nd year; IIa) beginning of heat injection 2nd year; IIb) after two years.

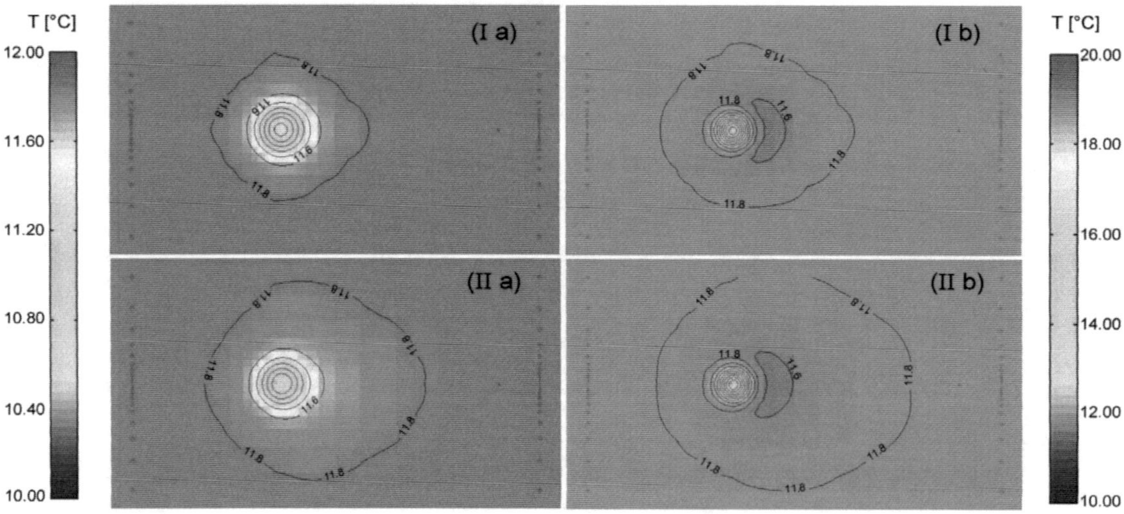

Figure 5: plan views of the temperature field in the aquifer for $v = 10^{-7}$ m/s Ia) beginning of heat injection 1st year; Ib) beginning of heat extraction 2nd year; IIa) beginning of heat injection 2nd year; IIb) after two years.

On the other side we find the $v = 10^{-5}$ m/s case (figures 7 and 8), where the high groundwater velocity brings away from the BHE most of the temperature perturbation due to the previous operating period. In these conditions, the benefits due to the use of the BHE in both seasons are less relevant. An intermediate behaviour is observed for $v = 10^{-6}$ m/s (figure 6): the effects of the previous season are still in the proximity of the BHE, although displaced downstream. In case $v = 10^{-6}$ m/s, apart from the beginning of the heat injection at the 1st year, two temperature plumes are generally visible, one due to the just finished operating period and another one due to the previous. Two temperature plumes are visible also for $v = 10^{-5}$ m/s, but only at the beginning of the heat injection periods. Since more energy is extracted then injected (figure 2), the cold plume is expected to develop more than the warm one. Therefore for large groundwater velocity as v = 10^{-5} m/s, the warm plume due to the previous heat injection is diluted, while the cold plume due to previous heat extraction is still visible. We can observe this specific situation better also in the vertical section view of this case (figure 8), where only one cold plume attends in Ia), while warm and cold plumes attend in Ib).

By comparing for each velocity the temperature fields Ia) and IIa) and, on the other side, Ib) and IIb), some considerations regarding the difference between the 1st and the 2nd year of operation may be derived. For v = 0 and v = 10^{-7} m/s the temperature field at the 2nd year appear very similar to the 1st year, but the contour corresponding to the unperturbed temperature 11.8°C is displaced further from the BHE. For v = 10^{-6} m/s, at the 2nd year the plume due to the just finished period is very similar to the 1st year, but the temperature plumes due to the past periods are more evident. Finally for v=10^{-5} m/s no significant difference may be found between the 1st and the 2nd year temperature field, in agreement with the energy data (figure 2). Therefore the situation after two years is still evolving for v = 0, 10^{-7} and 10^{-6} m/s, while it can be considered steady for v = 10^{-5} m/s.

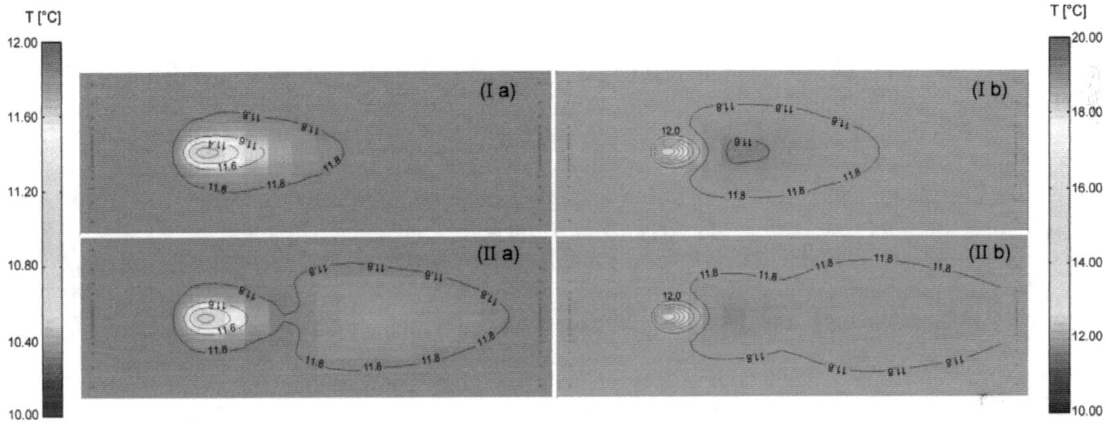

Figure 6: plan views of the temperature field in the aquifer for v = 10^{-6} m/s: Ia) beginning of heat injection 1st year; Ib) beginning of heat extraction 2nd year; IIa) beginning of heat injection 2nd year; IIb) after two years.

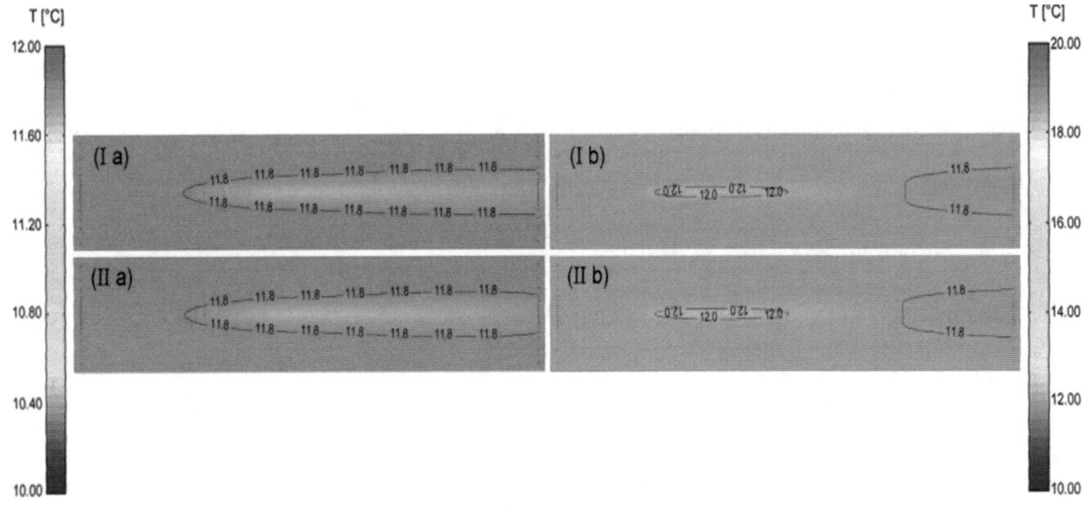

Figure 7: plan views of the temperature field in the aquifer for v = 10^{-5} m/s: Ia) beginning of heat injection 1st year; Ib) beginning of heat extraction 2nd year; IIa) beginning of heat injection 2nd year; IIb) after two years.

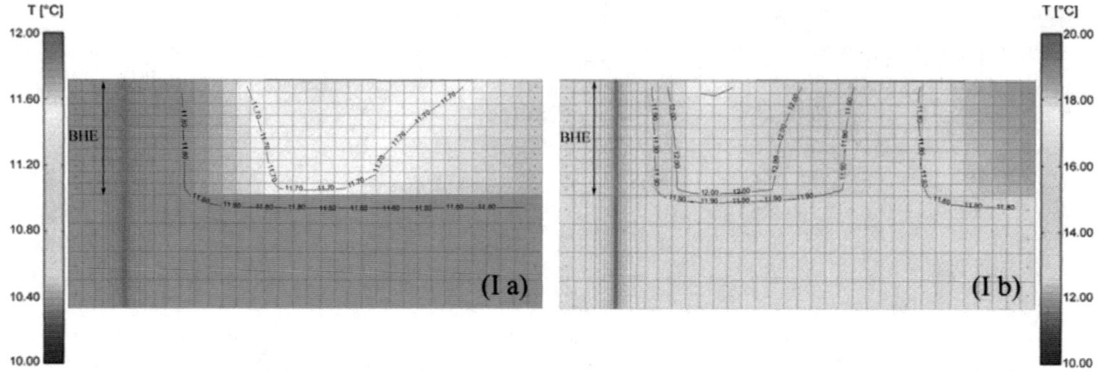

Figure 8: vertical section views of the temperature field in the aquifer for v = 10^{-5} m/s: Ia) beginning of heat injection 1st year; Ib) beginning of heat extraction 2nd year.

4. Conclusions

By using a numerical model of a BHE composed of a single U-pipe 100 m long in a sandy aquifer under different groundwater velocities the impact of the groundwater flow on the BHE energy performance has been assessed. For velocity lower than 10^{-6} m/s the heat transfer in the aquifer is mainly due to conduction. In turn for velocity equal and larger than 10^{-6} m/s the injected and extracted energy increase with respect to the null velocity case, up to 80 and 90% more for 10^{-5} m/s. Therefore further efforts are necessary to include groundwater influence in the design methods for BHEs. The case study, referring to a BHE operating in both seasons with a yearly unbalance between energy extraction and injection for two years, highlights the importance of long term evaluations, that however are characterised by a high computational effort.

References

[1] Hellstrom G, Mazzarella L, Pahud D 1995 Duct Ground Heat Storage Model – *Manual for Trnsys Version*

[2] Spitler JD 2000 GLHEPRO - A Design Tool For Commercial Building Ground Loop Heat Exchangers Proc Fourth Intern. *Heat Pumps in Cold Climates Conf.* (Aylmer, Québec)

[3] Hellstrom G and Sanner B 2001 Pc-programs and modelling for borehole heat exchangers design Proc. Intern. *Geothermal Days Germany* 2001 (Bad Urach, Germany)

[4] RETScreen International. Clean Energy Project Analysis: RETScreen Engineering & Cases Textbook. Minister of Natural Resources of Canada 2001

[5] UNI 11466 2012 Sistemi geotermici a pompa di calore - Requisiti per il dimensionamento e la progettazione (Milano: Ente Nazionale Italiano Normazione)

[6] Diao N, Li Q, Fang Z 2004 Heat transfer in ground heat exchangers with groundwater advection *Int. J. Thermal Sciences* **43** 1203-1211

[7] Molina-Giraldo N, Bayer P, Blum P 2011 Evaluating the influence of thermal dispersion on temperature plumes from geothermal systems using analytical solutions *Int. J. Thermal Sciences* **50** 1223-1231

[8] Wang H, Qi C, Du H, Gu J 2009 Thermal performance of borehole heat exchanger under groundwater flow: a case study from Baoding *Energy and Buildings* **41** 1368-1373

[9] Zheng C and Wang PP 1999 MT3DMS: a modular three-dimensional transport model for simulation of advection, dispersion and chemical reactions of contaminants in groundwater systems; documentation and users' guide. Contract Report SERDP-99-1

[10] Harbaugh AW, Banta ER, Hill MC, McDonald MG 2000 MODFLOW-2000, the U.S. Geological Survey modular ground-water model - User guide to modularization concepts and the Ground-Water Flow Process: *U.S. Geological Survey Open-File Report* 00-92

[11] Angelotti A, Alberti L, La Licata I, Antelmi M Energy performance and thermal impact of a borehole heat exchanger in a sandy aquifer: influence of the groundwater velocity, accepted for publication in *Energy Conversion and Management*. DOI: 10.1016/j.enconman.2013.10.018.

[12] Alberti L, Angelotti A, Antelmi M, La Licata I, Legrenzi C Low temperature geothermal energy: heat exchange simulation in aquifers through Modflow/MT3DMS codes *AQUA mundi* 2012 – Am05042: 039 - 051

Energy and economical comparison of possible cultures for a total-integrated on-field biodiesel production

G Allesina[1], S Pedrazzi[1], S Tebianian[2], A Muscio[1] and P Tartarini[1]

[1] University of Modena and Reggio Emilia, Department of Engineering 'Enzo Ferrari', Via Vignolese 905, 41125 Modena, Italy
[2]University of British Columbia, Department of Chemical and Biological Engineering, 2360 East Mall, V6T1Z3,Vancouver BC Canada

E-mail: giulio.allesina@unimore.it

Abstract.
This work is aimed at investigating the energy conversion effectiveness and the economical advantages of a total integrated solution for on-field biodiesel and electrical energy production. The system proposed here is based on the synergy of four sub-systems: a seed press for oil production, a downdraft gasifier, a biodiesel conversion plant and a Solid Oxide Fuel Cell (SOFC). Two possible culture rotations, suggested by literature review, were analyzed here from economical and energy balance points of view. Both the rotations were composed of oleaginous crops only, therefore the seeds collected from the different cultures were pressed, then the protein cake produced in the process was gasified in the downdraft reactor. The gasification process was modeled here, and its output suggested that, for a precise number of hectares, the syngas obtained through the cake gasification was enough for producing methanol required for oil-biodiesel conversion and feeding a 10-kW SOFC. The purge line in the methanol reactor was used in the SOFC as well. The system was simulated using ASPEN PLUS^{TM}and MATLAB^{TM}codes. Results of the SOFC and gasifier models underlined the capability of the fuel cell to work with this particular system, furthermore the whole system analysis suggested that the surface required for sustainability of the processes is a function of the rotation choice. In both cases little surfaces ranging from 11 to 21 hectares were found to be enough for system self-sustainability with a ROI under 7 years in all the operating conditions analyzed.

1. Introduction

Most of gasifiers for electrical and CHP applications are commonly coupled either with Organic Rankine Cycle (ORC) systems, Internal Combustion engines, Conventional and External Firing Gas Turbines (EFGT) or Stirling engines [1–6]. However, the efficiency of these systems is not as high as the one that can be obtained using advanced power generation systems such as Molten Carbonate Fuel Cells (MCFC) [7, 8] or Solid Oxide Fuel Cells (SOFC) [9–11]. For this kind of systems, a downdraft gasifier (DG) reactor is a suitable solution due to the low content of tar and particulate in its gas when compared with updraft, crossdraft or fluidized bed gasifiers [1, 2, 12, 13]. This feature optimizes the performance and the reliability of the SOFC subsystem that operates properly only when fed with clean syngas fuel [14].

This study starts from the feasibility analysis reported in [15], where a full integrated system for biodiesel production was proposed and its advantages were discussed. The previous work was based on a rotation of rapeseed-flax-sunflower, as suggested by [16]. All the crops composing this rotation could be used for oil production, but the major disadvantage of this solution was the low oil productivity of flax seeds. Furthermore flax oil is a valuable product and its use for biodiesel conversion is disadvantageous. For this reason, a new rotation was proposed and its feasibility and sustainability were discussed in [17]. This solution substituted flax with soy. All these crops are adequate to Mediterranean zones, therefore the simulations were based on data obtained from literature about average oil production for these cultures in the Emilia Romagna region of Italy as reported in [15, 18–27].

The system layout is sketched in figure 1. The approach suggested here consisted in cultivating every year all the surface (evaluated on economical end energy balance basis) with a different culture. The collected seeds were converted into oil and protein cake by a mechanical process. The protein cake obtained as a by-product of the oil extraction was gasified to syngas in a downdraft stratified reactor. A small part of the syngas was converted to methanol in a chemical reactor while the rest of the gas fed a 10 kW SOFC. The methanol was then used in the transesterification of the PVO to produce biodiesel. The gasifier model was based on: ultimate analysis of the protein cake, dimensions and physical properties of the cake pellets and geometrical parameters of the gasifier. The geometrical and physical properties of the protein cake were obtained through a series of experimental analyses, the results of these analyses are reported in table 1. The biodiesel production process consisted of two stages: in a first reactor the syngas was converted into methanol; then in a different reactor PVO and methanol were mixed for biodiesel production. The methanol production conversion reactions were simulated with an equilibrium-based approach whereas the biodiesel production was modeled by a stoichiometric approach with a known conversion typical of the catalytic reactions. SOFC modeling is well described in literature [28–30], the model proposed here was based on the works of Bang-Møller and Rokni [28] for the cell reactions and Rami Salah El-Emam for the reformer [29].

The whole system model was obtained coupling all the models of the sub-systems. It allowed to calculate the mass of biodiesel and glycerin obtained every year and the electrical energy produced by the SOFC connected to the gasifier. An economical analysis was made to evaluate the return of the investment and the economical differences between the rotations proposed here.

2. Gasifier mathematical model

For the gasification reactions modeling, a kinetic model was used in order to predict zone lengths and gas composition. This model was created starting from the method developed by Reed and Markson [31] to predict the flaming pyrolysis zone length l_p and the char reduction zone length l_c. The biomass properties used as input in this model are: moisture content F_m, biomass particle equivalent diameter d_p, area A_g and height H of the gasifier, density F_d of the biomass, and void fraction F_v in the biomass. Furthermore the model by Reed and Markson evaluates the heat transfer between particles located in the flaming pyrolysis layer using data by Reed and Markson themselves [31] or obtained from Huff [32]. One of the major output of this model is the char reduction time t_c. It was used here for coupling this model with the one presented and discussed by Wang and Kinoshita [33]. Wang's model consists of a kinetic system of differential equations that describes the mechanisms occurring in the char reduction zone. It starts from the evaluation of the effect of flaming pyrolysis on a particle homogeneously composed of equivalent biomass molecules $CH_\alpha O_\beta$. The system resolution requires the kinetic constants and biomass properties that were obtained from [34–36].

The resulting coupled model operates fixing a pressure p of gasification and the equivalence ratio ER of the reactor [1].

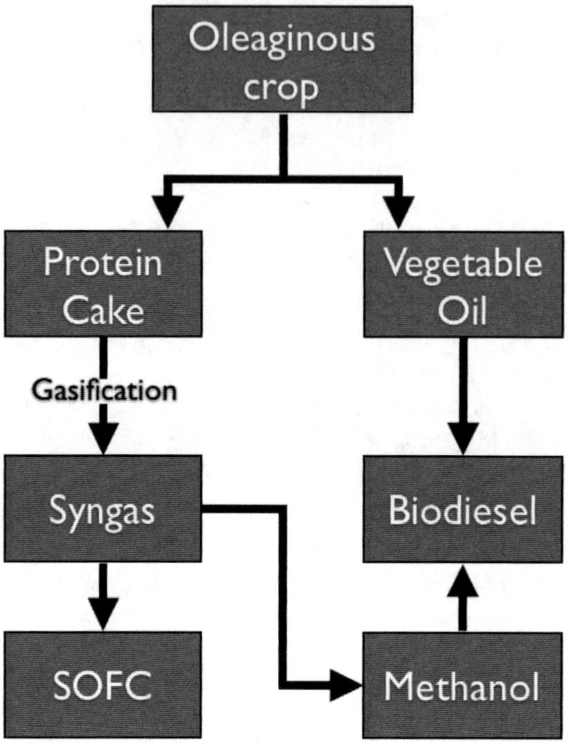

Figure 1. System layout

Table 1 lists the chemical, physical and geometrical properties of the protein cakes used as model input in this work: rapeseed, sunflower, soy and flax. Table 2 shows the geometrical and physical parameters of the gasifier. Furthermore, table 3 resumes the syngas compositions, HHV_{syngas} and gasifier cold gas inefficiencies η_{cold} for rapeseed, sunflower, soy and flax protein cakes. A complete discussion of the model used here can be found in [37].

Table 1. Protein cakes parameters

Parameter	Rapeseed protein cake	Sunflower protein cake	Soy protein cake	Flax protein cake
Equivalent cake diameter d_p	0.01 m	0.01 m	0.01 m	0.1 m
Density F_d	1.79 kg/m^3	1.79 kg/m^3	1.79 kg/m^3	1.79 kg/m^3
Void fraction F_v	0.48	0.48	0.48	0.48
Moisture	10.8%	10.8%	10.8%	10.8%
C (wt%)	43.2%	46.78%	55.89%	46.46%
H (wt%)	6%	6.425%	6.57%	5%
O (wt%)	44.27%	45.91%	28.25%	47.64%
N (wt%)	5.6%	0.88%	9.29%	0.6%
S (wt%)	0.93%	0 %	0%	0%
Ash (wt%)	7.20%	3.82%	5.36%	4.4%
$HHV_{bio,dry}$	15.37 MJ/kg	16.76 MJ/kg	21.34 MJ/kg	17.15 MJ/kg

Table 2. Parameters of the downdraft stratified gasifier

Parameter	Value
Gasifier diameter D_g	0.28 m
Gasifier area A_g	0.0616 m^2
Fixed bed height H	0.4 m
Flaming pyrolysis temperature T_s	1173 K
Char reduction temperature T	1073 K
Equivalence ratio ER [1]	0.35
Pressure in the gasifier p	1 bar

Table 3. Results of the gasification model

Variable	Rapeseed	Sunflower	Soybean	Flax
H_2 [%vol]	12.2 %	12.42 %	14.72 %	14.96 %
CO [%vol]	13.98 %	14.17 %	17.38 %	19.43 %
CH_4 [%vol]	6.01 %	5.92 %	4.77 %	3.11 %
CO_2 [%vol]	17.86 %	17.06 %	11.78 %	14.76 %
N_2 [%vol]	50.31 %	50.43 %	51.35 %	47.73 %
Syngas production per hectare	5582 kg	4328 kg	10353 kg	2012 kg
HHV_{syngas}	5.22 MJ/Nm^3	5.32 MJ/Nm^3	5.49 MJ/Nm^3	5.18 MJ/Nm^3
η_{cold}	96.67 %	96.41 %	93.76%	90.34%

3. SOFC model

The SOFC model has been developed by Bang-Møller and Rokni [28]. This model did not take into account the recirculation of the exhaust in the anode. To overcome this issue, the reforming model presented by Rami Salah El-Emam et al. [29] was used in this work. The reforming of the fuel occurs near the anode. Here take place both the reforming of the methane (equation 1) and the water gas shift of the carbon monoxide (equation 2). The electrochemical reaction takes place in the anode and in the cathode. At the anode, the hydrogen reacts with the oxygen ions to form water and electrons according to the equation 3. At the cathode, the oxygen of the inlet air reacts with the electrons from the anode (equation 4) to form oxygen ions that they are conveyed to the anode through the solid oxide electrolyte. Equation 5 resumes the overall electrochemical reaction.

$$CH_4 + H_2O \rightarrow CO + 3H_2 \tag{1}$$

$$CO + H_2O \rightarrow CO_2 + H_2 \tag{2}$$

$$H_2 + O^{2-} \rightarrow H_2O + 2e^- \tag{3}$$

$$\frac{1}{2}O_2 + 2e^- \rightarrow O^{2-} \tag{4}$$

$$H_2 + \frac{1}{2}O_2 \leftrightarrow H_2O \tag{5}$$

The mathematical modeling of reforming and electrochemical reactions are explained in [29] and [28] respectively. Using these models it was possible to calculate the electrical energy production and the electrical efficiency for a given syngas inlet flow with a specific gas composition. The SOFC model parameters adopted in the simulations are reported in table 4.

Table 4. SOFC model parameters

Description	Symbol	Value
Fuel utilization factor	U_f	0.85
Recirculation factor	r	0.2
Steam to carbon factor	STC	2.5
Cathode air factor	$vent$	1.1
Pressure ratio	PR	2.5
Operating temperature	T_{SOFC}	1073.15 K
Anode pressure loss	p_a	500 Pa
Cathode pressure loss	p_c	1000 Pa
Current density	i	3000 mA cm^{-2}
Active cell area	A_{cell}	81 cm^2
Cells for each stack	$n_{cell,stack}$	75 cells
Number of stacks	n_{stack}	8 stacks
Electro-chemical parameters	Reported in [28]	

4. Biodiesel production model

The modeling of syngas→methanol conversion and methanol+oil→biodiesel conversion were developed in ASPEN^{TM} Plus software. The model used the data obtained from literature about the productivity of the mentioned cultures together with the data output of the kinetic model of the gasification process for the four different protein cakes. Once the amount of vegetable oil collected from every hectare of the different cultures was evaluated, ASPEN software was used first to calculate the amount of methanol required for complete oil conversion and then the volume of syngas necessary for the production of that very amount of methanol.

The biodiesel production model from vegetable oil and methanol reaction was based on the transesterification of triolein [38] reported in equation 6, the reactor was configured RSOITC and it was based on stoichiometric approach with a defined conversion rate (95%). The model thermodynamic activity is set in the software to 'universal quasi-chemical' as suggested by the presence of highly polar components.

$$TRIOLEIN + 3CH_3OH \rightarrow^{KOH} 3METHYL\ OLEATE + GLYCEROL \quad (6)$$

The syngas to methanol conversion was based on the following reactions [39]:

$$CO + 2H_2 \rightarrow CH_3OH \quad (7)$$

$$CO_2 + 3H_2 \rightarrow CH_3OH + H_2O \quad (8)$$

$$CO_2 + H_2O \rightarrow CO_2 + H_2 \quad (9)$$

The methanol conversion modeling was based on equilibrium reactor in ASPEN plus (REQUIL). The syngas composition suggested by the gasification model showed that the CO and CO_2 were in excess compared to hydrogen, therefore the molar flow rate of methanol produced in the reactor was limited by the conversion of H_2. For the operating conditions of

Journal of Physics: Conference Series **501** (2014) 012034 doi:10.1088/1742-6596/501/1/012034

the reactor given in table 5 the hydrogen conversion was about 15%. The methanol production model parameters are reported in table 5). The whole system was resolved using 'Design Specification' function of Aspen Plus in order to obtain the flow-rate of syngas necessary for producing the methanol required for PVO conversion. Table 6 resumes the output of the ASPEN model together with the amount of syngas required for the process and the amount and the composition of the syngas purged after multiple recirculation. Figure 4 schematizes the process as reported in ASPEN^{TM} plus.

Table 5. Parameters of the biodiesel production model

Parameter	Value
Methanol reactor pressure	76 bar
Methanol reactor temperature	523.15 K
Recycle ratio	8
Biodiesel reactor pressure	1 bar
Biodiesel reactor temperature	343.15 K

Table 6. Results of the biodiesel production model for 1 hectare of soil

Variable	Rapeseed	Sunflower	Soy
Seeds	4000 kg	2800 kg	3700 kg
PVO	1840 kg	1260 kg	740 kg
Biodiesel	1465 kg	997 kg	583 kg
Glycerin	488 kg	332 kg	194 kg
Methanol required	204 kg	137 kg	80
Syngas required	4160 kg	2718 kg	1173
Syngas purged	3956 kg	2580 kg	1093
H_2 purged	4.47 %	4.42 %	4.60%
CO purged	11.22 %	11.41 %	14.24 %
CH_4 purged	6.86 %	6.78 %	5.69 %
CO_2 purged	19.88 %	19.45 %	14.01%
N_2 purged	57.41 %	57.76 %	61.30 %
HHV_{purged}	4.16 MJ/Nm^3	4.35 MJ/Nm^3	4.33 MJ/Nm^3

5. Results and discussion

5.1. Chemical balance

The first result that need to be discussed is the chemical balance of the system proposed. Figure 2 shows the differences in syngas production, syngas required for methanol production and syngas HHV. The value reported in figure 2 guaranteed, for all the cultures used in this study in biodiesel production, an excess of syngas. This gas excess is fundamental because it was the parameter used for evaluating the surface required for SOFC run over the year. More detailed data can be found in tables 6 and 7.

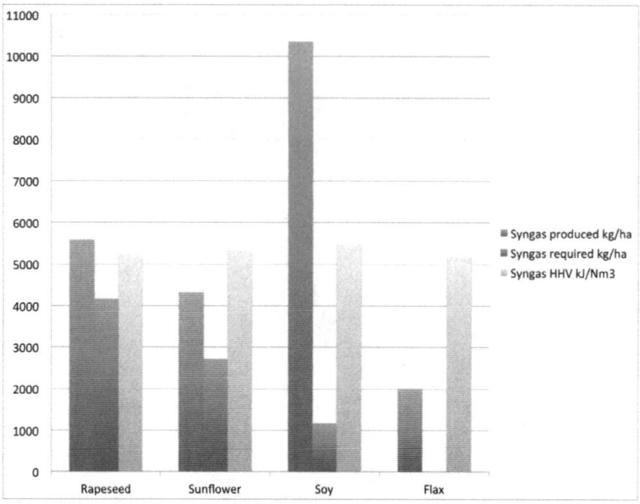

Figure 2. Syngas production, requirement and quality comparison

The heating values of the syngas obtained through gasification of the four protein cakes were founded to be all similar with an estimated value of about 5 MJ/Nm^3. Figure 2 together with the comparison reported in figure 3 underlined the differences between the soy and flax uses. Flax produced less protein cake and less oil, for this reason larger surfaces were required for reaching the sustainability of the system fixing the electrical power output to 10 kW. Soy was similar to flax from an oil production point of view but it had more protein cake that produced higher amount of syngas. Furthermore the proposed results outline two energy recovery features of the system proposed here: culture such as soy are characterized by high solid matter when compared with the oil production. With the approach proposed here this characteristic is not a disadvantage any more, because the gasification and electrical energy conversion of the solid matter compensate the low biodiesel production. Moreover in all the biodiesel conversion processes described here, even if the methanol reactor had a low conversion rate, the non-reacted gas chemical energy was not wasted because the purge line was connected with the SOFC. The waste heat coming from the methanol reactor and the SOFC could be used for several purposes such as heating up the biodiesel reactor, drying the cake and pre-heating the air entering the gasifier. If the heat produced by the processes will exceed the heat required for the mentioned processes, the system can be considered "high efficiency CHP", this solution guarantees a higher subsidy [40].

5.2. SOFC results

The results of the gasifier-SOFC system are reported in table 7, reporting the number of hours and the energy produced by one hectare of different crops. The data discussed here allowed the calculation of the number of hectares necessary as function of the number hours of plant running per day. The SOFC worked together with pure and purged syngas from the gasifier and methanol conversion reactors respectively. Because the higher the surface tilled, the less the specific cost of cultivation, the system considered only one crop a year. Furthermore, if the protein cake was properly stored, there was no need to use all the cake in the very year in which it was produced. For this reason the system was balanced working on the three year

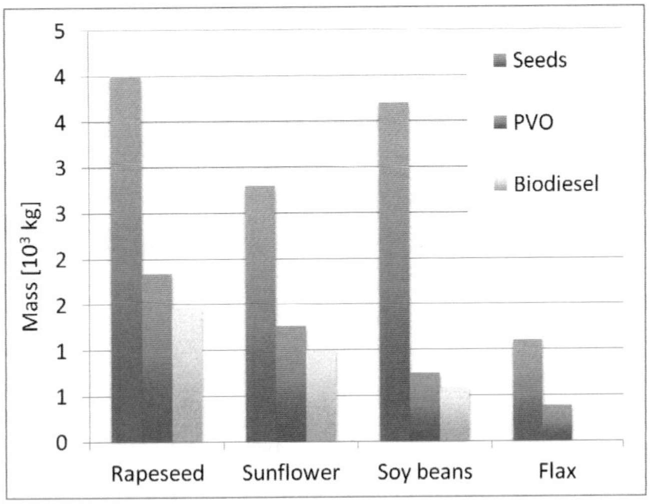

Figure 3. Productivity differences

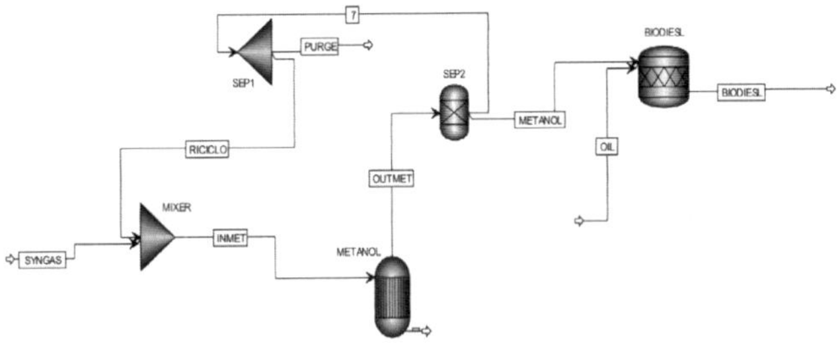

s

Figure 4. Biodiesel production model in $ASPEN^{TM}$ plus software

rotation instead of an year by year approach. Figure 5 and figure 6 show the different values in terms of hours of operation of the SOFC for the three different cultures and for the two rotations proposed as a function of the surface tilled. Horizontal dashed lines were placed at 8760 and 13140 total hours in a 3-year time interval, that is the same as 2920 and 4380 hours per year, 8 or 12 hours a day depending on the point of view. The higher the number of hours to run the plant per year, the higher the surface required to ensure the feeding of the gasifier. The line with the higher slope is the cumulative line. In case of 12 hours a day of plant run the minimum surface required is 11 hectares for the soy-based rotation and 21 hectares for the flax-based rotation.

5.3. Economical analysis

As reported in [15], the analysis was based on a few conservative assumptions:

- The cost for oil extraction was evaluated starting from technology review of seed presses available on the market. The average specific consumption is 0.06 kWh/kg of seeds [41]. This electrical energy was absorbed from the grid and not self-consumed from the power

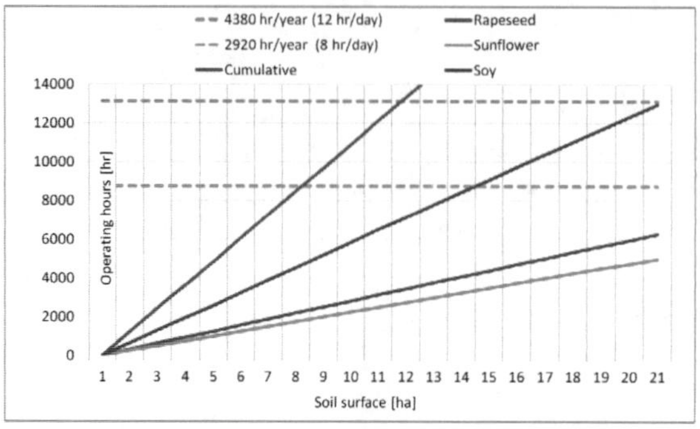

Figure 5. Hours of plant running per hectare of tilled surface with rotation A

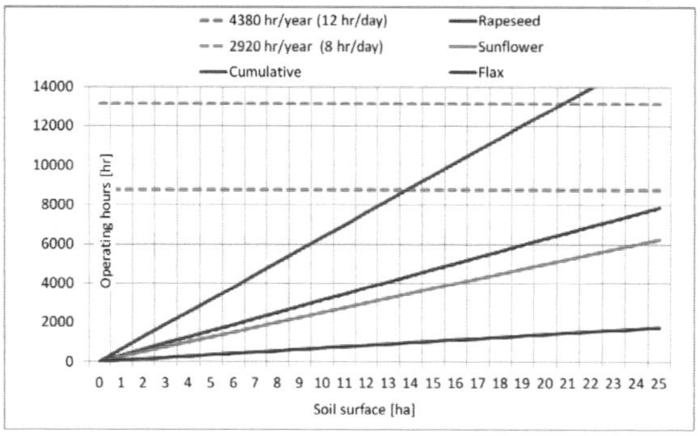

Figure 6. Hours of plant running per hectare of tilled surface with rotation B

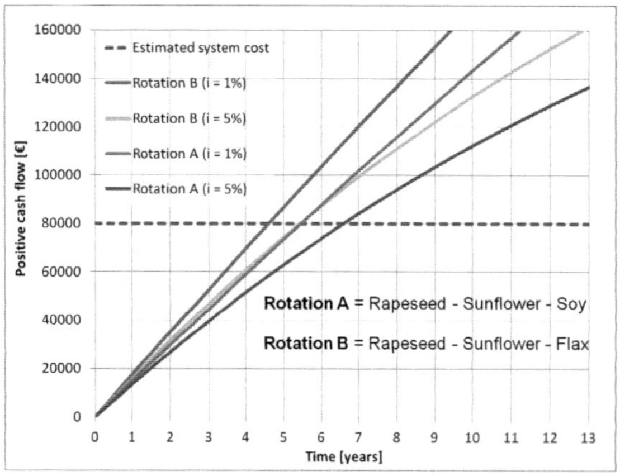

Figure 7. Net present value analysis

Table 7. SOFC output for 1 hectare of different cultures

Output	Rapeseed	Sunflower	Soy	Flax
Gasified protein cake kg/ha	2160	1540	2960	715
Operating hours at 10 kWel	314	250	650	70
Energy produced kWh/ha	3132	2503	6400	686
SOFC electrical efficiency	56.65 %	56.18 %	58.12 %	58.56 %

output of the SOFC. The considered cost for electricity at the current grid price for industrial uses was 0.1565 €/kWh$_{el}$ [42]).

- The costs for cultivating rapeseed, sunflower, flax and soy per hectare of soil are 680 €, 850 €, 500 € , 1245 € respectively. These costs were obtained as sum of seed, fertilization and farming costs [20, 27].
- This analysis did not take into account the electrical energy surplus obtainable through gasification of stalks and other vegetable byproducts of seeds crop. This surplus could be used for increasing the SOFC power output or together with the protein cake to increase the gasification performance.
- The income for sale of glycerin was 0.145 €/kg [43]).
- Other sources of electrical consumption were not taken into account in this study (i.e. compressor for methanol reactor pressurization).

The sale of biodiesel and electrical energy, the following parameters and subsidies were taken into account for 12 hours of plant running a day:

Biodiesel sale price set at 0.982 €/L (with biodiesel density < 0.90 kg/L), that is the current value on the Italian industrial market [44] and a feed-in tariff of electrical energy set at 0.297 €/kWh$_{el}$ for production from biomasses by means of small size plants thanks to a recent subsidy program in force in Italy (Ministerial Decree 06/07/2012 for the subsidy of electrical energy production from renewable sources different from photovoltaic systems [40])

For both the rotation proposed, this economical approach was found to yield a positive cash flow of about 45896 €/3year for rapeseed-sunflower-soy rotation, while the rapeseed-sunflower-flax rotation gives a cash flow of 54179 €/3year. On the other hand, the rotation with flax needed almost twice the surface required by the other rotation for being sustainable. Furthermore the rotation with flax needed to sell the flax oil, introducing the necessity to market an extra product when compared to the other solution. A conventional net present value analysis is showed in figure 7. It was based on an investment of about 80000 € (I_0) for the whole system and the capital cost was amortized within a period that range from 4.5 to 6.5 years depending on the rotation and the discount rate. Two possible discount rates (i) were considered in this study: 1% typical of alternative investments in government bonds (adjusted for inflation), and 5% typical of mortgage in Italy. The choice of the more suitable gasifier for this application, was determined by the characteristics of the protein cakes processed. For this non-conventional fuel an open core or an Imbert gasifier are good choices due to the capability of these reactors to process different kind of biomasses. The market offers different reactors for the power output size required in this work [45–47]. For cost estimations the All Power Labs GEK device was used as an example. The cost of the whole system was estimated on the following assumptions:

- Literature suggests a cost per kW of about 4000 € [48].

- A GEK 10 kW gasifier with bio-filter costs about 1500-2000 [46].
- The costs of the methanol and biodiesel production reactors and the seed-press was kept high (38000 €) in order to compensate possible extra costs for gas cleaning and SOFC maintenance [41, 49].

The following formula was used to calculate the net present value (NPV) at the N-th year as the sum of the discounted cash inflow in the years from 0 to N:

$$NPV = -I_0 + \sum_{n=0}^{N} \frac{(I_{bd+ee} - I_{seeds,cultivation} - I_{press\ el\ consumption})}{(1+i)^n} \tag{10}$$

It is interesting to find that 48% of all income of soy-based rotation comes from biodiesel, thus about 52% comes from electricity. In the flax-based rotation these values shift to 56% and 39% respectively (plus a 6% of income from the flax oil). Increasing the production of syngas and electrical energy through exploitation of stalks and other by products of seeds crop can thus improve significantly the cash flow.

6. Conclusions

The study discussed here demonstrated the capability of solid oxide fuel cells to be effectively used in systems such as the one proposed here. The presence of CO_2 and N_2 in the syngas seemed to affect the cell conversion efficiency within given limits, fixing the overall conversion rate to encouraging values. Furthermore two possible rotations are here modeled and discussed, substitution of flax with soy could drastically reduce the surface required for system self-sustainability without compromising the return of the investment. Future work will support the method proposed with accurate experimental campaign, trowing light on possible limits of the system such as capability to avoid ashes slagging in the gasifier or the effect of syngas composition fluctuations on methanol conversion rates. Once the system model is validated, different rotations can be simulated in order to maximize the profit.

References

[1] Reed T B and Das A 1988 *Handbook of Biomass Downdraft Gasifier Engine Systems* (The biomass energy foundation press)
[2] Mechanical Wood Products Branch F F D 1986 *Woodgas as engine fuel* vol ISBN 92-5-102436-7 (F.A.O.)
[3] Duvia A and Gaia M 2004 *Congress: Cogenerazione a biomassa mediante Turbogeneratori ORC Turboden: tecnologia, efficienza, esperienze pratiche ed economia.*
[4] Martelli F, Riccio G, Maltagliati S and Chiaramonti D 2000 *1st world Conference of Biomass, Sevilla*
[5] Naso V 1991 *La macchina di Stirling* (CEA)
[6] Souleymane C 2012 *Motori a combustione interna e turbine a gas di piccola taglia per gas di sintesi* Master's thesis Universit degli Studi di Padova
[7] Galeno G 2006-2007 *Modellizzazione di un micro cogeneratore basato sulla tecnologia MCFC accoppiata ad un gassificatore di biomassa* Ph.D. thesis University of Cassino, Italy
[8] Donolo G, Simon G D and Fermeglia M 2006 *Journal of Power Source* **158** 1282–1289
[9] Cordiner S, Feola M, Mulone V and Romanelli F 2007 *Applied Thermal Engineering* **27** 738–747
[10] Doherty W, Reynolds A and Kennedy D 2010 *Energy* **25** 4545–4555
[11] Achenbach E 1994 *Journal of Power Sources* **49** 333–348
[12] Devi L, Ptasinski K J and Janssen F 2003 *Biomass and Bioenergy* **24** 125–140
[13] Allesina G, Pedrazzi S and Cattini C 2011 *International Journal of Heat and Technology* **29** 151–156
[14] Hofmann P, Panopoulos K, Aravind P, Schweiger M S A, Karl J, Ouweltjes J and Kakaras E 2009 *International Journal of Hydrogen Energy* **34** 9203–9212
[15] GAllesina, Muscio A, Tebianian S, Pedrazzi S and Tartarini P 2013 *3th International Conference of Microgeneration and Related Technologies; Naples, IT*
[16] Lizarazu W Z and Monti A 2011 *Biomass and Bioenergy* **35**

Journal of Physics: Conference Series **501** (2014) 012034 doi:10.1088/1742-6596/501/1/012034

[17] Lizarazu W Z and Monti A 2010 Non-food energy crops rotation possibilities Tech. rep. Department of Agroenvironmental Science and Technology, University of Bologna (IT)
[18] Innocenti A, Poli M and Re L D 2008 *Rivista Agricoltura* 77–79
[19] ISMEA 2013 Commodity: Semi oleosi: Prezzi: Mercato internazionale Tech. rep. Istituto di servizi per il mercato agricolo alimentare
[20] Maisadour 2013 Seed cultivation costs Tech. rep. http://www.maisadour-semences.fr/it/
[21] di Statistica I I N 2013 Consultazione dati: Navigazione tra i dati: Coltivazioni: Seminativi: Coltivazioni industriali: Tav. c14/2012 – superficie (ettari) e produzione (quintali): arachide colza girasole Tech. rep. http://agri.istat.it
[22] online I A 2013 Atlante delle coltivazioni erbacee – piante industriali: Girasole – helianthus annuus l. Tech. rep. http://www.agraria.org/coltivazionierbacee/girasole.htm
[23] ISMEA 2013 Commodity: Semi oleosi: Prezzi: News mercati n. 32/2012 – settimana n.47 Tech. rep. ISMEA
[24] Colombo R and Poli M 2000 *Rivista Agricoltura* 33–35
[25] online I A 2013 Atlante delle coltivazioni erbacee – piante industriali: Lino- linum usitatissimum l. Tech. rep. Istruzione Agraria online
[26] Uzun B, Pötün A and Pütün E 2006 *Bioresource Technology* **97** 569–576
[27] Concezzi L 2007 La filiera del lino in umbria: una storia di successo con un futuro molto promettente Tech. rep. 3A-Parco Tecnologico Agroalimentare dellUmbria
[28] Bang-Moller C and Rokni M 2007 *Energy Conversion and Management* **51** 2330–2339
[29] El-Eman R, Dincer I and Naterer G F 2012 *International Journal of Hydrogen Energy* **37** 1689–1697
[30] Shirazi A, Aminiavary M, Najafi B, Rinaldi F and Razaghi M 2012 *International Journal of Hydrogen Energy* **37(24)** 19111–19124
[31] Reed T and Markson M 1983 *Progress in Biomass Conversion* **4** 219–254
[32] Huff E 1982 *Fundamentals of thermo-chemical biomass conversion: An international conference, Estes Park, CO*
[33] Wang Y and Kinoshita C 1993 *Solar Energy* **51** 19–25
[34] Laidler K J 1987 *Chemical kinetics* (New York: Harper sRow Publishers)
[35] Sharma A K 2011 *Energy Conversion and Management* **52** 1386–1396
[36] Everson R C, Neomagus H V J P, Kasaini H and Njapha D 2006 *Fuel* **85** 1076–82.
[37] Pedrazzi S, Allesina G and Tartarini P 2012 *International Journal of Heat and Technology* **30** 41–44
[38] Gerpen J V, Shanks B, Pruszko R, Clements D and Knothe G 2004 Biodiesel production technology Tech. rep. NREL Colorado
[39] N Dave G A F 1995 *Industrial and Engineering Chemistry Research* **34** 1037–1043
[40] Ministero dello Sviluppo Economico w D.m. 16/07/2012
[41] CRPA 2010 Presse a vite continua per la spremitura a freddo di semi e la produzione di olio vegetale Tech. rep. Cecchetti energia srl, La Mecanique Moderne
[42] Portal E E Electricity – industry (italy), http://www.energy.eu
[43] Bordin A 2008 *Biocombustibili e biocarburanti. Soluzioni, tecnologie, agevolazioni Data di Pubblicazione: 2008* (Ipsoa)
[44] http://wwwunionepetroliferait U P I Databook 2012 p. 103
[45] 2013 *Victory Gasifier, http://gasifier.wpengine.com/*
[46] 2013 *Gasifier Experimental Kit, http://www.gekgasifier.com/*
[47] Stak-Gasifier 2013 *Stak*
[48] Foger K 2011 Challenges in commercialising an ultraefficient sofc residential generator Tech. rep. Ceramic Fuel Cells Limited
[49] press cost S 2013 model clb-500-em, http://www.croplandbiodiesel.com

31st UIT (Italian Union of Thermo-fluid-dynamics) Heat Transfer Conference 2013 IOP Publishing
Journal of Physics: Conference Series **501** (2014) 012035 doi:10.1088/1742-6596/501/1/012035

Qualitative thermal characterization and cooling of lithium batteries for electric vehicles

A Mariani[1], F D'Annibale[1], G Boccardi[1], G P Celata[2], C Menale[3], R Bubbico[3] and F Vellucci[4]

[1] ENEA Laboratory of Thermal Fluid-Dynamics Applied to Energy Systems, Italy
[2] ENEA, Technical Unit for Advanced Technologies for Energy And Industries, Italy
[3] University of Roma "La Sapienza", Italy
[4] ENEA Laboratory of Low Environmental Impact Automotive, Italy

E-mail: mariani_a@enea.it

Abstract. The paper deals with the cooling of batteries. The first step was the thermal characterization of a single cell of the module, which consists in the detection of the thermal field by means of thermographic tests during electric charging and discharging. The purpose was to identify possible critical hot points and to evaluate the cooling demand during the normal operation of an electric car.

After that, a study on the optimal configuration to obtain the flattening of the temperature profile and to avoid hot points was executed. An experimental plant for cooling capacity evaluation of the batteries, using air as cooling fluid, was realized in our laboratory in ENEA Casaccia. The plant is designed to allow testing at different flow rate and temperatures of the cooling air, useful for the assessment of operative thermal limits in different working conditions. Another experimental facility was built to evaluate the thermal behaviour changes with water as cooling fluid. Experimental tests were carried out on the $LiFePO_4$ batteries, under different electric working conditions using the two loops.

In the future, different type of batteries will be tested and the influence of various parameters on the heat transfer will be assessed for possible optimal operative solutions.

1. Introduction

The maturation of the electric car market is strongly influenced by the performance of the energy storage systems. The development of new types of batteries is needed to improve the performance in terms of power and storable energy. In addition, these new types of batteries are required both to increase the reliability and to maintain the same performance for a longer time. In this respect, a key factor is their operating temperature; this value has to remain within a defined range that changes with battery type. Recently this problem has assumed a strategic importance after the inconvenience to the 787e airplane in Japan due, very probably, to an overheating of the lithium battery causing a smoke release in the cabin. The accumulators consist typically of modules constructed by assembling unitary elements (cells); in this configuration, even in a single module there is the possibility of a heat exchange unfavourable condition for some of its elements. Finally, considering that the available space for laying the modules is very limited, an usual situation in an electric car, the importance of an effective system of heat removal becomes very high.

Considering the use of air for battery thermal management may be the simplest approach, and air cooling systems are used for these vehicles because of cost and space limitations. Sun et al. [1]

Journal of Physics: Conference Series **501** (2014) 012035 doi:10.1088/1742-6596/501/1/012035

developed a three-dimensional thermal model to gain better understanding of thermal behavior of battery cells in a pack cooled with air under simulated driving cycles. The baseline battery pack includes 40 battery units and cooling plates and lower and upper cooling ducts. Simulations indicate that the geometry of cooling ducts can significantly affect the battery temperature uniformity across a pack. In particular, the maximum temperature variation of battery units in a baseline battery pack can be improved by approximately 70% by using a tapered upper cooling duct. However most of the past studies showed that the heat dissipation could not be significantly alleviated by air natural or forced convection, particularly in large-scale batteries. Heat transfer by liquid may be more effective. Heat transfer with liquid could be achieved either through discrete tubing around each module; with a jacket around the module; submerging modules in a dielectric fluid for direct contact; or placing the modules on a liquid heated/cooled plate (heat sink). If the liquid is not in direct contact with modules, such as in tubes or jackets, the heat transfer medium could be water/glycol or even refrigerants, which are common automotive fluids. If modules are submerged in the heat transfer liquid, the liquid must be dielectric, such as silicon-based or mineral oils, to avoid any electrical shorts. However, because of oil's higher viscosity and associated higher pumping power, a lower flow rate is usually used, making the oil heat transfer coefficient only 1.5 to 3 times higher than with air. Indirect-contact heat transfer liquids such as water or water/glycol solutions generally have lower viscosity and higher thermal conductivity than most oils, resulting in higher heat transfer coefficients. Although liquid cooling/heating is more effective and takes up less volume, it has its drawbacks. It could have more mass, has a potential for leaks, needs more components and could cost more [2]. In an interesting review about power battery thermal energy management by Rao and Wang [3] it is concluded that at stressful and abuse conditions, especially at high discharge rates and at high operating or ambient temperatures, traditional battery thermal energy management systems, such as air and liquid, may be not meeting the requirements. Pulsating heat pipe may be more effective but needs to be well designed. In addition, progress in developing new high temperature material is very difficult. PCM for battery thermal management is a better selection than others. Nevertheless, thermal conductivity of the PCMs such as paraffin is low and some methods are adopted to enhance the heat transfer of the PCMs. Duan and Naterer [4] used an electric heater to simulate the heat source of a battery cell. Two different PCM designs are investigated: one with a PCM cylinder surrounding the heater, and the other with PCM jackets wrapping the heater. Both configuration exhibit good effectiveness in maintaining the heater within a desired temperature range. Paraffin wax can be encapsulated in the graphite matrix to produce a composite with high thermal conductivity and high latent heat storage [5]. Although many thermal properties of PCMs for battery thermal management are discussed, very little results are available about mechanical properties of these material [3]. For these reasons we decided to test an air thermal management system with electric fans to study the effects of the fluid on the battery block.

2. Preliminary thermographic analysis

The thermal characterization of the single cell was made with a infrared camera coupled with some thermocouples: we are able to see the thermal field on the surface of the cell during charging and discharging electric phases.

The purpose is to identify critical hot points to place the thermocouple. In this manner the maximum temperature will be controlled avoiding the use of the infrared camera. The infrared camera is a FLIR S60, 320x240 pixel sensor and a temperature resolution < 0.1°C. The infrared movies during the transients are taken at constant intervals (from few second to few minutes depending of the supposed velocity of the transient), synchronizing the images obtained from the camera with the data of voltage and current recorded by the cycler. Recording various discharge and charge transient (discharge current from 100 to 500 A), from the thermography of the outer surface, we find the hottest point. This point is located nearby the anode electrode as shown in figure 1. The result in term of temperature are shown in the figures 2-9. In the top graph of each figure the maximum and the mean value of temperature of the external surface are reported, while in bottom graph the value of voltage and current are reported, the abscissa is the time. The hottest point is about 70°C during the 300 A discharge (figure 5 and figure 6).

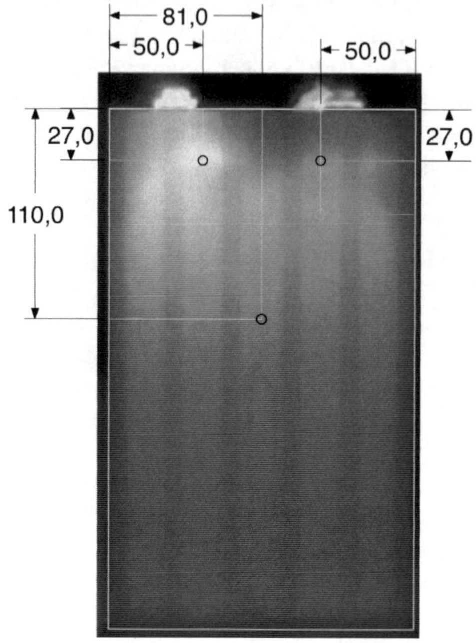

Figure 1 – Typical picture obtained with the infrared camera

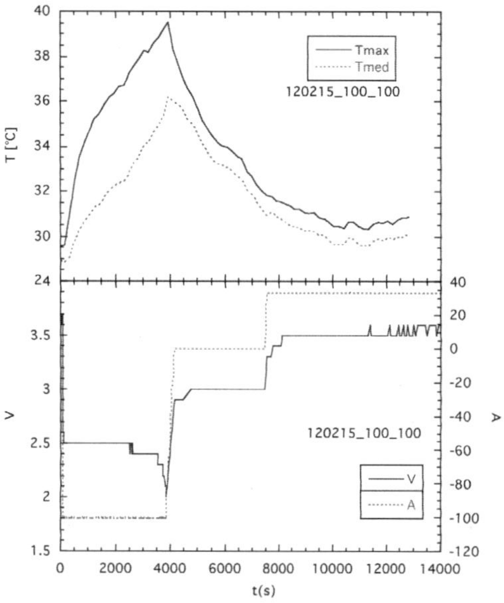

Figure 2. 100 A discharge for 3850 s, stop 3600 s and 33 A charge

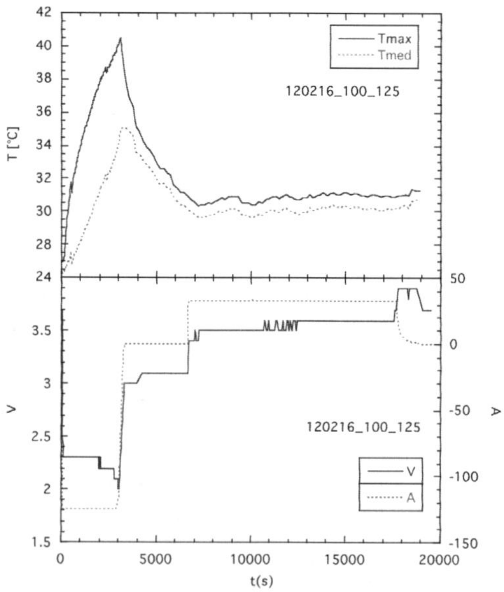

Figure 3. 125 A discharge for 3000 s, stop 3600 s and 33 A charge

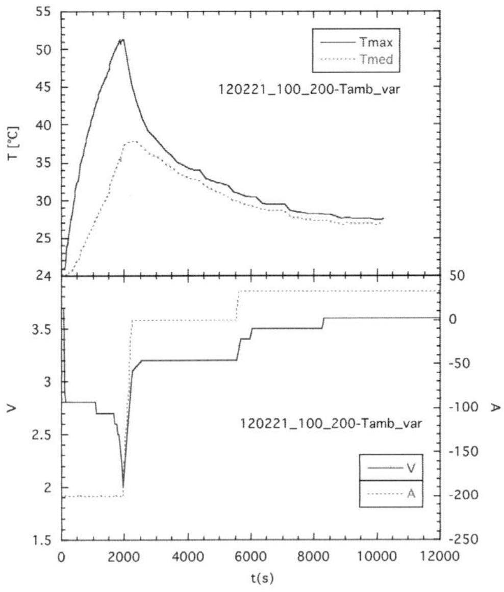

Figure 4. 200 A discharge for 2000 s, stop 3600 s and 33 A charge

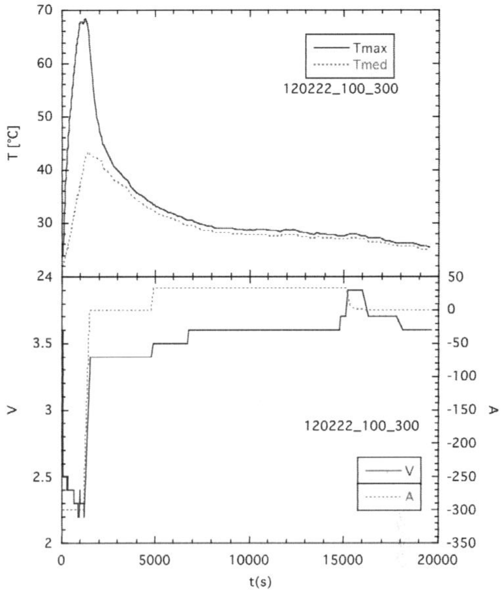

Figure 5. 300 A discharge for 1200 s, stop 3600 s and 33 A charge

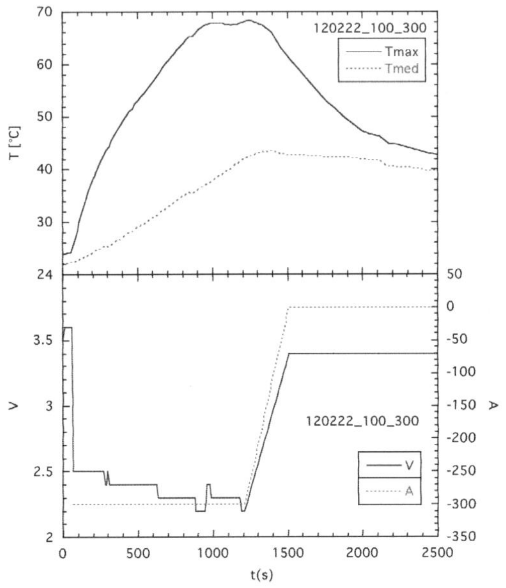

Figure 6. same run of figure 5, only discharge

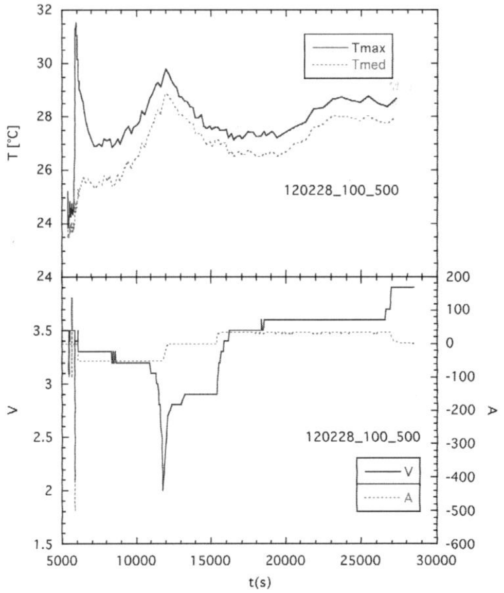

Figure 7. 500 A discharge for 30 s, stop 180 s, 50 A discharge for 5700 s 33 A charge

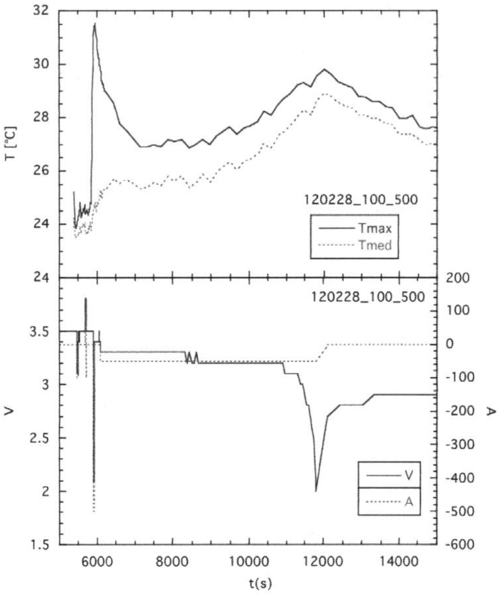

Figure 8. same run of figure 7, only 500 A discharge for 30 s, stop 180 s and 50 A discharge for 5700 s

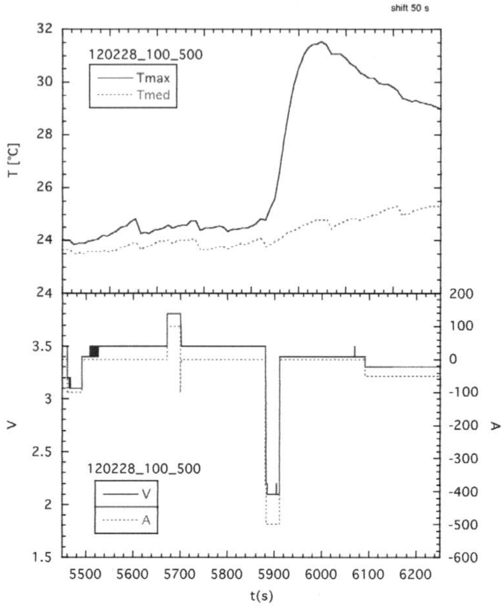

Figure 9. same run of figure 7, only 500 A discharge for 30 s and stop 180 s

3 Air cooling

3.1 BA.CO. (BAttery COoling) experimental loop

An experimental plant for cooling capacity evaluation of the batteries, using air as cooling fluid, was realized in the Laboratory of Thermal Fluid-Dynamics Applied to Energy Systems, and the Laboratory of Low Environmental Impact Automotive of ENEA Casaccia. The plant is designed to allow testing at different flow rate and temperatures of the cooling air, useful for the assessment of operative thermal limits in different working conditions. Another experimental facility was built to evaluate the thermal behaviour changes with water as cooling fluid.

3.2 Features demands on BACO:

The loop must cool 4 lithium batteries, the dimension of one battery are 163 x 278 x 51 mm.
The electric characteristics are:

Charge = 100 Ah
I_{max}=300 A
V_{nom} = 3.2 V
W_{nom} = V*I = 960 W
Internal resistance = 0.001 Ω

The maximum power to remove with the cooling fluid is:

$$W_{dis} = R_i I^2 = 90W \quad (9.4\% \text{ of } W_{nom}) \tag{1}$$
$$W_t = 4W_{dis} = 360\ W \tag{2}$$

Imposing the air heating up ΔT = 2°C, the mass flow rate is

31st UIT (Italian Union of Thermo-fluid-dynamics) Heat Transfer Conference 2013 IOP Publishing
Journal of Physics: Conference Series **501** (2014) 012035 doi:10.1088/1742-6596/501/1/012035

$$\Gamma = \frac{W_t}{c_p \Delta T} = 0.1793 \frac{kg}{s} \tag{3}$$

$$Q = \frac{\Gamma}{\rho} = 9052 \frac{l}{min} = 543 \frac{m^3}{h} \tag{4}$$

Using a gap of 1 cm, the passage area is $A_p = 6480\ mm^2$. Using equation 1 the air velocity is:

$$u = \frac{Q}{A_p} = 23.28 \frac{m}{s} \tag{5}$$

The pressure drop is (f calculated from the Pethukov correlation):

$$\Delta p = f \frac{L}{D} \frac{\rho u^2}{2} = 111 Pa \tag{6}$$

$$f = (1.82 \log_{10} Re - 1.64)^{-2} \tag{7}$$

The power of the fan is calculated using an efficiency $\eta = 25\%$:

$$P = Q \Delta p \eta = 67W \tag{8}$$

The maximum power in the pre-heater is calculated from equation 1 using $T_{max} = 40°C$ and $T_i = 20°C$:

$$W_s = \Gamma c_p \Delta T = 3600W \tag{9}$$

The heat transfer coefficient is calculated for a $\Delta T_{wall\text{-}fluid} = 20$ K:

$$q'' = \frac{W_e}{2 S_L} = 1021 \frac{W}{m^2} \tag{10}$$

$$h = \frac{q''}{\Delta T} = 51 \frac{W}{m^2 K} \tag{11}$$

The heat transfer coefficient calculated for the selected geometry vary from 36 W/m^2K (correlation of Bejan-Krauss, 12) to 70 W/m^2K (correlation. of Gnielinski, 13) and to 101 (correlation of Dittus-Boelter, 14)
The used correlation are reported hereunder:

$$h = 3.886 \left(\frac{u}{L}\right)^{0.5} \tag{12}$$

$$\mathrm{Nu} = \frac{\frac{f}{8} \mathrm{Pr}(\mathrm{Re} - 1000)}{1 + 12.7 \sqrt{f/8} (\mathrm{Pr}^{2/3} - 1)} \tag{13}$$

Using f from equation (7)

$$Nu = 0.023 Re^{0.8} Pr^{0.4} \tag{14}$$

In figure 10 is shown the diagram of the experimental loop. The characteristics of the loop are:

- Four axial fans, powered with a 24Vcc ebm-papst 4114NH3;

- An electric pre-heater to assure the desired inlet air temperature;
- An anemometer SCHMIDT ss20.500 to measure the air mass flow rate;
- A channel which length is sufficient to assure a velocity profile completely developed to guarantee a right operation of the anemometer;
- The test section with a thermocouple (to measure the inlet air temperature) and the module (four batteries with a gap of 1 cm and instrumented as shown in Figure 10);
- An air mixing and two thermocouple to measure the outlet mean air temperature;
- The cycler that charge and discharge electrically the module recording the supplied voltage and current;
- The acquisition system, which provides for the registration of the data and the adjustment of the inlet temperature (by controlling the voltage of the heater with a routine PID).

We used four fans mounted in pairs. The channel is a square with insulated panels made in polyurethane coated with aluminium foils. The preheater power is 6,6 Kw. The total length is about 5 m and the internal dimension are 10,6x10,6 cm with a calming length of 2,40 m.
The test section is a channel of 30,6x24,4 cm which allows the module. The air passed through the battery in longitudinal sense.
From the result of the previous thermographic campaign we insert the thermocouple in the batteries using 14 thermocouples positioned as shown in figure 11, all the batteries have a thermocouple in the hot point (27 mm under the negative pole) the two central batteries have the thermocouple also in the centre and under the positive pole.
A picture of the test section is in figure 12.
During the experimental runs the charge and discharge of the module is done changing the current and measuring the superficial temperature and the heat removed from the air.

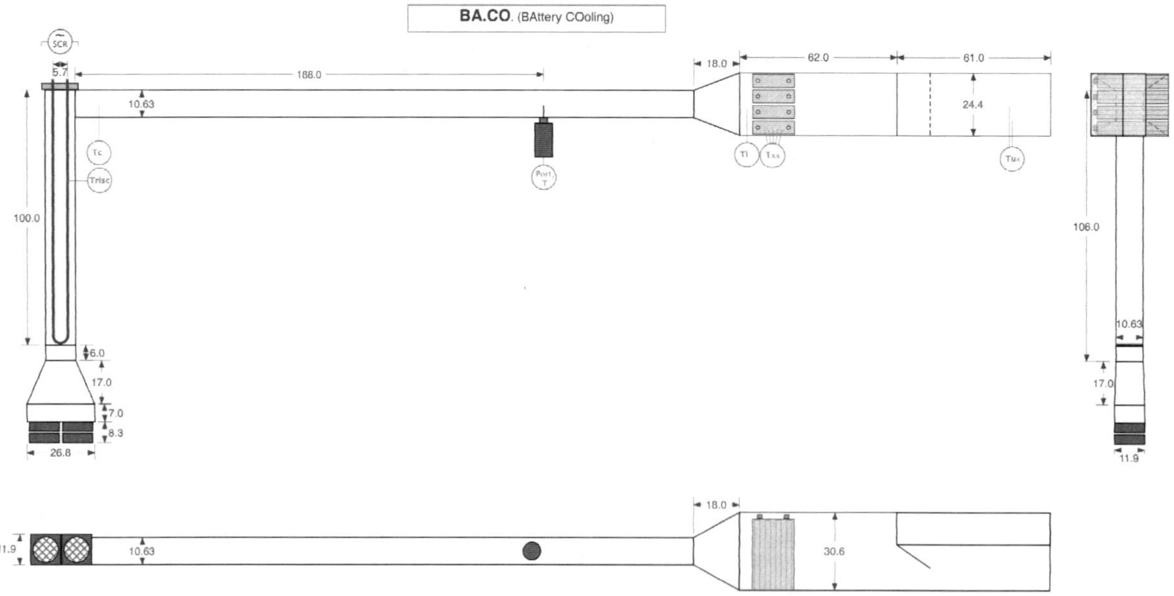

Figure 10. Diagram of the experimental loop and test section in air cooling

31st UIT (Italian Union of Thermo-fluid-dynamics) Heat Transfer Conference 2013 IOP Publishing
Journal of Physics: Conference Series **501** (2014) 012035 doi:10.1088/1742-6596/501/1/012035

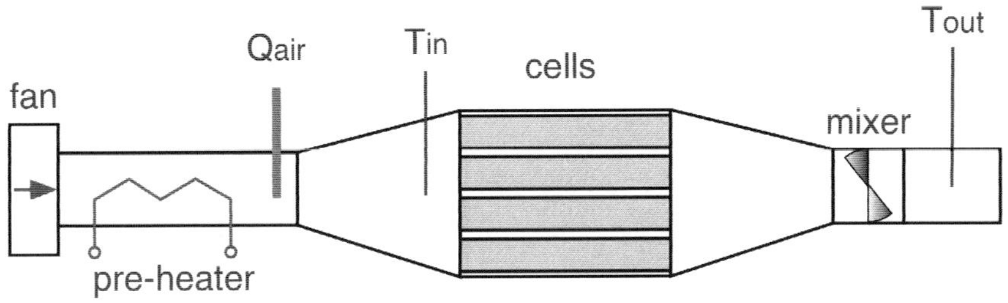

Figure 10.continue Diagram of the experimental loop and test section in air cooling

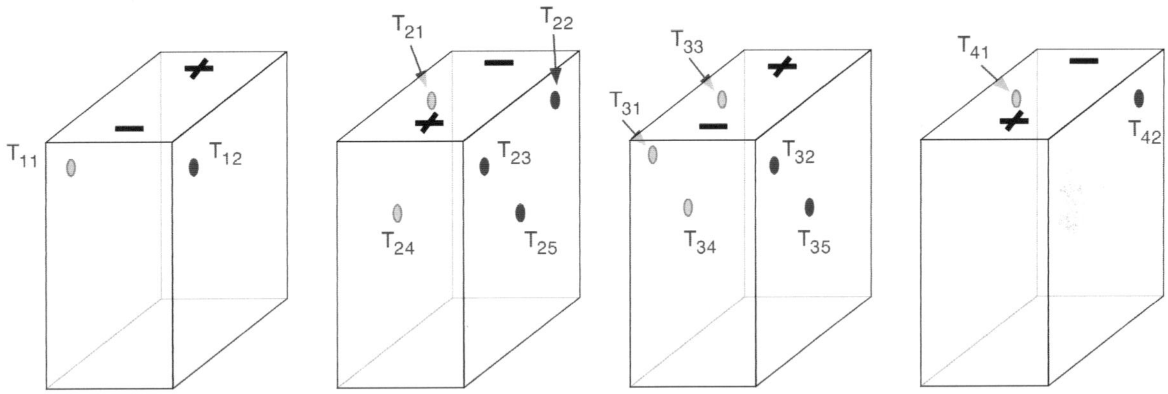

Figure 11. Arrangement of thermocouples on batteries

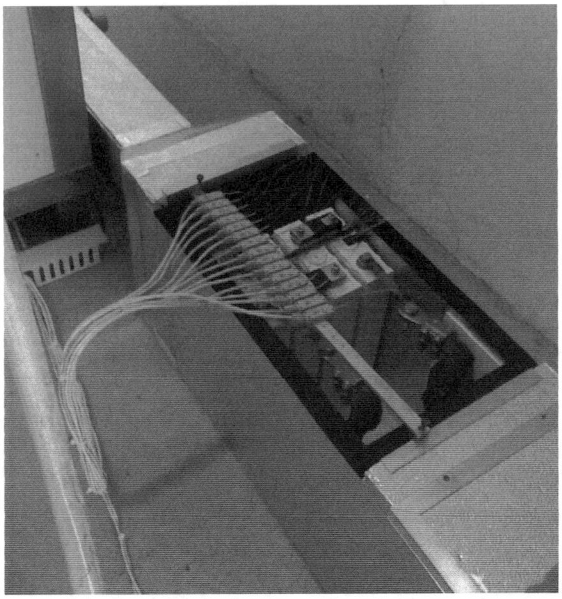

Figure 12. Photo of the module in air cooling loop

3.3 Results of air cooling

To test the batteries we used them in the worst case in charge and discharge. According to data provided by the manufacturer and confirmed by the previous tests, the worst case is when the charge has the value of 1C (a value of current equal to the nominal value of the batteries, 100A for our batteries) and the value of 3C in discharge (300 A), being this parameter the most critical.

So the test has this course: Starting with a completely discharged battery the cycler has a charge at 1C (constant current of 100 A until a tension of 15 V then decreasing tension until 1A, end of charge), after a pause of 60 min the cycler impose a discharge at 3C (constant current of 300 a, until a voltage of 10V, end of discharge)

In this first step we perform limited runs just to evaluate the cooling rate of the various system. For the air cooling we run only with a mass flow rate of 100 m^3/h (typical value of low efficiency and reduced dimension fan used in an electric car).

The results are shown in figure 13, for the charge, and figure 14 for the discharge. From the figures we can see that the temperature is well below the limit imposed. (26.5°C during the charge, figure 13, 36°C during the discharge, figure 14). We can say that air could perform properly the function of refrigerant.

4 Water cooling

4.1 Water cooling loop

The water cooling is more complicated than the air cooling due to hazard using an electric device. The good results of the air cooling to maintain the temperature in the operating range suggest us to reduce the number of tests using water as cooling fluid. Therefore the experimental loop was realized in a simplified form. We immerged the module in a pool of water with an adjustable input of water and an output situated at a certain height. This was possible due to the external hermetic housing of the batteries, figure 15. A picture is in figure 16.

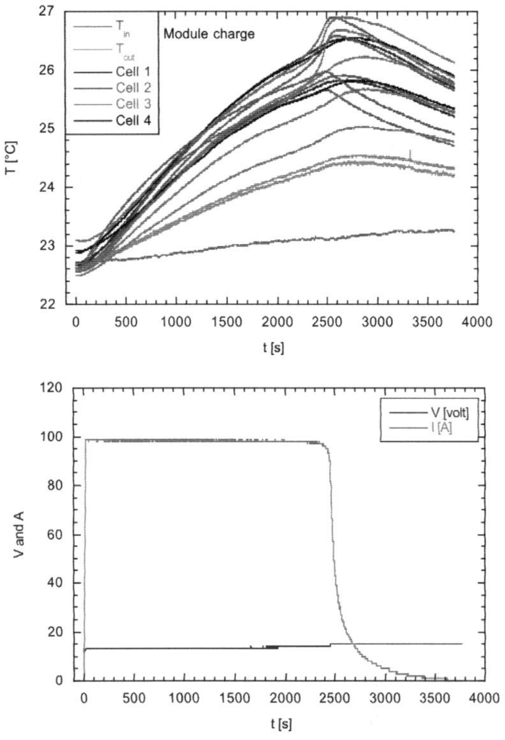

Figure 13. Charge of the module in the air

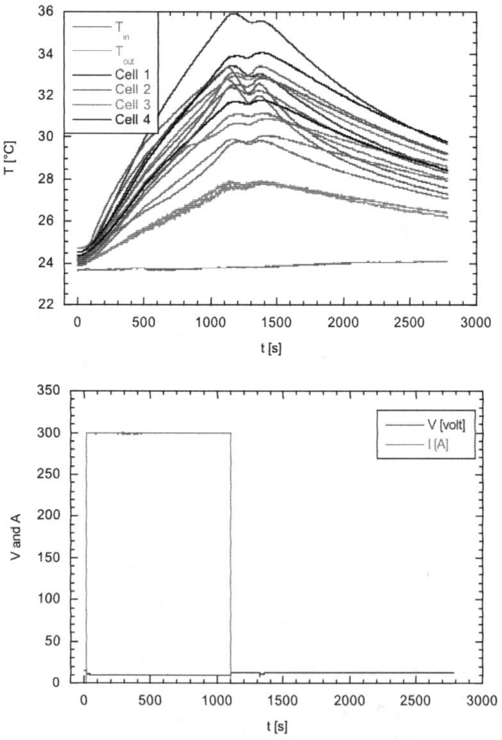

Figure 14. Discharge of the module in the air

The module is the same to the previous one used in the Ba.Co loop, the batteries have a gap of 1 cm. The water input is situated on the bottom of the module to avoid fluctuations in the level and damage batteries. The module is immersed in water up to 1 cm to the electric connectors. The water mass flow rate is regulated by a valve and connected to the domestic line. The water mass flow rate is calculated weighing the waste water in an electronic scale at fixed time and it is constant
The module instrumentation is the same of the air cooling tests.

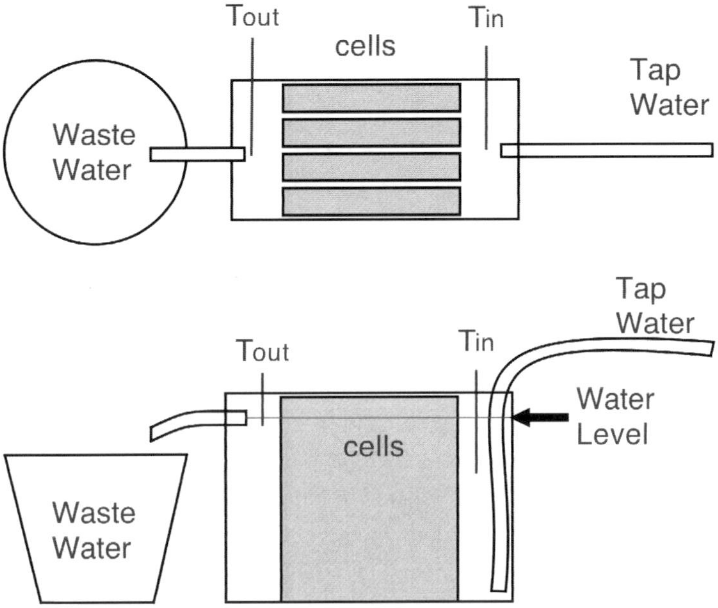

Figure 15. Schematic of the experimental loop in water cooling

Figure 16. Photo of water cooling loop

4.2 Results of water cooling
For the water cooling the tests are exactly the same of the air cooling. The electric cycle is the same in order of magnitude and timing.

The water mass flow rate is 100 l/h and it is obtained with a little circulation pump which can be easily mounted in an electrical car. The results are shown in figure 17, for the charge, and figure 18 for the discharge. From the figures we can see that the maximum temperature reached during the charge is 25°C, figure 17, and during the discharge it is 27.5°C, figure 18.

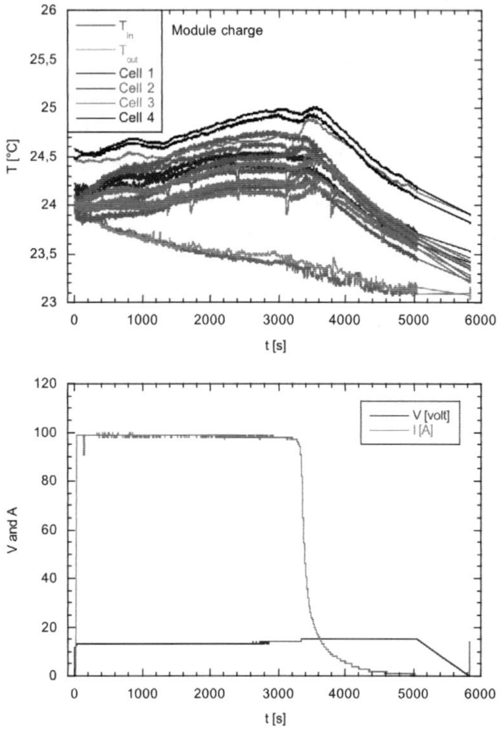

Figure 17. Charge of the module in the water

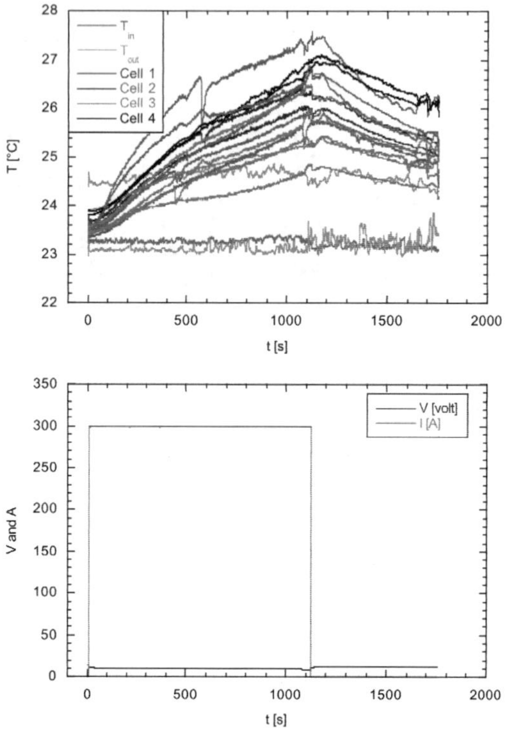

Figure 18. Discharge of the module in the water

5. Comparison air-water cooling

The results of this comparison are shown in figure 19 for the charge and figure 20 for the discharge.
In both figures, the blue lines indicated the temperature recorded by all thermocouples mounted on the battery 1 as shown in figure 11, similarly all the curves relating to the thermocouples respectively mounted on the battery 2, 3 and 4 have been indicated with the green, red and black respectively, (for instance the batteries 1 and 4 are external ones, the batteries 2 and 3 the internal one), always with reference to figure 11.
The analysis of the temperature profiles show that the maximum reached temperature is 25°C in the case of water cooling, 26.5°C in the case of air cooling during the charge, figure 19. These temperatures are well below the limit imposed and we argued that the module can resist to a different charge (es. 1,5÷2C), This fact seems to indicate a good response of this type of batteries to a rapid charge use.
In discharge 3C temperatures are higher: 27.5°C in the case of water cooling, 36°C in the case of air cooling, figure 20. Both cooling systems have proved effective, in fact, the temperatures were well below the limits stated by the manufacturer (+65°C) and above all well below the values recorded in analogous tests carried out at the stage of thermal characterization without cooling. The comparison of efficiency argues in favor of the cooling water.

31st UIT (Italian Union of Thermo-fluid-dynamics) Heat Transfer Conference 2013 IOP Publishing
Journal of Physics: Conference Series **501** (2014) 012035 doi:10.1088/1742-6596/501/1/012035

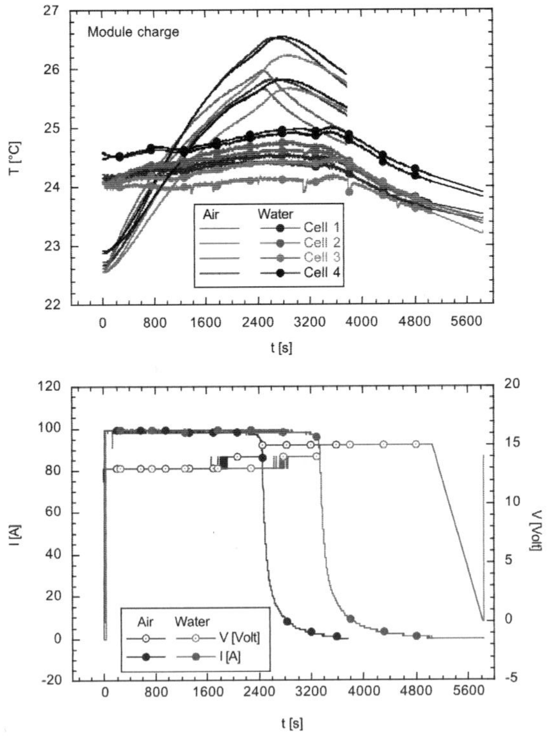

Figure 19. Comparison of air-water cooling in charge of the module

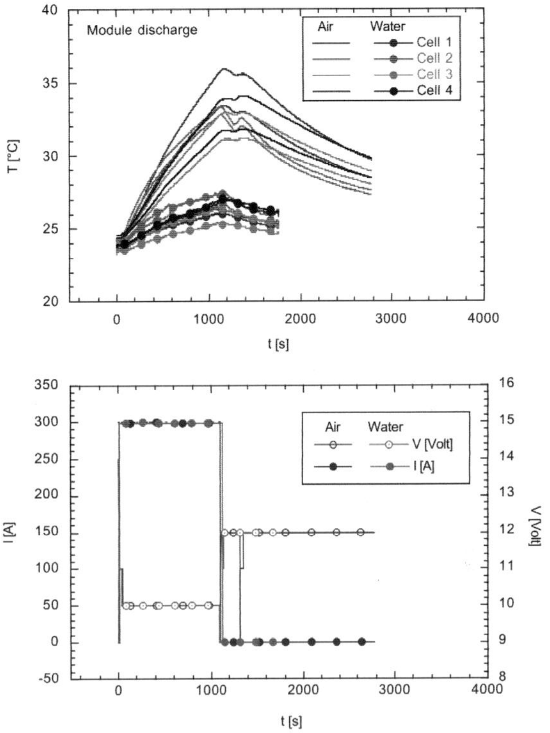

Figure 20. Comparison of air-water cooling in discharge of the module in the water

6. Conclusion

An experimental study was conducted to evaluate the possible cooling of the batteries employed as power supply in the electric car. The first step of this work was to ascertain the presence of hot spots on the single cell of the module used (four batteries linked together). This part was conducted using an infrared camera together with two thermocouples. The hot spot exits and is situated 27 mm under the negative pole. After that two different loops were designed and constructed to cool the module using air and water as cooling fluids. The two system are able to dissipate the heat produced. Obviously the water maintains a lower temperature, but the use of air is preferred because of its greater simplicity and the absence of risks during exercise.

Nomenclature

A Area [m^2]
c_p specific heat [J/kg K]
D diameter [m]
f_l Fanning factor
h heat transfer coefficient [W/m^2K]
I current [A]
k thermal conductivity [W/m K]
L length [m]
Nu Nusselt number $Nu = \frac{hD}{k}$
Pr Prandtl number $Pr = \frac{c_p \mu}{k}$
q" heat flux [W/m^2]
R resistance [Ω]
Re Reynolds number $Re = \frac{\rho u D}{\mu}$
S surface [m^2]
T temperature [°C]
u velocity [m/s]
V voltage [V]
W power [W]

31st UIT (Italian Union of Thermo-fluid-dynamics) Heat Transfer Conference 2013 IOP Publishing
Journal of Physics: Conference Series **501** (2014) 012035 doi:10.1088/1742-6596/501/1/012035

Greek Symbols

Γ mass flow rate [kg/s]

μ viscosity [N s/m^2]

ρ density [kg/m^3]

References

[1] Sun H, Wang X, Tossan B and Dixon R 2012 Three-dimensional thermal modeling of a lithium-ion battery pack, *J. of Power Sources* **206** 349– 356

[2] Pesaran A A 2001 Battery Thermal Management in EVs and HEVs: Issues and Solutions, *Proc. Advanced Automotive Battery Conference Las Vegas, Nevada* February 6-8

[3] Rao Z and Wang S 2011 A review of power battery thermal energy management, *Renewable and Sustainable Energy Reviews* **15** 4554– 4571

[4] Duan X and Naterer G F 2010 Heat transfer in phase change materials for thermal management of electric vehicle battery module, *Int. J. of Heat and Mass Transfer* **53** 5176–5182

[5] Kizilel R, Sabbaha R, Selmana J R and Al-Hallaj S 2009 An alternative cooling system to enhance the safety of Li-ion battery packs, *J. of Power Sources* **194** 1105–1112

Journal of Physics: Conference Series **501** (2014) 012036 doi:10.1088/1742-6596/501/1/012036

CO and NO emissions from pellet stoves: an experimental study

D Petrocelli and A M Lezzi

Dipartimento di Ingegneria Meccanica ed Industriale,
Università degli Studi di Brescia,
via Branze 38, 25123 Brescia

E-mail: domenico.petrocelli@ing.unibs.it

Abstract. This work presents a report on an experimental investigation on pellet stoves aimed to fully understand which parameters influence CO and NO emissions and how it is possible to find and choose the optimal point of working. Tests are performed on three pellet stoves varying heating power, combustion chamber size and burner pot geometry.

After a brief review on the factors which influence the production of these pollutants, we present and discuss the results of experimental tests aimed to ascertain how the geometry of the combustion chamber and the distribution of primary and secondary air, can modify the quantity of CO and NO in the flue gas.

Experimental tests show that production of CO is strongly affected by the excess air and by its distribution: in particular, it is critical an effective control of air distribution. In these devices a low-level of CO emissions does require a proper setup to operate in the optimal range of excess air that minimizes CO production. In order to simplify the optimization process, we propose the use of instantaneous data of CO and O_2 concentration, instead of average values, because they allow a quick identification of the optimal point.

It is shown that the optimal range of operation can be enlarged as a consequence of proper burner pot design.

Finally, it is shown that NO emissions are not a critical issue, since they are well below threshold enforced by law, are not influenced by the distribution of air in the combustion chamber, and their behavior as a function of air excess is the same for all the geometries investigated here.

1. Introduction

In Europe the use of woody biomass – of wood pellets, in particular – for residential heating has increased in recent years. In Italy, pellet heating technology was introduced in the '90s. In 2007 wood pellets accounted for 2.4% of woody biomass consumed for domestic use (Lombardy, 3.3%) [1]. Lombardy is the Italian region in which pellet appliances are more widespread: in 2007 they were the 6.9% of wood appliances (Italy, 3.9%) [1].

End-users and legislation have paid increasing attention both to energy efficiency and to pollutant emissions of wood-burning appliances. Depending on the country, some of the following pollutant substances are taken into account by legislation: carbon monoxide (CO), nitrogen oxides (NOx), organic gaseous carbon (OGC) and particulate matter (PM). In order to promote the use of efficient and clean woody biomass heating systems, several European governments have introduced subsidies or tax incentives which reduce, for end-users, the cost of purchase and installing of these devices. Focusing on Italy, subsidies have been introduced recently through the so-called "Conto termico" decree [2]. For pellet appliances it requires the following characteristics:

- pellet stoves: thermal efficiency greater than 85%; CO emission less than 250 mg/Nm^3 (200 ppm) at 13% of O_2; total particulate matter emission less than 40 mg/Nm^3 at 13% of O_2.
- pellet boilers: must comply with class 5 of European standard UNI EN 303-5:2012, that is thermal efficiency must be greater than 87% + $\log(P_n)$, where P_n is the nominal power of the boiler; CO emission less than is 250 mg/Nm^3 at 13% of O_2; total particulate matter emission less than 30 mg/Nm^3 at 13% of O_2.

The total particulate matter (TPM) is the quantity measured by a full flow dilution tunnel or, alternatively, by the measurement of the solid particles (PM), with gravimetric method, in combination with an OGC measurement, according the standard prEN 16510-1:2013. In "Conto termico" the subsidy is determined taken into account the value of TPM also: the smaller TPM, the larger the subsidy. For all these reasons wood-appliance manufacturers are increasing their efforts to design stoves and boilers with higher efficiency and lower pollutant emissions, investing in research and development. A review of current pellet stove technology can be found in [3–5].

Here we present the results of an experimental study on CO – mostly – and NOx emissions, which is part of a research program in collaboration with a pellet stove producer. The goal is to find methods to reduce emissions of stove models already on the market and to identify design criteria for new models to keep emissions low.

In [6] we described a preliminary investigation on a pellet stove focused on thermal efficiency. We also looked at CO emission. As expected they resulted strongly dependent on combustion air mass flow rate and on its distribution.

In this work we have extended the study, testing burning pots with different combustion air inlets and pellet stoves with different combustion chamber height. This with the aim to investigate how these geometrical features influence the relation between emissions and excess air. The influence of fuel (pellet) mass flow rate – that is to say, of heating power – has been considered also.

In pellet stoves combustion air control and fuel feeding systems cause significant fluctuations of air excess and pollutant emissions. Thus legislation, standards, and literature refer to mean values obtained over long time intervals of steady operations: so, for example, several hours of testing may be necessary to verify the effectiveness of a modification of the burner pot. In Sec. 4 we propose to use plots of instantaneous CO vs. O_2, in addition to those of average values. We show that such instantaneous plots – obtained in less than one hour of testing – can be successfully used to determine the effectiveness of modifications of the stove or to improve settings of the control systems, both in the producer's testing laboratory and during the periodic maintenance on an installed stove.

Emissions of PM are currently under investigation.

2. Characteristic of pellet stove emissions

The principal elements which constitute wood are C, H, O and N, furthermore, water is always present. The flue gas, produced by ideal combustion of wood with air, contains N_2, O_2, CO_2 and H_2O. Because of non-optimal combustion conditions, in real cases other substances are present: among others CO, NO, NO_2, H_2, hydrocarbons (HC), poly-cyclic aromatic hydrocarbon (PAHs) and particulate matter.

Carbon monoxide is a pollutant substance due to incomplete combustion. In many countries, it is the only pollutant limited by law because it is strongly related to the quality of combustion and then to the concentration of other products of incomplete combustion.

The main cause of incomplete combustion in grate-fired systems is poor mixing of air with combustible substances both in the fuel bed and in the combustion chamber. Advanced air supply systems and optimized grate systems can significantly enhance the mixing, reduce the excess air, improve the combustion process, and lower pollutant levels [3, 7–12]. Another key parameter for a complete combustion is the temperature of the combustion chamber. It must be high enough to allow for oxidation reactions to take place. Cold air streams or drafts, for instance as secondary air, or the contact between flame and cold walls can stop the reaction and, consequently, increase the concentration of unburnt in flue gas. Combustion quality is also influenced by the residence time of gases in high temperature zone: combustion at low temperatures requires a longer residence time [13, 14].

The value of NOx is the sum of NO and NO_2, but NO is the only one present in significant amount in pellet stove flue gas. Models of NO formation in combustion processes are available in literature [15–19]: NO can derive both from the nitrogen of combustion air and from the nitrogen present in fuel. It is well known that the principal mechanism of formation of NO in biomass combustion systems is the fuel-NO mechanism [7, 14–16, 20, 21], i.e. the NO formed from oxidation of nitrogen of the fuel.

The yield of NO from fuel-NO mechanism increases with excess air [17], conversely in the absence of oxygen the fuel-nitrogen forms N_2 rather than NO.

Other important substances present in pellet stove flue gas and considered by legislation are PM and HC. PM are largely investigated in literature. In works as [7, 9, 14, 22–33] the problem of emission of PM is discussed in depth, with particular attention to its causes and to particle size distribution. In our work PM emissions are currently under investigation, so we do not report on them here. However, according to the results reported in [23–25], we assume that optimization of the combustion process based on observation of CO-O_2 behavior would lead to a reduction of formation of PM also.

The same conclusion holds for HC emissions, in general, and OGC emissions, in particular (see prEN 16510-1:2013 for correlations between HC and OGC). As a matter of fact, explicit correlations between the concentration of CO and HC in flue gas can be found in [23, 24, 32, 34], for different type of biomass combustion systems: all of them are monotonically increasing functions of CO, so lower CO emissions correspond to lower OGC emissions. Furthermore [23, 24] report also a direct correlation between OGC and PM. Similar conclusions can be drawn from data collected in [12, 14, 27, 30, 31], although in these works explicit correlations are not presented. Other results regarding OGC can be found in [11, 14, 28, 29].

Other pollutants may be present in pellet stove flue gas, such as PAHs, SOx HCl, PCDD/PCDF (see [7, 12, 14, 23, 28–31, 35] for some experimental studies), in [35], in particular, a correlation between PAHs and CO concentration is presented. In our experimental work we are not considering them yet.

3. Experimental tests

For all tests reported here we used the experimental setup and procedure proposed by the standards [36] and used by many other authors [21, 27, 34, 37, 38]. The whole stove, including the hopper, is put on a platform scale in order to measure the consumption of pellets. The composition of flue gas is measured at the flue gas socket. The gas analyzer used is Testo 350 XL, equipped with cells to measure the concentration (volume fraction) of O_2, CO, NO, and NO_2 in dry gas, i.e. the flue gas without H_2O. Measurement accuracy is the following.

- Oxygen: $\pm 0,2\%$.
- CO: ± 10 ppm when concentration is below 100 ppm; $\pm 5\%$ of reading, otherwise.
- NO: ± 5 ppm when concentration is below 100 ppm; $\pm 5\%$ of reading, otherwise.

The CO_2 and N_2 concentration can be deduced from the stoichiometry of combustion, therefore it is possible to describe the process accurately. In this study we have tested 3 different pellet stoves: S1, S2, and S3. Each of them has a nominal heating power of 12 kW. Stove S1 is the base model, while stoves S2 and S3 have a combustion chamber shorter (S2) or taller (S3) than S1. Stove S1 have been used with different burner pots, to test the effect of changing combustion air distribution. The burner pot is a small holed basket (figure 1). Pellets are fed into the furnace by means of the so-called overfed burner system [4]. These stoves are equipped with an air flow

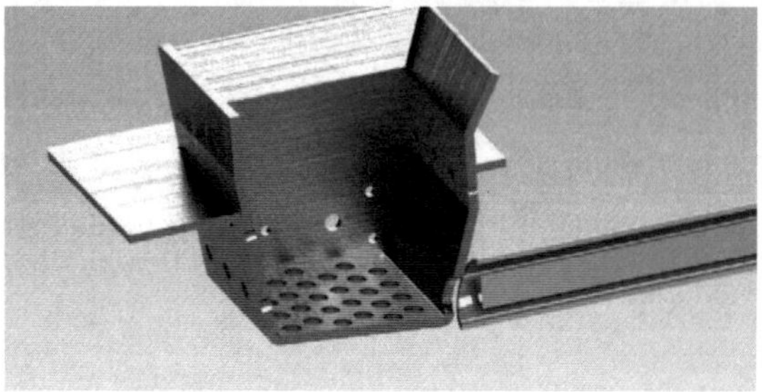

Figure 1. Typical geometry of burner pot.

meter which measures the mass flow of combustion air. Primary air enters through the bottom grate and it passes through the chars, while secondary air is supplied directly in the flame zone by means of holes on vertical walls of the burner pot. During the tests, performed with stove S1, five burner pots were used. They differ by the total inlet area, by the primary air inlet area, and by the ratio between the inlet area for primary air and secondary air (see table 1).

At the beginning of a test session the stove is lit, setting the mass flow rate of air and of fuel for the first test, but measurements are started only when the stove reaches steady state conditions, i.e. when the flue gas temperature at the exit of the stove does not change more than a few kelvin. For subsequent tests mass flow rate of air or of fuel are changed without turning off the stove, but only waiting for steady state conditions with the new settings. During the tests, each lasting at least 30 minutes, concentrations are recorded, acquiring a sample every 5 seconds, i.e. at least about 360 samples for each test. CO and NO emission data are plotted versus the concentration of O_2. In this work we present two types of data: time averaged values and instantaneous values. In figures 2 and 3, for examples, average concentration of CO or NO is plotted versus average concentration of O_2 and each single point represents one test. In figures 4, 5, 6, 7, 8, and 9, instead, a point represents the instantaneous concentration of CO or NO plotted as a function of excess air or O_2 concentration: one test is represented by a cloud of points.

In order to make a proper comparison of results, CO and NO are normalized at the same concentration of O_2, using the relation $X_{\mathrm{ref}} = X\,(21 - O_{2,\mathrm{ref}})\,/\,(21 - O_2)$ (see UNI EN 14785:2006). where X is the actual (averaged or instantaneous) concentration of CO or NO in ppm, O_2 the actual (averaged or instantaneous) concentration of O_2 in %, $O_{2,\mathrm{ref}}$ the reference concentration of oxygen, here 13% and X_{ref} the concentration of CO or NO at reference conditions, in ppm, i.e. the one shown in the diagrams.

4. Results

For a better understanding of our experimental results it is important to recall some characteristics of testing procedure due to the way the stoves, here tested, work. Combustion air mass flow rate, $\dot{m}_a$, is kept constant by a feedback control system which measures the velocity of air at the inlet and varies the rotation speed of the fan which evacuates the flue gas. With regard to the fuel (pellet) mass flow rate, $\dot{m}_f$, the typical feeding systems of pellet stoves are not continuous in time. A few pellets are released and fall into the burner pot with a time period of a few seconds: it is possible to change the fuel mass flow rate by changing this time period. Actually, the mass of pellet varies at each release, depending on the number and total length of the grains. Only fuel mass flow rate averaged over time intervals of some minutes can be considered roughly constant (nominal fuel mass flow rate).

Because of time variations of the instantaneous fuel mass flow rate, the amount of char and unburnt pellets in the burner pot vary as well, and instantaneous fuel mass flow rate doesn't

Journal of Physics: Conference Series **501** (2014) 012036 doi:10.1088/1742-6596/501/1/012036

Table 1. Air flow inlet areas for burner pots used in this study.

Burner pot	**Total area** mm^2	**Primary area** mm^2	**Sec./Tot. ratio**
A	1848	1710	0.074
B	2043	1710	0.163
C	1535	1358	0.115
D	1715	1358	0.208
E	1358	1358	0

Table 2. Mean value with indication of 5th (min) and 95th (max) percentile for burner pot geometry A and D (figure 2).

	O_2 (%)			**CO** (ppm)		
	mean	min	max	mean	min	max
A	11.7	10.1	13.5	159	83	313
A	12.2	10.0	14.0	183	87	383
A	13.9	12.1	15.5	357	106	753
A	14.4	12.6	16.5	523	123	1582
D	14.5	12.5	16.1	203	42	465
D	13.6	12.1	15.1	114	36	269
D	13.0	11.0	14.7	86	31	182
D	12.2	10.1	14.5	77	35	179

coincides with instantaneous burning rate, $\dot{m}_p$: only values averaged over time intervals longer than a few minutes are equal. Variation of burning rate at constant combustion air mass flow rate causes significant fluctuations of the air excess. If we express it in terms of O_2 concentration in flue gas, $\pm 2\%$ oscillations are commonly observed, even when the stove is nominally at steady state. The instantaneous burning rate of pellet, $\dot{m}_p$, can be estimated by means of equation 1 (see [39]),

$$\dot{m}_p = \dot{m}_a \left(\frac{79}{721} \frac{7 + 4y_{CO_2} + y_{O_2}}{1 - y_{CO} - y_{CO_2} - y_{O_2}} - 1 \right) \tag{1}$$

where y_X is the instantaneous volume fraction of O_2, CO_2 and CO in flue gas and $\dot{m}_a$ is the combustion air mass flow rate.

4.1. Influence of excess air and air distribution

In this section we report on how the amount and the distribution of air affect the combustion, therefore the production of CO and NO. Conclusions regarding the role of distribution will be qualitative. In order to put on a quantitative base the results shown here, we are planning a new series of tests on a modified stove. Primary and secondary air mass flow rates will be independent one from the other and will be controlled and measured separately.

The combustion air mass flow rate influences not only the excess air, but also the velocity of the air in the burning pot. A different distribution of air between primary and secondary air inlet, can change the local excess air and the fluid dynamics in the combustion chamber. Figure 2 shows the average concentration of carbon monoxide versus the average excess air for tests performed with different burner pots, while figure 3 shows results for NO concentration. Five different types of burner pot were used, table 1 reports their characteristics.

The different values of air inlet areas modify its distribution in the combustion chamber: burner pot B and burner pot D, for examples, allow a larger amount of air to pass above the char with respect to burner pots A and C, while in burner pot E air can pass through the bottom holes only. Results obtained with burner pot B and D are better than those obtained with burner pot A, C, and E, characterized by poorer performances. For burner pots B and D the ratio between secondary air inlet area and total air inlet area is greater then for the other burner pots. Therefore, a large amount of air entering from the top, i.e. in the flame zone, causes lower emission of CO. A significant secondary air flow allows the char and the first zone of flame – the one below the secondary air inlet – to reach higher temperatures, because of a lower local excess air, thus improving conditions to obtain complete combustion in this zone.

In addition, a strong secondary air flow produces a better mixing of gases in the upper zone of flame.

For all burner pots the quality of combustion strongly depends on the excess air: in particular, CO concentration is an increasing function of O_2 for the O_2 interval that could be investigated. These results agree with those in [23] and are compatible with results of [38, 40]. In the latter studies the relation between CO and O_2 admits a minimum (optimum): the absence of the decreasing branch of the curve in our results is due to the high concentration of O_2 obtained with stove S1 and we expect that for lower concentration of oxygen, CO concentration would increase, as instantaneous results for stoves S2 and S3 do confirm.

In any case, the important point is that for burner pots B and D low emissions were obtained over a rather large interval of O_2 concentration, whereas with burner pots A, C and E, the O_2 interval where CO is low seems to be narrower.

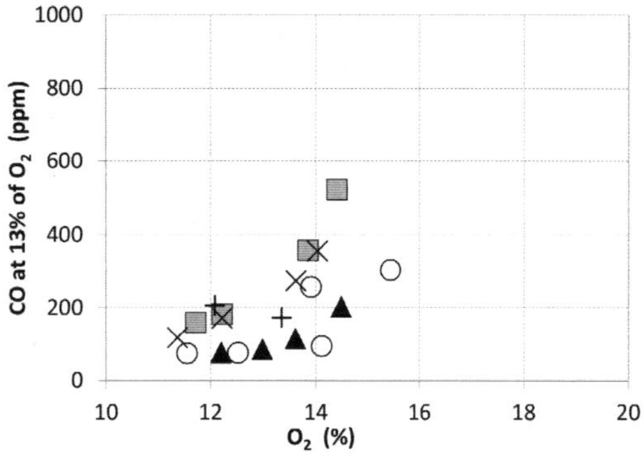

Figure 2. Average value of CO as a function of O_2 for different type of burner pots: □ A, ○ B, × C, ▲ D, + E.

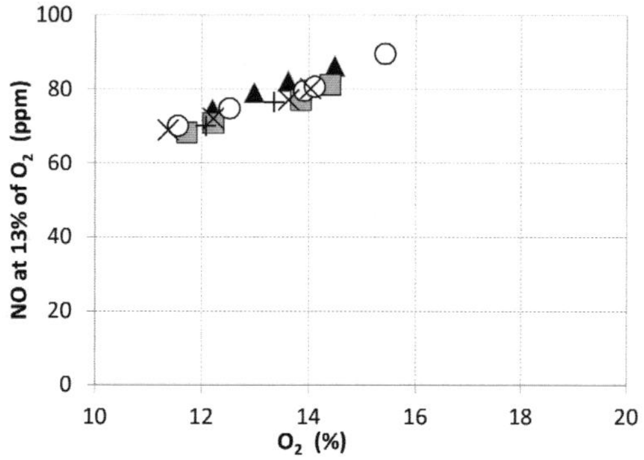

Figure 3. Average value of NO as a function of O_2 for different type of burner pots: □ A, ○ B, × C, ▲ D, + E.

The production of NO seems to be favored by an increase of air in the combustion chamber, but at the same time it is independent from the geometry of the burner pot. As a matter of fact NO emissions are the same for each type of burner pot: all points fall around the same curve. The trend of this curve is comparable with results of [15, 38, 40], in which NO concentration increase in flue gas is explained by the high conversion efficiency of fuel-NO mechanisms. With regard to the other production mechanisms of NO, the contribution of thermal-NO mechanism can be considered negligible since maximum temperatures, usually below 1000°C as shown in [6], are not high enough to produce significant amounts of NO with this mechanism.

The average concentration of CO and NO quantifies the total amount of substances emitted in the atmosphere during the test and for this reason it is used by standards to declare the emission of the stove, and in [22–25, 27, 37, 38] to report their results. As a matter of fact, during a test the instantaneous fluctuations are very high: the continuous variation of burning rate modifies the instantaneous excess air thus influencing the behavior of combustion and in particular the production of CO. This results in turn in significant oscillations of concentrations during the tests, as it can be seen in table 2. High oscillations of concentration values can be observed in the table 6 of [31] and in many test reports of [41], for instance.

As already mentioned, O_2 concentration values collected during a single test at steady conditions do span a quite large interval: about 4 or 5 percentage points. Upon plotting these instantaneous data on the O_2-CO plane, one can obtain a reasonable approximation of the CO vs. O_2 correlation for the stove configuration tested. As an example, consider data reported in figures 2 (average) and 4 (instantaneous). In order to compare performance of burning pots C and D using average data, 4 tests were necessary for each burner pot to span the range from

11% to 15% of O_2. That took more than two hours of testing. Using instantaneous data a single test (taking slightly more than half an hour of testing) was sufficient to show the better performance of burning pot D. In addition, it is evident from figure 4 that burner pot D is characterized by low and approximately constant CO emissions over a larger interval (12%,15%) of O_2 concentration.

Figures 4 and 5 show the results obtained with burner pot C and D. Results obtained with other burner pots are not shown only to avoid to crowd the plots with too many points. The

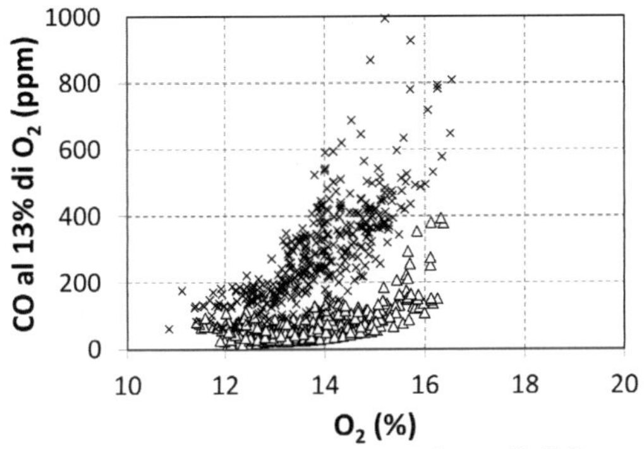

Figure 4. Instantaneous value of CO as a function of O_2 for different type of burner pots. × burner pot C, △ burner pot D.

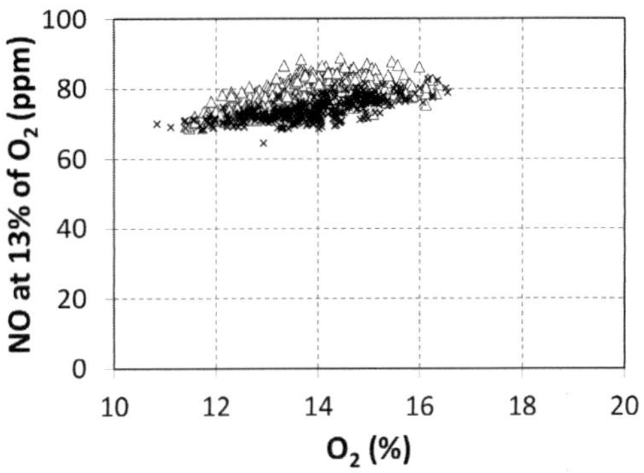

Figure 5. Instantaneous value of NO as a function of O_2 for different type of burner pots. × burner pot C, △ burner pot D.

conclusions that can be drawn from the instantaneous data agree with those obtained analyzing average values. From both points of view it is clear that the CO trend as a function of the excess of air is the same. For each burner pot it is possible to identify an optimal range of oxygen content, i.e. a value of excess air, which minimize the production of CO. In addition, burner pots with larger secondary air inlet area are characterized by a larger optimal range and by lower CO production levels.

4.2. Influence of fuel mass flow rate

The user of a pellet stove can choose and set the heating power level. As a consequence the electronic control system, present in all modern pellet stoves, sets both the fuel and the air mass flow rate. Since the combustion chamber and burner pot geometry is not adjustable, one could not expect to obtain optimal performances for all the available power levels. As a matter of fact, high CO concentration in flue gas at reduced power – about 30 % of nominal heating power, usually – is commonly observed: see, for example, the relation between O_2 in flue gas and input energy reported in [38].

Several tests have been performed at different heating power levels, to investigate this effect. In all cases stove S1 with burner pot A was used. At each power level tests were performed for different values of air mass flow rate.

Figure 6 shows instantaneous CO concentration versus O_2 concentration for four tests performed at the same heating power level, but for different air mass flow rates: these tests are useful to determine the optimal stove settings, that is the proper mass flow rate of air for a given heating level that minimizes CO emission. To interpret the data it is useful to recall that during a single test air flow rate is kept constant. Nevertheless, due to variation of the burning rate, O_2 concentration varies, causing the dispersion of points along the abscissa. Data seem to exhibit a regular behavior and to distribute along the same curve, independently from the particular test considered.

Figure 7 shows instantaneous CO concentration versus O_2 concentration for tests at different power levels. Despite a wide range of variation of fuel mass flow rate, all points fall onto the same

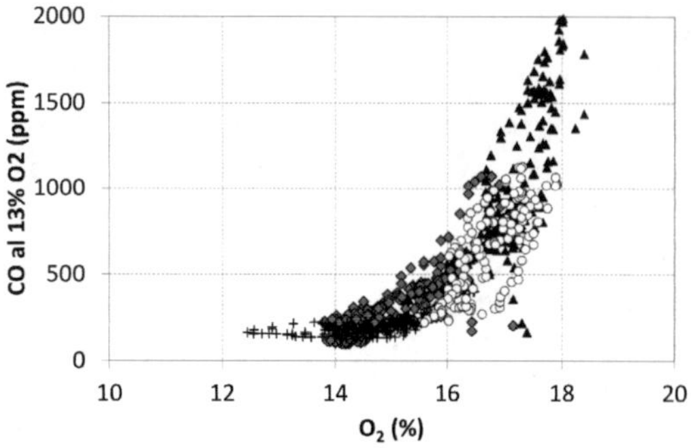

Figure 6. CO as a function of O_2 in flue gas for different mass flow of air at 4.8 kW. Velocity of the air measured at the air inlet (flow area diameter 47 mm): + 0.5 m/s, ◇ 1 m/s, ○ 1.5 m/s, ▲ 2 m/s.

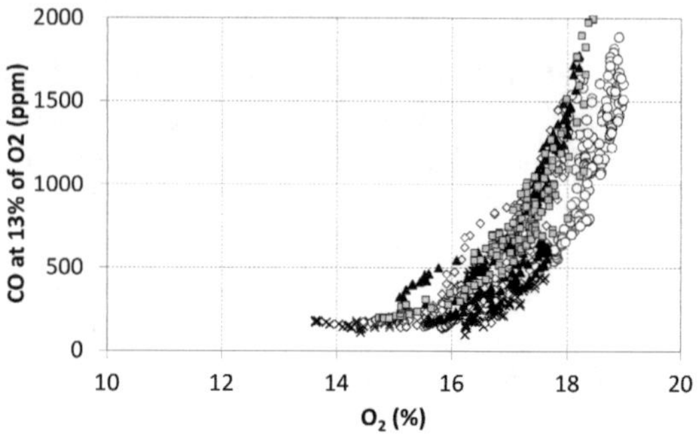

Figure 7. CO as a function of O_2 in flue gas for different mass flow rate of pellet. Power of: ○ 2.2 kW , ◇ 3.6 kW, □ 4.2 kW, ▲ 4.8 kW, × 5.8 kW.

region of plane O_2-CO without the possibility to clearly distinguish one case from the other. For this range of variation it exists a fixed concentration of oxygen in flue gas, hence a fixed excess air, that optimizes the combustion, reducing the production of CO: the minimum CO emission level is the same for any power level. In other tests, performed at 5.8 kW, the concentration of CO increases faster as the concentration of O_2 increases, but the optimal range and CO emission in the optimal range are the same. We conclude that for these stoves, at any power level, adjustment and control of the excess air through the air regulation system is sufficient to obtain a good combustion. High CO concentration usually observed at reduced power seem to be a problem related to the air suction system, rather than to low heating power levels. In designing the combustion air supply system it is necessary to guarantee a range of mass flow rates wide enough to ensure a good combustion both at nominal and at reduced heating power.

4.3. Influence of combustion chamber and bottom grate area dimensions

The size of the combustion chamber is an important parameter to obtain a good combustion: in particular, it should be tall enough to allow the full development of the flame. Tests performed using the same burner pot in stove S2, with short combustion chamber, and stove S3, with tall combustion chamber, are compared in figure 8. It is clear from the instantaneous data that combustion in stove S3 is better than in stove S2. During the tests O_2 content spanned an interval wide enough to show the minimum of the CO-O_2 curve. In the case of stove S2, the

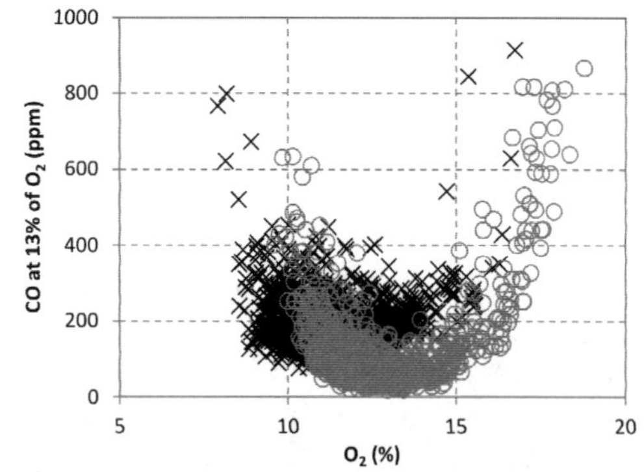

Figure 8. CO as a function of O_2 content in flue gas for short combustion chamber (×, S2) and long combustion chamber (∘, S3).

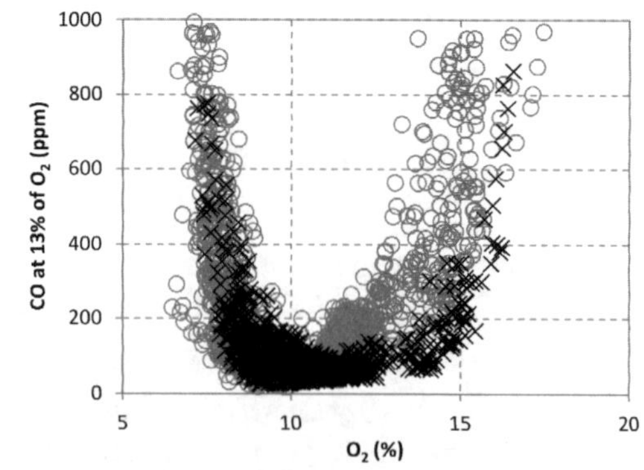

Figure 9. Before (∘ mean CO = 329 ppm) and after (× mean CO = 105 ppm) the optimization of combustion.

flame would be longer than the available space, so it hits cold walls, where the combustion of some substances stops. In stove S3 the flame can develop completely in its tall combustion chamber: this improves combustion yielding a lower level of CO emissions.

The deposition of ash on the grate is a characteristic issue of overfed burner combustor. It can be removed only by the air flow. Ash deposition can occur at reduced power level if the velocity of the air entering through the bottom grate is too low to remove the ash. At nominal heating power if excess air is too low, a thick layer of fresh pellet and char forms above the ash and inhibits its removal. In the worst case, a layer of sintered ashes (clinker) forms on the bottom of fuel bed, blocking the air flow through the bottom grate and deviating most of the air flow through the secondary air inlets. This process depends on the quality of pellet, also: in particular, on the deformation temperature of ashes. For this reason, during a long run test the behavior of combustion can vary, i.e. the CO-O_2 curve and optimal point of working change. Further tests must be carried out to study these phenomena in depth and, in particular, to find the optimal dimension of the combustion grate that ensures low pollutant emissions for long time periods (from few hours to some days).

5. Conclusions

CO production in pellet stoves strongly depends on the amount and distribution of air in combustion chamber. To ensure low emissions of CO and fulfill subsidy or tax incentive requirements, it is necessary to increase the ratio between secondary and total air mass flow rate. In addition, the combustion air supply system and its electronic control must be properly designed.

For each type of burner pot it is possible to find an optimal interval of excess air that reduces the emission of CO over a wide range of thermal power.

Emissions of NO depend only on the excess air. In any case, in all tests we ran it was well below limit enforced by legislation.

These conclusions can be drawn considering either data obtained as average over time intervals of steady operation or instantaneous data.

In particular, plots of instantaneous emissions versus O_2 concentration (i) allow us to find quickly the optimal point of working, just after one test, (ii) allow to obtain useful information from any phase of operation (start, stop, etc.) and from any thermal load, (iii) supply reliable information also for the setup after the installation – currently the stove setup is mostly based on the stove installer experience – (iv) in the future, if the installation cost of O_2 and CO transducer will be affordable for most stoves, they will be an important tool for the self-adjustment of the stove/boiler.

We have applied the results presented here to modify stoves with relatively high CO emissions. In figure 9, for example, the application of these solutions has made possible the reduction of mean concentration of CO and the enlargement of the excess air range corresponding to low CO emission.

Acknowledgments

The authors wish to thank Ravelli Group, its former R&D Manager, Mr. Mauro Brais and the technical department, for financial and technical support.

References

[1] ARPA 2008 Stima dei consumi di legna da ardere per uso domestico in Italia - Rapporto finale *Progetto APAT - ARPA Lombardia*

[2] DM 28/12/12 (Italia) 2012 Incentivazione della produzione di energia termica da fonti rinnovabili ed interventi di efficienza energetica di piccole dimensioni

[3] Fiedler F 2004 The state of the art of small-scale pellet-based heating systems and relevant regulations in Sweden, Austria and Germany *Renewable and Sustainable Energy Reviews*, **8**(3), 201–221.

[4] Obernberger I, Thek G 2006 Recent developments concerning pellet combustion technologies:a review of Austrian developments *Proc. 2nd World Pellets Conf.*, 31–40.

[5] Miguez J L, Moran J C, Granada E, Porteiro J 2012 Review of technology in small-scale biomass combustion systems in the European market *Renewable and Sustainable Energy Reviews*, **16**(6), 3867–3875.

[6] Petrocelli D and Lezzi A M 2012 Experimental study of a pellet stove *XXX UIT Heat Transfer Conf.*, Bologna, June 25–27 2012

[7] Yin C, Rosendahl L A, Kr S K 2008 Grate-firing of biomass for heat and power production *Progress in Energy and Combustion Science*, **34**(6), 725–754.

[8] Obernberger I, Thek G 2006 Recent developments concerning pellet combustion technologies: a review of Austrian developments *Proc. 2nd World Pellets Conf.*, 31–40.

[9] Wiinikka H, Gebart R, Boman C, Boström D, Nordin A, Öhman M 2006 High-temperature aerosol formation in wood pellets flames: Spatially resolved measurements *Combustion and Flame*, **147**(4), 278–293.

[10] Strehler A 2000 Technologies of wood combustion *Ecological Engineering*, **16**, 25–40.

[11] Scharler R, Obernberger I 2000 Numerical modeling of biomass grate furnaces *Proc. of the European Conf. on Industrial Furnaces and Boilers (INFUB)*, (x), 1–17.

[12] Brunner T 2009 Primary measures for low-emission residential wood combustion: comparison of old with optimised modern systems *Proc. of the 17th European Biomass Conf. & Exhibition, Hamburg.*

[13] Petterson E, Nordin A, Ohman M 1996 Effect of temperature and residence time on emissions of CO, THC, tars and NOx during pellet combustion *Proc. of Nordic Seminar on Thermochemical conversion of biofuels.*

[14] Boman C, Nordin A, Öhman M, Boström D, Westerholm R 2005 Emissions from Small-scale Combustion of Biomass Fuels: Extensive Quantification and Characterization *Energy Technology and Thermal Process Chemistry, Umea University.*

[15] Glarborg P 2003 Fuel nitrogen conversion in solid fuel fired systems *Progress in Energy and Combustion Science* **29** 89–113

[16] Klason T and Bai X 2007 Computational study of the combustion process and NO formation in a small-scale wood pellet furnace *Fuel* **86** 1465–1474

[17] Hill S and Douglas Smoot L 2000 Modeling of nitrogen oxides formation and destruction in combustion systems *Progress in Energy and Combustion Science* **26** 417–458

[18] Malte P C and Pratt D T 1974 The Role of Energy-Releasing Kinetics in NOx Formation: Fuel-Lean, Jet-Stirred CO-Air Combustion *Combustion Science and Technology* **9** 221–231 (*Preprint*)

[19] Fenimore C 1971 Formation of nitric oxide in premixed hydrocarbon flames *Symposium (International) on Combustion* **13** 373–380

[20] Klason T, Bai X S 2007 Computational study of the combustion process and NO formation in a small-scale wood pellet furnace *Fuel*, **86**(10-11), 1465–1474.

[21] Widmann E, Scharler R, Stubenberger G, Obernberger I 2004 Release of NOx precursors from biomass fuel beds and application for CFD-based NOx postprocessing with detailed chemistry *Proc. of the 2nd World Conf. and Exhibition on Biomass for Energy, Industry and Climate Protection*, X, 1384–1387.

[22] Wiinikka H, Gebart R 2004 Experimental investigations of the influence from different operating conditions on the particle emissions from a small-scale pellets combustor *Biomass and Bioenergy*, **27**(6), 645–652.

[23] Johansson L S, Leckner B, Gustavsson L, Cooper D, Tullin C, Potter A 2004 Emission characteristics of modern and old-type residential boilers fired with wood logs and wood pellets *Atmospheric Environment*, **38**(25), 4183–4195.

[24] Kelz J, Brunner T, Obernberger I 2012 Emission factors and chemical characterization of fine particulate emissions from modern and old residential biomass heating systems determined for typical load *Environmental Sciences Europe*, 1–18.

[25] Bäfver L S, Leckner B, Tullin C, Berntsen M 2011 Particle emissions from pellets stoves and modern and old-type wood stoves *Biomass and Bioenergy*, **35**(8), 3648–3655.

[26] Schmidl C, Luisser M, Padouvas E, Lasselsberger L, Rzaca M, Ramirez-Santa Cruz C, Puxbaum H 2011 Particulate and gaseous emissions from manually and automatically fired small scale combustion systems *Atmospheric Environment*, **45**(39), 7443–7454.

[27] Win K M, Persson T, Bales C 2012. Particles and gaseous emissions from realistic operation of residential wood pellet heating systems *Atmospheric Environment*, **59**, 320–327.

[28] Olsson M, Kjällstrand J, Petersson G 2003 Specific chimney emissions and biofuel characteristics of softwood pellets for residential heating in Sweden *Biomass and Bioenergy*, **24**(1), 51–57.

[29] Olsson M, Kjällstrand J 2006 Low emissions from wood burning in an ecolabelled residential boiler *Atmospheric Environment*, **40**(6), 1148–1158.

[30] Boman C, Nordin A, Westerholm R, Petterson E 2005 Evaluation of a constant volume sampling setup for residential biomass fired appliances: influence of dilution conditions on particulate and PAH emissions. *Biomass and Bioenergy*, **29**(4), 258–268.

[31] Boman C 2005 Particulate and gaseous emissions from residential biomass combustion (Doctoral dissertation, Umea University).

[32] Brunner T 2008 Evaluation of parameters determining PM emissions and their chemical composition in modern residential biomass heating appliances *Proc. of the International Conf. World Bioenergy*, 81–86.

[33] Johansson L S, Tullin C, Leckner B, Sjövall P (2003). Particle emissions from biomass combustion in small combustors *Biomass and Bioenergy*, **25**(4), 435–446.

[34] Ozil F, Tschamber V, Haas F, Trouvé G 2009 Efficiency of catalytic processes for the reduction of CO and VOC emissions from wood combustion in domestic fireplaces *Fuel Processing Technology*, **90**(9), 1053–1061.

[35] Khalfi A, Trouvé G, Delobel R, Delfosse L 2000 Correlation of CO and PAH emissions during laboratory-scale incineration of wood waste furnitures *Journal of Analytical and Applied Pyrolysis*, **56**(2), 243–262.

[36] EN 14785:2006 Residential space heating appliances fired by wood pellets - Requirements and test methods

[37] Persson T, Fiedler F, Nordlander S, Bales C, Paavilainen J 2009 Validation of a dynamic model for wood pellet boilers and stoves *Applied Energy*, **86**(5), 645–656.

[38] Dias J, Costa M, Azevedo J L T 2004 Test of a small domestic boiler using different pellets *Biomass and Bioenergy*, **27**(6), 531–539.

[39] Petrocelli D 2013 Indicazioni sperimentali per la progettazione di stufe a pellet ad alta efficienza. PhD thesis, Università degli Studi di Brescia

[40] Eskilsson D, Rönnbäck M, Samuelsson J, Tullin C 2004 Optimisation of efficiency and emissions in pellet burners *Biomass and Bioenergy*, **27**(6), 541–546.

[41] Josephinum BLT tests report
`http://www.josephinum.at/blt/pruefung/pruefberichte/feuerungen/pelletsfeuerungen.html`

AUTHOR INDEX

Alberti, L 333
Allesina, G 342
Angeli, D 215
Angelotti, A 333
Antelmi, M 333
Aquaro, D 310
Arnulfo, G 148
Arteconi, A 227
Barletta, A 20, 30
Bertani, C 121, 148
Bisetto, A 280
Boccardi, G 168, 354
Bortolotti, S 141
Bozzoli, F 10
Bubbico, R 354
Buonomo, B 39
Calabrese, N 168
Cammi, A 300
Caruso, G 181
Casagrande, D 87
Casano, G 111
Cattani, L 10
Celata, G P 168, 354
Celli, M 30
Ciofalo, M 249
Colombo, E 191
Colombo, L P M 61, 141
Corradi, C 10
Corticelli, M A 215
Cristofano, L 181
Crivellini, A 237
Croce, G 290
Cucchi, M 61
D'Alessandro, V 237
D'Annibale, F 354
De Salve, M 100, 131, 148
Del Col, D 280
Del Giudice, S 1
Di Benedetto, D 237
Di Liberto, M 249
Di Maio, D Vitale 181
Di Marco, P 77
Di Schio, E Rossi 20
Diani, A 158
Ferraro, G 39
Fichera, A 215
Filippeschi, S 77
Fiorina, C 300
Frano, R Lo 310
Fustinoni, D 61, 69
Geri, M 261
Giuliani, G 227
Gramazio, P 61, 69
Grassi, W 49
Guilizzoni, M 141
Inzoli, F 191
La Licata, I 333
Lampitella, P 191
Lezzi, A M 367
Luzzi, L 300
Malandrone, M 121
Manca, O 39
Mancin, S 158
Mancini, L 204
Marchesi, R 320
Mariani, A 354
Marinelli, L 39
Marotta, G 249
Menale, C 354
Mikityuk, K 300
Monni, G 100
Montelpare, S 237
Mordacci, M 10
Morini, G L 261, 270
Muhlen, L S W 320
Muscio, A 342
Najafi, B 320
Nardini, S 39
Ninokata, H 300
Niro, A 61, 69
Nobile, E 204
Nobili, M 181
Nonino, C 1
Orio, M 131
Pagano, A 215
Panella, B 100, 121, 131
Pavan, L 141
Pedrazzi, S 342
Petrocelli, D 367
Pietrasanta, L 77
Piller, M 87
Piva, S 111
Polonara, F 227
Poulikakos, D 280
Puccetti, G 270
Pulvirenti, B 270
Rainieri, S 10
Ranut, P 204
Ricci, R 237
Ricotti, M E 300
Rinaldi, F 320
Rossetto, L 158
Rovenskaya, O 290
Santini, M 87
Savino, S 1
Schena, G 87
Sotgia, G M 141
Spiga, M 261
Storesletten, L 20, 30
Tartarini, P 342

AUTHOR INDEX

Tartuferi, M. **227**
Tebianian, S **342**
Testi, D **49**
Tiwari, M K **280**
Torresin, D. **280**
Trinchieri, R **168**
Vellucci, F **354**
Vista, D Della **49**
Vitali, L **69**
Vocale, P **261**
Zummo, G. **168**